CADDIS LARVAE

CADDIS LARVAE

Larvae of the British Trichoptera

NORMAN E. HICKIN

RUTHERFORD · MADISON · TEANECK
FAIRLEIGH DICKINSON UNIVERSITY PRESS

 First American edition published 1968 by Associated University Presses, Inc., Cranbury, New Jersey 08512

Library of Congress Catalogue Card Number: 68-58408

6945
Printed in the United States of America

CONTENTS

ACKNOWLEDGEMENTS

My first acknowledgement is to Mr. N. D. Riley who, when Keeper of the Department of Entomology at the British Museum, suggested that I might occupy my leisure time by making a study of the immature stages of caddis flies. I am very grateful to him for the many years of pleasure that this pursuit has afforded me. Mr. Riley introduced me to Mr. Martin Mosely who, until his death, encouraged me greatly, identifying my bred insects and criticizing my first papers.

From that time Mr. D. E. Kimmins, also of the British Museum, has taken a great interest in my work and supported me by identifications and in other ways. Dr. E. B. Worthington interested himself in my early illustrations of caddis larvae and suggested ways in which I could effect improvements so that they could eventually be reproduced with advantage.

Dr. Georg Ulmer always wrote a personal letter to me when each of my papers was published. He was ever kind and thoughtful in his remarks. His death left a wide gap in my circle of trichopterist friends.

Those who have published papers on the biology of caddis flies have supported this present work by allowing me to reproduce illustrations and substantial text matter, and I offer my grateful thanks to Miss R. M. Badcock, Mr. W. V. Baker, Mr. A. Brindle, Dr. J. M. Edington, M. B. Frochot, Dr. H. M. Hanna, Mr. W. W. Macdonald, Mrs. J. C. Mackereth, Dr. G. Marlier, Dr. A. Nielsen, and Dr. G. N. Philipson.

Mr. F. D. Buck has prepared the index and made a number of helpful suggestions and I am very grateful to him for this. I also wish to thank Miss E. Evans who gave valuable assistance.

I wish especially to acknowledge the help of my friend and colleague Mr. Edward M. Buchan, who encouraged me to bring this work to a conclusion. I would like also to thank the Board of Directors of Rentokil Laboratories Limited who ensured its publication.

Lastly, I wish to pay a special tribute to the Royal Entomological Society of London who have very graciously given me permission to reproduce illustrations and to draw heavily on the text matter of the papers in the *Proceedings* and *Transactions* which appeared from 1942 to 1959 under my name.

PREFACE

This book is the outcome of work commenced in 1937, the object of which was to describe and illustrate the immature stage of the British caddis flies or TRICHOPTERA. From that date until 1959, the larvae of seventy-three species were collected and reared to the adult stage and illustrated descriptions of fifty-six species published in *Series A* of the *Proceedings of the Royal Entomological Society of London.*

The greater opportunities that now exist especially for young people to visit, observe and study the countryside under the expert guidance of trained helpers has, during the last few years, resulted in an increased awareness of many aspects of Britain's wild life. More especially there has been an increasing interest shown in the biology of our streams and ponds which the publication of Macan & Worthington's *Life in Lakes and Rivers*, Harris's *An Angler's Entomology* and Clegg's *The Freshwater Life of the British Isles* did much to sustain. It would be a poor observer of the life in our fresh water who was not aware, within a few minutes, of the caddis worms present in almost all freshwater habitats. So that for some years now many students of our freshwater fauna have been enquiring whether it would not be possible to bring together my papers and illustrations of caddis larvae that otherwise must be searched for amongst the heavy volumes of many years on the shelves of specialist libraries. In addition to the fifty-six descriptions of caddis larvae which constitute my own contribution to our knowledge of the British fauna, Mr. Allan Brindle has described four, Mr. W. V. Baker three, Dr. J. M. Edington twelve, Dr. H. M. Hanna four, Mr. Kimmins one, Mr. W. W. Macdonald three, Mrs. Jean Mackereth ten and Dr. G. N. Philipson three. Fifteen are redescriptions.

I have been able to incorporate their work and a number of illustrations into this book, but there has been a little difficulty owing to the variation in interpretation of the insect morphology by the different authors. At first an attempt was made to find a solution which would give the same word the identical meaning in every larval description, but this proved unwieldy. It is hoped that the reader will not be too exasperated by the differences in the nomenclature used by each author. Perhaps in the future new workers should follow the nomenclature for the head of the larva so admirably proposed and described by Miss R. M. Badcock (1961). They should study also the recent and most detailed work on the thorax and limbs of caddis larvae carried out by Dr. A. R. Tindall (1963).

Because the distribution of our British species of caddis fly is so little known, I have thought it worth while to include notes on this in the general text but under a separate heading for each species. In recent years there have appeared a number of useful lists for specific localities and a number of special records.

However, there are many counties for which there are few or no records of TRICHOPTERA. Surely there are many among the lepidopterists with their special means of attracting night-flying insects who could help in providing a much broader base for our knowledge of TRICHOPTERA distribution? Many of our species are known only from one, two or three localities and it is certain that this does not represent their complete distribution in our isles. It is virtually certain that species new to Britain will be collected in the next few years.

I have used the references to distribution given by Mosely (1939) but I have not made a detailed search into historical material. I have used the initials of the recorder. My principal information has been derived from papers the references to which will be found at the end of Chapter 5, and from my own collection, which includes the Peacey collection.

It was thought worth while to include some information on the adult flight period because so much more is known of this since the work of Crichton. The particular method of displaying it has been adopted so that it can be extended quite simply as more information is obtained, particularly concerning the rarer species. I have used only recent records where the actual date of capture was given. A knowledge of the adult flight period is, of course, of considerable value when searching for larvae of the particular species.

A chapter is devoted to each of the fourteen families. Some chapters are therefore small, indeed two refer only to single species, while Chapter 15 is concerned with the fifty-six species of the family LIMNEPHILIDAE. In each of Chapters 6 to 19 a broad description of the adult caddis fly is given, followed by a check-list of species in that family. This was thought to be preferable to listing all the species by way of an appendix. Mosely's handbook, *The British Caddis Flies, Trichoptera*, published in 1939, has stood as a landmark in the study of this group of insects. It has seldom been far from my right hand and I have drawn heavily upon it. Since its publication, however, there have been changes in nomenclature, a number of species new to Britain have been collected, and our knowledge of the distribution and dates of emergence of caddis flies has been extended. Some of this information has been included by way of introduction in these chapters and arranged in such a manner that the student of this group can add to it, either from his own collecting and observations or from that of others.

When Mosely first outlined to me the way he thought my work on the immature stages of the British TRICHOPTERA should be carried out, he stressed the obvious desirability of confining the work to actual British collected specimens. He thought that it might be found that some of our species were sub-specifically separate from the species found on the European mainland.

In addition, identifications based on European literature have been made from larval material which had proved to be erroneous. At the time of going to press, however, descriptions of larvae of only eighty-two species have been made in detail from British material. In this account, therefore, detailed descriptions of larvae are given only where they can be based on British material. Descriptions of larvae necessarily taken from European literature are given in very much briefer form. The reader will find three exceptions. The work of Nielsen on the HYDROPTILIDAE, Frochot on some species of the LIMNEPHILIDAE and Marlier on *Chimarra marginata* is included. In some families sufficient information is available to make possible the construction of keys. In other families, e.g. the LIMNEPHILIDAE and HYDROPTILIDAE, accurate keys are not yet a possibility. It is hoped that some students of entomology, looking through this book, will perceive the gaps in our knowledge and will be stimulated to remedy this state of affairs. The writer would be well pleased with his efforts if this were to come about.

PREFACE TO THE AMERICAN EDITION

This book on the larval stage of the Caddis Flies, *Trichoptera*, deals specifically with the British fauna, of which nearly two hundred species are known. Although this order of insects has not been as popular as, say, the *Lepidoptera* for entomological collections, nevertheless, it is thought that the total of this species is within approximately ten per cent of the actual total of indigenous British insects in this order. Few species are yet to be found in this highly populated island.

The Nearctic trichopterous fauna presents many similarities to the European fauna. On the other hand, there are important diversities and, indeed, some of the most recent work has pointed to relationships to the Asiatic fauna. Nevertheless, this present work is presented as a pattern in the study of the immature stages of this most widely distributed and ubiquitous order.

The role of freshwater biology, in which the aquatic larvae of the present study plays a part, has, during the past thirty years or so, increased immeasurably in importance. The economics of freshwater, feeding habits of fish, pollution of streams, ponds and lakes, water resources and water conservation, are all subjects within which falls an appreciation of this most numerous, interesting and often dominant group of aquatic animals.

In preparing this book for the North American edition, a few alterations in the original text have been made. A small number of references to conservation techniques, only applicable to Britain, have been deleted and the opportunity afforded to remedy a few inconsistencies in the text. A short section has been added incorporating references to American literature, and in some cases, a few brief notes have been added which it is hoped will be appreciated by the worker who finds interest in this group of insects for the first time.

Home Farm,
Fetcham,
Leatherhead, Surrey,
England.

5th July, 1968.

1

INTRODUCTION TO THE STUDY OF TRICHOPTERA. THEIR IMPORTANCE IN THE NATURAL ECONOMY

There can be few groups of insects more attractive to study than the TRICHOPTERA, or caddis flies. Whether one is observing or studying the adult flies or their immature stages, fresh water is not far away, be it mountain rill, broad slowly flowing river, pond or lake-side, and such situations invariably give pleasure in themselves. Apart from the delightful situations where caddis are found, this order of insects is of extraordinary biological interest. All but one of the British species have aquatic immature stages and almost every situation where fresh, unpolluted water flows or collects will be found to contain a caddis fauna, even on the surface of wet rocks, deep in caves and boggy seepages on mountain-tops!

Adult caddis flies are moth-like in appearance, but their wings, instead of being clothed with tile-like overlapping scales, are hairy. The name of the order TRICHOPTERA literally means 'hairy-winged', although many cases of scale-like hairs are known in the group. Some small moths can only with difficulty be separated from caddis flies. The larvae are known by a variety of common names such as caddis grubs, caddis worms or stick worms and, as already indicated, may be found in a very wide range of aquatic habitats.

The Derivation of 'Caddis'

Two distinct derivations and meanings are given for this word. In the first place the word, with many variants, is used for cotton or silk used as a padding, and this meaning is known to go back to 1400. It was then spelt 'cadaz', but forty years later, with the same meaning, it was spelt 'cadace'. The word then came to be applied to worsted yarn, the first-known reference in this context being in 1530. Very soon afterwards the word was used to describe tape or ribbon of the same material, and Shakespeare wrote in 1611 in *The Winter's Tale*, 'He hath Ribbons . . . Points . . . Incles, Caddysses'. This meaning of the word was used as late as 1887, and perhaps even later, but is not now in general use.

The second meaning of the word is given as of uncertain origin (but see later). This concerns its use in the sense that it is used in the title of this book, 'a larva . . . which lives in water, and forms for itself a curious cylindrical case of hollow stems, small stones, etc.; it is used as a bait for anglers'. This is taken from the *Oxford Dictionary*, but the writer hesitates to insert the missing words, which are 'of the May-fly and other species of *Phryganea*'. The May-fly belongs to the distinct order of insects EPHEMEROPTERA, whilst *Phryganea* is a single genus of one family (PHRYGANEIDAE) of caddis.

Many of the early references concern fishing. Thus Izaak Walton, in the *Compleat Angler*, 1653, states 'The May flie . . . is bred of the cod-worm or Caddis', and in the first book devoted to insects in the English language, Mouffets *Theatrum Insectorum*, published in 1658, is found, 'The great variety of those little cados worms whereof they come'.

In the *Oxford Dictionary* the connexion between the two meanings of the word is not apparent. My own explanation concerns the itinerant vendors of bits of stuff, braid and ribbons, who travelled the countryside calling at isolated cottages and hamlets. They pinned or stuck their wares on their coats as an advertisement for their business, and they were called cadice men. Is it not then very likely that the 'cadice-worm' in the streams and pools received its name from the cadice man who stuck all sorts of cloth and worsted braid (or caddis) on his coat?

Caddis larvae may be sought for under and on top of submerged stones, amongst water-weeds or swimming freely in the water. Whereas some make protective cases for themselves as small larvae, gradually enlarging the case as they grow, others are free-living, or make nets for trapping organisms and debris, and make a protective chamber for themselves only just prior to pupation. The larvae of one family of caddis make fixed tunnels on stones or submerged wood which may be several times their own length. Altogether 193 different species of caddis fly have been recorded as occurring in Britain and it is certain that a number of other species await discovery by assiduous collectors. As in most entomological textbooks caddis larvae are referred to as 'usually bearing a case', it is of interest to see how many of the British species belong to the different categories of larval habit mentioned above.

Altogether, 148 species make cases of various extraneous material such as grains, vegetable debris, pieces of cut leaf or of silk-like secretion; twenty-eight species construct bag-like nets usually communicating with a tube-like shelter which they usually inhabit; thirteen species live in silk-like fixed tubes to which sand-grains or vegetable particles may adhere and four species are entirely free-living, crawling slowly over the surfaces of rocks and stones. The net-spinners and those living freely make a pupal chamber not unlike the larval cases of related species except that they are usually fixed to a stone. Of the case-makers, we must keep in mind the twenty-nine species in the HYDROPTILIDAE which, as far as we know from those few species studied, are free-living during the first four instars and only make a case at the beginning of the fifth and final instar.

The Importance of Trichoptera in the Natural Economy

One is often asked by one's lay friends when the subject of TRICHOPTERA is being discussed and one is trying to describe their natural history, 'But what do they do?' or 'What good are they?' These are extraordinarily difficult questions to answer satisfactorily for persons who have little knowledge of the great interdependent reticulated pattern of life brought about by the existence of animals and plants. Caddis are often present in extraordinary numbers, yet they seldom, if ever, impinge on the consciousness of the ordinary individual. It is only occasionally that they occur as pests. In the larval stage the species *Limnephilus lunatus*, and perhaps other species, cause damage to watercress beds, chewing the stalks so that the shoots are cut and they drift downstream. The housewife sometimes finds the larvae in the watercress bunch she is washing. In Japan another species causes damage to the newly transplanted rice-plants.

The vast numbers of caddis flies that sometimes occur usually go unnoticed in Britain, but in America and Africa the very numbers of insects cause jamming of air-conditioning plant and other electrical equipment, and at the times of maximum emergence outdoor painting is impossible because of the numbers of caddis flies that would stick to the paint.

On the other hand, caddis larvae, pupae and adults are often present in significant proportions in the food of fish. This is a fact which can usually be assimilated and indeed it is the fisherman who has appreciated the existence of the caddis. Three hundred years ago our literature contained the expression 'our caddissed hooks', meaning that they were baited, thus showing that the importance of caddis larvae as providing sustenance for fish was well known. This subject is discussed further in Chapter 5. During the last two or three hundred years, however, there has evolved a method of fishing for trout, and to a lesser degree for other species also, which makes use of the trout's habit of taking from the water surface the adult stage of a number of insects with an aquatic larval stage such as mayflies, caddis flies and stone flies. Artificial flies are fashioned from fur, feathers and other materials on to a hook. The various patterns are given names and although many have no resemblance to living flies, others are tolerably good imitations. The dry-fly fisherman, as he is called, learns the identification and habits of many of the aquatic insects, and a considerable literature has evolved around his pursuit. Another method of fishing for trout is known as 'wet-fly' fishing, and in this case the active pupal stage of caddis as it is swimming to the surface is imitated.

Birds, bats and other insectivorous animals take their toll of the adult stage of caddis, as anyone who has observed swallows and bats hawking over rivers and lakes in the late summer evenings well knows.

As to what the larvae of caddis 'do' in their aquatic environment, this is considered in more detail in Chapter 5, where the biology of TRICHOPTERA is oultined and the work done on the feeding habits discussed. It can be said, however, that almost all forms of organic matter, living and dead, found in water, are utilized as foodstuff by TRICHOPTERA larvae. Some species are mainly herbivorous, some mainly carnivorous and the others omnivorous.

CLEGG, J. 1952. *The Freshwater Life of the British Isles*. Frederick Warne, London.
HARRIS, J. R. 1952. *An Angler's Entomology*. Collins New Naturalist Series, London.
HICKIN, N. E. 1952. *Caddis. A Short Account of the Biology of British Caddis Flies with Special Reference to their Immature Stages*. Methuen, London.
MACAN, T. T. & WORTHINGTON, E. B. 1951. *Life in Lakes and Rivers*. Collins New Naturalist Series, London.
MOSELY, M. E. 1939. *The British Caddis Flies, Trichoptera*. Routledge, London.

2

THE EVOLUTION OF TRICHOPTERA AND THEIR POSITION IN THE CLASSIFICATION OF INSECTS. THE FAMILIES OF TRICHOPTERA

All members of the insect order TRICHOPTERA are known as caddis flies and all caddis flies are classified in the order TRICHOPTERA. There is, in all probability, an element of convenience in this because a substantial proportion of the larvae of TRICHOPTERA do not construct cases to envelop their bodies but live in nets, silken tubes or are free-living. Thus, this group, by derivation of the word 'Caddis', do not truly belong.

The order TRICHOPTERA, in the most recent interpretation, follows the order LEPIDOPTERA, the butterflies and moths, and, in turn, is followed by the DIPTERA, the true two-winged flies. I must confess to some considerable difficulty in believing this myself. There seems to me so many reasons why the TRICHOPTERA were the forerunners of the LEPIDOPTERA rather than the reverse. One piece of evidence that has always weighed heavily with me concerns the caddis larva parasite *Agriotypus armatus* Curtis. This archaic hymenopteron is the only known parasite of the TRICHOPTERA, indeed it is confined entirely to the sub-family GOERINAE. LEPIDOPTERA, on the other hand, are parasitized by a large number of very much more highly developed hymenopterous species. Surely this was much less likely to have happened if the TRICHOPTERA were more highly developed, in any sense, than the LEPIDOPTERA?

The TRICHOPTERA was formerly included in the order NEUROPTERA together with the MECOPTERA and the NEUROPTERA as at present constituted. The first use of the word TRICHOPTERA is due to Kirby in 1813. Great difficulties have been found in deriving any satisfactory explanation of the evolution of the insect orders through this complex of groups which Tillyard called the panorpoid complex. One problem has been the primitive family of the LEPIDOPTERA, the MICROPTERYGIDAE. The adult stage of the latter possesses functional mandibles, the maxillae have the lacinia developed and the galea are not haustellate. The larva has a small transverse post-clypeus and bears eight pairs of abdominal legs, each of which terminates in a single hook. The pupa, also, possesses functional mandibles.

No other family of the LEPIDOPTERA has all these characters, and although it may be more usual to place this single family in a suborder

ZEUGLOPTERA, it must be borne in mind that Chapman (1917) and Hinton (1958) have put forward convincing arguments for considering this family as constituting a separate order, the ZEUGLOPTERA.

Our interest in the ZEUGLOPTERA lies in the belief that its members are more primitive than any of the LEPIDOPTERA and TRICHOPTERA. Possibly the TRICHOPTERA should be placed *within* the present LEPIDOPTERA, between the ZEUGLOPTERA and the two suborders MONTRYSIA and DITRYSIA. The LEPIDOPTERA and the TRICHOPTERA have certain venational characteristics in common and both appear to stem from the more primitive members of the MECOPTERA first known from the Lower Permian. Schwanwitsch, in his tabular classification, placed the TRICHOPTERA horizontally in the inferiores position.

Hinton, however, does not consider that modern MECOPTERA could have been ancestral to the TRICHOPTERA, but Badcock, from her study of musculature of trichopterous larvae, believes that the TRICHOPTERA could have been developed from unspecialized mecopterous ancestors.

The order TRICHOPTERA is defined as being composed of endopterygotous or holometabolous insects with membraneous wings more or less densely hairy (pubescent), which may be held tent-like over the body when at rest, giving a moth-like appearance. The fore-wings are more or less elongate and the hind-wings are broader, usually with a folding anal area. The venation is generalized and there are few cross-veins. The mandibles are vestigial or absent and the maxillae are single-lobed with elongate palpi. The labium has a median glossa with the palpi well developed. There are five tarsal segments. With but one exception in the British fauna, the larvae are aquatic. The abdomen of the larva is always terminated by a pair of hooked appendages. Whereas in some families the larva constructs a case of various materials taken from its environment, or secreted for the purpose, in which it lives, in other families the larva constructs a net or silken tunnel or it may live quite freely.

The pupae are exarate with strong functional mandibles and the wing tracheation is reduced.

After Klapálek had described the two distinct larval forms, campodeiform and eruciform, present in the TRICHOPTERA, Martynov (1924) classified the order into two suborders, the ANNULIPALPIA containing those families with campodeiform larvae and the INTEGRIPALPIA containing the eruciform. The ANNULIPALPIA is considered to be the most primitive suborder even though some considerations of the head structure, particularly the angle of head insertion, are somewhat anomalous. Nielsen, however, convincingly demonstrates that the eruciform larvae are derived from the campodeiform. Wesenberg-Lund, however, pointed out that the ANNULIPALPIA were not a truly coherent group, although the INTEGRIPALPIA formed a well-defined systematic unit. Nielsen assumes that there are several parallel developmental series in the ANNULIPALPIA and, in the case of *Hydropsyche*, this reaches a development by no means second to that of the most highly specialized eruciform larva.

It is generally considered that *Rhyacophila* represents the most primitive stage of development shown in recent caddis, and Nielsen puts forward convincing arguments for this. The GLOSSOSOMATIDAE were thought to be a sub-family of the RHYACOPHILIDAE, and Nielsen again has shown the great similarity, exhibited by pupae of the sub-family, to *Rhyacophila*, even though they are very much smaller. The GLOSSOSOMATIDAE are now considered a separate family. Eaton considered that the HYDROPTILIDAE were related to the GLOSSOSOMATINAE, a view which is generally accepted, and, more recently, Nielsen, while agreeing with this contention, has put forward arguments to show that there are affinities also with the INTEGRIPALPIA. He also states that the HYDROPTILIDAE exhibit a mixture of primitive and highly specialized features, and that the variation within the family is greater than is found in any other family of caddis flies.

In this present account the arrangement of the families caused some concern. It was at first decided to list them in the same order as in Mosely's handbook and it was not until late in 1965, just before the book went to press, that it was decided to follow the arrangement in Kloet & Hinck's *Check-list of British Insects*, Second Edition (Revised), 1964.

Mosely and Kimmins recognized eighteen families of TRICHOPTERA in the world fauna, and, of these, representatives of thirteen are found in Britain. The number of British species at present known in each family is listed below:

PHRYGANEIDAE	10	HYDROPSYCHIDAE	10
LIMNEPHILIDAE	56	POLYCENTROPIDAE	13
SERICOSTOMATIDAE	9	PSYCHOMYIIDAE	13
BERAEIDAE	4	PHILOPOTAMIDAE	5
MOLANNIDAE	2	RHYACOPHILIDAE	10
ODONTOCERIDAE	1	HYDROPTILIDAE	29
LEPTOCERIDAE	31		

More recently the GLOSSOSOMATIDAE have been separated from the RHYACOPHILIDAE, making fourteen families, and giving RHYACOPHILIDAE four species and GLOSSOSOMATIDAE six.

In 1966 Kimmins elevated the four sub-families of the SERICOSTOMATIDAE to family rank which arrangement has not been followed here.

The following families are not represented in Britain: PLECTROTARSIDAE, PHILANISIDAE, HELICOPHIDAE, CALAMOCERATIDAE, and PHILORHEITHIDAE.

BADCOCK, R. M. 1961. The Morphology of Some Parts of the Head and Maxillolabium in Larval TRICHOPTERA, with special reference to the HYDROPSYCHIDAE. *Trans. R. ent. Soc. Lond.* **113**, 217–48.

CHAPMAN, T. A. 1917. *Micropteryx* Entitled to Ordinal Rank: Order Zeugloptera. *Trans. ent. Soc. Lond.* **1916**, 310–14.

EATON, A. E. 1873. On the HYDROPTILIDAE, a Family of TRICHOPTERA. *Trans. ent Soc. Lond.* [**21**], 125–50.

HINTON, H. E. 1958. The Phylogeny of the Panorpoid Orders. *A. Rev. Ent.* **3**, 181–206.

KIMMINS, D. E. 1966. A Revised Check-List of the British TRICHOPTERA. *Ent. Gaz.* **17**, 111–120.

KIRBY, W. 1813. *Trans. Linn. Soc. Lond.* **11**, 88.

KLAPÁLEK, F. 1888. Metamorphose der Trichopteren. *Archiv der Naturwissenschaftliche Landesdurchforschung von Böhmen.* **6** (5).

MARTYNOV, A. V. 1924. TRICHOPTERA. *Practical Entomology*. Leningrad.

MOSELY, M. E. & KIMMINS, D. E. 1953. *The* TRICHOPTERA (*Caddis-flies*) *of Australia and New Zealand.* Trustees of the British Museum. London.

SCHWANWITSCH, B. N. 1943. *Nature, Lond.* **152**, 727.

TILLYARD, R. J. 1918–20. The Panorpoid Complex. A Study of the Phylogeny of the Holometabolous Insects with Special Reference to the Subclasses PANORPOIDEA and NEUROPTEROIDEA. *Proc. Linn. Soc. N.S.W.* **43**, 265–84, 395–408, 626–57; **44**, 533–718; **45**, 21–7.

WESENBERG-LUND, C. 1943. *Biologie der Süsswasserinsekten.* København.

3

THE LIFE CYCLE OF THE TRICHOPTERA

The Egg Stage

The eggs of caddis flies are laid either in water or near to it. In the former case the female enters the water and usually cements the eggs to rocks or, less commonly, water-plants. Ross considers that this is the case in the families RHYACOPHILIDAE, PHILOPOTAMIDAE, PSYCHOMYIIDAE, HYDROPSYCHIDAE and HYDROPTILIDAE. The eggs are laid in strings and grouped together to form irregular plates which may contain just a few eggs or as many as 800. The individual eggs are enveloped in a cement-like matrix. There are records of a number of observers having witnessed the unusual event of an aerial insect entering water, usually fast-flowing. Hora gives an account of a caddis fly in India in another family, the SERICOSTOMATIDAE, which crawls into the torrential water of a waterfall and lays its gelatinous egg masses on the actual lip of the fall. Needham (1918) also mentions that the females of some HYDROPSYCHIDAE enter the water and spread their eggs on the lee side of a stone in the gentler currents. However, Badcock (1953) has given a detailed account of a female *Hydropsyche angustipennis* which she observed to bend the tips of its antennae into the water after it had alighted on a projecting stone in an upland stream. It then suddenly flew up two or three feet, and after it had zigzagged a few yards it dived vertically into the stream. Miss Badcock was able to follow its course by its silvery appearance as it swam to the under-surface of an inclined submerged stone. After a few seconds the stone was lifted and the insect, which had already laid some eggs in a flat sheet, was captured. In captivity the insect again entered the water and laid 460 eggs whilst being continuously submerged for thirty-seven minutes. In the two accounts of her unique observation Miss Badcock discusses the insect's adaptations to this extraordinary behaviour, and the part played in respiration by the air trapped in the hairy folds of the wings and body.

In families of TRICHOPTERA other than those mentioned above, the female deposits the eggs in an ovoid or irregularly shaped mass. The eggs are enveloped in a greenish or bluish-green gelatinous material, and in some species, notably the sericostomatid *Brachycentrus subnubilus*, the egg-mass is carried about for a time by the female, extruded at the tip of the abdomen. Jacques has recorded his detailed observations on this species which, as well as having the common name of 'Grannom', is known also as the 'Greentail'.

Jacques (1960) states that within an hour or two of copulation the green egg-mass begins to appear at the abdominal vent and, by reason of the contraction forwards of the abdominal segments, it becomes almost fully exposed within a day or so. In fact about one-sixth of the egg-mass, which is now about 'twice as big as a pin-head', is held within the abdomen. A slight external force is required to eject it, and usually this is accomplished by what is known in angling circles as 'dipping'. In fine weather the female flies upstream, against the current, and from time to time 'dips', striking her abdomen against the pull of the surface film of the water. This she does until the egg-mass is finally lost in the stream. The number of eggs in the egg-mass varies from sixty to 700. It swells immediately on contact with water, doubling in size within an hour, and in twenty-four hours reaching its final size of a quarter of an inch. It becomes very glutinous so that as it travels downstream it becomes attached to rough surfaces such as wooden bridge piles. The numbers of egg-masses stuck on to such structures an inch or so below the water surface gave rise to the idea that in this species also the female crawls under water to lay her egg-mass, but this is not the case.

In many cases the egg-mass has no distinct shape but in the family PHRYGANEIDAE it is in the form of a ring which is often found hanging on aquatic plants or adhering to the surfaces of water plantain or water-lily leaves. It is soft and gelatinous in texture and the eggs may be seen clearly within it, arranged in transverse discs. The eggs of *Triaenodes bicolor*, and probably of other species in the same genus, are deposited in a disc and arranged spirally.

In the LIMNEPHILIDAE the egg-masses are deposited on plants and stones above the water and sometimes on the leaves and twigs of trees very high above it. The exact manner in which the hatched larvae actually make their way into the water has not been recorded. It is thought, however, that the egg-mass swells during rainfall and that subsequently the young larvae are washed into the water. This is a subject urgently requiring detailed observation and experiment.

The Larva

Trichoptera have truly aquatic branchio-apneustic or apneustic larvae, the whole larval state being spent in some aquatic habitat with but one exception, the species *Enoicyla pusilla*. The type of habitat ranges from torrential stream to stagnant ditch. A species found in fast-running streams is usually confined to this type of water, whilst those found in lakes and ponds are usually similarly restricted. Some species, however, have a rather wider range of habitat. Thus larvae of *Phryganea grandis* L. and *Mystacides nigra* L. are found on fairly open lake shores, weedy pools and also rivers. Oxygen for respiration is absorbed from the water, either by means of gills of various types or directly through the cuticle.

Three distinct types of trichopterous larva are fairly readily discernible. The eruciform larva (Fig. 1) is typical of the families LIMNEPHILIDAE,

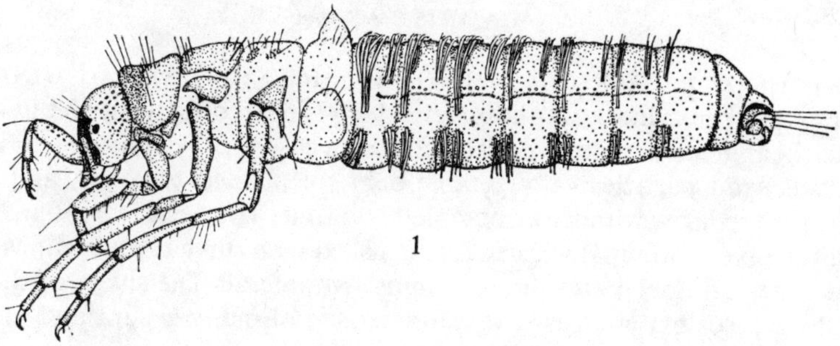

1

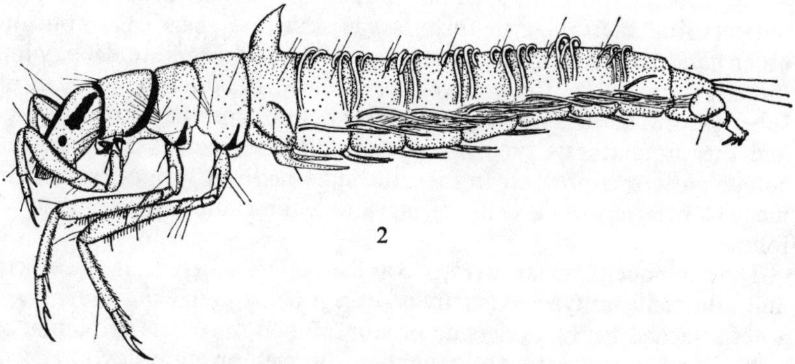

2

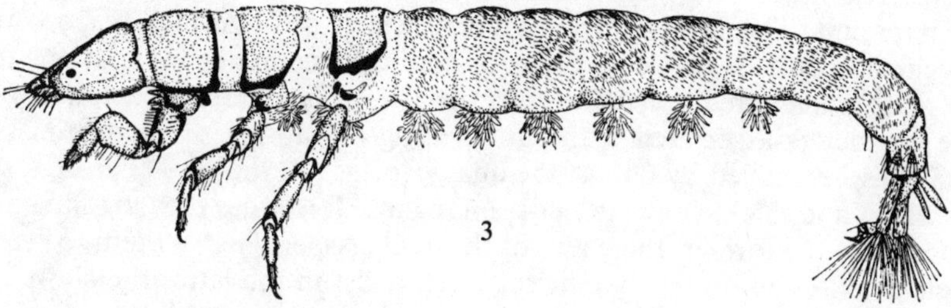

3

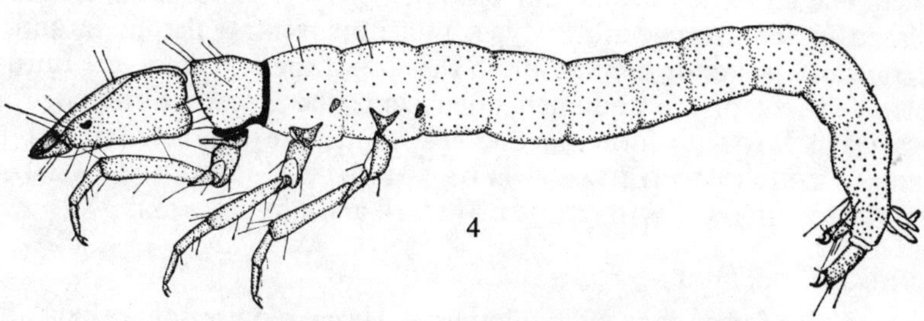

4

Fig. 1. Eruciform larva of *Limnephilus* sp.
Fig. 2. Suberuciform larva of *Phryganea* sp.
Fig. 3. Campodeiform larva of *Hydropsyche* sp.
Fig. 4. Campodeiform larva of *Philopotamus montanus*.

LEPTOCERIDAE, SERICOSTOMATIDAE, MOLANNIDAE and ODONTOCERIDAE. The cylindrical abdomen, which is most often whitish but sometimes light green in colour, is protected by a case made of various materials such as sand particles or vegetable debris cemented together, and when alarmed the larva withdraws completely into it. The head is hypognathous and the first abdominal segment is furnished with three protuberances, one being median dorsal and the remaining two lateral. The shape of the case or the type of material used is characteristic of many genera and in some particular instances even of species.

The suberuciform larvae (Fig. 2) of the PHRYGANEIDAE are in some respects similar to the eruciform larvae; thus they bear cases, but on the other hand the intersegmental grooves in the abdomen are deep, whilst in the eruciform larvae the grooves are shallow. In addition the head of the suberuciform larva is not hypognathous but is somewhat between this type and the prognathous type. Frequently the abdomen is coloured, and a colour pattern is present. In fact, the suberuciform type of larva is intermediate between the eruciform larva and the campodeiform type which follows.

The campodeiform larva (Figs. 3 and 4) differs widely from the eruciform and suberuciform types. Usually no case is constructed and the larva spins a net attached to the underside of stones or to the stems of plants. Small insects and crustaceans are caught in the web and are devoured by the larva. The net-spinning type of larva constructs a case immediately prior to pupation. The head is prognathous, which distinguishes it from the eruciform and suberuciform types. In habit the campodeiform larvae are more agile than are the eruciform larvae and, as is to be expected, many of the characters associated with case-bearing are absent or modified. The anal claspers, called by Orcutt the anal 'prolegs', in the case-bearing types are short and thick, whilst in net-spinning and free types they are long and slender. In the former, the claws of the anal claspers enable them to retain their position inside the cylindrical case, whilst in the latter the claws are for holding position on the threads of the web, much as do those of the spider. The uniformly cylindrical abdomen of the eruciform larva is often replaced in the campodeiform type by a dorsoventral flattening and an anterior and posterior diminution in the size of the segments, the third or fourth segment of the abdomen often being the largest. Campodeiform larvae (the 'thysanuriform' larvae of Orcutt) are characteristic of the families RHYACOPHILIDAE, HYDROPTILIDAE, PHILOPOTAMIDAE, HYDROPSYCHIDAE, POLYCENTROPIDAE and PSYCHOMYIIDAE.

Cuticular Formations

The cuticular formations of trichopterous larvae require some elucidation in order that the terms may bear some homology with the terms of previous authors. Siltala (1900–8) made a study of the chaetotaxy of the larva both in relation to the disproportionate spacing of the bristles, spines, hairs, etc., during larval growth, and also as regards variation in the different families. Orcutt *in* Betten (1934) gives some figures of the various cuticular outgrowths with a reference to Siltala's terminology in the text-figure legend.

Philipson (1961) has studied the detailed placement of bristles and gill arrangement of the fifth instar larva of *Potamophylax stellatus*. A study of this in a wide range of families to the same degree of exactitude is urgently required. In this general account, however, it has been thought best to include the information given by Philipson in Chapter 7.

Arising from the head, legs and anal segments of the larva are long stout bristles almost always dark in colour and with a light-coloured ring at the base which is probably a local thinning of the membrane, allowing the bristle to articulate. These bristles are typical sensillae and are the 'Borsten' of Siltala. The golden-yellow, thick, sword-shaped spines which are a characteristic feature of the ventral edge of the femur and tibia of many larvae are the 'Sporne' of Siltala. Sometimes the spines are hairy (Federsporne), as in the tarsus of *Holocentropus* McLachlan and other POLYCENTROPIDAE. Sometimes a series of very small spines are situated along the ventral edge of some of the leg segments—these are Siltala's 'Spörnchen'. Other outgrowths and modifications of the bristles, spines and claws are dealt with later in the text.

Head. The head is a hard sclerotized capsule roughly in the shape of a truncated cone with the base hemispherical, the latter being the proximal part of the head at its insertion into the membranes of the prothorax. Variation in the shape of the head amongst the different families is shown in Fig. 5. In *Rhyacophila* Pictet the head is much longer than broad, whereas in *Molanna* Curtis the head capsule is equally as long as broad. In some cases, e.g. *Oligotricha* Leach, the sides of the head (genae) are parallel, in others, e.g. *Silo* Curtis, the genae converge towards the mouthparts. The head capsule is made up of four distinct sclerites, firstly a three-sided sclerite, the clypeus, situated on the dorsal or anterior surface of the head with the vertex directed aborally and the base forming the anterior

Fig. 5. Types of larval head: A. *Limnephilus flavicornis*; B. *Rhyacophila dorsalis*; C. *Polycentropus flavomaculatus*; D. *Goera pilosa*.

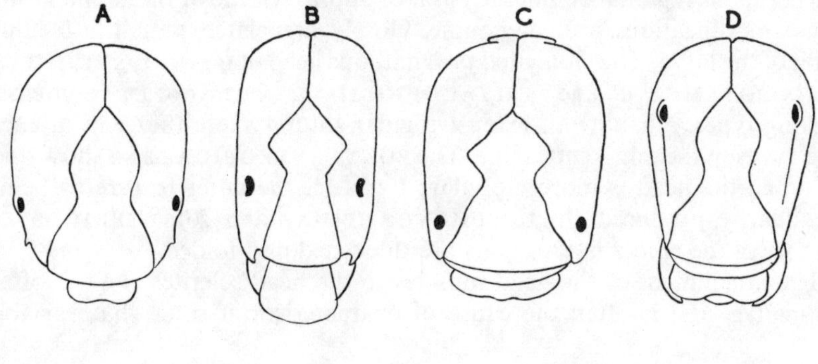

margin of the head capsule at the insertion of the mouthparts. There is some discussion, however, as to the true homology of this sclerite. Whereas Ulmer and Lestage together with most of the older authors have called this sclerite the clypeus, Orcutt prefers to call it the frons. Badcock, who has made a detailed study of the head of larval TRICHOPTERA, considers this to be the fronto-clypeus. This is the term used by Krafka, Nielsen, Mackareth and Hanna. Whilst accepting the correctness of this definition, I have retained the use of clypeus mainly because the two main continental works likely to be available to the student, Lestage and Ulmer, use this term.

In shape, the clypeus (Fig. 6) is very roughly triangular with the lateral margins incised to a greater or lesser degree and often with the anterior lateral angles either extending or contracting. The clypeus of American species of *Chimarrha* Leach is said to be asymmetrical. Bounding the clypeus are two large curved epicranial sclerites with a dorsal and a ventral suture. The dorsal suture extends from the vertex of the clypeus to the occipital foramen. The ventral suture extends from the submentum to the occipital foramen, but the two sclerites may be separated to a greater or lesser extent by the fourth sclerite, known variously in the literature as hypostome, gula and gular sclerite. I have preferred to use the last term. The gular sclerite has many modifications, its shape being of taxonomic importance. In Fig. 7 the range of shape of the gular sclerite is shown but, although examples are given of the form in several families, the figures are not meant to imply that the shape shown is strictly typical for that family, but merely that it does occur in that family.

It has been pointed out by Siltala (1907) that the gular sclerite is not strictly homologous throughout the Order. Siltala considered that this sclerite, in the case of the HYDROPSYCHIDAE, was the mentum. If the whole range is compared it is quite apparent that the hard corneous or sclerotized anterior margin of the gular sclerite seen in the PHRYGANEIDAE, LIMNEPHILIDAE, SERICOSTOMATIDAE and ODONTOCERIDAE may be part of the true submentum. Indeed, in the SERICOSTOMATIDAE on the one hand and the RHYACOPHILIDAE on the other, we see intermediate steps in the suppression of the gular sclerite and the elaboration of this corneous region forming the basal or aboral sclerite of the submentum. These modifications are, of course, closely associated with the feeding habit of the larva. The elongated prognathous head of HYDROPSYCHIDAE, PHILOPOTAMIDAE and RHYACOPHILIDAE, carnivorous, non-case-making types, all show an extensive gular suture where the gena of each side meets in the mid-ventral line. The POLYCENTROPIDAE also show this, although the head is more globular: I suspect the latter feature as being secondarily produced. In the PHRYGANEIDAE and some SERICOSTOMATIDAE the epicranial margins are thickened and folded.

The attachment of the head muscles to the head sclerites is quite often distinctive, and is often the cause of characteristic if somewhat variable

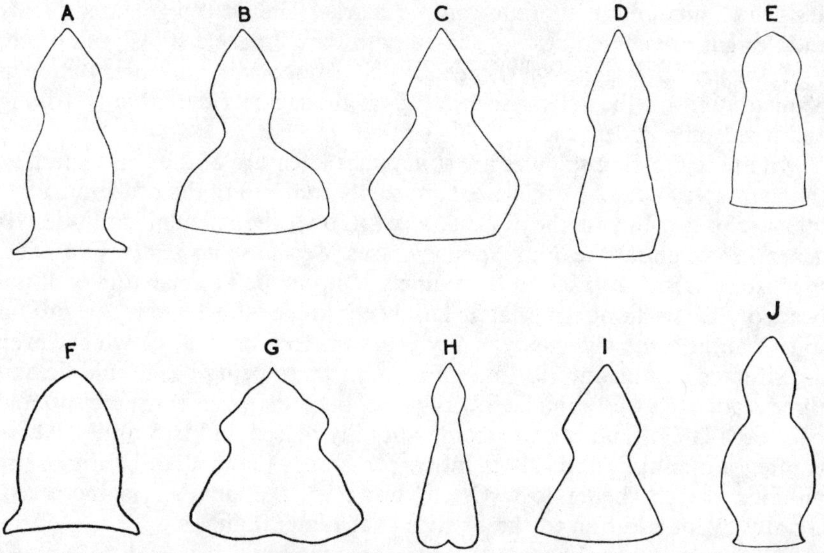

Fig. 6. Types of larval clypeus: A. PHRYGANEIDAE (*Oligotricha ruficrus*); B. LIMNEPHILIDAE (*Limnephilus flavicornis*); C. SERICOSTOMATIDAE (*Silo pallipes*); D. ODONTOCERIDAE (*Odontocerum albicorne*); E. LEPTOCERIDAE (*Setodes argentipunctella*); F. HYDROPSYCHIDAE (*Hydropsyche* sp.); G. POLYCENTROPIDAE sp.; H. PHILOPOTAMIDAE (*Philopotamus montanus*); I. RHYACOPHILIDAE (*Rhyacophila dorsalis*); J. SERICOSTOMATIDAE (*Lepidostoma hirtum*).

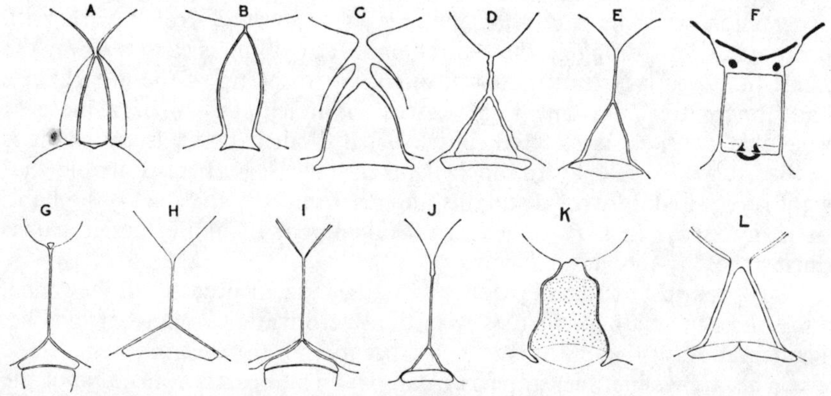

Fig. 7. Types of larval gular sclerite: A. PHRYGANEIDAE (*Oligotricha ruficrus*); B. LIMNEPHILIDAE (*Anabolia nervosa*); C. SERICOSTOMATIDAE (*Lepidostoma hirtum*); D. SERICOSTOMATIDAE (*Goera pilosa*); E. ODONTOCERIDAE (*Odontocerum albicorne*); F. LEPTOCERIDAE (*Mystacides nigra*); G. HYDROPSYCHIDAE (*Hydropsyche* sp.); H. POLYCENTROPIDAE (*Holocentropus dubius*); I. PHILOPOTAMIDAE (*Philopotamus montanus*); J. RHYACOPHILIDAE (*Rhyacophila dorsalis*); K. MOLANNIDAE (*Molanna angustata*).

designs of small elliptical light-centred marks. The dark pigmented bands and designs are sometimes associated with these 'muscle spots' and often, e.g. *Oligotricha* Leach, *Phryganea* L. and *Mystacides* Latreille, the deep pigmentation on the yellow sclerites gives the head of the larva a striking and handsome appearance.

Minute sclerotized spicules are sometimes a feature of the head sclerites. In the LIMNEPHILIDAE they are apparently confined to the head and a few other sclerites, but in the HYDROPSYCHIDAE the spicular condition is general throughout the body. In a few cases (*Lepidostoma hirtum* Fabricius, for instance) the head sclerites are finely sculptured. The chaetotaxy of the head of the trichopterous larva has been studied by Ulmer. According to this author, the clypeus typically bears thirteen bristles, of which seven are situated on the anterior margin, one of these being in the middle and three associated with each lateral angle; of the remainder, three are situated near each lateral margin, one being aborally placed and two orally. There is often a diminution in this number, sometimes the median bristle of the anterior margin being lost. Often, however, the number is increased, usually by an addition to the bristles in the lateral angles.

The eyes of all trichopterous larvae are small and it would not appear that sight plays an important part in their natural economy. Observation of larvae in aquaria gives this impression; thus in the chapter 'Habits of the Trichoptera' by M. B. Davis in Betten's *Report on the Trichoptera of New York State*, 1934, she states that larvae of *Hydropsyche* Pictet were unconscious of the presence of their prey until it actually touched them. Whilst this is being witten a small number of larvae of *Phryganea grandis* L. are in front of me in an aquarium. The method of progression when feeding is by making jerky forward strokes with both prothoracic legs, dragging the case behind. If anything is caught up by the sickle-like legs, it is immediately brought up to the tip of the mouthparts and apparently 'felt', 'smelt' or 'tasted'. If it happens to be a piece of vegetable debris it is rejected, but if a soft-bodied insect is caught it is devoured whilst held by the locked prothoracic legs. It is probable that sight is required only for distinguishing between light and dark—the habit of many larvae (e.g. *Potamophylax* Wallengren) of hiding under stones during the day-time is well known.

The eyes appear as small patches of black pigment situated on the genae, sometimes in small convexities with the sclerotization somewhat hyaline. Pigmentation of the chitin is usually absent from the area adjacent to the eye. This area sometimes appears as an oval white patch as in some of the POLYCENTROPIDAE. On the other hand, the eyes of *Silo pallipes* Fabricius are more deeply set into the head. Wesenberg-Lund has pointed out the relation between the feeding habits and the position of the eyes. Carnivorous larvae have eyes situated close to the anterior margin of the genae, whilst in herbivorous types they are set much farther back. Omnivorous larvae have the eyes intermediate in position.

Except in the LEPTOCERIDAE, the antennae are never conspicuous. In many cases, antennae, in the fully grown larvae, are atrophied; they are always present, however, in the young larvae but degenerate at a later stage. In the LEPTOCERIDAE the antenna consists of a proximal bulbous and a long, filiform, distal segment furnished with a sensory bristle. In several other families the distal segment is short but there are usually some sensory bristles or rods which may be quite long, as in *Agraylea* Curtis and *Oxyethira* Eaton, or finger-like, as in *Agapetus* Curtis.

The fairly typical mandibulate 'biting' mouthparts (Fig. 8, A–L) show a range of form which, if not very diverse, is at least quite important for determination of genera. A large membranous anteclypeus partially folds over the labrum. The anteclypeus is normally colourless, transparent and devoid of marking, but in the POLYCENTROPIDAE it is divided into four dark areas by three longitudinal white lines. The labrum (Fig. 11, A–D) is almost always yellowish, transversely elliptical, with an incision in the

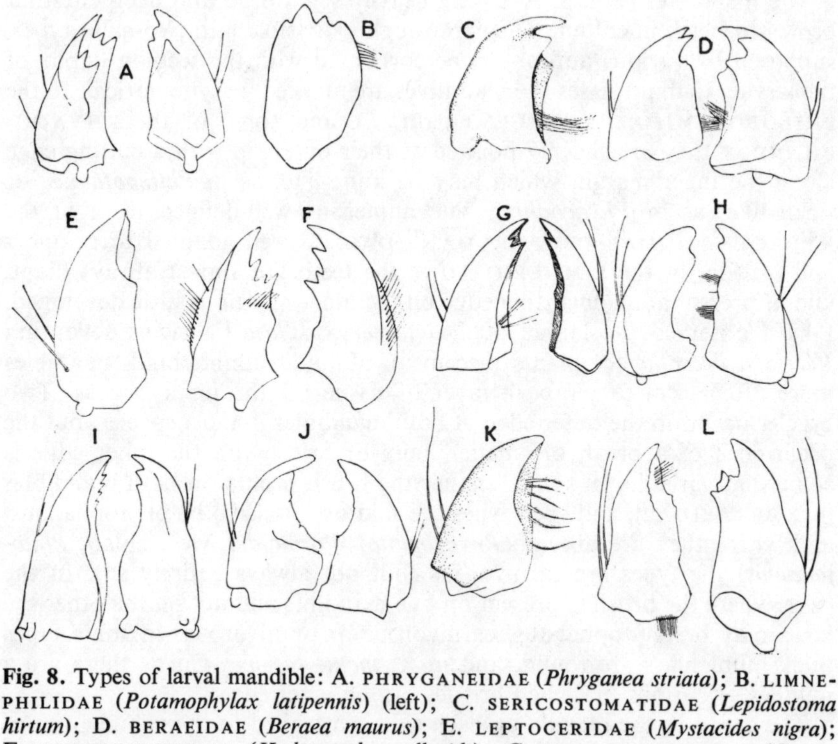

Fig. 8. Types of larval mandible: A. PHRYGANEIDAE (*Phryganea striata*); B. LIMNEPHILIDAE (*Potamophylax latipennis*) (left); C. SERICOSTOMATIDAE (*Lepidostoma hirtum*); D. BERAEIDAE (*Beraea maurus*); E. LEPTOCERIDAE (*Mystacides nigra*); F. HYDROPSYCHIDAE (*Hydropsyche pellucida*); G. POLYCENTROPIDAE (*Cyrnus flavidus*); H. PSYCHOMYIDAE (*Psychomyia pusilla*); I. PHILOPOTAMIDAE (*Wormaldia subnigra*) (from below); J. RHYACOPHILIDAE (*Rhyacophila dorsalis*); K. RHYACOPHILIDAE(*Agapetus fuscipes*); L. HYDROPTILIDAE (*Oxyethira costalis*). Fig. 8, D, F, G, H, I and L after Lestage.

middle of the anterior margin. The incision varies in extent and is absent in *Rhyacophila* and *Hydropsyche*. However, in the PHILOPOTAMIDAE are seen some remarkable deviations from the typical labrum. In *Philopotamus* Leach it is elongated longitudinally and has the anterior lateral areas extended into prongs.

In *Chimarra* the labrum is intermediate between this extreme and the typical form. Some black bristles and yellow spines are variously placed on the labrum; often one or two pairs of curved spines are situated on the anterior margin directed towards the median incision. In some species a cushion-like protuberance covered with minute spines is situated immediately posterior to the incision and probably has a sensory function. The ventral surface of the labrum has hairy tracts, and further spines are generally present. In the HYDROPTILIDAE the labrum is often asymmetrical and furnished with a beak and sometimes other adaptations, probably of use in puncturing algal filaments, the contents of which are sucked out.

The mandibles (Fig. 8, A–L) are heavily sclerotized and deep chestnut brown to black in colour. They are roughly divisible into two main types, supposed by various authors to be correlated with the feeding habits of the larvae. In many cases the mandibles are markedly asymmetrical. In the PHILOPOTAMIDAE, POLYCENTROPIDAE and some of the RHYACOPHILIDAE they are sharply pointed at their outer tip with a cutting edge along the inner margin which may be knife-like, as in *Philopotamus*, or sickle-like, as in *Rhyacophila*. The numerous well-defined teeth in the HYDROPSYCHIDAE and PHRYGANEIDAE are well adapted to gripping and tearing. In the LIMNEPHILIDAE the teeth are almost always blunt, and, if present at all, a cutting edge on the inner margin is little developed. I have observed the larvae of *Chaetopteryx villosa* Fabricius devouring *Veronica beccabungae* plants piecemeal, although limnophilid mandibles more often seem to scrape a layer of tissue off the plant organs. Two bristles occur on the outer edge of both mandibles in most species, but the occurrence of a brush of smaller hairs or bristles on the inner edge is somewhat variable. In some families the brush occurs on both mandibles (e.g. BERAEIDAE), and such types are said by Siltala to be phytophagous, and where they are absent (*Rhyacophila*, *Wormaldia* McLachlan, *Philopotamus*) the types are carnivorous (but not always entirely so). In the cases where the brush is present on one mandible and not on the other the larva may be phytophagous, carnivorous or omnivorous. In some types thick blunt hairs are found, and in *Agapetus fuscipes* Curtis there are a number of curious branched bristles on the inner edge.

The cardo of the maxilla on each side is attached to the submentum of the labrum at its base, and most figures in the literature show the maxillae and labium undivided. The maxilla (Fig. 9, A–L) consists of cardo, stipes, maxillary lobe (the inner lobe) and the palp. Cardo and stipes are both broad structures, sometimes hairy, often bearing bristles, and sclerotized bands are often conspicuous. Great variation attends the shape of the maxillary lobe where one or, in some cases, both endites are suppressed. In some families—PHRYGANEIDAE, ODONTOCERIDAE, HYDROPSYCHIDAE, POLYCENTROPIDAE and RHYACOPHILIDAE—the lobe is long, almost as long as the maxillary palp, and finger-like with the tip furnished with sensory papillae and rods. In some cases, as in *Holocentropus* McLachlan, the lobe is furnished with some inwardly directed yellowish spines, but more often, as in *Oligotricha* and *Odontocerum* Leach, the spines arise from the broadened proximal base of the maxilla. *Lepidostoma hirtum*

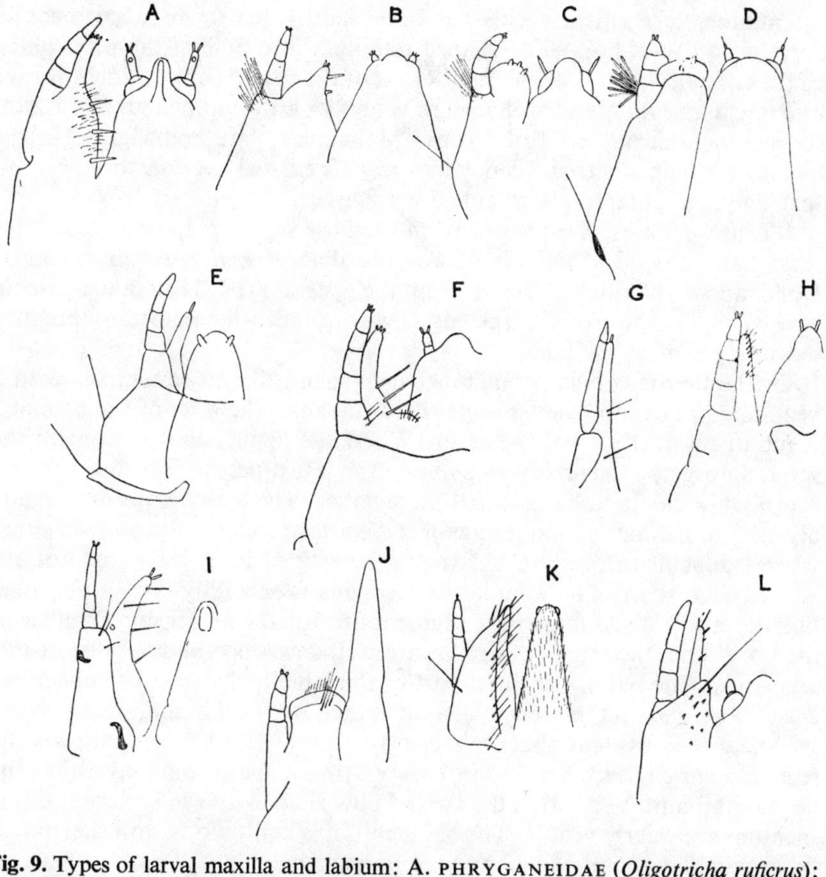

Fig. 9. Types of larval maxilla and labium: A. PHRYGANEIDAE (*Oligotricha ruficrus*); B. LIMNEPHILIDAE (*Limnephilus flavicornis*); C. SERICOSTOMATIDAE (*Lepidostoma hirtum*); D. SERICOSTOMATIDAE (*Goera pilosa*); E. BERAEIDAE (*Beraea maurus*); F. ODONTOCERIDAE (*Odontocerum albicorne*); G. LEPTOCERIDAE (*Oecetis ochracea*); H. HYDROPSYCHIDAE (*Hydropsyche pellucida*); I. POLYCENTROPIDAE (*Holocentropus dubius*); J. PSYCHOMYIIDAE (*Tinodes waeneri*); K. PHILOPOTAMIDAE (*Philopotamus montanus*); L. RHYACOPHILIDAE (*Rhyacophila dorsalis*).

Fabricius and *Goera pilosa* Fabricius both show flattened cushion-like lobes with the associated spines reduced to warty knobs. In *Tinodes waeneri* L. a similar cushion-like lobe is found but the spines are apparently suppressed. The maxillary palps are strong and sclerotized and consist usually of four segments inwardly curved. In the LEPTOCERIDAE there are often only three segments apparent, but in some other species of this family, the fourth (proximal) segment is just visible as a sclerotized patch on the outer side near the base of the maxillary lobe. Thus *Leptocerus aterrimus* Stephens and *Setodes argentipunctella* McLachlan both have four-segmented maxillary palps, whilst *Triaenodes bicolor* Curtis has only three-segmented palps. Some doubt exists amongst authors on the question of the number of segments. This has been mainly due to the occurrence in some groups of a horseshoe-shaped sclerite at the base of the palp which could or could not be considered as part of the palp. It appears that we are looking at an evolutionary change as yet only partly completed. I have not, so far, collected material of *Oecetis* McLachlan, but from figures in the literature it appears that a similar process is carried even farther here, so that only two complete segments are apparent.

A brush of hairs often arises from the base of the palp and sometimes these hairs are thick and blunt. The maxillae of *Philopotamus montanus* Donovan are strikingly different from the general type. They take the form of a pair of scoop-like sheaths edged with spines which close round, and seemingly protect the labium.

The labium is conical, sometimes hemispherical and sometimes with a beak-like projection (the ligula) at its distal end. The duct of the spinning organ opens at the tip of the ligula. Where the ligula is not prominent the spinning organ opens midway between the labial palps. This distal region is probably the fused glossae and paraglossae. The labial palps are usually distinct, sometimes prominent as in *Oligotricha ruficrus* Scopoli; in other species difficult to discern, but they are vestigial in at least some of the PSYCHOMYIIDAE. The number of segments is generally two, a proximal bulbous and a distal finger-like segment furnished with sensory papillae at the tip. (Often the sensory papillae are well developed and may be of two segments.) The basal, proximal part of the labial palp must be considered a segment and not merely a protuberance of the labium, as in a few instances it is evident that it is clearly separated by a separate sclerite (e.g. *Oligotricha ruficrus*). PSYCHOMYIIDAE have a characteristic long acicular labium. Very often the curved bow-shaped internal sclerites of the mentum are clearly visible. The submentum is contiguous with the gula at its proximal extremity as mentioned previously.

Thorax. Larvae of Trichoptera are endopterygotous, there being no external sign of developing wings. The three leg-bearing segments are quite distinct, although in some instances the pronotum is closely associated with the aboral region of the head. This is the case in *Oligotricha ruficrus*,

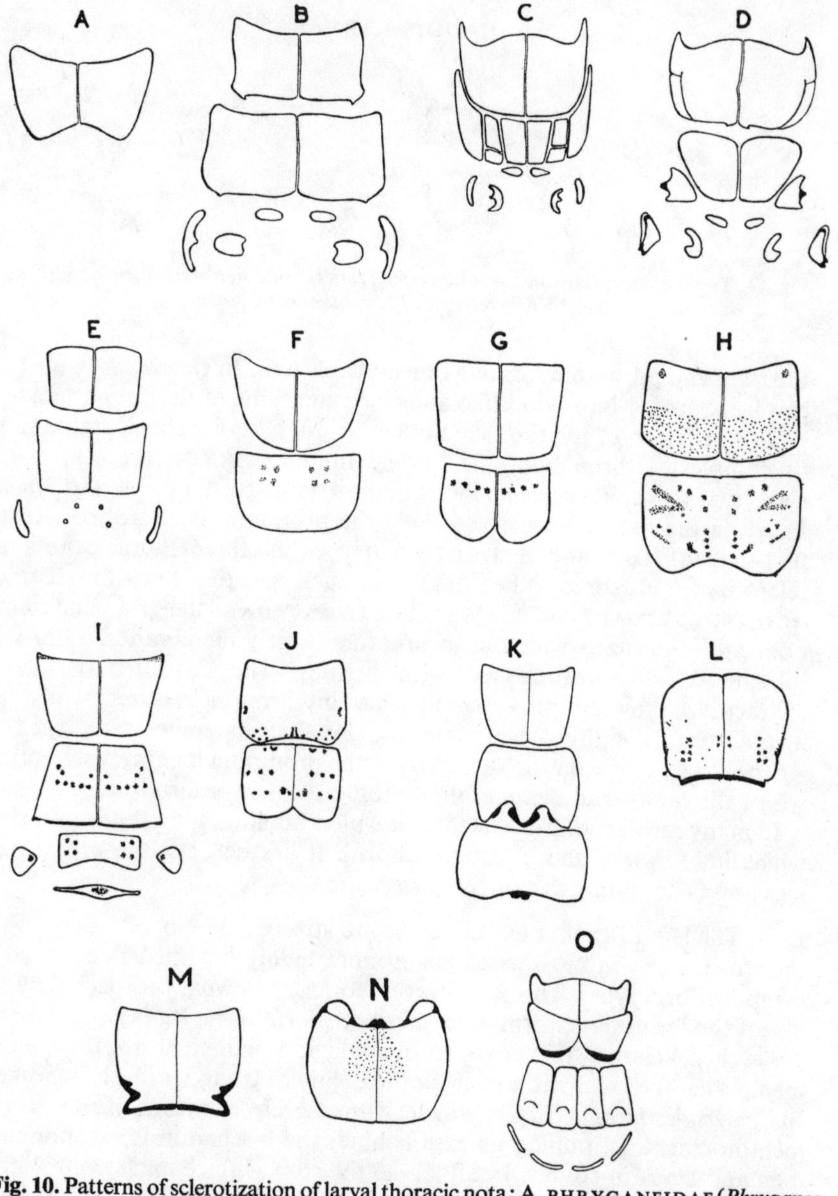

Fig. 10. Patterns of sclerotization of larval thoracic nota: A. PHRYGANEIDAE (*Phryganea grandis*); B. LIMNEPHILIDAE (*Limnephilus flavicornis*); C. SERICOSTOMATIDAE, GOERINAE (*Silo pallipes*); D. SERICOSTOMATIDAE, GOERINAE (*Goera pilosa*); E. SERICOSTOMATIDAE, LEPIDOSTOMATINAE (*Lepidostoma hirtum*); F. SERICOSTOMATIDAE, SERICOSTOMATINAE (*Sericostoma personatum*); G. BERAEIDAE (*Beraeodes minuta*) after Struck; H. MOLANNIDAE (*Molanna angustata*); I. ODONTOCERIDAE (*Odontocerum albicorne*); J. LEPTOCERIDAE (*Triaenodes bicolor*); K. HYDROPSYCHIDAE (*Hydropsyche* sp.); L. POLYCENTROPIDAE (*Holocentropus dubius*); M. PHILOPOTAMIDAE (*Philopotamus montanus*); N. RHYACOPHILIDAE (*Rhyacophila dorsalis*); O. SERICOSTOMATIDAE (*Brachycentrus subnubilus*).

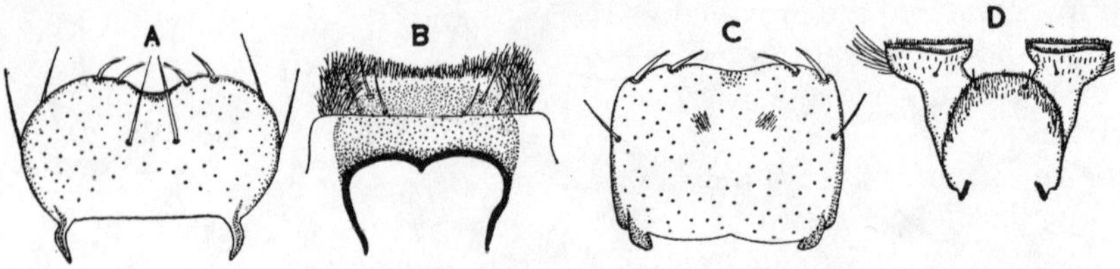

Fig. 11. Types of larval labrum: A. *Oligotricha ruficrus;* B. *Agapetus fuscipes*; C. *Setodes argentipunctella*; D. *Philopotamus montanus*.

the head and pronotum forming one movable unit. In *Goera pilosa* the head can be retracted into hood-like anterior extensions of the pronotum.

Sclerotization of the thoracic nota (Fig. 10) shows a considerable range of complexity throughout the Order. In the PHRYGANEIDAE, POLYCENTROPIDAE, *Rhyacophila* Pictet in the RHYACOPHILIDAE, and *Tinodes* Leach in the PSYCHOMYIIDAE, only the pronotum is sclerotized. In the HYDROPTILIDAE and HYDROPSYCHIDAE all three thoracic nota are sclerotized. In some other families, such as the LIMNEPHILIDAE, ODONTOCERIDAE, and the SERICOSTOMATIDAE, the pro- and mesonota are sclerotized with the metanotum largely membranous, but with sclerotized patches furnished with bristles. The LEPTOCERIDAE are characterized by the pro- and mesonotum being sclerotized whilst the metanotum is entirely membranous. Finally, in some cases such as *Agapetus* C. in the RHYACOPHILIDAE, the pronotum is entirely sclerotized whilst the meso- and metanotum are only partially sclerotized.

In many case-bearing larvae there is a membranous finger-like appendage which is known as the 'prosternal horn'. It lies between the prothoracic legs and is an outgrowth of the prosternum.

Legs. The legs of trichopterous larvae are strongly developed, those of the eruciform types in most cases being more highly modified than those of campodeiform types. This is, however, contrary to what is expected by the use of the terms 'eruciform' and 'campodeiform'. The leg is supported by an arched sclerite, the coxopleurite which is adherent to the pleural membrane, the trochantin bearing the condyles being variously modified. In *Triaenodes bicolor* Curtis, which swims largely by means of the fringed metathoracic legs, pulling its case behind, the trochantin is exceptionally long and black in colour. In all larvae the coxa is thick and strong with a deep incision on the dorsal surface of the distal margin of the sclerite which is covered with a thin membrane and allows the trochanter to move in an upward direction. The trochanter consists almost always of two divisions. In *Athripsodes aterrimus*, however, the trochanter of the metathoracic legs consists of three divisions, but this is not general throughout the LEPTOCERIDAE, although there is a furrow in the metathoracic trochanter in some other species as though there were an incipient division. The divisions of the trochanter do not appear to exercise any separate movement and the proximal division is nearly always considerably smaller than the distal. Sometimes, as in the LEPTOCERIDAE, the trochanters are of considerable length. Very seldom does the proximal segment bear many bristles.

The femur is always the largest single sclerotized segment of the leg and often bears one or two black bristles dorsally, of considerable length, and also a series of spines on the ventral edge, which, in eruciform and suberuciform larvae, are often strong and of two distinct sizes, so that when the tibia and tarsus are pulled against the femur by the strong muscles which are present, the prey is very efficiently trapped. It is peculiar that this mechanism is not developed in the campodeiform types which are looked upon as typically carnivorous; the legs of the latter forms are quite innocent of such strong spines. The femur is inserted deeply into the distal segment of the trochanter, and previous authors have differed widely in making out the structures whereby these segments are attached. Siltala considers that the femora of meso- and metathoracic legs in the LEPTOCERINAE are divided into two segments. A. W. Orcutt adds to this and states that in *Mystacides* and possibly other LEPTOCERINAE the trochanters are divided into three segments. From studies of *Athripsodes aterrimus*, *Triaenodes bicolor*, *Mystacides nigra* L., *M. azurea* and *Setodes argentipunctella* my view is that in some cases the trochanters are divided into two and in others into three. Where the division is into two parts the distal part may show a further division. The femur is always of one segment, but in this sub-family (LEPTOCERINAE) there is a considerable amount of articular movement between the femur and the distal segment of the trochanter. (In other groups the amount of movement is very limited.) Apparently there is a thin, light but strong membranous sheath forming a hinge between the segments; where this sheath joins the femur it gives the misleading appearance of a divided femur. Again, the ventral sclerotized wall of the distal segment of the trochanter extends upwards and inwards and forms an anvil against which the proximal end of the inserted femur can be levered by means of the muscles. It should be noted that in the LEPTOCERINAE the legs are very long and slender.

The tibia and tarsus both consist of a single sub-cylindrical sclerite, but in *Mystacides* both show an incipient but not quite complete division in the metathoracic legs. In some campodeiform types the tarsus, and in addition sometimes the tibia, is covered with short hairs or spines which are said to serve as a brush for cleaning the nets. In connexion with this observation, however, it should be mentioned that nets are frequently found which are completely silted up with fine mud. In *Holocentropus* there is a tuft of branched hairs at the distal extremity of the tarsal sclerite. The tarsus is terminated by a strong claw furnished with a spine at its base. (The spine in *Molanna* is longer than the claw.) The modifications of claw and spine offer useful characters for determination. The claw is perhaps least modified in some of the campodeiform types such as *Rhyacophila*.

Orcutt has called attention to the play of functional adaptation in the legs of trichopterous larvae. Thus in *Brachycentrus* Curtis, the femora of the prothoracic legs are short but very thick and strong. This is related to the habit of this species of anchoring the case to a frond of weed and seizing small insects and vegetable debris as they are carried past the opening of the case by the current. The long fringed metathoracic legs of *Triaenodes bicolor* with well-developed trochantin have already been alluded to in the text as serving for swimming. The short legs of most campodeiform net-spinning types, such as *Hydropsyche*, are other examples of this functional adaptation. Generally, in campodeiform types, the legs are approximately equal in size, whereas in eruciform types the prothoracic legs are always shortest, usually held bunched up under the head, whilst the metathoracic legs are longest, but in some cases, e.g. some LIMNEPHILIDAE, the mesothoracic legs are slightly longer.

Abdomen. The abdomen is almost entirely membranous, very little sclerotization being present on this part of the body. In eruciform larvae the abdomen is usually uniformly cylindrical or tapering to a small degree

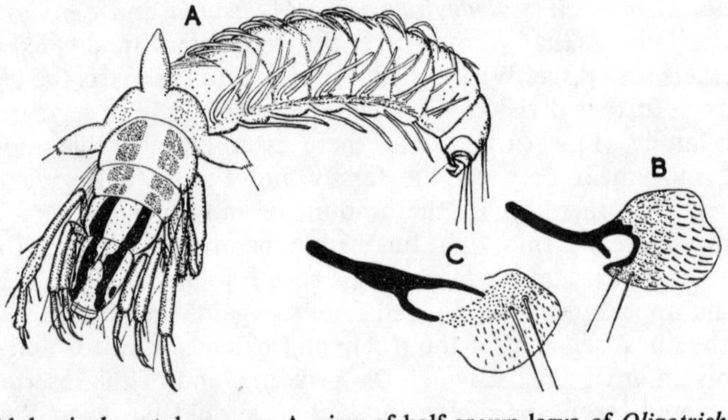

Fig. 12. Abdominal protuberances: A. view of half-grown larva of *Oligotricha ruficrus* showing protuberances; B. lateral sclerite on protuberance of *Mystacides nigra*; C. lateral sclerite on protuberance of *Leptocerus aterrimus*.

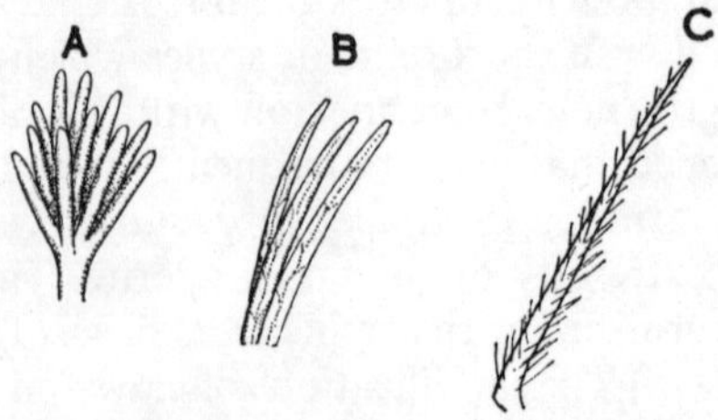

Fig. 13. Types of gill: A. branched tracheal gill (HYDROPSYCHIDAE); B. blood gills (POLYCENTROPIDAE); C. simple tracheal gill (PHRYGANEIDAE).

only. It is white or cream in colour but may be light green and with no colour pattern on the dorsal surface. There may, however, be a shield-shaped patch, sometimes sclerotized and furnished with bristles, and dark in colour on the dorsal surface of the ninth segment. The shape of the abdomen of eruciform larvae is, of course, related to the case-bearing habit of this type of larvae. The abdomen is very seldom completely withdrawn from the case; in many species this may only happen if an accident occurs. In those species which cut off the tail end of the case when the head end is increased in size, the anal claspers still retain a grip on the rim of the head end of the case. The intersegmental grooves in the abdomen of eruciform types are shallow, whereas in suberuciform and campodeiform types they are deep. In some campodeiform forms (POLYCENTROPIDAE) the abdomen is somewhat dorsoventrally compressed, but most are usually thickest at about the third segment; other campodeiform forms (PHILOPOTAMIDAE and HYDROPSYCHIDAE) have cylindrical abdomens. In HYDROPTILIDAE the abdomen is very distended at the fourth and fifth segments and is so reminiscent of some coleopterous eruciform larvae that it may be mistaken for a larva of that form. Campodeiform and suberuciform larvae are not usually white (*Philopotamus* is an exception). They may be green (*Rhyacophila* and *Hydroptila*), reddish-brown (POLYCENTROPIDAE), or greyish-brown (*Hydropsyche* Pictet), and a colour pattern may be present on the dorsal surface (*Oligotricha*).

Most eruciform and suberuciform larvae bear three protuberances or 'tubercles' on the first abdominal segment. One is disposed dorsally in the middle of the segment and the remaining two laterally. In species in which they are most developed (such as *Oligotricha* and *Phryganea*) they are fleshy and finger-like and often bear sensory bristles or patches of small spicules (Fig. 12, A). They are absent in free-living larvae, and often in eruciform larvae the median dorsal protuberance is absent. In some species, e.g. *Athripsodes aterrimus*, the lateral protuberances are not well developed, but the patches of sclerotized spicules are conspicuous and a spur-like black sclerite may be present (Fig. 12, B and C).

Gills. The aquatic habitat of the larvae has led to a closed tracheal system and to gill development. It is agreed with Orcutt that in many cases a large proportion of the oxygen absorption for respiration must be cuticular, yet gills of some type are almost always present. Several authors, such as Siltala and Orcutt, have made mention of localized patches of membrane on the coxa of each leg for oxygen absorption which occur in *Hydropsyche* and some POLYCENTROPIDAE, whilst in *Ithytrichia* Eaton the same function is undertaken by sac-like diverticulae on the pleural region of the third to eighth segments. Amongst the more normal types of gill occurring in the Order there are simple filiform gills which may be covered with small black hairs (Fig. 13, C) as in *Oligotricha,* and may vary in length from a small conical process to a long filament as long as, or even longer than, the abdominal segment from which it arises. On the other hand, many larvae bear compound gills (some species of *Rhyacophila* and *Hydropsyche*) which have the appearance of tufts of small gills arising from a common stalk or stalks (Fig. 13, A). The number of gills present varies to a great extent throughout the Order, and indeed small variations are found within the species. Thus it is not wise to place undue reliance on the number of gill filaments in determining species, especially where closely related species are said to differ in respect of single gills. In general the gills occur as longitudinal series with dorsal, lateral and ventral rows. Transversely the gills occur at either the anterior or lateral, or both margins. The occurrence of gills in the different species is shown by Lestage (*in* Rousseau) in tabular form. Gills are usually absent from the first abdominal segment but extend from the second to the eighth segment. Very seldom do tracheal gills occur on the ninth segment. All gills are directed posteriorly.

Blood gills (which are non-tracheated) occur in the Trichoptera only as evaginations of the anal region and are few in number, never exceeding six (Fig. 13, B). They are usually eversible: this is brought about by pressure exerted by the abdominal segments, and the retraction is carried out by an inserted branched muscle arising from the body wall. Many species have the habit of moving the abdomen in an undulating rhythmic manner in the vertical plane. This is most probably an accessory aid to respiration, causing a flow of water over the gills and the surface of the body. Campodeiform types such as some of the POLYCENTROPIDAE, suberuciform, for example, *Phryganea striata* L. and eruciform (*Limnephilus flavicornis*), all show this habit.

In all case-bearing larvae there occurs a tract of dark hairs running along the pleural region on both sides of the abdomen. This is known as the 'lateral line'. In its most highly developed condition it extends from the third abdominal segment to the eighth but is often lacking in some of the hinder or fore segments. Generally the lateral line is made up of two distinct lines of hairs and has been likened to a fold or wrinkle.

It should be mentioned, however, that some free-living larvae, such as some of the POLYCENTROPIDAE, show a hairy tract running along the pleural region of the abdomen. These hairs are long and not nearly so numerous as those in the lateral line of the case-bearing larvae. Probably associated with the 'lateral line' is a series of very small sclerotized points or tubercles which are reddish-brown in colour. These occur in a small linear series just dorsal to the lateral line. There may be eight or less tubercles in each segment on each side in species where the tubercles are fully developed.

Anal Claspers. All trichopterous larvae bear a pair of claws at the posterior end of the abdomen (Fig. 14, A). These are situated on variously segmented appendages or pygopods which arise from the ninth abdominal segment. In case-bearing larvae these appendages are short and bulbous with the proximal segments much enlarged, often fused to form a tenth abdominal segment. With the free-living larvae the appendages are long and slender. These claws are very strong and when hooked into the silken lining of the case it is often impossible to remove the larvae without rupturing the abdomen. Even when the larva is cutting away part of the hinder end of the case, it retains a grip on the silken lining of the case by means of the claws. In the free-living types the anal claws serve to grip the silken net when the larva is running backwards, which many of them do over their

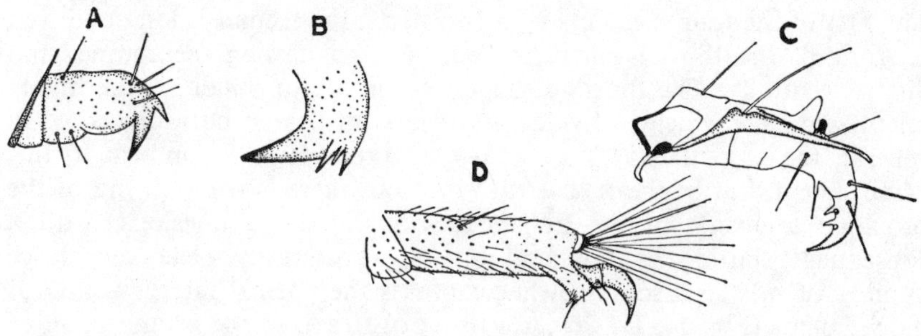

Fig. 14. Types of anal clasper: A. claw of anal clasper of *Limnephilus lunatus*; B. tip of claw of clasper of *Mystacides nigra*; C. anal clasper of *Rhyacophila dorsalis*; D. anal clasper of *Hydropsyche* sp.

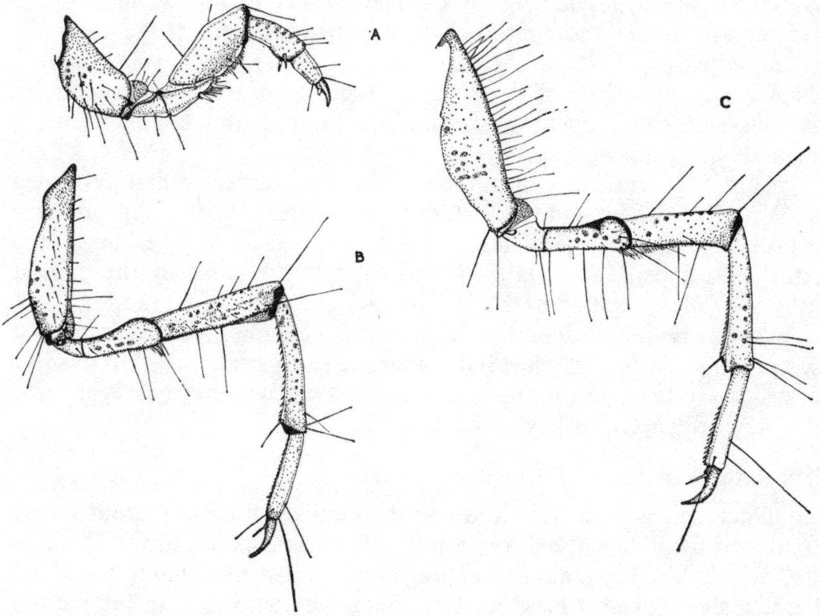

Fig. 15. Legs of trichopterous larva (*Limnephilus flavicornis*): A. prothoracic; B. mesothoracic; C. metathoracic.

nets or down their silken tubes. The tips of the claws are directed anteriorly and often bear two or three smaller auxiliary hooks (Fig. 14, B) and sometimes rows of small spicules. Much discussion has taken place on the homology of the segments of the claw-bearing appendages. Embryological studies have apparently not been carried out, but Siltala has put forward his views based on the positioning of bristles, and Orcutt (in Betten) has discussed them. Briefly, Siltala considers the typical trichopterous anal appendage (as found in certain POLYCENTROPIDAE) as being divided into four segments and as borne on the ninth abdominal segment. The first of the four segments is in two parts, the second is long and well sclerotized, the third is small, of soft interconnecting membrane, and dilated ventrally. The fourth segment, according to Siltala, is the highly sclerotized claw which is divided and bears bristles on both divisions. In general, the PSYCHOMYIIDAE and PHILOPOTAMIDAE conform to this arrangement, but in the HYDROPSYCHIDAE there is some fusing of the proximal segments of the appendages as the larva develops, with a consequent shifting of the anal opening posteriorly. The very large number of bristles also somewhat confuses the normal interpretation of the segments (Fig. 14, D). In the RHYACOPHILIDAE the arrangement of the bristles is again rather confusing, but in the HYDROPTILIDAE changes take place progressively during the larval life, but a tenth segment is formed during the later stages. In *Rhyacophila* a finger-like appendage arises from each pygopod (Fig. 14, C). The tenth abdominal segment of the case-bearing larvae is composed of the first two segments of the anal prolegs, and the two sclerites at the base of the claw (one usually being lunate), which are characteristic of these larvae, represent the other segments. In these larvae the anal opening is situated at the posterior line of fusion in the made-up tenth segment.

The above general account of the trichopterous larva was first published in 1946. Since then Badcock, Hanna, Macdonald and Philipson have discussed the morphological nomenclatures of parts of the larval head used in the 1946 paper. In the latter it was thought advisable by the present author to follow the important earlier works, notably of Ulmer.

There has been no attempt to incorporate a common nomenclature for the particular parts of the head of the larva as this might have led to some confusion. Instead, the individual author's own wording has been used when their descriptions have been quoted.

Keys to the Families of Trichopterous Larvae

The difficulties in presenting a key to the families of trichopterous larvae are numerous. Firstly it will be seen that the classification of the Trichoptera, which has been almost entirely (except by Martynov) based on characters of the adult flies, certainly does not conform with the natural groups which I have found amongst the larvae. The SERICOSTOMATIDAE is one example of an assemblage rather than a natural group, and *Agapetus*

Curtis in the RHYACOPHILIDAE and *Ecnomus* McLachlan in the PSYCHOMYIIDAE, as well as several others, occupy purely arbitrary positions. Ulmer, in his researches on the Trichoptera in Baltic Amber, pointed out in 1912 the polyphyletic nature of the SERICOSTOMATIDAE. Some authors, such as Martynov, have attempted to rectify this by alterations of the classification but inevitable complications have occurred, and as the object of this work is to present descriptions and keys which the general biological worker and freshwater biologist can use, the key has been kept as simple as possible. When this work was first put to press, the classification adopted by Mosely in his *British Caddis Flies*, a classification based upon the work of McLachlan and Ulmer was followed, Mosely's book being the only one likely to be available referring only to the British species. However, the order of families was revised in accordance with the second edition of Kloet and Hincks check list (1964) and even more recently it was revised again in the light of the check-list of Kimmins (1966).

Although much help has been derived from the keys of Ulmer, Lestage (*in* Rousseau), and Krafka (*in* Betten), an entirely different approach has been made in the present key. It has been usual to make an initial splitting on the considerations of whether the larvae are of the campodeiform or eruciform types. To the general entomological student, the TRICHOPTERA present some difficulties here. Not only do some species (PHRYGANEIDAE) exhibit both campodeiform and eruciform characteristics, but several genera, such as *Agapetus*, are difficult to differentiate in this regard. Accordingly characters have been chosen which are positive, such as the presence and shape of sclerotized areas on the thoracic nota, the clypeus and gular sclerite. Confirmatory differentiating evidence is also found in the nature of the case (in the case-bearing types), but such evidence must bear a purely secondary role, as a larva after collection often leaves its case and may invade the case of another. The present author has even found this in nature, having taken a half-grown specimen of *Phryganea striata* from the case of a fully grown example, and it is quite possible that different species may occupy each other's cases when driven out of their own.

The marking of the sclerites and the position of bristles have not been used as determining characters on account of variation of the former and the possible breakage of the latter. Larvae are often found on examination to have several bristles broken off close to the cuticle and only the circular pit is left.

It has not been possible at the present stage entirely to differentiate between larvae of the BERAEIDAE and those of the SERICOSTOMATINAE. Ulmer suggested that the BERAEINAE should be split off from the MOLANNINAE and placed with a nearer relationship to the SERICOSTOMATIDAE, and Mosely has described the genus *Beraeodina* which is intermediate between these two families. Thus the larval characters tend to confirm this view.

(a) *Key in which Initial separation is based on Degree of Sclerotization* (Note: Figures and letters in parentheses refer to the text-figures to which reference is made as a guide in using the key. More detailed figures are given in the short separate papers, references to which are given on pages 50 et sqq.)

1. Metanotum entirely membranous 2
   Metanotum sclerotized wholly or in part 10
2. Mesonotum membranous 3
   Mesonotum sclerotized 7
3. Pronotum sclerotized with heavily sclerotized concave anterior margin, protuberant lobes on anterior margin absent (10, A). Larva of suberuciform type. Large medium and two lateral lobes on first abdominal segment. Gular sclerite narrow longitudinally with adjacent areas of genae thickened into lobes and completely dividing genae (7, A). Larva bears a case made of vegetable fragments, cut leaves, stalks, etc., arranged spirally PHRYGANEIDAE
   Pronotum sclerotized but not heavily sclerotized at anterior margin (3 and 4). Larvae of campodeiform type. Do not bear cases. Gular sclerite transverse and not dividing genae 4
4. Labrum not sclerotized, elongated with transverse ciliated anterior margin (6, A). Clypeus extremely narrow (6, H) PHILOPOTAMIDAE
   Labrum sclerotized and transverse 5
5. Labium long, slender and pointed, projecting beyond the maxillary lobe (9, J). Gills absent except 5 anal gills. Lateral line absent PSYCHOMYIIDAE (except *Ecnomus*)
   Labium not projecting beyond the maxillary lobe 6
6. Clypeus with sinuate margin and excised anterior margin (6, G). Abdomen brownish or reddish. Abdominal gills (except anal gills) absent. Gular sclerite simple (7 H) POLYCENTROPIDAE
   Clypeus with single excision only on lateral margins (6, I). Anterior margin not excised. Abdomen green, sometimes with tufted gills. Gular sclerite with transverse anterior sclerotized lobe (7, J) RHYACOPHILIDAE (except *Agapetus*)
7. Mesonotal sclerite divided into four. Gular sclerite trapezoidal in shape and completely divides genae (7, K). Larval case shield-shaped, formed of lateral wing-like extensions of the tubular centre MOLANNIDAE
   Mesonotal sclerite entire or divided into two sclerites only by a longitudinal suture 8
8. Brush of hairs absent on inner (concave) side of right mandible LEPTOCERIDAE
   Brush of hairs present on inner (concave) side of right mandible 9
9. Brush of hairs present on outer (convex) side of mandible BERAEIDAE (*Beraeodes* only)
   Brush of hairs absent from outer (convex) side of mandible BERAEIDAE (*Beraea* only) and SERICOSTOMATIDAE (SERICOSTOMATINAE)
   Note: In the absence of detailed studies of all the British species it is not possible to differentiate between *Beraea* and SERICOSTOMATINAE.
10. Metanotum entire or divided into 2 or 4 sclerites 11
    Metanotum divided into 6 sclerites 12
11. Metanotum entire or divided into 2 sclerites only 13
    Metanotum divided into 4 sclerites 14
12. Four median sclerites very small, reduced to small patches, from each of which springs a single bristle (10, E). Clypeus with an additional excision on each side anteriorly (6, J). Genae with small inwardly directed lobes adjacent to gular sclerite (7, C). Larval case generally of quadrangular cross-section throughout its length SERICOSTOMATIDAE (LEPIDOSTOMATINAE)

Four median sclerites each furnished with more than one bristle. Clypeus with a single lateral excision only. Inwardly directed lobes of genae adjacent to gular sclerite absent. Gular sclerite narrow or wide. Larval case never of quadrangular cross-section throughout its length 15

13. Abdomen cylindrical and slender 16

Abdomen thick, widest at about the fifth abdominal segment, whitish and dark spicules absent. Tufts of long black hairs on anal claspers absent. Length of larva up to 8 mm. Makes a hemispherical case of sand grains, small stones or diatoms, with both openings on the ventral surface, or of secretion only, seed-like with small amount of added matter 17

14. Metanotum with two large parallel transverse sclerites, anterior of which is broad with posterior margin excised. Posterior sclerite narrow, thickened at the centre (10, I) ODONTOCERIDAE (*Odontocerum albicorne* is only British representative)

Metanotum without two large parallel transverse sclerites 18

15. Gular sclerite narrow, widest at the centre and excised on lateral margins anteriorly (7, B). Mesonotum divided by longitudinal suture. Anterior projections of the lateral sclerites of the mesonotum (10, B) LIMNEPHILIDAE

Gular sclerite wide, widest at the anterior margin. Not excised on lateral margins (7, L). Mesonotum divided into 8 sclerites. Anterior projections of the lateral sclerites of the mesonotum present (10, C) SERICOSTOMATIDAE (*Silo* in GOERINAE)

16. Tufts of long black stiff hairs present on anal claspers (14, D). Body dark greyish or greyish-brown covered with dark-coloured spicules, legs short (3). Length of larva 10–20 mm. Do not bear a movable larval case but inhabit a fixed case adjacent to a small silken net attached to small stones or vegetable debris HYDROPSYCHIDAE

Tufts of long black stiff hairs on anal claspers absent. Mentum very wide, ligula acute PSYCHOMYIIDAE (*Ecnomus* only)

17. Legs long, metanotal sclerites restricted to small patches more or less distinct. Case of sand-grains, etc., hemispherical with both openings on ventral surface. Larva of eruciform type. Relative size of abdomen to thorax normal. Length of larva up to 9 mm. RHYACOPHILIDAE (*Agapetus*)

Legs short, metanotal sclerites large. Case sometimes seed-like, often with rounded ends forming lateral shields to the openings. Larva of campodeiform type. Abdomen relatively large to size of thorax. Length of larva up to 7 mm. HYDROPTILIDAE

18. Metanotal sclerites in two groups of two. Mesonotum of four sclerites, the outer of which are triangular with heavily sclerotized black projections outwardly directed. Head strongly retractile into pronotum, the latter furnished with wing-like anterior lateral projections (10, D). Posterior margin of pronotum not strongly excised SERICOSTOMATIDAE (*Goera* only in GOERINAE)

Metanotal sclerites grouped in arc with concavity directed anteriorly. Mesonotal sclerites four in number, quadrangular and transverse. Posterior margin of pronotum strongly excised (10, O) SERICOSTOMATIDAE (BRACHYCENTRINAE)

Brindle considers that the key given above, in which the initial separation concerns the degree of sclerotization of the thoracic nota, can be misinterpreted. He believes that a key in which the initial separations are based on the type of case produced by the larva would be easier to use. Such a key is given below. It should be borne in mind, however, when using it that many larvae leave their cases when taken out of water. Take care, therefore, that the case has been assigned to the correct larva.

(b) *Key in which Initial Separation is based on Type of Case*

1. Larvae in a transportable case 2
— Larvae without a transportable case 13
2. Cases almost hemispherical, both openings on the flattened ventral surface; cases of comparatively large pieces of stone, rough, fastened to the upperside of stones in streams and rivers RHYACOPHILIDAE (GLOSSOSOMATINAE)
— Cases not hemispherical 3
3. Cases very small, of fine mineral particles, greatly flattened or flask-shaped; larvae small, campodeiform, some segments of abdomen much wider than thorax; each thoracic nota with a single sclerotized plate; in static or lotic water HYDROPTILIDAE
— Cases not greatly flattened or flask-shaped; larvae usually larger, eruciform or suberuciform, no segments of abdomen wider than thorax; metanotum at least without a single sclerotized plate 4
4. Cases shield-shaped, composed of sand-grains, and formed of a tube with lateral extensions; head with two dark bands; in static water MOLANNIDAE
— Cases not shield-shaped, but usually more or less tubular, less commonly of quadrangular or triangular section (cf. *Glyphotaelius*) 5
5. Larvae sub-eruciform with only the pronotum sclerotised; cases of vegetable material, tubular, hardly tapering posteriorly, open at both ends, typically composed of uniform lengths of material arranged spirally; sometimes a hollow stem is used (*Agrypnia*) and the case of *Trichostegia* is not spirally made; in static or slowly moving water PHRYGAENIDAE
— Larvae eruciform with at least both pro- and meso-nota sclerotized (the latter slight in some Leptoceridae); cases variable but with one exception not spirally made. (Exception: *Triaenodes*, in which the case is very thin, long and tapers posteriorly) 6
6. Pronotum produced at anterior corners 7
— Pronotum not produced at anterior corners 8
7. Cases tubular or elliptical, of mineral particles, with larger pieces of stones arranged down either side; tibiae without spines on processes; pronotum with median longitudinal suture only; eyes lateral; in rivers and streams SERICOSTOMATIDAE (GOERINAE)
— Cases tubular, of secretion only, anchored at the distal end; tibiae with spines on processes; pronotum with longitudinal and transverse sutures; eyes dorsal; in rivers SERICOSTOMATIDAE (BRACHYCENTRINAE)
8. Clypeus with distinct anchor-like mark; thoracic nota with sclerotized plates; case of mineral particles, smooth, tubular, slightly curved; in rapid streams ODONTOCERIDAE
— Clypeus otherwise; thoracic nota not with such an arrangement of plates 9
9. Prosternal process present 10
— Prosternal process absent 11
10. Cases quadrangular in section, at least partially (*Crunoecia*, *Lepidostoma*); if round in section (*Lasiocephala*) then mesonotum not with a complete sclerotized plate; in lotic water SERICOSTOMATIDAE (LEPIDOSTOMATINAE)
— Cases variable, usually tubular, sometimes triangular in section, rarely quadrangular; mesonotum always with a complete sclerotized plate and metanotum typically with three pairs of small sclerotized plates; in lotic or static water LIMNEPHILIDAE
11. Head uniformly dark brown or blackish or with light spots; thoracic nota densely hairy; abdomen wide, cylindrical, not tapering posteriorly, ninth segment almost as wide as eighth; cases of mineral particles; in rivers SERICOSTOMATIDAE (SERICOSTOMATINAE)
— Head lighter or with a distinct pattern; thoracic nota not densely hairy;

abdomen cylindrical, narrower, tapering, ninth segment usually much narrower than eighth; cases of fine or coarse mineral particles, or vegetable material, or of secretion only 12

12. Head either round and unicolorous reddish or slightly elongated, black and yellow; antennae short, inconspicuous; right mandible with brush of setae on inner side; cases of fine mineral particles; in lotic or static water BERAEIDAE

— Head elongated, with or without distinct pattern; antennae long or very long, conspicuous; no brush of setae on right mandible; cases variable; in lotic or static water LEPTOCERIDAE

13. Larvae free-living amongst stones in rapid streams and rivers; lateral gill tufts on abdomen RHYACOPHILIDAE (*Rhyacophila*)

— Larvae not free-living, but in nets or tubes; no lateral gill tufts on abdomen 14

14. Larvae in tubes or tunnels of silk, covered with mineral particles or other debris, on submerged stones or wood; anal appendages short (only the pronotum sclerotized except *Ecnomus*); in lotic or static water PSYCHOMYIIDAE

— Larvae not in tubes or tunnels but in silken nets, anal appendages long 15

15. Larvae with ventral gill tufts; often dark coloured and communal; in rapid streams HYDROPSYCHIDAE

— Larvae without gill tufts ventrally 16

16. Labrum soft, whitish; head elongated, parallel-sided, unicolorous red or yellow; abdomen whitish or yellowish; in lotic water PHILOPOTAMIDAE

17. Labrum sclerotized, yellow or brown; head not elongated, more rounded; yellowish or brown, often with darker spots or bands; abdomen reddish, pink, yellowish, or greenish, etc., never white; in lotic or static water POLYCENTROPIDAE

(c) *Key Based on that of Bertrand* (1954)

Modified to accommodate only families known to be British.

1. Larvae-free, naked, often in webbing, nets or galleries, but never found in a mobile case, even when mature; flattened prognathous head; well-developed anal claws, often very long, always clearly visible 2

— Larvae never free (except when young), always enclosed in a mobile case of varied form and structure (rarely entirely of silk), more often covered with diverse materials, mineral or vegetable, also shells, which are sometimes replaced by pieces of vegetable debris; head variable, rarely flattened and prognathous, more often hypognathous, sometimes of the intermediate type; anal claws short or very short, hardly visible 7

2. Sclerotized pronotum, mesonotum and metanotum 3

— Only pronotum sclerotized 4

3. Body more or less curved with gills in bunches on the ventral surface HYDROPSYCHIDAE

— Body not curved, without gills, in spun galleries PSYCHOMYIIDAE partim (ECNOMINAE)

4. Labrum normal, sclerotized 5

— Labrum soft, triangular PHILOPOTAMIDAE

5. Anal claws long, gills absent 6

— Anal claws shorter, robust, occasionally lateral gills arranged in a variety of ways RHYACOPHILIDAE partim (RHYCOPHILINAE)

6. Labium with ligular extension projecting beyond the maxillary palps PSYCHOMYIIDAE partim (PSYCHOMYINAE)

— Labium does not possess the preceding characteristic POLYCENTROPIDAE

7. Head prognathous, gills absent, very rarely non-tubular gill expansions; case never cylindrical or conical 8

— Head not distinctly prognathous; only pronotum entirely sclerotized; gills long and slender (fringed), the lateral ones spread out, thorax bearing a prosternal horn; abdomen with tubercles on the first segment and well-developed lateral lines; larval case cylindrical composed of vegetable matter PHRYGANEIDAE

— Head always hypognathous sclerotization of the dorsal surface of the thorax variable; gills not possessing preceding characteristics, occasionally ramified, arranged in groups or even isolated filaments or in small groups (two or three); occasionally absent; case extremely variable 9

8. Only pronotum entirely sclerotized; case short, rounded on top, flat underneath, 'saddle'-shaped; average size or small RHYACOPHILIDAE partim (GLOSSOMATINAE)

— Pronotum, mesonotum and metanotum sclerotized, occasionally with small sclerotized patches on the abdomen; small or very small in size HYDROPTILIDAE

9. Pronotum and mesonotum sclerotized, metanotum always with small sclerites; gills variable, more or less numerous; thorax bearing a prosternal horn, abdomen with tubercles on the first segment and well-developed lateral lines; case extremely variable, tubular (occasionally flat or of triangular section by addition of materials) (mineral, vegetable or both) LIMNEPHILIDAE

— Not as above; sclerotization of the dorsal surface of the thorax variable; tubercles and lateral lines occasionally only slightly developed or even absent, gills present or absent; larval case extremely variable, occasionally not tubular 10

10. First segment of abdomen furnished with three tubercles generally well developed, median tubercle very rarely absent 11

— First abdominal segment having no tubercles; gills in groups or absent; prosternal horn absent; occasionally projection on median and posterior tibiae BRACHYCENTRIDAE

11. Metanotum with two sclerotized patches; the anterior one large, the posterior short and transverse; no prosternal horn; gills in groups; cylindrical larval case ODONTOCERIDAE

— Not as above 12

12. Legs unequal, the posterior ones almost three times as long as the anterior; mesonotum sclerotized and metanotum membraneous; no prosternal horn; tibiae having no projection; cylindrical larval case LEPTOCERIDAE

— Not as above 13

13. Larval case shield-like and flat, more or less broad; mesonotum divided into four plates; legs long; lateral lines present; no prosternal horn MOLANNIDAE

— Larval case always of the tubular type, very variable (occasionally with 'wings' but formed only by the addition on the sides of more or less voluminous surplus materials) 14

14. First abdominal segment furnished with two tubercles only; median tubercle absent, the lateral, however, are moderate only or sometimes prominent and even pointed; larval case tubular, cylindrical or often quadrangular, not furnished with 'wings' formed of miscellaneous debris LEPIDOSTOMATIDAE

— First abdominal segment having three tubercles 15

15. Mesonotum entirely sclerotized; gills in groups or absent; legs occasionally very unequal; larval case cylindrical BERAEIDAE

— Mesonotum slightly sclerotized or even with simple coloured patches; gills not in groups, formed by a small number of filaments SERICOSTOMATIDAE

The Pupa

Trichopterous pupae in the swimming phase (when they have cut their way out of the cocoon and are swimming in search of a suitable place for metamorphosis) constitute an important item in the food of fish, especially trout. Cast pupal skins on stones, or marginal vegetation, or floating on the surface of the water are often, however, the only indication to the biologist, fisherman or other investigator of a recent 'hatch' of caddis fly. In this paper, therefore, special importance has been given to descriptions of those features of the pupae it is possible to examine in the cast skins. In addition, as many of these latter features as possible are incorporated into the key to the families.

In the section on the larvae above, descriptions were made of the fully grown larvae in the feeding condition (in those case-bearing types when the fore part is still able to emerge from the case in order to forage). This section will thus commence with a description of the cocoon, the prepupal resting phase of the larva, and continue with descriptions of the pupa.

The material on which this work is based has been obtained by the author during the rearing of Trichoptera from larvae to adults. Pupae have been preserved in the fluid advocated for this purpose by M. E. Mosely—one part of commercial formalin to nineteen parts of water—having first been placed in 'collecting fluid' made up of two parts of 2 per cent solution of formalin added to one part of 90 per cent alcohol. Pupal skins were dehydrated in alcohol and mounted in Canada balsam.

All Trichoptera possess *pupae liberae*[1] and thus a general resemblance to the adult is shown by all pupae. The pupal integument is, for the most part, colourless and rather loosely envelops the formed imago lying beneath. The general shape of the pupa is, thus, that of the adult insect, and the sizes of the different parts of the pupa bear the same relation to each other as in the adult. There are, however, some features of the pupal integument of exceptional interest. The mandibles are large, well formed and functional, whereas those of the adult, on the other hand, are either absent or mere vestiges and are not functional. Moreover, the shape of the pupal mandibles is quite unlike that of the larval mandibles. The dorsal hook-bearing plates form a series of horny sclerites, variously shaped and situated, and used for gripping the sides of the case when emerging from it. Mandibles and dorsal hook-bearing plates are yellow, light brown or chestnut, and partly often black in colour in contrast to the third special feature of the pupal integument, the anal processes, which are of the same colour and texture as the general pupal integument. In some families of Trichoptera the pupal integument fits fairly closely the underlying genitalia,

[1] The attention of readers is drawn to the paper by H. E. Hinton, 'On the Function, Origin and Classification of Pupae' (1949. *Proc. Trans. S. Lond. ent. nat. Hist. Soc.* **1947–1948,** 111–54). Hinton proposes the name *pupa dectica* for exarate pupae (*pupae liberae*) where functional mandibles exist, as in the Trichoptera. The development of pupal types in relation to their escape from the pupal cell is discussed.

in other families two processes emerge more or less elongate and usually furnished with bristles. Because the pupal integument is almost transparent the underlying genitalia can almost always be seen, so that species determination can be made. This cannot, of course, be done with a cast skin.

The Cocoon

All trichopterous pupae are protected by a cocoon. In case-bearing larvae this is made from the larval case by blocking in various ways the front and rear openings. The cocoon thus made is anchored by secreted cement to a stone, water-weed or some other submerged object. In some species the larva shortens the case before plugging, by cutting off one or both ends.

The plugging or sealing of the openings of the case is accomplished in diverse ways. In many species, e.g. *Limnephilus flavicornis*, *Triaenodes bicolor*, a plug of cut pieces of water-weed, tufts of alga or moss, or pieces of vegetable debris, is pulled in either at one end or at both ends. This not only prevents the ingress of marauding carnivorous insects but often serves to hide the general shape of the case which might otherwise show up in its environment because it is very characteristic. Some species fill in the opening with a web of silk which is perforated in a characteristic manner, presumably to allow the passage of water into the cocoon for the gaseous exchange needed in respiration. This is correlated with the habit of the pupa of rhythmically undulating the abdomen to maintain a flow of water over the surface of the abdomen. Most often the pupal gratings are situated at the extreme ends of the case, but in some species they are fixed some way inside it. In *Brachycentrus subnubilus* (Fig. 16, c), the pupal grating consists of about eight small perforations in the centre of the disc. In *Sericostoma personatum* the grating consists of an oblong slit (Fig. 16, f), whereas in species of LIMNEPHILIDAE it consists of a number of perforations left in the weaving of algal filaments, fibres, etc. (Fig. 16, g). In *Chaetopteryx villosa*, the number of perforations is large, whereas in *Limnephilus rhombicus*, it is small. Small stones are sometimes woven into the grating, as in *Stenophylax sequax* (Fig. 16, c and d), so that when the pupa cuts the grating away the stones come away with it. Species inhabiting fast-running streams commonly plug the fore and hind end of the case with small pebbles, e.g. *Goera pilosa* (Fig. 16, h), *Sericostoma personatum* and *Odontocerum albicorne*, but small pores allowing a flow of water through the case occur between the larger sand-grains of which they are made. Pores such as those just described occur also in *Agapetus fuscipes*. In this species anterior and posterior openings occur on the ventral surface of the case and thus, before pupation when the case is fixed down on to a stone with silk, both openings are closed. Thienemann reports that the terrestrial species, *Enoicyla pusilla*, makes a simple membrane with no apertures. *Hydropsyche* constructs a pupal case by blocking in with small stones the shelter at the side of the net which is used during the larval stage. Completely free-living species such as *Rhyacophila* and *Philopotamus* construct

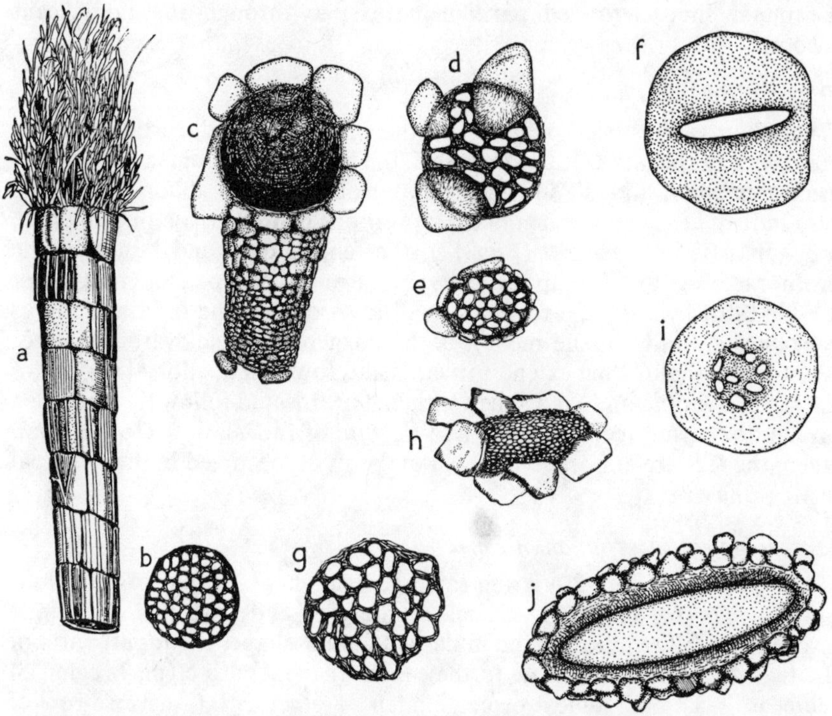

Fig. 16. Types of case-plugging to form cocoon. a. *Phryganea varia* F., case with plug; b. *Phryganea varia* F., sieve membrane (posterior end); c. *Stenophylax sequax* McL., with sieve membrane cut off by pupa; d. *Stenophylax sequax* McL., anterior sieve membrane cut off by pupa; e. *Stenophylax sequax* McL., posterior sieve membrane; f. *Sericostoma personatum* Spence, anterior sieve membrane; g. *Limnephilus flavicornis* F., anterior sieve membrane; h. *Goera pilosa* F., case with plug stones; i. *Brachycentrus subnubilus* Curt., anterior sieve membrane; j. *Rhyacophila dorsalis* Curt., case taken from stone, from below.

cases of small stones fastened together with silk on to the side of a stone (Fig. 16, j). In *Rhyacophila*, however, a characteristic cocoon is secreted within the pupal case and entirely separate from it. This is membranous and tough, brownish in colour and semi-translucent.

In *Holocentropus* the cocoon is made by the secretion of a plentiful supply of silk to which is attached parts of decaying leaves of water-plants.

In the HYDROPTILIDAE the pupal cocoon is made by filling the apertures of the larval case with a silken plug.

When the pupa emerges from the larva where a posterior sieve membrane is present, the larval skin breaks up into a number of pieces and is then kicked out of the pupal case through the pores in the membrane.

In *Rhyacophila*, however, where a membranous cocoon closely encompasses the pupa, the larval skin remains in various positions close to the pupa, the sclerotized parts being visible through the translucent cocoon.

Prepupal Resting Phase

When the case has been plugged or otherwise completed, certain changes take place in the larva. It becomes stiff, the head and abdomen losing their usual flexibility, and the intersegmental grooves of the abdomen become very indistinct. The legs then occupy positions characteristic of this phase only. In *Mystacides nigra* (Fig. 17), the femur, tibia and tarsus of the prothoracic legs are held up vertically; in the meso-legs coxa and trochanter are held downwards, the femur is held upwards and the tibia and tarsus backwards; whilst in the meta-legs the coxa points backwards, the trochanter and femur then extend forwards and towards the dorsal surface of the larva, the tibia and tarsus then turn backwards and follow the abdomen on the dorsal surface. (Lestage gives a figure of the larva of *Oecetis furva* where the legs are almost certainly in the position occupied in the prepupal resting phase.)

Morphological Description of Pupa

Head. The head (Fig. 20) is the same general shape as that of the adult, but whereas in some organs, such as the eyes and antennae, the pupal integument follows closely the underlying adult shape, the mouthparts of the two stages are dissimilar. In the centre of the fronto-clypeal region of the head is a small conical projection. In *Mystacides* a transverse row of three equal projections occurs. It was formerly thought that adult caddis flies rarely drank or sipped nectar, but Crichton, in describing the mouthparts, has called attention to the widespread occurrence of a protruscible haustellum capable of performing such a function. The mandibles, however, are non-functional. Those of the pupae, on the other hand, have the important function of cutting the exit hole in the pupal case. In addition, the bristles of the labrum, assisted in some cases by the mandibles, serve to keep the anterior sieve membrane of the pupal case free from silt and debris, thus allowing an uninterrupted flow of water into the case. (This is related to the same function carried out by the anal processes or bristles on the posterior lobes of the abdomen in respect to the posterior membrane.) The bristles of the labrum have also been said to have secretory functions.

Eyes. These are a prominent feature of the head of the pupa, and are the adult eyes covered by the thin, transparent pupal integument (Fig. 20). In the fresh pupa they are light cream in colour but become black some time before emergence.

Antennae. The pupal antennae follow closely the length, shape, number of segments, etc., of the underlying adult antennae. They are, however, loosely held down in a position different from that found in the adult. In the pupa, after skirting the eyes dorsally, they usually follow the anterior margin of the anterior wings. Thereafter, what occurs depends largely on the length of the antennae. In some species (Fig. 19, b and c) they are approximately equal in length or only slightly shorter than the wings, e.g. SERICOSTOMATIDAE, RHYACOPHILIDAE, LIMNEPHILIDAE and PHRYGANEIDAE. Sometimes they are much shorter, e.g. HYDROPTILIDAE, and in some cases they are extremely long and slender (Fig. 19, a), being twice or three times the length of the anterior wing, e.g. LEPTOCERIDAE. In many species of this family the tips of the antennae are wound round the posterior abdominal segments, and a pair of lappets bearing forwardly directed bristles are situated on the ninth segment, e.g. *Mystacides azurea* (Fig. 18, c). These keep the wound antennae in position and prevent them from slipping off and blocking the posterior sieve membrane of the pupal case. In *Athripsodes aterrimus* (Fig. 18, b) lappets are present, but each antenna is loosely looped over on to the opposite side of the abdomen and not wound round the body. In *Triaenodes bicolor* (Fig. 18, d) the antennae loosely adhere together and are wound round the abdomen in a clockwise direction for almost two complete loops.

Labrum. This is a distinct, usually well-sclerotized part. It differs widely in shape in the different families, and there is variation also in the manner in

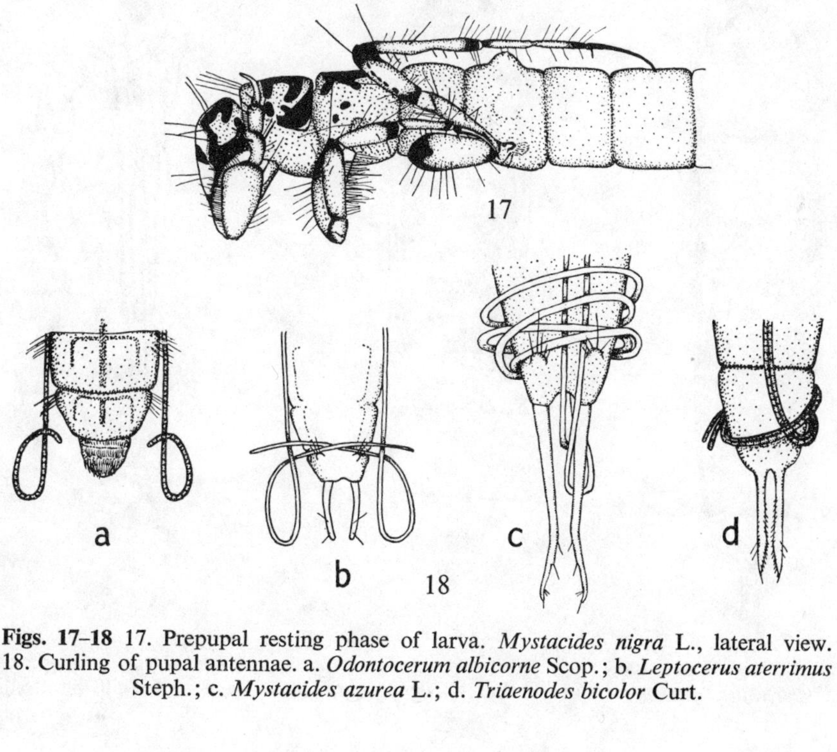

Figs. 17–18 17. Prepupal resting phase of larva. *Mystacides nigra* L., lateral view. 18. Curling of pupal antennae. a. *Odontocerum albicorne* Scop.; b. *Leptocerus aterrimus* Steph.; c. *Mystacides azurea* L.; d. *Triaenodes bicolor* Curt.

which it projects from the head capsule. Bristles are an important feature of the labrum, not only in their placing but also in the nature of the bristles themselves.

In many pupae the labrum lies flat, in the same plane as the fronto-clypeus. In other pupae it projects outwards from the head. In general the labrum is disc-like, wider at the base, but in some families it is narrower at the base—this is found in the PHRYGANEIDAE (Fig. 21, a) and in *Mystacides* (Fig. 21, i), whilst in *Hydroptila occulta* (Fig. 21, k) it is almost perfectly oval. Those species with the labrum wider at the base also show some thickening of this region and often a folding (*Odontocerum* (Fig. 21, g) and *Rhyacophila* (Fig. 21, j)), which sometimes gives a two-lobed effect. In *Lepidostoma hirtum* (Fig. 21, e) the base is extended laterally.

A characteristic feature of the bristles of many of those species which construct the membranes in the openings of the cocoon is their long sturdy nature and the bending or hooking at the tip. In *Lepidostoma hirtum* the tips of the bristles form a distinct loop. The function attributed to these bristles is the cleaning of the membranes of debris and silt in order to allow

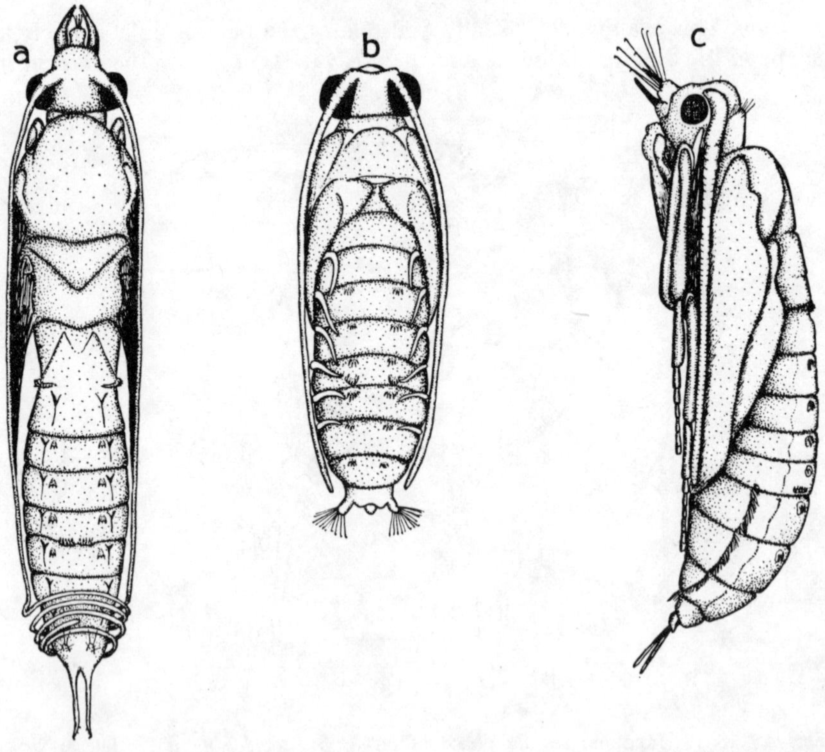

Fig. 19. a. Dorsal view of pupa of *Mystacides azurea* L.; b. Dorsal view of pupa of *Plectrocnemia conspersa* Curt.; c. Lateral view of pupa of *Goera pilosa* F. (gills omitted).

the uninterrupted flow of water through the cocoon for gaseous exchange. The bristles are usually in fairly well-defined groups, a pair of distal and a pair of proximal groups on each side. In *Limnephilus flavicornis* (Fig. 21, b) the distal groups are hooked, whilst, on the other hand, in *Lepidostoma hirtum* the proximal groups are hooked. In *Stenophylax sequax* (Fig. 21, d) neither group is hooked, although the bristles are long and strong, and project upwards from the surface of the labrum, but in *Athripsodes aterrimus* (Fig. 21, h) and *Hydroptila occulta* (Fig. 21, k) a single pair of bristles only is present, situated on the anterior margin, minute in the latter species.

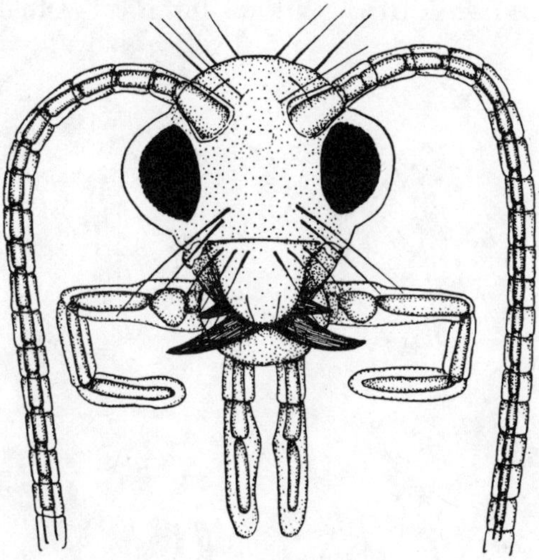

Fig. 20. Head of pupa of *Rhyacophila dorsalis* Curt., from the front.

Mandibles. The mandibles show a wide range of form, and are highly characteristic of their groups. They are well developed, robust and well adapted to their function of sawing through the cocoon and thus releasing the pupa for its swim to the surface in order that metamorphosis into the aerial adult can take place. The mandibles are usually sickle-shaped with the base broad. Teeth are usually present and are of two distinct sizes. A few large teeth (nearly equalling in size the apex of the mandible) may be present, or the teeth may be small (they may be so small as scarcely to be distinguishable under a one-sixth-inch microscope objective). In these forms the inner edges of the mandible resemble a saw. In some species a combination of the two types of teeth may be present. A large number of species have two bristles on the outer convex side of the mandibles near the base. Articulation of the mandible with the head is by means of a spherical condyle on the base, usually between the centre and the outer edge, more rarely on the outer angle and the inner angle of the base of the mandible which makes contact with the sclerotized ring which surrounds the mouthparts (Fig. 21, g).

The mandibles of the PHRYGANEIDAE are characterized by having a slender inward-curving apical part with a stout base, and the bristles on the outer face are situated on a conical projection. The teeth on the inner edge are very numerous but can hardly be distinguished with a one-sixth-inch objective. But Ulmer gives the mandibles of *Oligotricha clathrata* as rudimentary. They are small and obtuse. Although I have twice searched the Staffordshire locality for this species I have not been fortunate enough to come across it. The mandibles of RHYACOPHILIDAE, PHILOPOTAMIDAE and HYDROPSYCHIDAE are all furnished with large teeth. In

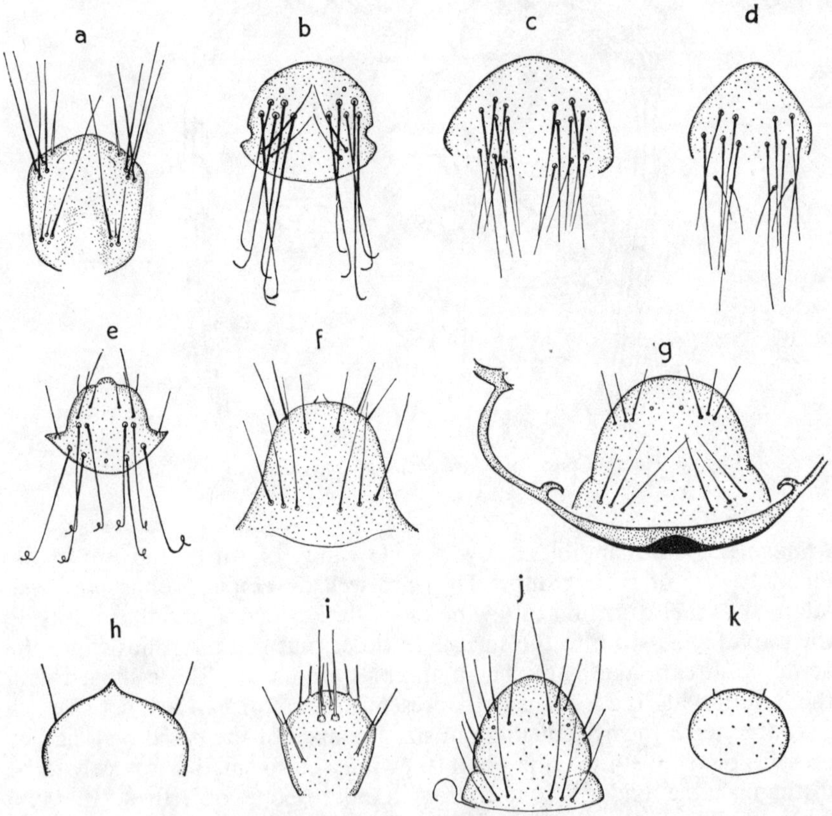

Fig. 21. Pupal labra. a. *Phryganea varia* F.; b. *Limnephilus flavicornis* F.; c. *Limnephilus extricatus* McLach.; d. *Stenophylax sequax* McLach.; e. *Lepidostoma hirtum* F.; f. *Molanna angustata* Curt.; g. *Odontocerum albicorne* Scop.; h. *Athripsodes aterrimus* Steph.; i. *Mystacides azurea* L.; j. *Rhyacophila dorsalis* Curt.; k. *Hydroptila occulta* Eaton.

RHYACOPHILIDAE they are asymmetrical, two teeth being present on the left mandible and three on the right. Fig. 22 shows the range of shape found in pupal mandibles.

Maxillary Palps. These are well defined in the pupa. Five segments are present in the maxilliary palp in the female in all families, but males of the family PHRYGANEIDAE have four-segmented palps, and males of LIMNEPHILIDAE and SERICOSTOMATIDAE have three-segmented palps. Males of the remaining families have five-segmented maxillary palps similar to those of the females.

Labial Palps. These are always of three segments.

Wings. These are held closely pressed to the pleural region of the thorax, and extend usually as far as the sixth segment, the tips of the forewings almost meeting in the mid-ventral line. The wings are, of course, much smaller than the adult wings. The veins of the wing can often be made out after extraction of the wing from the pupal integument, but it would be unwise to rely on venation of the pupal wing as a determining characteristic. The wing venation cannot, however, be made out from the empty wing case.

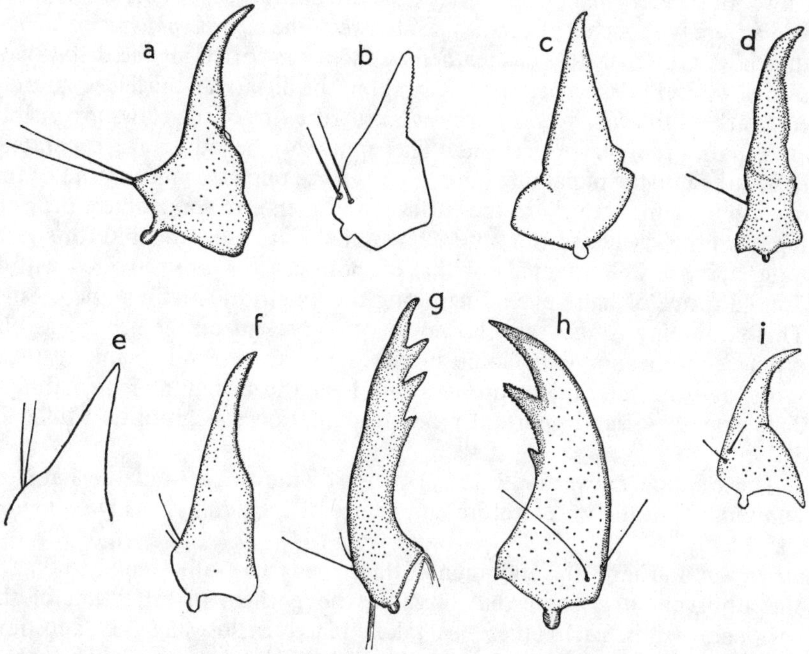

Fig. 22. Pupal mandibles. a. *Phryganea varia* F.; b. *Limnephilus extricatus* McLach.; c. *Goera pilosa* F.; d. *Molanna angustata* Curt.; e. *Odontocerum albicorne* Scop.; f. *Athripsodes aterrimus* Steph.; g. *Philopotamus montanus* Don.; h. *Rhyacophila dorsalis* Curt.; i. *Hydroptila occulta* Eaton.

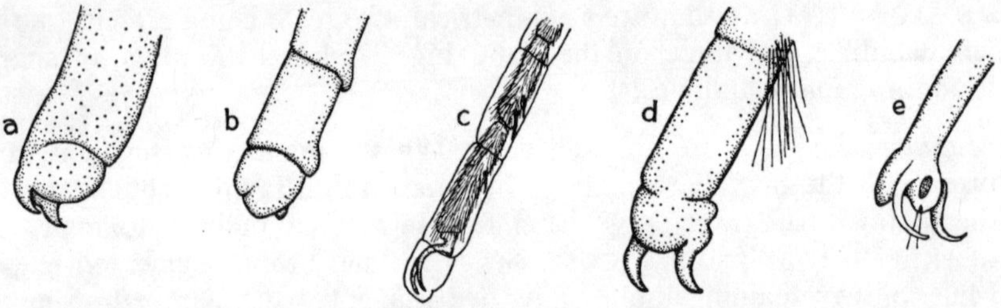

Fig. 23. Tarsal claw of mesoleg of pupa. a. *Phryganea varia* F.; b. *Limnephilus flavicornis* F.; c. *Chaetopteryx villosa* F.; d. *Lepidostoma hirtum* F.; e. *Rhyacophila dorsalis* Curt.

Legs. The legs of the pupa are free, the meso-legs entirely so; pro- and meta-legs have the coxa, trochanter and femur loosely adherent, but tibia and tarsus are free. The pupal integument follows fairly closely the shape of the adult legs and the segments correspond, but a separate segment is occupied by the tarsal claw and the number of segments of the tarsus is thus six. The tarsal claw segment shows a wide variation in shape, the pupal integument following to a greater or lesser degree the shape of the claws. In *Rhyacophila dorsalis* the claws are entirely separately ensheathed and a pore with a pair of bristles lies between them. In *Lepidostoma hirtum* the claws are separately ensheathed to the extent of about the distal two-thirds, whilst in *Phryganea varia* only about the distal one-third is separately ensheathed. In *Limnephilus flavicornis* only the tips of the claws are visible as a pair of blunt projections. The spurs of the adult are separately ensheathed by the pupal integument. When the pupa cuts its way out of the cocoon it swims actively to the surface, and thence to some object projecting from the surface of the water. The middle legs only are used for swimming and are well adapted for this purpose, as they are provided with a double fringe of hairs extending along the length of the tibia and tarsus. The 'swimming fringes' are, however, often present on the other legs but are never so pronounced. Some species, however, which live amongst wet moss, and thus do not require to swim, have the swimming fringe absent. This example of an important type of adaptation was given by Müller in 1879.

The relation between the tarsal claws of the pupae and the habit of climbing from the water before emergence of the adult is easily seen.

Abdomen. In shape, the abdomen of the pupa is generally similar to that of the adult but it is somewhat larger as no permanent retraction of the segments within each other has taken place at this stage. In addition, processes on the terminal segment often considerably extend the length. Nine segments are present. In those species where a lateral line is present in the larva, it is present also in the pupa, but usually the hairs comprising it are longer and darker in colour than those in the larva. The lateral line

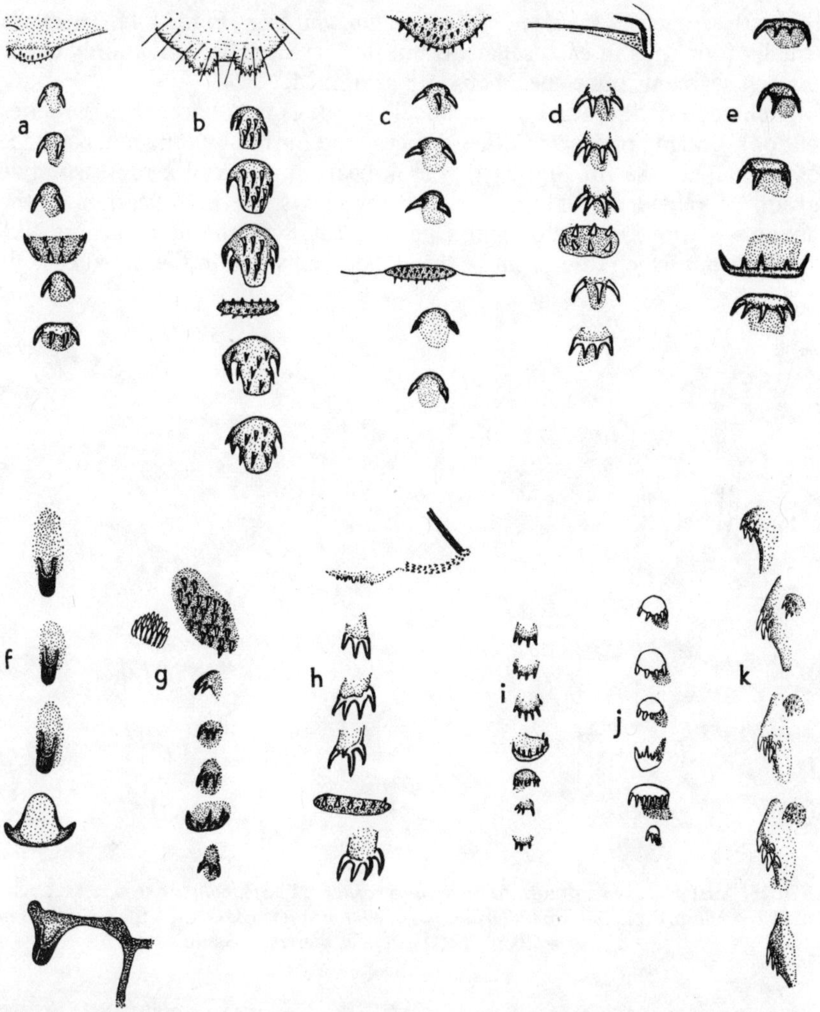

Fig. 24. Hook-bearing plates of abdomen. a. *Limnephilus extricatus* McLach.; b. *Stenophylax sequax* McLach.; c. *Goera pilosa* F.; d. *Lepidostoma hirtum* F.; e. *Molanna angustata* Curt.; f. *Odontocerum albicorne* Scop.; g. *Athripsodes aterrimus* Steph.; h. *Mystacides azurea* L.; i. *Plectrocnemia conspersa* Curt.; j. *Philopotamus montanus* Don.; k. *Hydroptila occulta* Eaton.

in the pupa differs also from that of the larva by the posterior part curling under the posterior segments, and almost meeting at the mid-ventral line.

A number of skeletal rods are often visible in the pupal skin. These take the form of a series of umbrella ribs in each segment, often forked at the anterior end. They are quite hard, stiff and dark in colour and in some species some connexion between the rods and the hook-bearing plate on the sixth segment is seen, e.g. *Odontocerum albicorne* (Fig. 24, f). There are usually four rods in each segment, but in *Phryganea varia* two only occur in each segment. (One specimen only examined.)

A characteristic feature of the dorsal surface of the abdomen is the series of hook-bearing plates and other devices used for holding the inner surface of the pupal case. In Fig. 24 the hook-bearing plates of a representative group of pupae are shown. Usually there is some modification of the dorsum of the first abdominal segment, but it is absent in genera like *Rhyacophila* which do not move their abdomen when in the pupal case. It

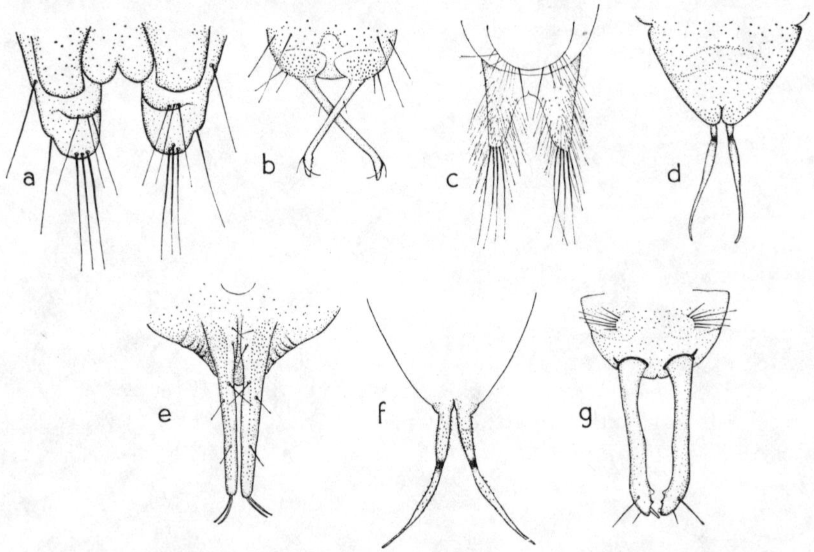

Fig. 25. Anal processes of pupa. a. *Phryganea varia* F.; b. *Stenophylax sequax* McLach.; c. *Goera pilosa* F.; *Lepidostoma hirtum* F.; e. *Molanna angustata* Curt.; f. *Odontocerum albicorne* Scop.; g. *Athripsodes aterrimus* Steph.

is thought that this device is for gripping the inner lining of the pupal case whilst the remainder of the abdomen is rhythmically undulated. At first sight it would appear that the abdominal hook-bearing plates are 'gin-trap' mechanisms similar to those described by Hinton (1946) in certain beetle pupae. Closer observation, however, shows that the posterior plate on the fifth segment is thrown out of action (by the telescoping of the segment) when the anterior plates are forced downwards. The hook-bearing plates thus serve only a locomotory function. There is a wide diversity in the form of this gripping device. In one genus, *Athripsodes*, it is not unlike, in general shape, the hook-bearing plates situated on the third segment onwards, except that it is much larger and contains many more hooks. Probably associated with the hook-bearing plates on the first abdominal segment in

Athripsodes is a group of finger-like organs inclined towards the plate on each side. In *Limnephilus* the first abdominal segment extends backwards on the dorsal side into a pair of bulbous projections beset with short blunt spines. In *Mystacides* the bulbous projections are present to a lesser degree, but they are joined on the outside by a curved spinous bar which in turn is joined to a sclerotized bar pointing obliquely forwards. In *Lepidostoma* a narrow transverse band extends across the segment, ending in the pleural region with a stirrup-shaped part, both arms of which are covered with small spines.

The second abdominal segment seems always to be devoid of hook-bearing plates, the series proper usually commencing on the third segment towards the anterior margin and being repeated in the same place on the fourth, fifth and sixth segments. An additional plate on or near the posterior edge of the fifth segment is almost always present. The hook-bearing plates consist of sclerotized patches from which emerge a number of short hooks. The latter point backwards in the case of the anterior plates, and forwards in the case of the posterior plate on the fifth segment. The hook-bearing plates show a certain range of complexity, but it is almost a rule that the posterior plate on the fifth segment is larger and bears more hooks than the other plates.

In some genera the plate on the third segment may be absent, as in some species of *Limnephilus*. The series may terminate on the sixth segment (*Athripsodes*, *Odontocerum*, *Molanna*, *Mystacides*, etc.) or may terminate on the seventh segment (some species of *Limnephilus*, *Lepidostoma*, *Goera*, etc.). In *Plectrocnemia* it terminates on the eighth segment. *Odontocerum albicorne* (Fig. 24, f) is unique in possessing sac-like organs attached to the anterior plates. The posterior plate of the fifth segment only is provided with hooks. These are two in number. In *Hydroptila occulta* there is a single pair of anterior plates on the third segment, two pairs on the fourth, fifth and sixth, and one pair on the seventh. The posterior plate on the fifth segment is absent.

Anal Processes. Considerable variation in shape of the posterior end of the abdomen is found in the different families. In the pupae from campodeiform types of larvae (RHYACOPHILIDAE, PHILOPOTAMIDAE, HYDROPTILIDAE, POLYCENTROPIDAE, PSYCHOMYIIDAE) the pupal integument more nearly follows the shape of the underlying adult genitalia, although in HYDROPSYCHIDAE there is a pair of two-segmented blunt appendages. In the PHRYGANEIDAE the pupal integument is a simple envelope to the genitalia. On the other hand, in all the other families there is an elaboration, to a greater or lesser degree, of the last abdominal segment to form a pair of acicular processes. These are single-pointed and devoid of bristles in *Odontocerum albicorne* (the sole British species of the ODONTOCERIDAE) and GOERINAE. Emphasizing the diverse assemblage in the SERICOSTOMATIDAE the anal processes in the LEPIDOSTOMATINAE are short, triangular, confluent at their bases and very hairy—in contrast to the GOERINAE. In the LIMNEPHILIDAE, MOLANNIDAE and LEPTOCERIDAE they are long and slender, but usually blunt at the tip, with a few short bristles and sometimes some small hooks. A diversity of form is found in the BERAEIDAE, whilst in the genus *Beraeodes* the anal processes are long and slender with a few long hairs. In *Beraea* they are short, broad at the base and furnished with a large number of shorter hairs.

The function of these processes is generally thought to be the cleansing of the orifices in the plug of the cocoon—similar, in fact, to the large bristles on the front of the head. The movement of the abdomen during the current-forming rhythmic undulations would cause the anal processes to clear silt, etc., from the orifices. This function fits in well with the facts. Thus the processes are absent in those cases where well-defined terminal orifices are not present.

Gills. In those families with campodeiform larvae, gills are usually absent in the pupa even though they may be present in the larvae (e.g. *Rhyacophila*). In the POLYCENTROPIDAE, however, abdominal filiform gills are present although absent in the larva, and in the HYDROPSYCHIDAE tufted gills are present in the pupa similar to those of the larva, but generally not so numerous. In all the families with eruciform and sub-eruciform larvae, gills are present in the pupa if they are present in the larva (which is more usually the case), but almost always some slight reduction in numbers or disposition occurs.

Key to the Families of Trichoptera based on the Pupal Skin

1. Mandibles with several large teeth — 2
   Mandibles with no teeth, small or minute teeth only — 4
2. Mandibles very hairy at the base, anal processes long, very hairy, sclerotized apices — HYDROPSYCHIDAE
   Mandibles with two bristles only at the base, anal processes short, not hairy and without sclerotized apices — 3
3. Mandibles sharply bent near the base, symmetrical — PHILOPOTAMIDAE
   Mandibles not bent, asymmetrical — RHYACOPHILIDAE
4. Mandibles with two bristles emerging from conical projection on outer edge of base — PHRYGANEIDAE (except *Oligotricha clathrata*)
   Mandibles without conical projection on outer edge of base from which bristles emerge — 5
5. Anal processes absent or very short and broad at the base — 6
   Anal processes long and slender — 9
6.[1] Small auxiliary dorsal hook-bearing plates on abdominal segments 3, 4 and 5 (see Fig. 24, k). Pupae very small — HYDROPTILIDAE
   Auxiliary dorsal hook-bearing plates not on abdominal segments 3, 4 and 5. Pupae medium size — 7

[1] From this point onwards the key should be used with some reserve, as only representative species in the families have been examined.

7. Mandibles with long, thin apical part curved with a few teeth at the apex or strongly curved at the junction of the thin part with the base, and curved again near the apex without teeth PSYCHOMYIIDAE
Mandibles without long thin apical part 8
8. Hook-bearing plates on first abdominal segment bracket-shaped and joined by a transverse sclerotized bar. Hook-bearing plates on segments 3, 4, 5 ant. and 5 post., 6 and 7 SERICOSTOMATIDAE (LEPIDOSTOMATINAE)
Hook-bearing plates absent on first and seventh abdominal segments POLYCENTROPIDAE
9. Dorsal hook-bearing plates sac-like except that on the posterior fifth segment which has two hooks. Hook-bearing plate on sixth segment continuous with a sclerotized bar
ODONTOCERIDAE. *Odontocerum albicorne* Scopoli. Sole British species
Dorsal hook-bearing plates not sac-like—that on sixth segment not continuous with a sclerotized bar 10
10. Antennae long, curved around posterior abdominal segments with forwardly-directed lappets on ninth segment LEPTOCERIDAE
Antennae not curved around posterior abdominal segments, lappets on ninth abdominal segment absent 11
11. Posterior margin of first abdominal segment with two bulbous projections beset with spicules or with two well-defined spicule-bearing patches 12
Posterior margin of first abdominal segment without spicule-bearing projections or patches 13
12. Pair of bulbous projections with spicules on posterior margin of first abdominal segment LIMNEPHILIDAE
Bulbous projections on posterior margin of first abdominal segment absent
SERICOSTOMATIDAE (GOERINAE, BRACHYCENTRINAE, SERICOSTOMATINAE)
13. Lateral line present MOLANNIDAE
Lateral line absent BERAEIDAE

For references to publications dealing with the British Species see pages 50–52.

The works which have been consulted are listed here, but for an almost complete list of references on Trichoptera the reader is referred to Fischer's Trichopterorum Catalogus (see below). References in List 1 deal solely with British specimens. List 2 gives references to sources other than British and List 3 presents references dealing with North American Fauna. In general the references are restricted to publications giving some account or figures of the larvae.

List 1. *References to Publications Dealing with the British Species*

HICKIN, N. E. 1942–59. Larvae of the British Trichoptera.

1. *Stenophylax stellatus* Curtis. 1942. *Proc. R. ent. Soc. Lond.* (A) **17**, 9–11.
2. *Triaenodes bicolor* Curtis. 1942. *Proc. R. ent. Soc. Lond.* (A) **17**, 12–13.
3. *Rhyacophila dorsalis* Curtis. 1942. *Proc. R. ent. Soc. Lond.* (A) **17**, 14–16.
4. *Philopotamus montanus* Donovan. 1942. *Proc. R. ent. Soc. Lond.* (A) **17**, 16–17.
5. *Odontocerum albicorne* Scopoli. 1942. *Proc. R. ent. Soc. Lond.* (A) **17**, 119–22.
6. *Silo pallipes* Fabricius. 1942. *Proc. R. ent. Soc. Lond.* (A) **17**, 123–6.
7. *Phryganea grandis* L. 1942. *Proc. R. ent. Soc. Lond.* (A) **17**, 134–7.
8. *Limnephilus flavicornis* L. 1943. *Proc. R. ent. Soc. Lond.* (A) **18**, 6–10.
9. *Anabolia nervosa* Curtis. 1943. *Proc. R. ent. Soc. Lond.* (A) **18**, 11–14.
10. *Lepidostoma hirtum* Fabricius. 1943. *Proc. R. ent. Soc. Lond.* (A) **18**, 15–17.
11. *Holocentropus dubius* Rambur. 1943. *Proc. R. ent. Soc. Lond.* (A) **18**, 19–21.
12. *Limnephilus lunatus* Curtis. 1943. *Proc. R. ent. Soc. Lond.* (A) **18**, 66–8.
13. *Mystacides nigra* L. 1943. *Proc. R. ent. Soc. Lond.* (A) **18**, 69–71.
14. *Limnephilus vittatus* Fabricius. 1943. *Proc. R. ent. Soc. Lond.* (A) **18**, 72–4.
15. *Goëra pilosa* Fabricius. 1943. *Proc. R. ent. Soc. Lond.* (A) **18**, 75–7.
16. *Agapetus fuscipes* Curtis. 1943. *Proc. R. ent. Soc. Lond.* (A) **18**, 78–80.
17. *Brachycentrus subnubilus* Curtis. 1943. *Proc. R. ent. Soc. Lond.* (A) **18**, 81–3.
18. *Leptocerus aterrimus* Stephens. 1943. *Proc. R. ent. Soc. Lond.* (A) **18**, 106–8.
19. *Setodes argentipunctella* McLachlan. 1943. *Proc. R. ent. Soc. Lond.* (A) **18**, 109–11.
20. *Neuronia ruficrus* Scopoli. 1944. *Proc. R. ent. Soc. Lond.* (A) **19**, 12.
21. *Glyphotaelius pellucidus Retzius*. 1946. *Proc. R. ent. Soc. Lond.* (A) **21**, 61–5.
22. *Molanna angustata* Curtis. 1946. *Proc. R. ent. Soc. Lond.* (A) **21**, 55–60.
23. *Plectrocnemia conspersa* Curtis. 1947. *Proc. R. ent. Soc. Lond.* (A) **22**, 114–117.
24. *Limnephilus extricatus* McLachlan. 1948. *Proc. R. ent. Soc. Lond.* (A) **23**, 12–13.
25. *Limnephilus rhombicus L.* 1948. *Proc. R. ent. Soc. Lond.* (A) **23**, 54–56.
26. *Chaetopteryx villosa* Fabricius. 1948. *Proc. R. ent. Soc. Lond.* (A) **23**, 59–61.
27. *Halesus digitatus* Schrank. 1949. *Proc. R. ent. Soc. Lond.* (A) **24**, 56–59.
28. *Tinodes waeneri L.* 1950. *Proc. R. ent. Soc. Lond.* (A) **25**, 67–70.
29. *Lype reducta* Hagen. 1950. *Proc. R. ent. Soc. Lond.* (A) **25**, 71–74.
30. *Tinodes pallidula* McLachlan. 1950. *Proc. R. ent. Soc. Lond.* (A) **25**, 103–6.
31. *Stenophylax vibex* Curtis. 1950. *Proc. R. ent. Soc. Lond.* (A) **25**, 107–10.
32. *Sericostoma personatum* Spence. 1951. *Proc. R. ent. Soc. Lond.* (A) **26**, 93–6.
33. *Polycentropus flavomaculatus* Pictet. 1952. *Proc. R. ent. Soc. Lond.* (A) **27**, 86–8.
34. *Tinodes assimilis* McLachlan. 1952. *Proc. R. ent. Soc. Lond.* (A) **27**, 89–90.
35. *Tinodes unicolor* Pictet. 1953. *Proc. R. ent. Soc. Lond.* (A) **28**, 36–8.
36. *Phryganea varia* Fabricius. 1953. *Proc. R. ent. Soc. Lond.* (A) **28**, 39–40.
37. *Setodes tineiformis* Curtis. 1953. *Proc. R. ent. Soc. Lond.* (A) **28**, 74–6.
38. *Athripsodes bilineatus L.* 1953. *Proc. R. ent. Soc. Lond.* (A) **28**, 111–13.
39. *Mystacides longicornis L.* 1953. *Proc. R. ent. Soc. Lond.* (A) **28**, 114–16.
40. *Micropterna sequax* McLachlan. 1953. *Proc. R. ent. Soc. Lond.* (A) **28**, 163–5.
41. *Stenophylax latipennis* Curtis. 1954. *Proc. R. ent. Soc. Lond.* (A) **29**, 55–8.
42. *Rhyacophila septentrionis* McLachlan. 1954. *Proc. R. ent. Soc. Lond.* (A) **29**, 59–61.
43. *Granmotaulius ato-manius* Fabricius. 1954. *Proc. R. ent. Soc. Lond.* (A) **29**, 89–92.

44. *Lype phaeopa* Stephens. 1954. *Proc. R. ent. Soc. Lond.* (A) **29**, 93–5.
45. *Limnephilus centralis* Curtis. 1954. *Proc. R. ent. Soc. Lond.* (A) **29**, 96–8.
46. *Micropterna lateralis* Stephens. 1954. *Proc. R. ent. Soc. Lond.* (A) **29**, 145–6.
47. *Triaenodes conspersa* Rambur. 1954. *Proc. R. ent. Soc. Lond.* (A) **29**, 153–5.
48. *Crunoecia irrorata* Curtis. 1954. *Proc. R. ent. Soc. Lond.* (A) **29**, 172–3.
49. *Trichostegia minor* Curtis. 1954. *Proc. R. ent. Soc. Lond.* (A) **29**, 174–6.
50. *Phryganea obsoleta* McLachlan. 1955. *Proc. R. ent. Soc. Lond.* (A) **30**, 55–8.
51. *Phryganea striata L.* 1955. *Proc. R. ent. Soc. Lond.* (A) **30**, 55–8.
52. *Enoicyla pusilla* Burmeister. 1958. *Proc. R. ent. Soc. Lond.* (A) **33**, 176–8.
53. *Beraeodes minuta L.* 1959. *Proc. R. ent. Soc. Lond.* (A) **34**, 83–9.
54. *Beraea maurus* Curtis. 1959. *Proc. R. ent. Soc. Lond.* (A) **34**, 83–9.
55. *Beraea pullata* Curtis. 1959. *Proc. R. ent. Soc. Lond.* (A) **34**, 83–9.
56. *Ernodes articularis* Pictet. 1959. *Proc. R. ent. Soc. Lond.* (A) **34**, 83–9.

BADCOCK, R. M. 1961. The Morphology of some Parts of the Head and Maxillo-Labium in Larval TRICHOPTERA, with special Reference to the HYDROPSYCHIDAE. *Trans. R. ent. Soc. Lond.* **113**, 217–48.

BAKER, W. V. 1963. Some Notes on the Taxonomy of the Last Instar Larvae of British Species of the Sub-family GOERINAE (TRICHOPTERA:SERICOSTOMATIDAE). *Proc. R. ent. Soc. Lond.* (B) **32**, 171–4.

BRINDLE, A. 1960. The Larva of *Wormaldia occipitalis* (Pict.) (TRICHOPTERA, PHILOPOTAMIDAE). *Entomologist's Rec. J. Var.* **72**, 144–7.

BRINDLE, A. 1960. The Larva of *Neureclipsis bimaculata* (L.) (TRICHOPTERA, POLYCENTROPIDAE). *Entomologist's Rec. J. Var.* **72**, 244–5.

BRINDLE, A. 1960. The Larva of *Psychomyia pusilla* (F.) (TRICHOPTERA, PSYCHOMYIIDAE). *Entomologist's Rec. J. Var.* **72**, 265–7.

BRINDLE, A. 1960. The Larva of *Hydropsyche angustipennis* Curtis (TRICHOPTERA, HYDROPSYCHIDAE). *Entomologist's Rec. J. Var.* **72**, 267–70.

BRINDLE, A. 1961. The Larval Taxonomy of the British Trichoptera. *Entomologist's Rec. J. Var.* **73**, 114–25.

BRINDLE, A. 1961. A Family Key to the Pupae of the British Trichoptera. *Entomologist's Rec. J. Var.* **73**, 156–62.

BRINDLE, A. 1964. The Larval Taxonomy of the British Trichoptera. *Entomologist's Rec. J. Var.* **76**, 194–8.

EDINGTON, J. M. 1964. The Taxonomy of British Polycentropid Larvae (TRICHOPTERA). *Proc. zool. Soc. Lond.* **143**, 281–300.

FISCHER, F. C. J. 1960–7. *Trichopterorum Catalogus.* Nederlandsche Entomologische Vereeniging. Amsterdam.

HANNA, H. M. 1956. The Larva of *Notidobia ciliaris* (L.) (TRICHOPTERA:SERICOSTOMATIDAE). *Entomologist's Gaz.* **7**, 77–81.

HANNA, H. M. 1956. The Larva of *Limnephilus politus* McLachlan (TRICHOPTERA: LIMNEPHILIDAE). *Entomologist's Gaz.* **7**, 139–44.

HANNA, H. M. 1956. The Larva of *Limnephilus marmoratus* Curtis (TRICHOPTERA: LIMNEPHILIDAE). *Entomologist's Gaz.* **7**, 194–8.

HANNA, H. M. 1957. The Larva of *Agrypnia pagetana* Curtis (PHNYGANEIDAE). *Entomologist's Gaz.* **8**, 110–14.

HANNA, H. M. 1960. The Larva of *Halesus radiatus* (Curtis) (TRICHOPTERA:LIMNEPHILIDAE). *Entomologist's Gaz.* **11**, 153–9.

HICKIN, N. E. 1941. Some Records of *Trichoptera* in the Lake District. *Entomologist's mon. Mag.* **77**, 252.

HICKIN, N. E. 1944. *Neuronia clathrata* Kol. (*Phryganeidae, Trichoptera.*) Chartley Moss and the *Caddis Fly. Entomologist* **77**, 20.

HICKIN, N. E. 1946. Larvae of the British *Trichoptera. Trans. R. ent. Soc. Lond.* **97**, 187–212.

HICKIN, N. E. 1946. *Neuronia ruficrus* (Scop). (*Trich. Phryganeidae*) and *Phalacrocora replicata* (L.) Dipt. Tipulidae in a high Tarn. *Entomologist's mon. Mag.* **82**, 11.

HICKIN, N. E. 1949. Exhibit. *Proc. R. ent. Soc. Lond.* (C) **14**.

HICKIN, N. E. 1949. Pupae of the British Trichoptera. *Trans. R. ent. Soc. Lond.* **100**, 275–89.

HICKIN, N. E. 1952. *Caddis*. A short account of the biology of British Caddis flies with special reference to the immature stages. Methuen & Co. Ltd. London.

HICKIN, N. E. 1953. An Aquatic Larva. Suitable for Laboratory work. *Nature, Lond.* **172**, 784.

HICKIN, N. E. 1953. *Mystacides nigra L.* Courtship of a Caddis Fly. *Entomologist's mon. Mag.* **89**, 71.

HICKIN, N. E. 1953. *Neuronia ruficrus* Scopoli. In Sussex. *Entomologist's mon. Mag.* **89**, 168.

HICKIN, N. E. 1957. Studying *Caddis. Country-side* **18**, n.s., 99–103 (Autumn).

KIMMINS, D. E. 1943. A Note on the Caddis Fly, *Mystrophora intermedia* Klapálek (TRICHOPTERA). *Proc. R. ent. Soc. Lond.* (A) **18**, 96–8.

KIMMINS, D. E. 1966. A Revised Check-list of the British Trichoptera. *Entomologist's Gaz.* **17**, 111–20.

KLOET, G. S. and HINCKS, W. D. 1964. *A Check-list of British Insects*. Second edition revised. Royal Entomological Society of London.

MACDONALD, W. W. 1950. The Larvae of *Mystacides azurea* L., *Cyrnus flavidus* McLachlan and *Oxyethira simplex* Ris. (TRICHOPTERA). *Proc. R. ent. Soc. Lond.* (A) **25**, 19–27.

MACKERETH, J. C. 1954. Taxonomy of the Larvae of the British Species of the Genus *Rhyacophila* (TRICHOPTERA). *Proc. R. ent. Soc. Lond.* (A) **29**, 147–52.

MACKERETH, J. C. 1956. Taxonomy of the Larvae of the British Species of the Subfamily GLOSSOMATINAE (TRICHOPTERA). *Proc. R. ent. Soc. Lond.* (A) **31**, 167–72.

MIALL, L. C. 1905. *The Natural History of Aquatic Insects*.

MORTON, K. J. This author contributed a large number of papers on Trichoptera from 1883 to 1930, many of them containing descriptions of the immature stages. A small selection of these papers is given here.

MORTON, K. J. 1884. On the Larva, etc., of *Beraeodes minuta* L. *Entomologist's mon. Mag.* **21**, 125–6.

MORTON, K. J. 1888. The Larva and Case of *Ithytrichia lamellaris* Eaton, with references to other species of HYDROPTILIDAE. *Entomologist's mon. Mag.* **24**, 171–5.

MORTON, K. J. 1888. The Larva, etc., of *Philopotamus. Entomologist's mon. Mag.* **25**, 89.

MORTON, K. J. 1890. Notes on the Metamorphoses of Two Species of the Genus *Tinodes*. *Entomologist's mon. Mag.* (2) **1**, 38–42.

MORTON, K. J. 1890. Notes on the Metamorphoses of British LEPTOCERIDAE. *Entomologist's mon. Mag.* **26**, 127–31, 181–4, 231–6.

MORTON, K. J. 1904. The Preparatory Stages of *Adicella filicornis* Pict. *Entomologist's mon. Mag.* (2) **15**, 82–4.

MOSELY, M. E. 1921. *The Dry-Fly Fisherman's Entomology*.

MOSELY, M. E. 1926. *Insect Life and the Management of a Trout Fishery*.

MOSELY, M. E. 1939. *The British Caddis Flies*.

PHILIPSON, G. N. 1953. The Larva and Pupa of *Hydropsyche instabilis* Curtis (TRICHOPTERA, HYDROPSYCHIDAE). *Proc. R. ent. Soc. Lond.* (A) **28**, 17–23.

PHILIPSON, G. N. 1953. The Larva and Pupa of *Wormaldia subnigra* McLachlan (TRICHOPTERA:PHILOPOTAMIDAE). *Proc. R. ent. Soc. Lond.* (A) **28**, 57–62.

PHILIPSON, G. N. 1961. The Fifth Instar Larva of *Potamophylax stellatus* (Curtis) (LIMNEPHILIDAE:TRICHOPTERA). *Hydrobiologia* **18** (4), 321–6.

TINDALL, A. R. 1960. The Larval Case of *Triaenodes bicolor*. *Proc. R. ent. Soc. Lond.* (A) **35**, 93–6.

TINDALL, A. R. 1963. The Skeleton and Musculature of the Thorax and Limbs of the Larva of *Limnephilus* sp. *Trans. R. ent. Soc. Lond.* **115**, 409–77.

List 2. The following references give a review of the literature on the immature stages of TRICHOPTERA *from sources other than British.*

BADCOCK, R. M. 1952. Observation of Egg-laying Under Water of the Aerial Insect *Hydropsyche angustipennis* (Curtis) (TRICHOPTERA). *Nature, Lond.* **170**, 40.

BADCOCK, R. M. 1953. Observation of Oviposition Under Water of the Aerial Insect *Hydropsyche angustipennis* (Curtis) (TRICHOPTERA). *Acta Hydrobiol.* **5** (1–2), 222–5.

BERTRAND, H. 1954. Tableaux de Determination (Ordres et famillies) des Larves et Larves-nymphes aquatiques d'insectes. *Cahiers des Naturalistes. Bull. Des. N.P.* n.s. **9**, 3–18.

BETTEN, C. 1934 (with chapters by B. L. Kjellgren, A. W. Orcutt and M. B. Davis). The Caddis Flies or Trichoptera of New York State. *Bull. N.Y. St. Mus.* No. **292**.

CRICHTON, M. I. 1957. The Structure and Function of the Mouthparts of Adult Caddis Flies (TRICHOPTERA). *Phil. Trans. R. Soc.* **241**, 45–91.

ESBEN-PETERSEN, P. 1916. *Vaarfluer, Danmarks Fauna* **19**.

HAGEN, H. A. This author was responsible from 1851–87 for a large number of papers on Trichoptera, including the immature stages. The material described was collected from all parts of the world.

HAGEN, H. A. 1866. Description of a Genus of Caddis Flies, of which the Larvae Construct Cases, known as *Helicopsyche*. *Entomologist's mon. Mag.* **2**, 252–5.

HINTON, H. E. 1946. The 'Gin-traps' of some Beetle Pupae; a protective device which Appears to be Unknown. *Trans. R. ent. Soc. Lond.* **97**, 473–96, 27 figs.

HORA, S. H. 1930. Ecology, Bionomics and Evolution of Torrential Fauna, with Special Reference to the Organs of Attachment. *Phil. Trans. R. Soc.* (B) **218**, 171–282.

JACQUES, D. 1960. The Life and Death of an Angler's Fly. *Anglers Annual* 1961. Heathcote Press, London. Published Nov. 1960.

KLAPÁLEK, F. Much of our knowledge of the Trichoptera and their systematics is due to the work of this author, who published from 1888 to 1913 a series of papers dealing mainly with Hungarian and Balkan species.

KLAPÁLEK, F. 1889. The Metamorphoses of *Apatania muliebris* McLach., a chapter in Parthenogenesis. *Entomologist's mon. Mag.* **25**, 241–2.

KRAFKA, J. 1915. A Key to the Families of Trichopterous Larvae. *Can. Ent.* **47**, 217–22.

LESTAGE, J. A. 1921. Trichoptera in Rousseau: *Les Larves et Nymphes Aquatiques des Insectes d'Europe*.

LESTAGE, J. A. 1922. Note sur la Ponte immergée des *Micrasema* (TRICHOPTERA). *Annls Biol. lacustre* **11**, 152–62.

MARTYNOV, A. B. This author wrote mainly on Russian species and most of his papers are in Russian.

MCLACHLAN, R. Over a period of more than forty years (1862–1903) McLachlan contributed to the literature of the Trichoptera. His best-known work is given below.

MCLACHLAN, R. 1874. *A Monographic Revision and Synopsis of the Trichoptera of the European Fauna.* Pt. 1. Further parts and supplements published in 1874, 1875, 1876, 1877, 1878, 1879, 1880 and 1884.

MORETTI, G. 1937. Studi sui Tricotteri: V. Tricotteri cavernicoli. *Boll. Zool. agr. Bachic. Torino* **7** (1936), 229–39.

MORETTI, G. 1940. Studi sui Tricotteri: XIII. I Tricotteri della Sardegna. *Memorie Soc. ent. ital.* **19**, 259–91.

MORETTI, G. 1942. Studi sui Tricotteri: XIV. Comportamento del *Triaenodes bicolor* Curt. (Trichoptera—LEPTOCERIDAE) nelle risaie a vicenda e in allevamenti sperimentali. *Boll. Lab. Zool. agr. Bachic. Milano* **11**, 3–45.

MORETTI, G. 1942. Partecipazione dei Tricotteri alle biocenosi costiere del Lago di Fovel. *Memorie Ist. ital. Idrobiol, Milan* **1**, 295–306.

NIELSEN, ANKER. 1942. Über die Entwicklung und Biolgie der Trichopteren. *Arch. Hydrobiol. Plankton* (Suppl.) **17**, 255–631.

NIELSEN, ANKER. 1943. Postembryonale Entwicklung und Biologie der Rheophilen Köcherfliege *Oligoplectrum maculatum* Fourcroy. *Biol. Meddr., Kbh.* **19** (2), 87 pp.

NIELSEN, ANKER. 1943. *Apatidea auricula* Forsslund from a Norwegian Mountain Lake. *Ent. Meddr.* **23**, 18–30.

ROSS, H. H. 1944. The Caddis Flies, or TRICHOPTERA of Illinois. *Natural History Survey Division. State of Illinois.*

ROUSSEAU, E. 1921 (the chapter on Trichoptera is the work of Lestage). *Les Larves et Nymphes Aquatiques des Insects d'Europe.*

SCHWANWITSCH, B. N. 1943. *Nature, Lond.* **152**, 727.

SILFVENIUS, A. J. (SILTALA). Numerous papers were published from 1900 to 1908 on the metamorphoses of Trichoptera, mainly in *Acta Soc. Fauna Flora fenn.*, or *Meddn Soc. Fauna Flora fenn.* (Note: The author's change of name from Silfvenius to Siltala took place in 1906.)

SNODGRASS, R. E. 1935. *Principles of Insect Morphology.*

ULMER, G. The work of this author has already been referred to in the text. Much of his work is collected together in the following two works.

ULMER, G. 1903. Über die Metamorphose der Trichoptera. *Abh. naturw. Ver. Hamburg* **18**, 1–154.

ULMER, G. 1909. Trichoptera in *Süsswasserfauna Deutschlands* **5–6**.

WESENBERG-LUND, C. J. This author wrote a number of papers on the biology of Caddis larvae from 1908 to 1915, mainly in *Int. Rev. Hydrobiol.* The following paper is selected.

WESENBERG-LUND, C. J. 1911. Biologische Studien über Netzspinnen der Campodeoide Trichopterenlarven. *Int. Rev. Hydrobiol.* (*Biol. suppl.* 3) **4**, 1–64.

WESENBERG-LUND, C. 1943. *Biologie der Süsswasserinsekten.* Copenhagen.

List 3. *References to Publications Dealing with the North American Fauna. This list contains the more recent publications with some important exceptions. These are older papers to which reference is most often given by modern workers. More complete lists, including the extensive literature of* Banks, *are to be found in* Betten, *1934, and* Denning, *1956. In addition the reader may again be referred to* Fischer *1960–1967.*

ANDERSON, N. H., 1967, Life Cycle of a terrestrial caddisfly, *Philocasca demita* (TRICHOPTERA: LIMNEPHILIDAE) in North America. *Ann. Entomol. Soc. Am.* **60** (2): 320–323.

BETTEN, C., 1913, An interesting feature in the venation of *Helicopsyche*, the MOLANNIDAE, and the LEPTOCERIDAE. *Ann. ent. Soc. Amer.* **6** (1): 65–73.

BETTEN, C., 1934, The Caddis Flies or TRICHOPTERA of New York State. Chapters by Kjellgren, B. L., Orcutt, A. W., and Davis, M. B., pp. 1–576. New York State Museum Bulletin No. 292. University of the State of New York. Albany, N. Y. Amongst the essential literature for a study of North American caddis flies. An extensive bibliography is included.

BETTEN, C., and MOSELY, M. E., 1940, The Francis Walker Types of TRICHOPTERA in the British Museum. *Brit. Mus.* (*Nat. Hist.*), London. pp. 1–248, 122 figs.

BETTEN, C., 1950, The Genus Pycnopsyche (TRICHOPTERA). *Ann. ent. Soc. Amer.* **43** (4): 508–522.

BLICKLE, R. L., and MORSE, W. J., 1954, New Species of HYDROPTILIDAE (TRICHOPTERA). *Bulletin* of the *Brooklyn Entomological Society.* **69** (5): 121–127.

For references to publications dealing with the North American Fauna see page 453.

4

COLLECTING, PRESERVING AND MOUNTING TRICHOPTERA. REARING LARVAE

Collecting

Having arrived at the water's edge, the plan for collecting will depend largely on the nature of the water. As a general rule it is far better to get into the water and search for the larvae by sight than to collect by means of haphazard sweeps of the net. This latter method, however, may on occasion have to be the only choice when one is collecting in deep muddy pools or deep, swiftly flowing rivers. In streams, shallow rivers, pools and lake-sides, however, it is best to wade into the water, the feet clad in wellington boots or waders. If, on a hot summer day, a bathing costume is one's only dress, then wear strong pumps on the feet to guard against cuts from broken glass or tins. Then wade slowly against the current and examine all submerged accumulations of vegetable detritus such as bits of twig and leaf that get swept into side pools of relatively stiller water. A pond-net comes into only occasional use—to sweep among the submerged roots of vegetation bordering the bank or to catch what might be shaken from water-weeds. The main work, however, is to turn over the bottom stones, examining the top and the under-surface as well as the concavity left by the stone in the sand or silt. The top surface of the stone may bear the long fixed silken tunnels of a psychomyiid, and if the surface of the stone is rough they may be picked out by the eye only with difficulty.

In times of drought these tunnels may even be out of the water, but with the larvae living in their closely fitting tubes. It must be the rule of the collector not to leave any place where he collects any the worse for his having been there. Apart from stones, submerged branches or trunks of trees should be rolled over and inspected. The fixed tunnels of the psychomyiid *Lype reducta* are found on small branches submerged in small streams. Many limnephilids are sometimes found in crevices in submerged bark especially when they are searching for a place for anchorage of the case before pupation. It will be found also that some species appear to 'congregate'; often two or three dozen will be found under one stone or branch, yet perhaps none will be found for several yards, under apparently similarly attractive objects. Net-spinning caddis larvae will often be met with in this type of locality and special care will have to be taken when transferring them to the collecting tin in order not to injure them. The object should be to collect about two dozen larvae of the same species all from the same few square yards so that the chances of collecting mixed species is minimized.

Many special habitats will require rather special collecting techniques but much can be left to the judgement and ingenuity of the field naturalist. It is hoped, however, that the reader will accept from the writer a guiding principle for collecting, that is to use sight as far as possible, employing it before using haphazard methods. Even when it is necessary to use a pond-net, appreciate exactly where it is being used so that it will be known from near which species of plants the larvae have been obtained. Much useful information can be collected in this way. For example, the larvae of *Limnephilus extricatus* seem always to be found among the submerged roots of hairy willow herb and those of *Holocentropus dubius* are associated with water lilies and *Potamogeton*.

One of the most important points about collecting is that the collector will learn about the habits of the insects, especially their exact habitat preferences. Although well-known localities for insect collecting generally, or more particularly for caddis, will obviously be visited, the student of the TRICHOPTERA should endeavour to collect in out-of-the-way places where little in the way of collecting has ever been done. Many important records await the keen collector when he begins to get a 'feel' for knowing the likely locality for rare species. On the other hand, it is rewarding to look up the old (and the new) records of capture of adult caddis flies and then endeavour to search out the breeding-place in order to obtain some larvae. A note of warning, however, ought to be sounded here. Make sure that you have permission to collect. If the locality is a nature reserve or is controlled by the Forestry Commission or other authority, some authorization will be required.

Examining likely localities for the terrestrial species *Enoicyla pusilla* should not be neglected. At present known in Britain only from a few acres in Worcestershire, it is almost certain to be found elsewhere. In Chapter 15 details are given of its present habitat. Perhaps the two other species in the genus will be found in our own countryside.

The last note on collecting is for the town or city dweller. Every opportunity should be taken for examining water in tanks, ornamental garden pools and pools and lakes in parks. Caddis will often be found to be present in remarkable numbers, and a 'pure culture' of one species in quantity will sometimes be found.

Rearing

For many species of larvae that breed in standing water, almost any sort of container, provided it will hold water, will suffice to maintain them, from a one-pound jam-jar to a garden pond. That is not to say that all these species can be reared even if neglected, although the large limnephilid

Grammotaulius atomarius is often found in field ditches under very harsh conditions of stagnation.

For each container arrangements must be made for oxygenation. This can be done by tying some water-weed to a stone in order to sink it and maintain it in an upright position. In addition some sand and pebbles obtained from the collecting locality should be introduced. This is for a foot-hold for the larvae and also to provide a similar microflora on which they can graze. A few sticks or corks should be placed in the container so that the emerging pupa can crawl out of the water for its metamorphosis. Some species, however, are able to carry out their final metamorphosis on the water surface and then skate on the surface until they can find a solid object on which to climb. When the larvae have been introduced into the aquarium (be careful not to overstock) its aperture must be covered with gauze or some other material to prevent the escape of the adults.

A rubber band cut from an old motor-tube is excellent for holding the gauze. Finally the aquarium must be placed where subdued sunlight can reach it, in order that oxygen will be released from the water-weed, and where it is relatively cool. It must not be placed in the direct sun as the small volume of water would quickly warm up and kill the larvae.

Do not forget to examine the water-weed and stones very minutely to guard against introducing other caddis larvae which would cause confusion. If larvae die, take them out of the aquarium as soon as possible in order not to contaminate the water.

It is essential to label the aquarium and its contents and keep exact records. There are, no doubt, several ways of doing this but the writer has kept records for many years in the manner described later in this chapter.

Caddis fly larvae taken from swift-flowing water generally cannot be reared in the usual type of still-water aquarium or container unless some method is adopted which keeps the water in constant motion. Philipson (1953) described a simple stirring equipment for one or more two-pound glass jam-jars in which small stones and pebbles were placed. In these he reared successfully the net-spinning species *Hydropsyche instabilis* and *Wormaldia subnigra*. In these cases oxygenation of the water was not required but it is possible to provide sufficient water movement by a fairly powerful air-jet for net-spinning species to settle down and spin their nets. In both methods, of course, the containers must be covered to prevent escape of the adults.

Two additional methods of rearing caddis larvae are applicable to species found in fast-moving water. In the one, a continual current of water is circulated from container to container either being constantly fed from a tap or being pumped back to a tank. The other method consists of gauze containers placed in appropriate positions in a natural stream. These must be made to a robust pattern and in addition must be placed where there is no risk of animal or human interference. Wiggins (1959), gives a design of an apparatus for the large-scale rearing of caddis larvae involving small cells of aluminium 'screening' placed in sets in frames which are held in tanks.

Transporting

Much can go wrong in transporting the captured caddis larvae back to the laboratory or home. The most important point may sound something of a paradox, but in the ordinary way they should not be transported in water. The collecting jars or tins each with their complement of larvae should be tightly filled, preferably with a stiffish-stemmed water-weed such as *Elodea*, *Myriophyllum* or *Callitriche*; failing these, liverworts or moss. However, do not use *Sphagnum* or other soft mosses as they tend to form a solid ball which will kill all larvae imprisoned with it. Fill the collecting tins with water then leave them for a few minutes so that the larvae may gain a foothold on to the weed; then drain them out, fit the lids and pack them in the coolest and least jolting situation in the luggage for the journey home. On arriving home or at the laboratory, empty the tins into separate trays or bowls of water, and if it is possible to introduce individual aeration, do so. Then one may return in an hour's time and make an inspection and transfer the lively ones to their rearing quarters. Those that have not survived may be preserved, as described later.

If it is necessary at any time to send living caddis larvae by parcel post, pack them in a similar manner, then wedge the tin with rolls of stiff paper into a strong wooden box. Of course, time your packing so that the larvae catch the afternoon parcel collection as this may save a day in transit.

A second method of transporting caddis larvae, which is really only practical if a motor-car or boat is used, is to place all the larvae in a large container such as a milk-churn or 5-gallon drum. When the collecting is completed, and this is rushed through as quickly as possible, one gets home to cold aerated water with all possible speed. This ensures that the water in the drums remains relatively cool. It is impossible to transport living caddis larvae in de-oxygenerated warm water, and both the methods described guard against this.

Both methods of caddis larvae transport described above have been successfully employed by the writer in the tropics. On one occasion two species of larvae were collected at 9,000-feet altitude in Kenya, then taken a fourteen hours' very hot train journey in vacuum-flasks packed with weed to Jinja in Uganda, where they were eventually bred out. At another time two species collected in Uganda were flown home to England and bred out in aquaria in Surrey.

Finally a word about the tube-making larvae in the family PSYCHOMYIIDAE. The stone or wood to which the tubes are fixed should be brought home entire, in a polythene bag. This is easily accomplished in the boot of a motor-car but rather tiresome in a rucksack, but if it can be done it will give good results.

Where to Collect

The life-cycle of the different species of caddis flies is passed in an extraordinarily diverse range of aquatic habitats, so that the adult caddis may be sought for around almost every situation where there is fresh water. In addition it seems virtually certain that the rare, newly discovered, leptocerid species *Triaenodes reuteri*, is able to breed in water which is slightly brackish, and although this is the only British record of such a habitat, there are many records of caddis fly larvae living in brackish water around the European mainland. It would be very worth while, therefore, to explore estuarine areas which may not have had much attention from TRICHOPTERA collectors. Although the larvae of a number of species are found in stagnant ditches (such as the Limnephilid *Grammotaulius atomarius*) and even evil-smelling bogland (such as *Limnephilus incisus*), water polluted with industrial waste seldom contains caddis larvae. Except for polluted water, then, the vicinity of virtually every source of fresh water may be examined for caddis flies. Around pools, tarns and stream, on hill and mountain-tops, the underside of stones and rocks should be examined as well as the bases of grass tufts and indeed any situation which offers shelter. Here such species as *Phryganea obsoleta* rest out of the wind which may blow them far from their breeding-place. The bark of trees should be inspected, and although most caddis species possess a shape, wing pattern and colour which merges into the background, with a little practice in training the eye to pick out the shape of a caddis fly, proficiency will be acquired. The phryganaeid species *Trichostegia minor* usually rests in the crevices of bark of trees surrounding the pool in which its larval stage is passed. I have only found it in the bark of oak-trees but no doubt it may also be found on other tree species. The psychomyiid caddis *Tinodes waeneri* is often found nesting on the underside of oak-leaves near the water in which it breeds, and this sort of situation is commonly utilized by many species.

Caddis flies are thought to be attracted by aromatic odours, and it is certain that many species are found resting amongst pine-needles. The phryganaeid *Oligotricha clathrata* was usually collected from pine-trees which grow in and around Chartley Moss, the better known of its two British localities. The area is now a nature reserve.

Stonework of bridges and culverts is a favourite hiding-place for caddis flies, and they often fit themselves into thin crevices in the mortar. It is sometimes possible to examine the underside of a bridge, out of the direct sunlight, as many species prefer the shade. On the other hand, the tiny HYDROPTILIDAE are often seen in bright sunlight, making quick runs with great agility over the stonework. Woodwork too should be searched and boat-houses, sheds and fencing. All unpolluted moving water from slow-flowing rivers and streams to torrents and waterfalls supports a large number of TRICHOPTERA species; some are found only in the most swiftly flowing currents.

Lakes, pools and ponds may contain a large number of species, but when a visit is made to such localities the area around the outflow stream is often of particular interest, as is also the boggy ground from which a seepage trickle emerges. The actual amount of water in which some species can breed is sometimes amazingly small. The larvae of the psychomyiid species *Tinodes assimilis* are found only in the thin film of water on wet rocks, and one locality is known where about three square feet of wet rocks at the bottom of a wall in the middle of a town has supported a colony for many years.

All the situations for collecting so far described have been near water, but caddis flies are often found far from it. This may be due to their having been attracted to a source of light away from their normal breeding-place, and some species have been collected a mile or more away from the latter. On the other hand, they may have been carried involuntarily with produce from some riverside situation. Wherever one is, it is wise to carry a small box in the pocket so that such wanderers can be collected.

When to Collect. There is no period in the year when collecting may not take place. Adults of some species, such as a number of Limnephilids including the terrestrial species *Enoicyla pusilla*, may be collected until the middle of November.

Larvae of almost all species may be collected throughout the year, the adult emergence period usually being so prolonged that even during this period larvae and pupae may be found. Generally speaking, collecting larvae is only impossible when the weather is so inclement as to make it so; freezing and flooding alone bringing collecting to a standstill.

Collecting by Means of Light. A very convenient method of collecting caddis flies is to attract them to a source of light. It is known that a very large number of species may be concentrated in this way and, further, that, whilst a large number of species may be attracted to light in the near vicinity of water, a small number will fly a long distance to light. Perhaps the last statement needs some clarification. Certain species are found at a source of light a long way from their breeding-point, but whether they usually fly over a wide area and at some point in this wide-ranging flight come within the range of the light-source, or whether they are attracted directly from the immediate vicinity of the breeding-point, has not been determined.

It is always worth while looking around lights of all types near water; caddis flies will often be found resting in an exposed situation a few feet from the light-source. Two types of light-source, however, have received special attention as an aid to collecting caddis flies, the mercury vapour lamp and the Tilley lamp; they are briefly described below.

The Mercury Vapour Lamp. Crichton has given an account of the working of a light-trap situated close to a pond of eight and a half acres extent near

Reading in Berkshire. This was a Robinson-type trap with an 80-watt mercury vapour lamp connected to an a.c. main with a choke and Venner time-switch installed at the supply end. This trap was operated over several years every third night from dusk till dawn. Air and water temperature were recorded during this period, as well as humidity, rainfall and the general state of the weather. A total of 66,000 caddis flies were collected in 1954 and 43,000 in 1955, and the number of species amounted to seventy, representing just over one-third of the species recorded in Britain. From the information obtained by Crichton in this way it has been possible to produce a substantial amount of information on the relative abundance of the species caught, their emergence periods and the correlation of the latter with meteorological conditions. Additional information on the use of light-traps for collecting caddis flies and other insects, particularly in relation to the collection of biological information as well as the insects themselves, can be found in the papers of C. B. Williams (1939, 1940, 1948, 1951), Corbet & Tjønneland (1955), Robinson & Robinson (1950), and Anne Marshall (1939). It should be noted, however, that the killing of very large numbers of caddis flies is to be deprecated and should never be tolerated in the interests of collecting. Those responsible for the conducting of scientific experimental work which will inevitably bring about the death of thousands of insects should weigh very carefully the probable value of the results against this.

The Tilley Lamp. For some years Brindle has made a study of the biological information obtained by collecting caddis flies attracted by Tilley lamps, and he has recorded the methods used and his results in four short papers. He used two 300-candle-power Tilley lamps on a white sheet positioned close to the water and as nearly level with the water surface as possible. Tilley lamps use paraffin vapour under pressure ignited under an incandescent mantle. The sheet should touch the water if possible. This is because many species skim and skate about on the water surface, sometimes in swarms prior to mating. This is termed 'night swarming' by Brindle. The attractive power of the Tilley lamp, however, is restricted to about 20 feet radius so that it is necessary to set up a number of collecting stations around or along a body of water. The lamps should then be moved from time to time in order to collect all the species likely to have emerged, as many species are known to be very local in their range, seldom moving many feet from their breeding-place.

Killing

When the caddis fly is in the net or in the collecting-box the next step is to decide whether we wish to retain the specimen. It is wrong to dump the contents of the net straight into the killing-jar without inspection. The caddis should first be examined carefully after getting it into a glass tube. With a little experience this may be accomplished without difficulty, but some species have the habit of shooting about in the net in unforeseen directions. It is not for nothing that one species possesses the common name 'The Caperer'. There is neither sense nor ethics in killing more specimens than are actually required for the purposes of our study. On the other hand, we may wish, on the basis of the specimen, to record its existence in a county in which it had not been reported previously, or an extension to its known adult emergence period, or even perhaps as a species new to Britain. Then the specimen must be retained, killed and dealt with in a careful manner, as we will describe later. In addition, we may be forming a collection as a basis for identification requiring about half a dozen specimens of each sex.

These are the only reasons which should prompt us to kill our specimens. If we are neither going to have time nor energy to label carefully and preserve with precision our captures, then they should be released close to the point of capture so that the breeding stock in the particular environment is not depleted. Generally we never know upon what critical knife-edge the number of the population is balanced.

Having satisfied ourselves that we wish to kill the specimen we transfer it to the killing-jar. This jar may be constructed in a number of ways but in each the specimen is introduced to a poisonous substance. The container is usually of strong glass, short and squat with little shoulder and a very wide mouth. It should be so robust that it is not easily damaged or broken. A well-fitting cork bung is generally used although for convenience or necessity the writer has used a one-pound honey-jar with a metal screw-cap and cork insert. It is obviously desirable that the specimens when killed should be in a soft relaxed condition, and remain so for a few days, and in this regard one toxic substance, the liquid ethyl acetate, is perhaps the best. The killing-jar container, however, must be provided with an absorbent surface, and this is usually done by mixing plaster of paris with water into a thick paste and pouring it immediately into the jar, making sure the surface is as flat and smooth as it is possible to make it. The plaster layer should extend to about an inch from the bottom, and when it has set hard it is ready for use. When actual collecting is to commence, the killing-jar is primed with a small quantity of ethyl acetate. This is a very mobile liquid so that extreme care must be exercised in order that none of it runs down the inside wall of the jar. A few circles of clean white blotting-paper already cut to size are placed on top of the plaster in order to mop up any moisture there may be. If a number of specimens are likely to be collected, then, in addition, some crumpled tissue paper should be introduced so that the caddis flies get wedged into the folds, and thus damage to antennae, wings and legs is prevented. The inside of the jar should be inspected from time to time to ensure that excess moisture has not caused the caddis flies to stick to the glass sides. This may cause considerable damage to antennae, wings and legs.

When the insects are dead they may be stored for a few days in small tin boxes or glass tubes in which is also placed a piece of blotting-paper or other absorbent material moistened with ethyl acetate. It is most important that at the time the dead insects are emptied from the killing-jar, the essential data (locality of capture, date, collector's name, etc.) is immediately written out and placed *inside* the appropriate container in which the insects are next placed. In every subsequent handling or manipulation of the specimen the data label must accompany it and be so attached that there is no possibility of specimen and label becoming parted or mixed up in any way.

There are a few further points concerning the killing-jar that need to be pointed out.

It is best to protect the killing-jar from damage by getting it fitted into a leather case, preferably with a lining of sponge rubber. The leather case may then be carried by means of a strap over the shoulder or round the neck in the same way as a camera. This will be especially useful on a river-bank or in other situations near deep water where on occasion it may be necessary to leave both hands unencumbered.

As mentioned previously, ethyl acetate is a very mobile liquid and needs to be contained in a well-stoppered bottle. This also should be protected against damage and a good way to do this is to keep it in a circular tin lined with sponge rubber.

Other toxic agents are often described, such as potassium cyanide and bruised laurel leaves. The former is so poisonous to human beings, and in addition does not have the advantage of keeping the insects in a relaxed condition as does ethyl acetate, that its use for killing caddis flies is not recommended. Bruised laurel leaves have much to commend them as a lethal and relaxing agent for caddis flies but not everyone has the natural product growing in a convenient position.

Making a Dry Collection of Caddis Flies

'The first task in the scientific study of any group of insects is to recognize, define and name the species, thus providing a series of pegs upon which to hang all subsequent data.' This was stated by N. D. Riley over twenty-five years ago in the introduction to Mosely's handbook. There is no doubt that the building up of a dry, mounted collection of caddis flies is of great value in learning to identify them in the field. Whilst a fluid collection is also of importance in other ways, a caddis fly when wetted loses its natural appearance.

The conventional method of mounting caddis flies for a dry collection is to pin and set them in a way similar to that used for butterflies and moths. The legs of caddis flies are, however, of great importance in identification. The spurs on the tibia of each leg on one side have to be counted so that the legs must always be displayed. Unfortunately, caddis flies when dry are exceptionally fragile, the slightest jar or knock will often break off an antenna, a leg or perhaps the whole abdomen. The greatest care, therefore, must always be taken to ensure that this never happens. One cannot take the same liberties with a set caddis fly as one can with a butterfly.

When a study of the TRICHOPTERA is initiated it is very desirable to have the identification of a few species confirmed by an authority, and this should serve as the basis of the collection. In addition to the full data explained elsewhere, an extra label with the determination and the authority's name, e.g. *Phryganea grandis* Linné det. M. E. Mosely, should be added to the pin.

It is assumed that the reader has some experience or at least acquaintance with the method of setting butterflies. If this assumption is not justified then he is referred to Oldroyd's *Collecting, Preserving and Studying Insects*, published by Hutchinson, London, in 1958.

When caddis flies are being set the following recommendations are made. Always pin the specimens that are to be mounted when freshly killed, and set within two or three hours if possible or two or three days at longest. Do not attempt to relax dried caddis flies. Use setting-boards with deep grooves so as not to damage legs. It is a good plan to line the groove with cellophane as well as to cover the top surface of the board so that there are no rough projections on which the tarsal claws of the legs can catch and hold. Otherwise many legs may be lost when the insects, if very dry, are removed from the boards. Mosely also recommended cellophane as cover strips for the wings when setting.

Many of the smaller caddis flies will require staging, that is, the use of very small pins, known as points, or 'minuten', which pin the specimen on to a 'stage' consisting of a small piece of polyporus. This latter is then pinned with a large and stout pin into the cabinet drawer. The labels are transfixed by the larger pin. Not only does staging make the pinning of small caddis flies possible, but when the insect has to be moved in the collection, this is done by holding the larger pin and thus the shock of removal from position is somewhat absorbed. Caddis flies are generally so fragile that larger specimens can, with advantage, be staged.

Making a Fluid Collection of Caddis Flies and their Immature Stages

Although a mounted, dry collection of caddis flies has the advantage of giving a good impression of the shape and colour of the wing, neuration and length of antennae, it has a serious disadvantage when the specific determination requires a critical examination of the genitalia, which is so often the case. The serious collector will therefore require, in addition to his dry, mounted collection, a collection in fluid. The genitalia of the specimens preserved in fluid will remain, to a great extent, with the constituent parts in their natural shape.

With regard to egg-masses, larvae and pupae, there is no satisfactory method of dry mounting, so that in any case these will be collected and preserved in fluid.

Mosely recommended a 'collecting fluid' consisting of two parts of 2 per cent formalin and one part of 90 per cent alcohol, in which the insects are first wetted and killed, and then, as soon as convenient, transferred to 2 per cent formalin solution. Mosely considered that this mixture did not discolour the thick white abdomens of many limnephilid larvae. Many workers, however, believe that 75 to 80 per cent ethyl alcohol has many advantages over the use of formalin. The latter will harden the specimens, and also its strong odour affects the nose and eyes. On the other hand, it does not evaporate so rapidly as ethyl alcohol and is not inflammable. Ethyl alcohol is inflammable so that precautions should be taken to ensure that it does not accidentally ignite.

Glass tubes, $\frac{1}{2}$ inch $\times$ 2 inches in dimension, will be found convenient for storage of the specimens. Several specimens, either adults or larvae, may be introduced into each tube, providing they are of the same species and from the same locality, of course. This size tube is large enough for an adequate label. The latter should be written in soft pencil or in indian ink but allow this sufficient time to dry before inserting it into the fluid-filled tube with the specimens. Each and every tube must contain a label bearing the minimum data requirements, i.e. locality of capture, inclusive of parish and county, date of collection and name of collector. Beware of abbreviations! They are usually so ambiguous that they are best avoided altogether. If the name is known without question it may be inserted in the tube as an additional label with the name of the identifier—also as '*Phryganea grandis* L. det. M. E. Mosely'.

It will not of course be possible to write all the information concerning collection on the individual small labels, so collectors use a notebook in addition, but this must never take the place of full data labelling of the specimens. The notebook should merely augment the information on the label. The writer has used systems of numbering for his collection and a separate notebook for each year's collecting. Each batch of caddis larvae collected and believed to be of the same species is given a four-figure number, the first two figures of which represent the number of years collecting. As an example, in the notebook for 1954 all the numbers given to batches of larvae commence with 17, as this was the seventeenth year in which I had collected. Every tube of larvae bears such a four-figure number in characters as large as the tube allows. They are all in black indian ink on white paper. In the notebook very exact details of collecting are given, the locality being given in very much more precise terms than is possible on the label.

The small tubes are stored in larger vessels such as straight-sided wide-mouthed glass jars. Each tube is plugged with a wad of cotton-wool already dipped into alcohol and placed in the storage jar upside-down. A layer of cotton-wool is placed in the bottom of the storage jar to prevent glass hitting glass and thus cracking the tube. The storage jar is then partially filled with alcohol.

The conventional type of storage jar, as used in museums, has a ground-glass stopper, and is expensive, but other types of jars and bottles will do just as well. The writer's collection is contained in relatively small jars of 4 fl. oz. capacity and fitted with plastic screw-caps and rubber inserts. The small-size jar is used because each contains only a single species, but adults reared from the same batch are placed in the same storage jar although in different tubes. A regular examination of storage jars in the collection must be maintained so that the alcohol may be topped up if necessary. They must not be allowed to dry up.

Another series of numbers has been used by the writer for the series of short descriptions of larvae after their identity has been established by rearing. There is no special significance in this numbering. It merely relates to the order in which the descriptions were published, and the larvae associated with the published papers are identified by a large number printed in dark blue on a light blue background. These were the papers published by the Royal Entomological Society of London, commencing with Larvae of the British Trichoptera, 1: *Stenophylax stellatus*, in 1942, (the name has since been changed to *Potamophylax latipennis*), and ending with 53–56, the BERAEIDAE, in 1959.

Oldroyd (1958) has given an account of making a 'spirit collection' in the work referred to on page 64 which contains much useful information.

CORBET, P. S. & TJØNNELAND, A. 1955. The Flight Activity of Twelve Species of East African Trichoptera. *Univ. Bergen Årb. naturv. R.* 1955 (9), 44 pp.

CRICHTON, M. I. 1960. A Study of Captures of Trichoptera in a Light Trap near Reading, Berkshire. *Trans. R. ent. Soc. Lond.* **112**, 319–44.

HICKIN, N. E. 1952. *Caddis*. Methuen, London.

HICKIN, N. E. 1957. Studying Caddis. *Country-side* **18** (new series No. 3), 99–103.

HICKIN, N. E. 1962. Trichoptera from a Kenya High Altitude Stream. *Proc. R. ent. Soc. Lond.* (A) **37**, 129–34.

MARSHALL, ANNE C. 1939. A Qualitative and Quantitative Study of the Trichoptera of Western Lake Erie (as indicated by light trap material). *Ann. ent. Soc. Amer.* **32**, 665–88.

PHILIPSON, G. N. 1953. A Method of Rearing Trichopterous larvae Collected from Swift-flowing Water. *Proc. R. ent. Soc. Lond.* (A) **28**, 15–16.

ROBINSON, H. S. & ROBINSON, P. J. M. 1950. Some Notes on the Observed Behaviour of Lepidoptera in Flight in the Vicinity of Light Sources, together with a description of a light-trap designed to take entomological samples. *Entomologist's Gaz.* **1**, 3–20.

WIGGINS, G. B. 1959. A Method of Rearing Caddis Flies (Trichoptera). *Can. Ent.* **91** (7), 402–5.

WILLIAMS, C. B. 1939. An Analysis of Four Years' Captures of Insects in a Light Trap. Part I. *Trans. R. ent. Soc. Lond.* **89**, 79–132.

WILLIAMS, C. B. 1940. An Analysis of Four Years' Captures of Insects in a Light Trap. Part II. *Trans. R. ent. Soc. Lond.* **90**, 227–306.

WILLIAMS, C. B. 1948. The Rothamsted Light Trap. *Proc. R. ent. Soc. Lond.* (A) **23**, 80–5.

WILLIAMS, C. B. 1951. Comparing the Efficiency of Insect Traps. *Bull. ent. Res.* **42**, 513–17.

5

THE BIOLOGY OF THE BRITISH TRICHOPTERA. FEEDING, CASE AND NET CONSTRUCTION. GENERAL HABITS

In this chapter, which is concerned exclusively with the British fauna, the greatest importance has naturally been placed on the work of British authors, but the student should also consult the works of other authors listed in the bibliographies.

Food

Up to the present time the most complete study of the food of caddis larvae has been made by Slack. He collected between ten and twenty specimens of twelve representative species occurring in the upper reaches of the River Test flowing through the chalk of northern Hampshire. The contents of the alimentary canal were examined and identified as far as possible. Most of the species examined were found to be primarily phytophagous, although only three species, *Limnephilus rhombicus*, *Stenophylax* sp. and *Silo nigricornis*, were entirely so. Of the remaining species, *Glyphotaelius* sp., *Anabolia nervosa*, *Halesus* sp., *Sericostoma personatum*, *Odontocerum albicorne*, *Leptocerus* sp., *Molanna angustata* and *Hydropsyche* sp. were omnivorous to varying degrees, and *Rhyacophila dorsalis* was predominantly carnivorous. Plant material found in the gut of these species consisted of both living and dead particles of higher plants as well as diatoms. *Silo nigricornis* was found to be a diatom feeder exclusively, obtaining the diatoms from the river-bed. The diatoms found in the gut of *Limnephilus rhombicus*, on the other hand, were species epiphytic on such plants as *Ranunculus fluitans*, *Hippuris vulgaris*, *Sparganium* sp. and *Fontinalis* sp., portions of which were also found in the gut. The portions of animals found consisted of Copepods (*Glyphotaelius* sp.), *Baetis* nymphs (*Anabolia nervosa*) and the molluscan *Limnaea* (*Molanna angustata*). In addition several species were found to have been feeding on TRICHOPTERA larvae (*Glyphotaelius* sp., *Odontocerum albicorne* and *Rhyacophila dorsalis*). Dipterous larvae (*Simulium* and *Chironomus*) were also found in the gut of this latter species.

There are a number of records of several species attacking and devouring much larger animals. An instance is recorded in Switzerland of fifteen larvae of *Halesus tesselatus* seizing and eating the amphibian *Triton alpestris*, and Slack mentions a larva of *Halesus sp.* seizing a living trout alevin and killing and eating it. The author has recorded instances of trichopterous larvae feeding on the dead bodies of birds (song-thrush and coot) and a mammal (mouse). It seems evident that authors previous to Slack (Martynov, Wesenberg-Lund and Rousseau) believed that the feeding habits of caddis larvae were much more sharply defined than Slack found was actually the case. For example, Siltala states that the larvae of RHYACOPHILINAE and POLYCENTROPOPIDAE are carnivorous, the HYDROPSYCHIDAE PHRYGANEIDAE, MOLANNIDAE and ODONTOCERIDAE omnivorous and the remainder phytophagous.

Larvae of the terrestrial *Enoicyla pusilla* feed on dead oak-leaves with some moss and algae. The larvae remove the softer epidermal tissue from the leaves, producing a reticulated net-like pattern of veins. Van der Drift & Witkamp have studied the mechanical and chemical action resulting from the feeding habit of this species and calculated, that where it occurred, 9 per cent of the total leaf fall was consumed by *E. pusilla* alone. They also found that as a result of the chewing action of the larvae, an average oak-leaf is broken down into about three thousand faecal pellets containing about ten million particles. This has the result of increasing the volume twofold and the exposed surface area fifteenfold. Evaporation from the faecal pellets is only a third that of the whole leaves. Only about 7 per cent of the leaf material is assimilated by the larvae.

The author has recorded also the change in behaviour of certain caddis larvae such as *Phryganea* sp., which, browsing on plant material, quite suddenly change to a predatory habit, making sudden catching movements with their strongly armed prothoracic legs if water currents are produced.

The HYDROPTILIDAE have been specially studied by Nielsen, and he found that larvae of the genera *Agraylea*, *Hydroptila*, *Oxyethira* and *Orthotrichia*, all of which occur in Britain, feed by sucking the cells of filamentous green algae, and their mouthparts are adapted to this function. Larvae of the genus *Ithytrichia* feed on diatoms and their mouthparts are adapted for this special purpose.

In conclusion it might be stated that the larvae of TRICHOPTERA, by virtue of their large numbers and, in many cases, large size, play an important role in freshwater by cutting up vegetable material for case construction, and for food, removing diatoms and algal felt from the substrate, and feeding on a wide range of freshwater animals, including species in their own order. On the other hand, larval, pupal and adult caddis flies form a not inconsiderable content in the diet of freshwater fish and in some cases birds and bats.

Case Construction

Hanna (1960) recognized eight methods adopted by larvae of TRICHOPTERA in starting their cases. These are as follows:

1. *The Tunnel Method.* The larva collects a pile of portions of dead leaves by means of the claws of its pro- and mesothoracic legs. It then crawls beneath the pile and draws beneath itself a dead leaf, holding it in the angle of the femora and tibiae of the first two pairs of legs. The remainder of the portions of dead leaves are then joined together to form a tunnel. *Trichostegia minor* practises this method.

2. *The Dorsal Plate Method.* The larva lies on its back and pieces of leaf are picked up by the first two pairs of legs. The leaf fragments are pushed backwards by the metathoracic legs and joined together with silk, which then forms a plate over the abdomen. At the front end of the plate a square is formed encircling the thorax by first fixing a leaf portion to the front end of the plate with silk, adding one to each side and then to the ventral surface. In *Chaetopteryx villosa* and *Molanna angustata* the dorsal plate is formed of sand-grains.

3. *The Ventral Plate Method.* A mass of material collected by the larva is joined together by silk to form a plate beneath the abdomen. The larva then adds stalks at right-angles to the edges of the plate by turning first to one side, then to the other. A square is then formed by the larva turning on its back and adding material to the roof. The square is then slipped over the head. More stalks are then fixed to the anterior end. *Limnephilus marmoratus* constructs its case in this fashion.

4. *The Square Method.* The initial plate is made in much the same way as in the Tunnel Method. The larva then crawls under the heap and builds a square by adding material to each side and to the ventral surface. Some larvae, however, crawl over the heap of material, and after adding material to each side form a dorsal surface.

5. *The Belt Method.* When the larva has secreted silk over a heap of material which has been drawn together by the claws of the pro- and mesothoracic legs it crawls over it and, lying on its back, draws the two sides together with the mesothoracic claws and joins them with silk. As a variant to the above, however, some larvae crawl under the heap and join it together ventrally. In the case of three species *Beraeodes minuta*, *Athripsodes aterrimus* and *Triaenodes bicolor*, they sometimes fail to join the two ends together to form the belt but then fix the plate by means of silk threads to the sternum of the first abdominal segment. The larva then bends its abdomen forwards under the thorax until the anal claws hook the plate, and its end is then moved sideways in a dorsal direction a few times until it has been transferred to the side of the thorax. The plate is long enough to hang on both sides of the thorax. The prothoracic legs pull the two ends together and join them with silk to form the belt. Other species using this method are *Limnephilus flavicornis*, *L. lunatus*, *Anabolia nervosa*, *Potamophylax radiatus*, *Brachycentrus subnubilus*, *Athripsodes aterrimus*, *Mystacides nigra* and *M. longicornis*.

6. *The 'T' Method.* The caseless larva secretes silk over a heap of sand-grains and small stones collected together by the claws of the pro- and mesothoracic legs. This same material is then added first to one side and then to the other to form a T-shaped structure. The larva then lies on its back, pulls the two ends of the cross-plate together with the mesothoracic legs, and joins them with silk. *Limnephilus flavicornis* constructs its case in this fashion on occasion.

7. *The Burrowing Method.* The larva burrows into the sand, using the first two pairs of legs. The head and thorax only penetrate gradually, leaving the abdomen nearly vertical, protruding from the substrate. Burrowing then stops for about twenty minutes whilst the abdomen is seen to rotate slowly. At this stage a ring of sand-grains and small stones adheres to a ring of silk. The larva then burrows completely and a loose case of silk threads with adhering sand-grains and small stones has been formed. The larva then constructs a definitive case at the front end of the provisional one so formed. Examples of species using this method of case construction are *Drusus annulatus*, *Sericostoma personatum*, *Notidobia ciliaris*, *Odontocerum albicorne*, *Athripsodes cinereus* and *Oecetis lacustris*.

8. *The Triangle Method.* The larva holds on to a small stone by the anal claws. It then fixes a small stone to the sterna of the meso- and metathorax, and additional stones to each side of these segments, thus completing a triangle. The larva then proceeds to build the anterior part of its case in front of this girdle by fixing stones to the sides and then constructing the floor. When this is completed the larva reverses itself inside the case, cuts away the triangle at the posterior end and this completes a normal case. *Glossosoma conformis* constructs its case in this fashion.

Tobias (1962) has collected together a large number of illustrations of caddis cases.

Net-spinning

Larvae of caddis in the families POLYCENTROPIDAE, PHILOPOTAMIDAE and HYDROPSYCHIDAE do not construct transportable cases but spin nets, snares and webs of a silk-like secretion in which to live and trap their food. Whitehead has given an account of the net-spinning of *Hydropsyche angustipennis*. This species spins a web of fine silk amongst plant debris under partly submerged stones in clear, swift streams. The web is tunnel-shaped with the opening at the upstream end, and at the other end and generally at the side is placed an outlet which is covered by a well-formed net of rectangular mesh. The larva normally lives in the web, which is closely woven, offering considerable resistance to the water flow, but the net meshes, which average 0.3 by 0.13 mm. in size, allow the water to flow out with ease. In this way small insect larvae, worms and an assortment of vegetable matter get trapped on the outflow net meshes and it is on this that the larva feeds.

Whitehead found that the nets varied in size according to the space to be covered, but usually measured from 5 to 7 mm. across. There are irregular quadrilateral and triangular fibres across the middle of the net which form a seam and apparently causes the net to bulge somewhat. The net is composed of two kinds of fibres, those forming the shorter sides of the mesh, which are constant in width (21 μ), and those that form the long sides, which are not of constant width, being about 15 μ in the thinnest part, but where they cross the thicker fibres they are much wider. Whitehead suggests that the short sides of the rectangular mesh are spun by the larva first whilst the fibres of the long sides are produced afterwards and thicken out before they harden. On the other hand, Wesenberg-Lund (1913) considers that a special cement is used as an adhesive at the junction.

In the PHILOPOTAMIDAE, *Wormaldia subnigra* and *W. occipitalis* spin long loosely spun silken tubes open at the end, which is placed to receive the full force of the water current. The tubes are attached near the open end to a stone so that the hinder end of the tube waves freely in the stream. Philipson (1953) was able to get larvae in aquaria to construct their retreats and feed when the stirring rate was increased to 130 r.p.m., although they were not able to construct them at a stirring rate of 65 r.p.m., although larvae of *Hydropsyche instabilis* were successful in making their nets at this slow rate of stirring.

Rate of Growth

Hanna has studied the rate of larval growth of seven species of British TRICHOPTERA. He found that there was little growth in winter even in the species *Potamophylax latipennis* and *Agapetus fuscipes*, found in a stream where the temperature remained constant at about 10° C. throughout the year. The larvae of *Limnephilus flavicornis*, *L. lunatus*, *Potamophylax latipennis* and *Brachycentrus subnubilis* feed mainly on vegetable detritus, but diatoms, desmids and *Chlorococcales* are eaten in large numbers. In winter the great reduction in this microflora may be the cause of the reduction in growth. This suggestion had been made previously in connexion with the species *L. politus* and *L. marmoratus*. The larvae of *B. subnubilis* grew mainly in May and passed the winter as fully or nearly fully grown larvae. In *A. fuscipes*, also, the larvae grew rapidly between the middle of April and the middle of May.

Effect of Water Flow

Philipson has studied the effect of water flow and oxygen concentration on six species of caddis fly larvae and was able to demonstrate a number of interesting phenomena. The species chosen for study showed a distribution which can be correlated with the rates of flow of water in various regions of the Blyth in Northumberland, where the research took place. *Rhyacophila dorsalis*, *Hydropsyche instabilis* and *Wormaldia subnigra* were typical of the swifter-flowing reaches of the river, whilst *Anabolia nervosa* and *Polycentropus flavomaculatus* were typical of the slower-flowing parts. On the other hand, *Potamophylax latipennis* was found generally distributed throughout the river. In an experimental channel Philipson found that when larvae of all these species were freely exposed to the mechanical force of the water current, the two species which were least able to withstand the force of the current (*Polycentropus flavomaculatus* and *Anabolia nervosa*) were found typically in the slower flowing waters of the river. The former species, however, is known from fast-flowing water, but they must be established in their retreats to be able to withstand the force. The two species found in swift water, *H. instabilis* and *W. subnigra*, will not spin their typical nets in still water aquaria, but if the water is stirred mechanically then they will do so. *H. instabilis* will spin nets in slower-flowing water than will *W. subnigra*, but both species require the current of water not only to bring food into the net but also to enable them to construct it. Larvae of *Polycentropus flavomaculatus* were able to spin their snares in still water.

Philipson determined the minimum oxygen concentration required for each species, this being the concentration bringing about immobilization of the larva, and found that this was relatively high for the species living in swift water (*R. dorsalis*, *W. subnigra* and *H. instabilis*). When the water was stirred, however, the minimum oxygen concentration was similar to that obtained for the other two species. In *H. instabilis*, *Polycentropus flavomaculatus*, *Potamophylax latipennis* and *Anobolia nervosa* larvae, when the oxygen concentration decreased there was an increase in the number of abdominal undulatory movements. In the case of *H. instabilis* and *P. flavomaculatus*, however, if the water was stirred the rate of these undulatory movements was slowed down. Munro Fox & Sidney (1953) examined the influence of dissolved oxygen on the respiratory movements of caddis larvae, and found, in the case of the species *Limnephilus flavicornis*, that a diminution in dissolved oxygen accelerates the abdominal undulatory movements but an increase in carbon dioxide had no such effect.

Scott (1958) also examined the distribution of caddis larvae in relation to the current velocity of the water in the River Dean, Cheshire. It was found that in respect of the current velocity of the water in which they are found, caddis larvae can be divided into three groups as follows.

1. *Current Velocities below 20 cm./sec.* This group includes *Potamophylax cingulatus*, *P. latipennis* and *Odontocerum albicorne* which were found on the bed of the river usually below stones. They fed mainly on dead plant and animal material lying on the river-bed. Such food material tends to congregate in areas of low current velocity, especially under stones. Scott considers that in this group, larval distribution is directly related to food supply.

2. *Current Velocities from 20–40 cm./sec.* This group includes *Glossosoma conformis* and *Ecclisopteryx guttulata*, and they were most commonly found on the stone faces other than the ventral face, although 36 per cent of *E. guttulata* was found on the substrate below the stones. Both species are algal feeders and it is suggested that the distribution on the illuminated stone surfaces is directly related to this. *G. conformis* did not take in mineral particles with its food although *E. guttulata* did so.

3. *Current Velocities above 40 cm./sec.* This group includes *Hydropsyche fulvipes* and *Rhyacophila dorsalis*, and they occurred most commonly on the ventral faces of stones. They feed mainly on immature insects, which are found in large numbers in these situations. *Hydropsyche fulvipes*, of course, is unable to construct its net at low current velocity. Scott found also that the size of the stones was a factor governing their use by caddis larvae. With the exception of *Glossosoma conformis* the larval population increased with increasing size of substrate unit. *R. dorsalis* larvae, for example, reached their greatest population density on the larger stones, and this was obviously related to the greater occurrence of moss on such stones. *G. conformis*, on the other hand, was found most commonly on the medium-sized stones.

The distribution of caddis pupae in relation to water-current velocity resembled, in general, that of the larvae, except that in *Glossosoma conformis* and *Silo pallipes* prior to pupation the larvae migrated into faster currents, and it is suggested that this might be an adaption to summer drought.

Length of Pupal Stage

Hanna found that at room temperature in the laboratory, the pupal period lasted 23 to 26 days for *L. flavicornis*, 19 to 23 days for *L. lunatus*, 21 to 24 days for *P. latipennis*, 16 to 18 days for *B. subnubilus* and 14 to 16 days for *A. fuscipes*.

As some overwintering pupae of several species such as *A. fuscipes* and *P. latipennis* are sometimes found, it can be assumed that a small proportion of pupae remain in this stage for several months.

Flight Activity

Day Swarming. The swarming of caddis flies generally close to and over water during the day, and more especially towards dusk, is a well-known phenomenon. These swarms are thought to consist predominantly of males. Brindle recognizes two types of the day swarm, the stationary swarm and the mobile swarm. In the former, the individual insects fly rapidly to and fro but the swarm as a whole remains stationary, often related to some fixed outstanding object on the river-bank. Mosely describes this for the species *Ecclisopteryx guttulata*, and Brindle relates how, in the case of *Hydropsyche angustipennis* the swarm disperses suddenly and as suddenly is re-formed. *Mystacides longicornis* and *M. azurea* appear to produce only stationary swarms.

In the mobile swarm, in addition to the to-and-fro movement of the individuals, the swarm as a whole moves over the water surface. *Molanna angustata* is a species apparently always forming mobile swarms. Some species such as *Athripsodes cinereus*, however, may form either type of swarm. Brindle considers that high temperature and humidity favour swarming but the argument is less convincing with regard to humidity. The positioning of the swarm in relation to slight air-currents carrying or likely to carry the female gland odour seems likely to account for the swarming close to the water surface or around an outstanding object.

Night Flight. The most detailed study concerning the nocturnal flight activity of caddis flies is that of Corbet & Tjønneland (1955). Although the work was carried out in Uganda, the results are so interesting as to merit mention in this account of the British trichopterous fauna. Twelve species were investigated by making catches during successive ten-minute periods throughout the night in a light-trap. The nocturnal activity was found to be mainly bimodal. In six species the main flight occurred at dusk, after which numbers declined gradually throughout the night but showed an abrupt rise at dawn. In two species, whilst still being bimodal, the main activity was at dawn, whilst in three species only the dusk flight could be discerned, and finally, in a day-flying species, no recognizable pattern could be demonstrated. Wind at low velocities had no marked effect on the activity of the larger species but caused fluctuations in the case of smaller species. It was suggested that most species emerge at dusk and the main flight occurs shortly afterwards, and sexual activity takes place then.

Brindle (1958) also carried out light-trapping by means of Tilley lamps during three nights (separated by a week and a month), and counted the catch for each quarter of an hour. He found that a peak of flight activity occurred at the second or third hour after dusk, followed by a steady decline until the sixth hour. This was at dawn on the first two occasions (29–30 June and 5–6 July), but on the third night (4–5 August), a smaller peak of activity occurred at the seventh hour.

Brindle suggested that his results showed that the night activity depended on the length of night—the short night producing a unimodal flight and the longer night a bimodal flight, the peaks being separated by about five hours. It should be noted that in Corbet & Tjønneland's results the peaks were separated by ten hours, obviously because of the tropical location. Brindle distinguishes three separate types of flight:

(1) *Dusk Flight*. This is a short, intense flight beginning just before dusk and continuing for a short time afterwards. *Mystacides longicornis* and *Athripsodes cinereus* exhibit this type of flight.

(2) *Night Flight*. Brindle describes this as a long, more or less continuous flight in which the numbers build up much more slowly than in the dusk flight. Species exhibiting this type of flight are *Tinodes waeneri*, *Molanna*

angustata, *Psychomyia pusilla*, *Phryganea obsoleta*, *Cyrnus trimaculatus*, *C. flavidus* and *Polycentropus flavomaculatus*.

(3) *Intermittent Flight*. This is not continuous and the mass flights shown by other species were not in evidence. The first arrivals at the light-trap appeared two or three hours after dusk. Species in this group are *Hydropsyche instabilis*, *Rhyacophila dorsalis*, *Limnephilus rhombicus*, *L. centralis*, *L. extricatus* and *L. sparsus*.

Effect of Weather on Activity

Crichton argues that because caddis flies are not short-lived, fluctuations in the numbers captured at light-traps must be due to variations in their activity and not to sudden emergence of short-lived species. Williams, who operated light-traps at Rothamsted for many years, but whose captures were 97 per cent DIPTERA and LEPIDOPTERA, calculated that, on successive nights, an increase in the minimum temperature of 2.8° C. approximately doubled the catch. The nights of big catches, in these orders, were those of higher temperatures, little or no wind and a high barometric pressure. Rainfall, provided the above conditions were operating, appeared to have no effect.

Crichton, during his extensive light-trapping operations in 1957, showed that his large catches were made on the warmer nights and that with few exceptions the largest collections were made on rainy nights. Indeed, during eleven nights of heavy catches of LIMNEPHILIDAE during the late autumn, rain, sometimes heavy, fell on all but one night. Crichton's experiences led him to state his belief in the possibility that rain acts as a releasing stimulus to mating and egg-laying, and that there may be a necessary connexion between rainfall and their choice of egg-laying sites out of water.

This belief was exemplified by Wesenberg-Lund's observation, which Crichton recalls, of sheltering under a tree during a heavy rainstorm and noticing sticky drops containing limnephilid larvae falling from the tree, and also that the newly emerged larvae were swimming about in the wet gelatinous masses and in the water below.

Emergence Period

The work principally of Crichton and Hanna has built up a fair picture of the emergence period or seasonal occurrence of a large number of species, but even so there is much scope for work in elucidation of complete life histories. The first-named author divides species into three groups, as follows:

(1) *Summer Species*

These consist of the species captured in maximum numbers from May to August. This group can be further subdivided into those species with long and those with a relatively short period of flight activity.

Short Flight. The PHRYGANEIDAE generally belong to this group. *Phryganea grandis* is at its maximum emergence at the beginning of July whilst *P. striata* is most abundant towards the end of May. *P. obsoleta*, however, has been recorded as emerging, *en masse*, as late as 23 August, and all stages of larvae were present at the same time in the lake from which emergence was taking place. This was in an upland lake over 1,200 feet above sea-level. The impression has always been strong in the present author's mind that in this species, and perhaps others, some proportion at least of the individuals may take two years for completion of the larval stage. At least one species of LEPTOCERIDAE, *Leptocerus tineiformis*, is normally restricted to the month of May.

Long Flight. With the exception mentioned above, the commoner LEPTOCERIDAE probably all belong to this group, the main emergence being in July, although in some species some individuals are encountered at the end of May and as late as September. *Mystacides longicornis*, on the other hand, appears to emerge steadily during the whole of this period. Many species in the HYDROPSYCHIDAE, POLYCENTROPIDAE and HYDROPTILIDAE also exhibit a long summer flight period.

(2) *Autumn Species*

The main emergence period for many species of LIMNEPHILIDAE is during September and October. In the extensive light-trapping carried out by Crichton, relatively large numbers of the following species were collected only during the months of September, October and November: *Ironoquia dubia*, *Limnephilus politus*, *L. decipiens*, *L. griseus*, *L. bipunctatus*, *L. affinis*, *L. centralis*, *L. vittatus*, *Anabolia nervosa*, *Halesus radiatus*, *H. digitatus* and *Chaetopteryx villosa*. This is not to suggest that adults of these species are restricted to the autumn months. Indeed, many of the species listed above are not uncommonly found during the summer. The terrestrial species *Enoiclya pusilla*, also a limnephilid, is found in the adult stage only in October and November. Not all LIMNEPHILIDAE, however, can be described as autumnal species. Although it is perhaps unwise to place much reliance on emergence dates from aquarium-bred specimens, the author nevertheless records collecting a large number of *Grammotaulius atomarius* which were apparently fully-fed as early as the end of January, and they all emerged from aquaria during the first half of May. This large species often inhabits field-side ditches which must often contain little water during the late summer. This would seem to pose many difficulties for a large agile larva during such a period.

Length of Life Cycle

The great majority of TRICHOPTERA in the British fauna complete the life-cycle in one year. It has been suggested, however, that some species of caddis produce more than one brood during the year. Almost for certain, but still lacking absolute proof, the psychomyiid *Tinodes waeneri* and perhaps other species in the same family, and some hydroptilids such as *Oxyethira costalis* and *Agraylea multipunctata*, are examples. As Hanna has collected the larvae and pupae of the rhyacophilid *Agapetus fuscipes* in every month of the year and the adults in every month except January and February, he thought it almost unquestionable that in this species also two or perhaps more broods each year are produced. Mackereth, however, working in the Lake District, considers that this species undergoes one generation in a year. It is worthy of note that all the species mentioned are small in size.

Mention has already been made of the suggestion that perhaps in a few species, but including *Phryganea obsoleta*, the life-cycle may extend beyond one year in particular cases. Mackereth, studying the caddis larvae of a small Lake District beck, thought that her data for collections of *Wormaldia occipitalis* suggested a two-year life-cycle. She also discovered larvae of *Rhyacophila dorsalis* overwintering in the pupal case but could not be sure whether this was the first or second winter of the larval life.

Epiphytes and Epizoites

The rather slow speed of locomotion of many Trichopterous larvae render them somewhat vulnerable to a wide range of epiphytes and epizoites, but there has been surprisingly little recorded on this subject. Baker (1961), however, noted the sponge *Ephydatia fluviatilis* (L) encrusting the cases of a limnephilid. Several cases were also covered with algae of the genera *Cladophera*, *Stipitococcus*, *Synedra*, *Lyngbya*, *Oedogonium*, *Gloeochaeta*, *Pseudochaete* and *Characium*, as well as by members of the CYANOPHYCEAE. He also found epizoic PROTOZOA of the PERETRICHA group on nearly every larva collected, some of which were almost completely covered. One species was identified as *Epistylis brachiopyla* Perty.

Baker has also recorded adult and larval mites of the species *Atturus scaber* Kramer infesting larvae of *Goera pilosa*. This was the only species of caddis larva infested by this mite in the particular habitat so that there is the possibility that it may be specific to *G. pilosa*.

Dispersal

The upstream migration of ovigerous female caddis flies, shown by Roos (1957), explains the colonization and re-colonization of streams and rivers. It does not explain how caddis flies are able to colonize or re-colonize static water.

Crichton (1961), however, considers that there is widespread dispersal from the breeding-site in the case of the LIMNEPHILIDAE. He makes this claim largely on the small percentage of recoveries obtained from released marked specimens. Small-scale dispersal by limnephilids has been demonstrated by Crichton by running two light-traps simultaneously, one at the water's edge and one 100 metres away and screened from the water.

Actually, the distant trap caught more caddis flies than the one at the water's edge.

Hickin (1950) has recorded *Stenophylax vibex* coming to lighted windows of a house from water just over a mile away. *S. permistus* also is known to range fairly widely away from its breeding-site.

Upstream Migration

In streams and rivers, caddis larvae are likely to move some way downstream during their long larval life, especially as the late autumn, winter and spring, the times of floods, are spent in this stage. Roos (1957) was able to show, by means of trapping-experiments on a river in northern Sweden, that there is an upstream migration. This occurs when the females are ready for oviposition; they then fly strongly upstream even against a feeble wind. This occurred in the species *Cheumatopsyche lepida* and *Rhyacophila nubila*, and upstream distances of flight up to five kilometres were recorded.

Length of Adult Life

Crichton (1960) has established by the method of recapture of marked specimens that limnephilid caddis flies, in natural conditions in the field, live for several weeks. Döhler (1914), by supplying them with water and sugar solution, was able to keep specimens of *Limnephilus flavicornis* alive for from thirty to 105 days. Such a length of adult life, combined with the usual three or four months of emergence period for the species, makes the chances almost certain of the species encountering one or more warm wet nights for maximum activity, thus ensuring continuity.

Courtship

Hickin (1953) has recorded an observation of courtship behaviour in the species *Mystacides nigra*. At 8.30 a.m. at the end of June several pairs flew away from a zigzagging swarm on a river. They flew in tandem with similar wing-beats, the male behind grasping the female with the forelegs. On landing on river-side herbage the male still grasped the female whilst he twisted his abdomen around, searching for the vaginal aperture until copulation was effected, which took about one minute.

Sex Ratio

Crichton sexed a very large number of Trichoptera obtained in his light-trap experiments. As the trap was used every third night from April to November it might be expected that differences in the proportions of males and females due to the known usual first emergence of the males would be averaged out. On the other hand, as pointed out by Crichton (1960), relative proportions of the sexes of the insects captured in the trap will depend on their differential activity, behaviour and reaction to the source of light. This differential sex behaviour may vary between the different species and may of course be dependent on other factors, such as climate. The proportion of sexes caught varied greatly. Up to 25 May, that is, the early part of the swarming period of the first generation of *Oxyethira costalis*, about 85 per cent of all individuals were males, whereas after this date the proportions of females rarely fell below 70 per cent. Contrasting with this, however, is the case of another hydroptilid, *Orthotrichia tetensii*, in which the catch always consisted of about 95 per cent females. The hydropsychids *H. instabilis* and *H. angustipennis* showed a female percentage of 91 and 65 respectively, whilst *Leptocerus tineiformis* showed a preponderance of 80 per cent females. The trapping of emerging adults in floating traps of *Athripsodes aterrimus* by Morgan and Waddell (1961) gave a female percentage of 37, whereas by light-trapping the same species Crichton found only 21 per cent females.

Crichton's table of sex ratios of TRICHOPTERA as found from light-trapping the individuals is given below.

TABLE 1

Sex proportions of TRICHOPTERA, of which more than 100 specimens were captured in the four years, arranged in order of ascending percentage of females. All specimens captured were sexed, except those species marked *, where only a sample of the catch was sexed.

Species	♂♂	♀♀	Percentage ♀♀
Limnephilus sparsus	99	8	7
L. auricula	307	29	9
Anabolia nervosa	94	14	13
Glyphotaelius pellucidus	330	50	13
Phryganea grandis	98	16	14
Halesus digitatus	465	82	15
Mystacides longicornis	3,317	842	20
Athripsodes aterrimus	106	29	21
**Agraylea pallidula*	383	105	22
**A. multipunctata*	463	130	22
Limnephilus vittatus	161	47	23
L. flavicornis	720	214	23
Athripsodes senilis	189	67	26
Limnephilus lunatus	2,454	876	26
Stenophylax permistus	137	63	31
Oecetis ochracea	155	92	37
O. lacustris	336	357	52
Athripsodes cinereus	394	426	52
A. dissimilis	137	190	58
Tinodes waeneri	156	219	58
Ecnomus tenellus	197	290	60
**Oxyethira costalis*	6,359	9,811	61
Hydropsyche angustipennis	519	957	65
Leptocerus tineiformis	533	2,174	80
Hydropsyche instabilis	10	105	91
**Orthotrichia tetensii*	99	1,908	95

BADCOCK, R. M. 1955. Widespread Distribution of our Allegedly Rare Caddis, HYDROPSYCHE FULVIPES (Curtis) TRICHOPTERA, HYDROPSYCHIDAE. *Entomologist's mon. Mag.* **91**, 30–1

BAKER, W. V. 1961. Some Observations on the Associations of Trichopterous Larvae. *Entomologist's mon. Mag.* **97**, 192.

BRINDLE, A. 1956. The Ecology of North-East Lancashire TRICHOPTERA (with special reference to the use of light as a collection method). *Entomologist's Gaz.* **7**, 179–84.

BRINDLE, A. 1957. The Effect of Temperature and Humidity on the Flight of Trichoptera. *Entomologist's mon. Mag.* **93**, 63–6.

BRINDLE, A. 1957. Notes on the Use of Light for Attracting Trichoptera. *Entomologist's mon. Mag.* **93**, 127–9.

BRINDLE, A. 1958. Night Activity of Trichoptera. *Entomologist's mon. Mag.* **94**, 38–42.

BROWN, J. M. 1938. Some Derbyshire Caddis Flies. *Naturalist, Lond.*, 87–82 (J. M. B.).

BROWN, J. M. & WHITEHEAD, H. 1938. The Trichoptera or Caddis-Flies of Yorkshire. *Naturalist, Hull*, 315–19.

CORBET, P. S. & TJØNNELAND, A. 1955. The Flight Activity of Twelve Species of East African Trichoptera. *Univ. Bergen Årb.*

COWLEY, J. 1951. Some Recent Records of Somerset Insects. *Proc. Somerset. archaeol. nat. Hist. Soc.* **96**, 219

CRICHTON, M. I. 1959. Attacks by Birds on Caddis Flies. *Bird Study* **6**, 22–5.

CRICHTON, M. I. 1960. A Study of Captures of Trichoptera in a Light Trap near Reading, Berkshire. *Trans. R. ent. Soc. Lond.* **112**, 319–44.

CRICHTON, M. I. 1961. Observations on the Longevity and Dispersal of Adult LIMNEPHILIDAE (TRICHOPTERA). *XI Int. Kong. Ent. Wien.* **1**, 366–71.

CRICHTON, M. I., BAKER, B. R. & HANNA, H. M. 1956. Records of Trichoptera from the Reading Area. *Entomologist's mon. Mag.* **92**, 31–5.

CRICHTON, M. I., BAKER, B. R. & HANNA, H. M. 1957. Further Records of Trichoptera from the Reading Area. *Entomologist's mon. Mag.* **93**, 241–3.

CRICHTON, M. I. & BAKER, B. R. 1959. Records of Trichoptera from the Reading Area, 1957–8. *Entomologist's mon. Mag.* **95**, 85–7.

DÖHLER, W. 1914. Beiträge zur Systematik und Biologie der Trichopteren. *Sber. naturf. Ges. Lpz.* **41**, 28–102.

ERICHSON-JONES, J. R. 1958. A Further Study of the Zinc-polluted River Ystwyth. *J. Anim. Ecol.* **27**, 1–14

FEBBER, J. 1908. Die Trichopteren von Basel und Umgebung mit Berichtsichtigung der Trichoptera—fauna der Schweiz. *Arch. Naturgesch.* **74**, 199.

FOX, M. W. 1957. The Trichoptera of Skokholm. *Entomologist's mon. Mag.* **93**, 40–1.

FOX, H. MUNRO & SIDNEY, J. 1953. The Influence of Dissolved Oxygen on the Respiratory Movements of Caddis Larvae. *J. exp. Biol.* **30** (2), 235–7.

GLEDHILL, T. 1960. The EPHEMEROPTERA, PLECOPTERA and TRICHOPTERA caught by Emergence Traps in Two Streams during 1958. *Hydrobiologia* **15** (1–2), 179–88.

GRENSTED, L. W. 1935. The TRICHOPTERA of the Oxford District. *Trans. Soc. Br. Ent.* **2**, 68–72.

HANNA, H. M. 1957. Observations on Case-Building by the Larvae of *Limnephilus politus* McLachlan and *L. marmoratus* Curtis (TRICHOPTERA: LIMNEPHILIDAE). *Proc. R. ent. Soc. Lond.* (A) **32**, 47–52.

HANNA, H. M. 1957. A Study of the Growth and Feeding Habits of Four Species of Caddis Flies. *Proc. R. ent. Soc. Lond.* (A) **32**, 139–46.

HANNA, H. M. 1959. The Growth of Larvae and their Cases and the Life Cycles of Five Species of Caddis Flies. *Proc. R. ent. Soc. Lond.* (A) **34**, 121–9.

HANNA, H. M. 1960. Methods of Case-building and Repair of Larvae of Caddis Flies. *Proc. R. ent. Soc. Lond.* (A) **35**, 97–106.

HICKIN, N. E. 1944. Chartley Moss and the Caddis Fly *Neuronia clathrata* Kol. (PHRYGANEIDAE, TRICHOPTERA). *Entomologist* **77**, 20–1.

HICKIN, N. E. 1946. *Neuronia ruficrus* (Scop.) (TRICH. PHRYGANEIDAE) and *Phalocrocera replicata* (L.). (DIPT. TIPULIDAE) in a High Tarn. *Entomologist's mon. Mag.* **82**, 111.

HICKIN, N. E. 1952. *Caddis.* Methuen, London.

HICKIN, N. E. 1953. An Aquatic Larva, Suitable for Laboratory Work. *Nature* **172**, 874.

HICKIN, N. E. 1965. *Forest Refreshed.* Hutchinson, London.

HICKIN, N. E. & KIMMINS, D. E. 1949. [Exhibit of Adults and Larvae of PSYCHOMYIIDAE.] *Proc. R. ent. Soc. Lond.* (C) **14**, No. 11.

HOLMES, P. F. 1963. Insects of the Malham Tarn Area (*Trichoptera* by P. F. Holmes). *Proc. Leeds phil. lit. Soc.* (*Scientific Section*) **9** (2), 31–5.

IMMS, A. D. 1957. *A General Textbook of Entomology.* Methuen, London. Ninth Edition revised by O. W. Richards & R. G. Davies.

KIDD, L. N. 1963. A Note on the Distribution of *Limnephilus xanthodes* McLach. (TRICHOPTERA). *Entomologist* **96**, 246–7.

KIMMINS, D. E. 1943. A List of the TRICHOPTERA (Caddis Flies) of the Lake District, with Distributional and Seasonal Data. *J. Soc. Br. Ent.* **2**, 136–57.

KIMMINS, D. E. 1943. A Note on the Caddis Fly, *Mystrophora intermedia* Klapálek (TRICHOPTERA). *Proc. R. ent. Soc. Lond.* (A) **18**, 96–8.

MACKERETH, J. C. 1954. Taxonomy of the Larvae of the British Species of the Genus *Rhyacophila* (TRICHOPTERA). *Proc. R. ent. Soc. Lond.* (A) **29**, 147–52.

MACKERETH, J. C. 1960. Notes on the TRICHOPTERA of a Stony Stream. *Proc. R. ent. Soc. Lond.* (A) **35**, 17–23.

MARTYNOV, A. B. 1930. On the Trichopterous Fauna of China and Eastern Tibet. *Proc. zool. Soc. Lond.* V: 65–112.

MORGAN, N. C. & WADDELL, A. B. 1961. Diurnal Variation in the Emergence of some Aquatic Insects. *Trans. R. ent. Soc. Lond.* **113** (6), 123–34.

MOSELY, M. E. & KIMMINS, D. E. 1953. *The* TRICHOPTERA (*Caddis Flies*) *of Australia and New Zealand.* Trustees of the British Museum, London.

NIELSEN, A. 1948. Postembryonic Development and Biology of the HYDROPTILIDAE. *Biol. Skv.* **5** (1), 1–200.

PEACEY, A. F. 1953. NEUROPTERA and TRICHOPTERA at Symond's Yat. *Entomologist's Rec. J. Var.* **65**, 263–4.

PELHAM-CLINTON, E. C. 1964. The Distribution in Scotland of *Limnephilus xanthodes* McLachlan (TRICHOPTERA, LIMNEPHILIDAE). *Entomologist* **97**, 67.

PHILIPSON, G. N. 1953. A Method of Rearing Trichopterous Larvae Collected from Swift-flowing Waters. *Proc. R. ent. Soc. Lond.* (A) **28**, 15–16.

PHILIPSON, G. N. 1954. The Effect of Water Flow and Oxygen Concentration on Six Species of Caddis Fly (TRICHOPTERA) Larvae. *Proc. zool. Soc. Lond.* **124**, 547–64.

PHILIPSON, G. N. 1955. TRICHOPTERA in Northumberland. *Entomologist's mon. Mag.* **91**, 3.

PHILIPSON, G. N. 1957. Records of Caddis Flies (TRICHOPTERA) in Northumberland. *Trans. nat. Hist. Soc. Northumb.* **12** (3) (New Series), 77–92.

PHILIPSON, G. N. 1962. Further Records of TRICHOPTERA in Northumberland. *Entomologist's mon. Mag.* **98**, 67.

ROOS, T. 1957. Studies on Upstream Migration in Adult Stream-dwelling Insects. I. *Rep. Inst. Freshwat. Res. Drottningholm* **38**, 169–93.

ROUTLEDGE, G. B. 1933. The NEUROPTERA &c., and TRICHOPTERA of Cumberland, Westmorland and North Lancashire. *Trans. Carlisle nat. Hist. Soc.* **5**, 44–60.

SCOTT, D. 1958. Ecological Studies on the TRICHOPTERA of the River Dean, Cheshire. *Arch. Hydrobiol.* **54** (3), 340–92.

SLACK, H. D. 1936. The Food of Caddis Fly (TRICHOPTERA) Larvae. *J. Anim. Ecol.* **5**, 105–15.

TINDALL, A. R. 1962. Specific Gravity of the Cases of Larvae of *Limnephilus* and *Triaenodes* (TRICHOPTERA). *Nature, Lond.* **196**, 690–1.

Tobias, W. 1962. Die Gehäusebauten de Köcherfliegen (Trichoptera) unter Berückrichtigung der bis 1961 erschienenen Literatur. *Hydrobiologische anstalt der Man-Planck-Gesellschaft*.

Van der Drift, J. & Witkamp, M. 1960. The Significance of the Breakdown of Oak Litter by *Enoicyla pusilla* Burm. *Archs néerl. Zool.* **13** (4), 486–92.

Wesenberg-Lund, C. J. 1913. Fortpflanzungsverhältnisse: Paarung und Eiablage der Süsswasserinsekten. *Fortschr. naturw. Forsch.* **8**, 161–286.

6

RHYACOPHILIDAE

The adults of species in the family RHYACOPHILIDAE are moderate in size, the length of the anterior wing being from 12 to 13 mm. The antennae are usually shorter than the wings and the basal segment is short. Ocelli are present and the two first segments of the maxillary palps are short, whilst the terminal segment is not articulated. The wings are elongate and the anterior pair sometimes bear a fold or callosity at the base. The tibial spur formula is 3, 4, 4, the middle legs of the female are not dilated and the discoidal cell of both wings is open.

Following Kloet & Hincks (1964), British representatives of this family are classified in one genus in the sub-family RHYACOPHILINAE. Kimmins (1966) includes the GLOSSOSOMATIDAE in this family.

RHYACOPHILINAE
RHYACOPHILA (Pictet)
1. *dorsalis* (Curtis)
2. *munda* McLachlan
3. *obliterata* McLachlan
4. *septentrionis* McLachlan

This genus is one of the largest in the TRICHOPTERA, 250 species already being known. More than a third of them are from North America.

Distribution

Rhyacophila dorsalis. Throughout the British Isles and very abundant (M.E.M.); Berks. (M.I.C. *et al.*); Northumberland (G.N.P.); Cardigan (J.R.E.J.); Yorks. (P.F.H.); common, generally distributed (D.E.K.); Cumberland, Durham, Mid-Yorkshire, Pembroke, Shetlands, south Northumberland, Westmorland (J.C.M.).

Rhyacophila munda. Local, frequently taken in company with *dorsalis* (M.E.M.); Yorks. (P.F.H.); locally common west Britain (D.E.K.); Pembroke, Westmorland (J.C.M.).

Rhyacophila obliterata. Common in alpine and sub-alpine districts (M.E.M.); Northumberland (G.N.P.); Yorks. (P.F.H.); locally common in hilly districts, west and north Britain (D.E.K.); Westmorland, Mid-Yorkshire (J.C.M.).

Rhyacophila septentrionis. Very local. Yorks. (M.E.M.); Yorks. (P.F.H.); Yorks., Lanark, Haddington (D.E.K.); Glos. (N.E.H.); Yorks. (J.C.M.).

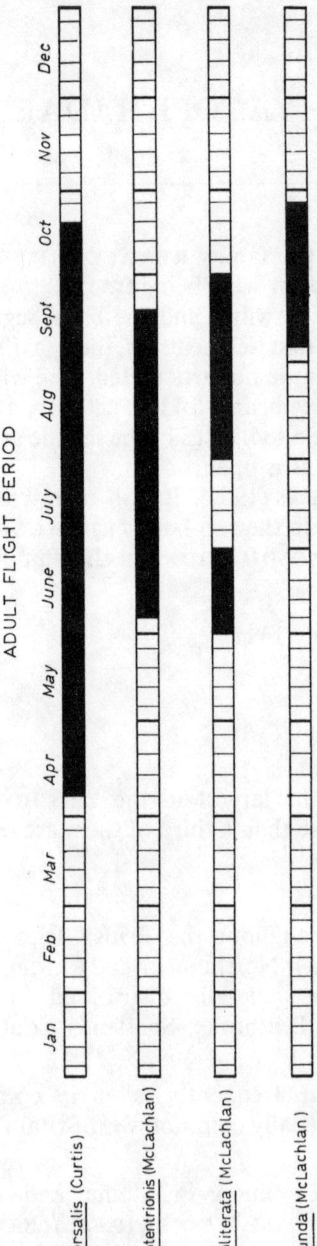

RHYACOPHILIDAE. *Adult Flight Periods taken from published records of individual dates.*

General Description of Larvae

The larvae are campodeiform, living entirely free. The larva is widest at about the second abdominal segment.

The mouthparts are prominent with a sclerotized labrum and a hypostome is present, which is small and triangular. The antennae are rudimentary.

The pronotum is entirely sclerotized with the meso- and metanotum generally soft. The legs are almost equal in length and width and furnished with robust claws which are sometimes short with a robust basal spur.

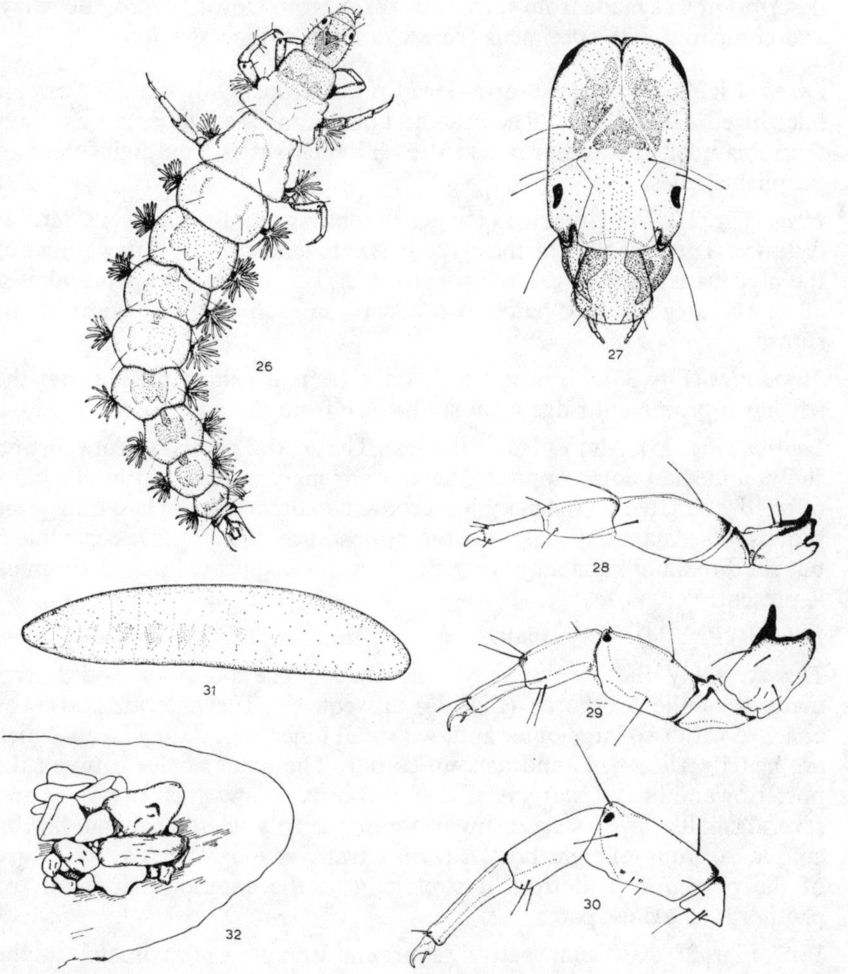

Figs. 26–32. 26. *Rhyacophila dorsalis*, larva; 27. *R. dorsalis*, head; 28. *R. dorsalis*, prothoracic leg; 29. mesothoracic leg; 30. metathoracic leg; 31. *R. dorsalis*, cocoon; 32. *R. dorsalis*, pupal case.

The abdominal segments are well marked, the first having no lateral or dorsal protuberances. Prosternal horn and lateral lines absent. Gills in the form of tufts; six anal gills always present. Anal claws free and well developed.

Detailed Descriptions of Larvae

Rhyacophila dorsalis (Curtis)

This species is commonly found crawling about on the under-surface of stones at the bottom of quickly running streams and brooks. The following description was made from specimens taken from Dowles Brook, Bewdley, and compared with specimens from Cranham, Gloucestershire.

Larva (Fig. 26). Campodeiform. Head procentrous. Only the prothorax is sclerotized. Gills present. The head and prothorax is a light creamy yellow, the remainder of the thorax and the abdomen varies from light green to purplish shades.

Head (Fig. 27). Very long with the genae running parallel for a considerable distance. The oral part of the clypeus is attenuated. At the aboral end of the clypeus is an inverted heart-shaped dark marking occupying almost all of the area. A dark patch also occurs on each of the genae near the suture.

Mandibles (Fig. 33). Asymmetrical. Each bears a pair of bristles, but the left has a prominent ridge a short distance from the outer border.

Labrum (Fig. 35). Almost hemispherical. The central part is sclerotized, and shows a median notch opposite the anterior margin. Apart from two pairs of bristles situated on the margin there are two other pairs in two transverse groups. The maxillary palp has the appearance of being five-segmented, but the proximal segment is very short. A one-segmented palpiform mala is present.

Labium (Fig. 34). Very small.

Thorax. Only the prothorax is sclerotized. The posterior and lateral margins of the prothorax (Fig. 37) are convex. The anterior margin is concave with two large outer and two small inner articulating facets which are heavily sclerotized and dark in colour. The outer angles between the posterior and lateral margins of the prothorax, from which two large and several smaller bristles arise, are also more heavily sclerotized and dark in colour. A group of eight bristles form a transverse band across the centre of the prothorax and from it stretching to the anterior margin of the prothorax is a dark patch.

Legs (Figs. 28–30). Comparatively short and stumpy. Approximately all the same size. Tarsal claws small with a small spur in the meso- and metathoracic legs. Trochanter is one-segmented. No spines on ventral margin of tarsus and tibia. Three bristles occur on ventral surface at distal end of proximal tarsal segment in meso- and metathoracic legs. A dark spot occurs on each side of the tibia at the distal end near the dorsal surface in meso- and metathoracic legs. A tuft of gills occurs at the bases of meso- and metathoracic legs.

Abdomen. The width of the abdominal segments decreases progressively from the anterior to the posterior end. Each of the first eight segments bears a large filamentous tuft of gills and four bristles on the dorsal surface. Each segment is strongly convex in dorsal and lateral regions. The ninth segment has no gills and has a dark heavily sclerotized transverse band in a median position. The anal hooks (Fig. 36) are two-segmented and long. The proximal segment bears a long bristle and the distal segment several smaller bristles and spines. Auxiliary hooks are present external to the main hooks but are one-segmented.

Pupation. Just before pupation the larva constructs a chamber of large fragments of stone which it lines with silk (Fig. 32). A tough chestnut-coloured pupal case or cocoon (Fig. 31) is formed in which the larva pupates.

Mackereth was able to compare larvae of all four British species of *Rhyacophila*, first by extracting last instar larval skins from pupal cases

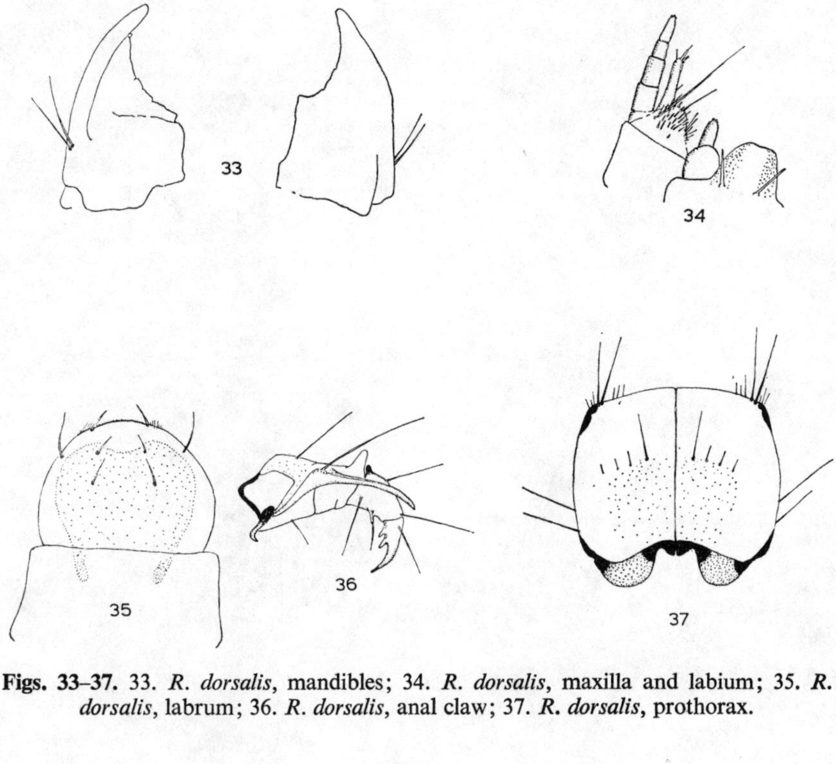

Figs. 33–37. 33. *R. dorsalis*, mandibles; 34. *R. dorsalis*, maxilla and labium; 35. *R. dorsalis*, labrum; 36. *R. dorsalis*, anal claw; 37. *R. dorsalis*, prothorax.

from which adult males had emerged; then, when good distinguishing characters had been established, she examined whole larvae.

With regard to this species, Mrs. Mackereth (1954) calls attention to the following points:

The head, which tapers slightly towards the anterior end and has a slight constriction about the level of the eyes, is creamy yellow. There is a characteristic dark brown marking at the aboral end of the fronto-clypeus, the shape of a spade or heart in a pack of cards. At the posterior end of the genae there are pale indistinct dots on a dark brown area (Fig. 38). The anterior edge of the fronto-clypeus is straight with minute projections at each end.

The pronotum is sclerotized and creamy yellow, with an irregular brown patch extending from the posterior margin to slightly beyond the mid-line. This patch is covered with small darker dots which vary in intensity in different specimens, but which are indistinct in most fully grown larvae.

The colour of the meso- and metathoracic segments and the abdomen is variable; most specimens are yellowish-green ventrally and darker

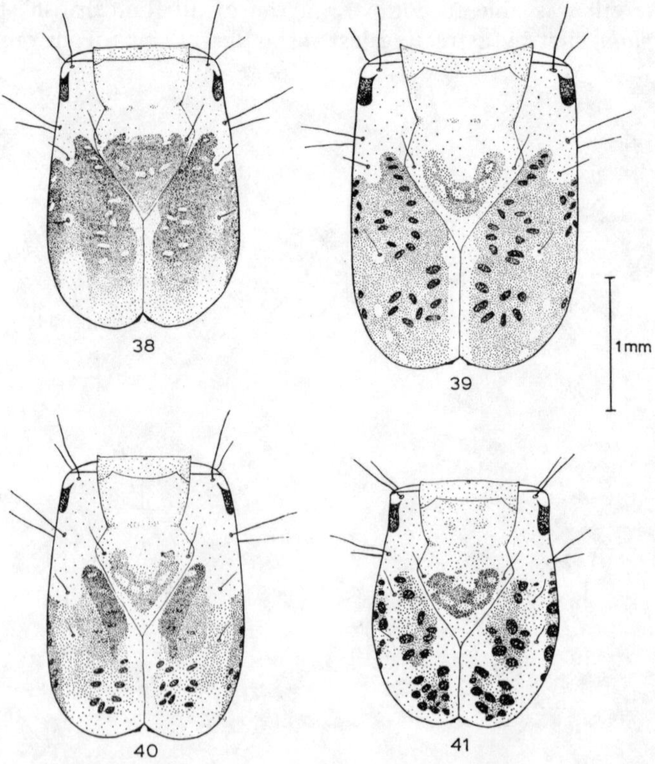

Figs. 38–41. Heads. 38. *Rhyacophila dorsalis*; 39. *R. obliterata*; 40. *R. septentrionis*; 41. *R. munda*. Mackereth.

purplish-green dorsally. On each side of these segments there is a many filamented gill (Fig. 42). The number of filaments varies in different specimens; in last instar larvae there are between eight and thirty. The number of filaments in each gill on each side of a segment is rarely the same, and the difference is as much as six in some specimens.

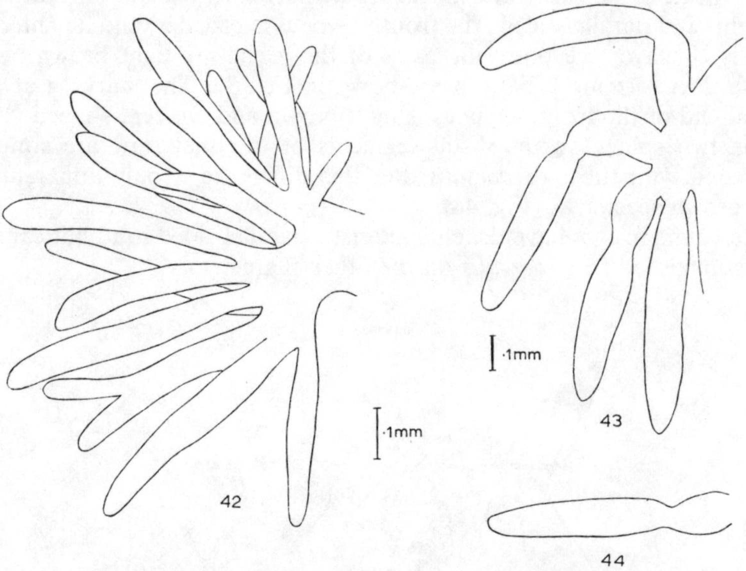

Figs. 42–44. Gills. 42. Thoracic gill of *R. dorsalis*; 43. Abdominal gill of *R. munda*; 44. Thoracic gill of *R. munda*. Mackereth.

The legs are short and robust. All the segments bear black hairs of varying lengths, and each claw has a short spur attached to the inner side.

The long curved anal claw is made up of two segments; the proximal segment has a long thread-like bristle on its inner surface; the distal segment bears two, and occasionally three, robust spines (Fig. 45). Attached to the outer side of the anal claw there is an auxiliary spine, which is long and curved (Fig. 47).

Rhyacophila munda McLachlan

Mackereth, whose specimens were obtained from Westmorland and Pembroke, comments on the larvae as follows:

R. munda is very distinct from the other three species of the genus in having the gills on each side of the meso- and metathorax with a single filament (Fig. 44), and those on the abdominal segments with four (Fig. 43). The auxiliary spine is very short as Fig. 46 shows. The head (Fig. 41) is even more constricted at the eyes than that of *R. dorsalis*. The head pattern is bold, the basic colour being orange-yellow. The abdomen is light green ventrally and darker green dorsally.

Rhyacophila obliterata McLachlan

Mackereth (1954) makes the following remarks concerning this species:

The head of *R. obliterata* differs from that of *R. dorsalis* in that it is straight- and parallel-sided; the fronto-clypeus is broader, and its anterior edge is concave; the posterior parts of the genae are light brown, with brown dots forming a pattern as shown in Fig. 39. The marking at the aboral end of the fronto-clypeus is light brown and crescent-shaped.

The two spines on the distal segments of the anal claw are stumpy compared with those of *dorsalis*, but this difference is only apparent in fully grown specimens (Fig. 48).

The young instars have the characteristic parallel-sided head but cannot be distinguished from *dorsalis* on the other characters.

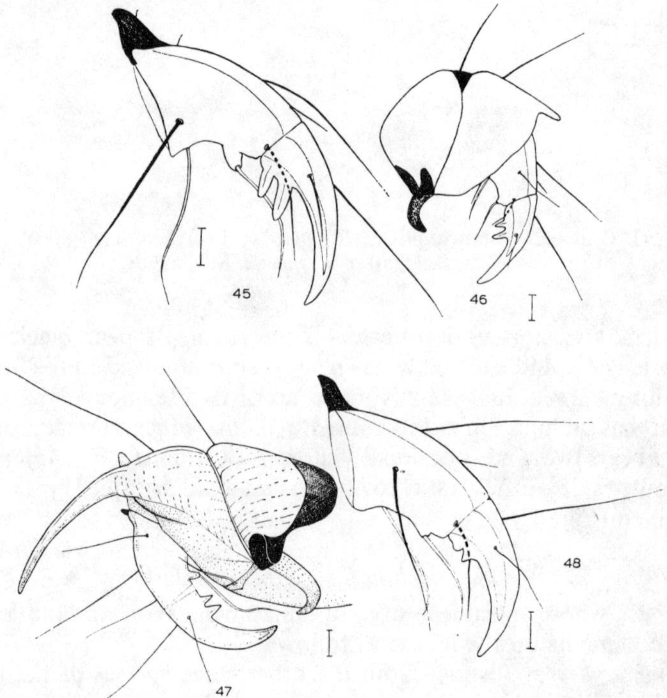

Figs. 45–48. 45. Anal claw of *R. dorsalis*; 46. Anal claw and auxiliary spine of *R. munda*; 47. Anal claw and auxiliary spine of *R. septentrionis*; 48. Anal claw of *R. obliterata*. Mackereth.

Rhyacophila septentrionis McLachlan

I am very grateful to Mr. Arthur Peacey for conducting me, on 5 April 1953, to the secluded locality near Bisley, Glos., where in previous years he had collected the adults of this very local species. A small, but fast-flowing rivulet ran down to the main stream of the Battlescombe Valley. The water was so heavily charged with lime that everything in the stream was encrusted; even the bag-like nets of Philopotamids were thick and brittle.

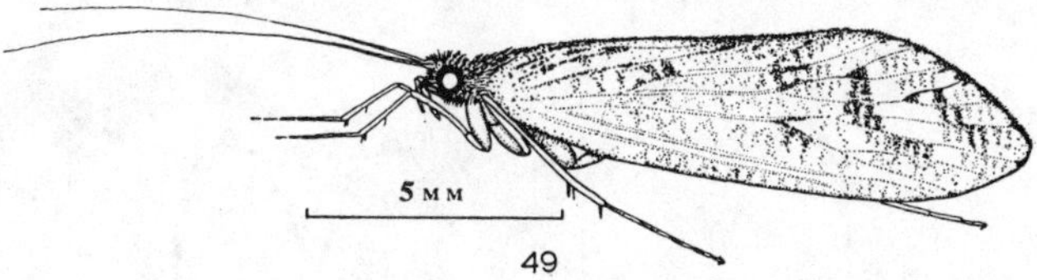

49

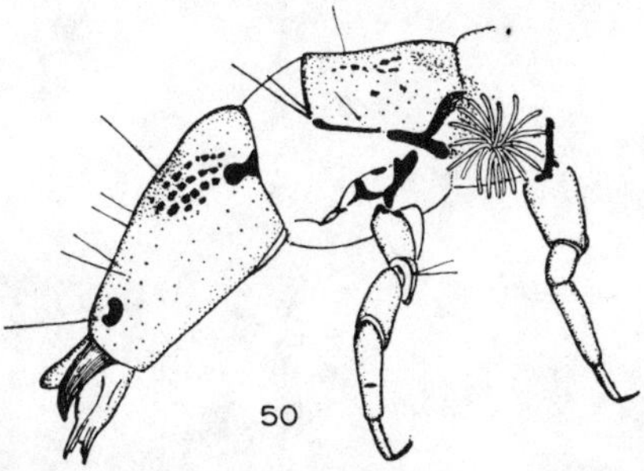

50

Figs. 49–50. *Rhyacophila septentrionis* McLachlan. 49. Adult; 50. Larva, side view of head and thoracic segments.

Twelve larvae and two pupae of a Rhyacophilid were quickly found under stones where the stream ran fastest, and the subsequent comparison of these with Nielsen's description and figures showed them to be *Rhyacophila septentrionis* McLachlan. My drawing of the adult was made from specimens taken by Mr. Peacey from this same locality, and the wing pattern was checked with a specimen from the British Museum (Natural History), kindly made available to me by Mr. D. E. Kimmins.

The larva of *R. septentrionis* is very like that of *R. dorsalis*. There are, however, several well-defined differences, some of which can be picked up with a $\times$ 10 lens, thus enabling identification in the field. All my specimens of *septentrionis* were golden-olive in colour, contrasting with the bright apple-green which is usual for *dorsalis*. In viewing the head from above, that of *septentrionis* is not nearly so heavily marked as *dorsalis*, and a pattern of brown spots occurs aborally and also laterally. This is absent in the latter species. In viewing the head from beneath, the aboral margins of the genae just before they become contiguous show a small fold, which is golden-brown in *dorsalis*, but jet black in *septentrionis* (Figs. 51–56).

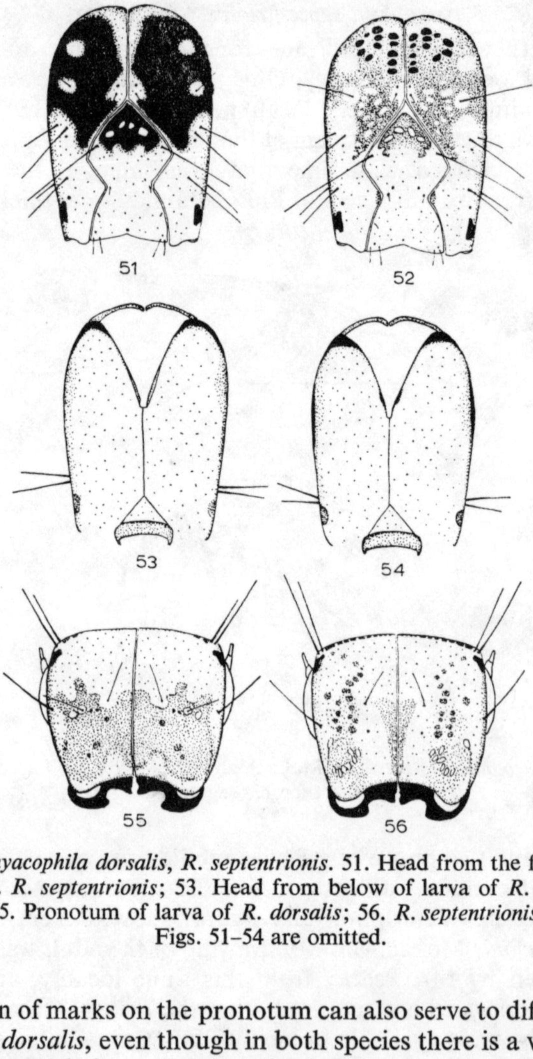

Figs. 51–56. *Rhyacophila dorsalis*, *R. septentrionis*. 51. Head from the front of larva of *R. dorsalis*; 52. *R. septentrionis*; 53. Head from below of larva of *R. dorsalis*; 54. *R. septentrionis*; 55. Pronotum of larva of *R. dorsalis*; 56. *R. septentrionis*. Mouthparts in Figs. 51–54 are omitted.

The pattern of marks on the pronotum can also serve to differentiate this species from *dorsalis*, even though in both species there is a wide variation in marking, not only in pattern but also in intensity. The larvae of *dorsalis* from which I made my original description were very lightly marked, so that I include drawings made from larvae of the more typical darker form. These latter were collected from the Dowles Brook at Furnace Mill, Wyre Forest, Salop, on 8 April 1953.

In *septentrionis* there is a triangular mark along the median suture with horn-shaped patterns of spots on each side. (The light areas between the former and the latter can easily be made out with a ×10 lens.)

In *dorsalis*, the dark lobar mass is continuous with a pair of projections on each side directed anteriorly, with a larger median projection between them. The anterior margin of the pronotum is more heavily sclerotized in my specimens of *septentrionis* than in *dorsalis*.

Mackereth (1954) makes the following remarks concerning *R. septentrionis:*

This species is difficult to distinguish from *R. dorsalis*. The head (Fig. 40) tapers slightly towards the anterior end, but is not constricted at the level of the eyes. The brown marking at the aboral end of the fronto-clypeus is crescent-shaped as in *obliterata* in all the British specimens seen. Of the Belgian specimens, some had this crescent shape, but darker ones had the heart-shaped marking as in *dorsalis*. According to Nielsen, this marking can be heart-shaped or crescent-shaped. At the posterior end of the genae there are dark brown dots; these are evident on both the British and Belgian specimens, and are also illustrated by Nielsen and Ulmer. No such dark dots have ever been seen on a specimen of *dorsalis*, and this does therefore provide a clear distinction between the two.

Key to Larvae of Rhyacophilidae

1. Gills on meso- and metathorax with one filament, those on the abdominal segments each with four filaments (Figs. 43, 44). Auxiliary spine on anal claw very short (Fig. 46) *munda* McLachlan
— Gills on meso- and metathorax and abdomen with many filaments (Fig. 42). Auxiliary spine on anal claw long and curved (Fig. 47) 2
2. Sides of head straight. Anterior edge of fronto-clypeus concave (Fig. 39) *obliterata* McLachlan
— Sides of head converging. Anterior edge of fronto-clypeus not concave (Figs. 38 and 40) 3
3. Markings on posterior dorsal side of genae dark brown with indistinct light dots. Sides of head converging and constricted at the level of the eyes (Fig. 38) *dorsalis* (Curtis)
— Markings on posterior dorsal side of genae brown with distinct darker brown dots forming a pattern (Fig. 40). Sides of head converging but not constricted at the level of the eyes (Fig. 40) *septentrionis* McLachlan

HICKIN, N. E. 1942. *Proc. R. ent. Soc. Lond.* (A) **17**, 14–16.
HICKIN, N. E. 1954. *Proc. R. ent. Soc. Lond.* (A) **29**, 59–61.
KIMMINS, D. E. 1943. A List of the TRICHOPTERA (Caddis Flies) of the Lake District with Distributional and Seasonal Data. *J. Soc. Br. Ent.* **2**, 136–57.
KIMMINS, D. E. 1966. A Revised Check-list of the British TRICHOPTERA. *Entomologist's Gaz.* **17**, 111–20.
KLOET, G. S. and HINCKS, W. D. 1964. *A Check-list of British Insects.* Second edition (Revised) Part 1. Royal Entomological Society of London.
LESTAGE, J. A., *in* ROUSSEAU, E. 1921. *Les Larves et Nymphes Aquatiques des Insectes d'Europe*, 415–419.
MACKERETH, J. C. 1954. *Proc. R. ent. Soc. Lond.* (A) **29**, 147–52.
MARLIER, G. 1943. Les métamorphoses de *Rhyacophila dorsalis* Curtis et de *Tinodes assimilis* MacLachlan. *Bull. Mus. Hist. nat. Belg.* **19**, 1–8.
NIELSEN, A. 1942. Über die Entwicklung und Biologie der Trichopteren. *Arch. Hydrobiol.* Suppl. **17**, 255–631.

ROUSSEAU, E. 1921. *Les Larves et Nymphes Aquatiques des Insectes d'Europe*, 415.
SILTALA, A. J. 1906. *Acta Soc. Fauna Flora fenn.* **27**, 146.
ULMER, G. 1903. *Metamorph. Trichopt.* 127.
ULMER, G. 1909. *Die Süsswasserfauna Deutschlands* **5–6**, 217–20.
WESENBERG-LUND, C. 1911. *Int. rev. Hydrobiol.* (Biol. Suppl.) **3**, 1–64.

7

GLOSSOSOMATIDAE

Kimmins (1966) includes *Glossosoma* and *Agapetus* in the RHYACOPHILIDAE but because of the distinct larval characteristics they are here given family status as in Kloet & Hincks (1964). The adults of species in this family are moderate to small in size. In *Glossosoma* the length of the anterior wing is up to 9 mm. but in *Agapetus* the anterior wing measures up to 5.5 mm. at most. The tibial spur formula is 2.4,4 and the middle legs of the female are dilated. The discoidal cell of the fore-wing is closed but that of the hind-wing may be open or closed.

GLOSSOSOMATIDAE
GLOSSOSOMA Curtis
1. *conformis* Neboiss=*boltoni* McLachlan *nec* Curtis
2. *intermedium* (Klapálek)
3. *boltoni* Curtis=*vernale* (Pictet)

AGAPETUS Curtis
4. *ochripes* Curtis=*comatus* (Pictet)
5. *delicatulus* McLachlan
6. *fuscipes* Curtis

NOTE: The species *intermedium* was placed in the genus *Mystrophora* by Klapálek. Kloet & Hincks showed that this name was preoccupied and erected the genus *Mystrophorella* to contain it. Kimmins (1965) merged this with *Glossosoma*. Neboiss (1963), examining the Curtis types, has rectified errors made by McLachlan.

Distribution

Glossosoma conformis. Rather more local than *boltoni*, *conformis* on small tributary streams (M.E.M.). Berks. (M.I.C. *et al.*); Northumberland (G.N.P.); Cardigan (J.R.E.J.); widely distributed in rivers and streams, west and north Britain, south Ireland (D.E.K.); Westmorland, Cumberland (J.C.M.);
Glossosoma intermedium. Very local, Lake District (Coniston Water, Windermere, Hayes Water), streams with stony bottoms, near inflow into lakes (D.E.K.); Cumberland, Westmorland (J.C.M.).
Glossosoma boltoni. Rather local, on small streams, west and north Britain

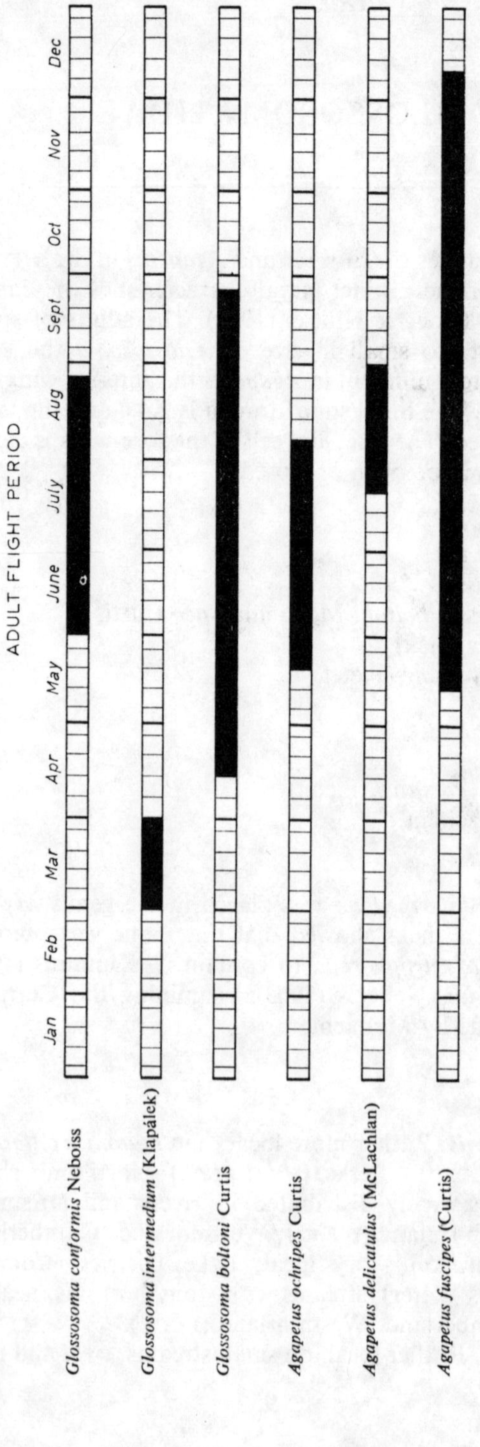

GLOSSOSOMATIDAE. *Adult Flight Periods taken from published records of individual dates.*

Agapetus ochripes. Very widely distributed and abundant (M.E.M.); Oxon. (L.W.G.); Northumberland (G.W.A.); widely distributed throughout the British Isles (D.E.K.); Mid-Yorkshire, Surrey (J.C.M.).
Agapetus delicatulus. Local. By no means so abundant where it occurs as the preceding species (M.E.M.); rather local and mainly western in distribution (D.E.K.); Cumberland, Mid-Yorkshire, Westmorland, Lanark (J.C.M.).
Agapetus fuscipes. Very widely distributed throughout the British Isles and one of the most abundant of the caddis flies (M.E.M.); Berks. (M.I.C. *et al.*); Northumberland (G.N.P.); Yorks. (P.F.H.).

General Description of the Larvae

The campodeiform larvae are widest at about the fifth abdominal segment. They construct mobile larval cases which are never tubular. The pronotum is entirely sclerotized and, whereas in the genus *Glossosoma* the meso- and metanota are entirely membraneous, in *Agapetus* these are sclerotized patches. In common with the larvae of RHYACOPHILIDAE the legs are almost equal in length and width and furnished with robust claws which are sometimes short with a robust basal spear. The abdominal segments are well-marked, the first having no lateral or dorsal protuberances. Prosternal horn and lateral lines are absent. Thoracic and abdominal gills are, however, absent. Six anal gills present.

Anal claws sometimes partly incorporated with the tenth abdominal segment.

Detailed Descriptions of the Species

Mrs. Mackereth (1965) studied the six British species in the GLOSSOSOMATIDAE by the method she adopted for the RHYACOPHILIDAE, dissecting

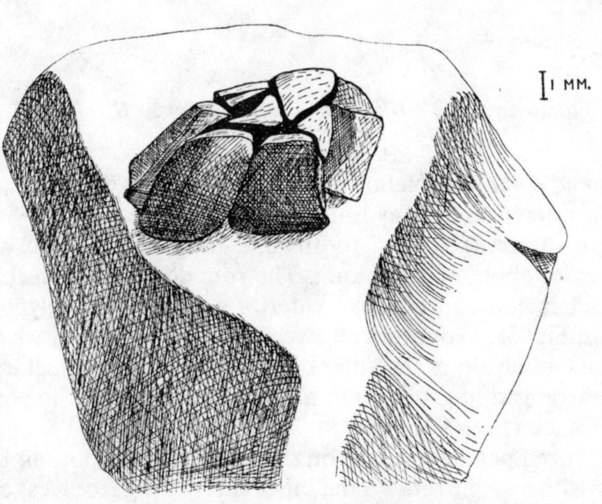

Fig. 57. Pupal case of *Glossosoma intermedia* Klapálek.

out last-instar larval skins from pupal chambers from which male adults had emerged. When good separating characters had been found then whole larvae were collected and examined. The descriptions of larvae of the genus *Glossosoma* which follow are those due to Mackereth with Kimmins's original descriptions of the larva of *intermedium* with the illustrations which he provided. In the case of the genus *Agapetus* the description of *fuscipes* by Hickin is given with the comparative notes made by Mackereth concerning the three species and her key to all species of the sub-family. Alterations have, of course, been made in accordance with current species nomenclature. The larvae of all species are small, being from 4 to 10 mm. in length. They inhabit elliptical cases made from coarse sand-grains or fragments of stone. They are all confined to rapid stony streams or rivers, but *A. fuscipes* is also found in lakes. Pupation takes place in a thin light-brown parchment-like cocoon within the stone case, which is securely fastened to a large stone with silk-like secretion (Fig. 57).

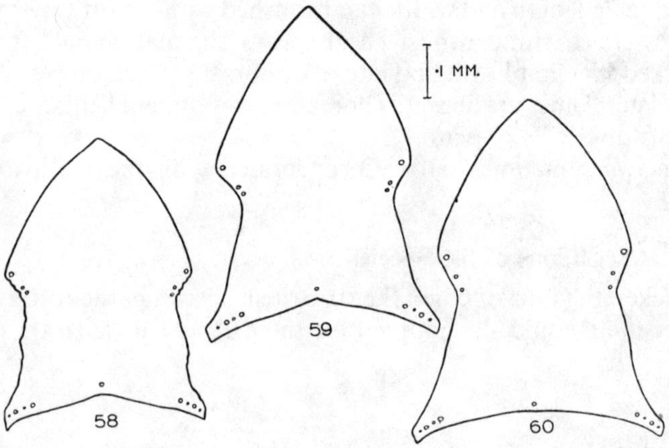

Figs. 58–60. **Fronto-clypeus.** 58. *Glossosoma intermedium*; 59. *G. boltoni*; 60. *G. conformis*.

Glossosoma conformis Neboiss—*boltoni* McL., 1879, *nec* Curtis, 1834

Mackereth's description is as follows:

Fully grown larvae from 8 to 10 mm. long and 1.5 to 3 mm. broad. Elliptical cases about 10 mm. long. The rounded head is dark brown and almost black in some specimens. Anterior edge of fronto-clypeus concave, as shown in Fig. 58. Pronotum also very dark brown, and with a single row of long black setae along the anterior edge and numerous setae posteriorly which are arranged in groups laterally. The band along the posterior edge is entirely black (Fig. 76).

No sclerotized patches present on the meso- or metanotum. Legs robust. Each claw with a single seta on the inner side arising from a short footstalk (Fig. 81).

Abdomen yellow or pinkish-yellow with short hairs arranged dorsally and ventrally on each segment. Anal claw short and robust, with one auxiliary spine attached to the outer edge.

The mandibles from the last larval skin, found in the pupal case, were all short, and very much as Ulmer and Rousseau depicted them, but the mandibles of newly moulted larvae were long and serrated. Nielsen (1942) states in his description of *A. fuscipes* that the claws and mandibles are longer than other authors have described them. It seems safe to surmise that the claws and mandibles get worn down as the larva crawls about and scrapes the stones for food.

Glossosoma intermedium (Klapálek)

Although first recorded as British by Morton in 1925, this species was not collected again until 1942, when Kimmins found a single dead larva at the original stream, a stream running into the western shore of Coniston Water in the Lake District. *G. intermedium* occurred where the beck runs through pastureland with scattered trees, bushes and clumps of rushes on its margins and also where it runs through a small wood. Small and medium-sized stones with silt in the quieter portions formed the stream bottom (Figs. 61–65).

Kimmins's description runs thus:

Larva. Closely resembling that of *Glossosoma.* In comparison with the figures given by Lestage (1921), the mandibles (Fig. 61) are broader at the

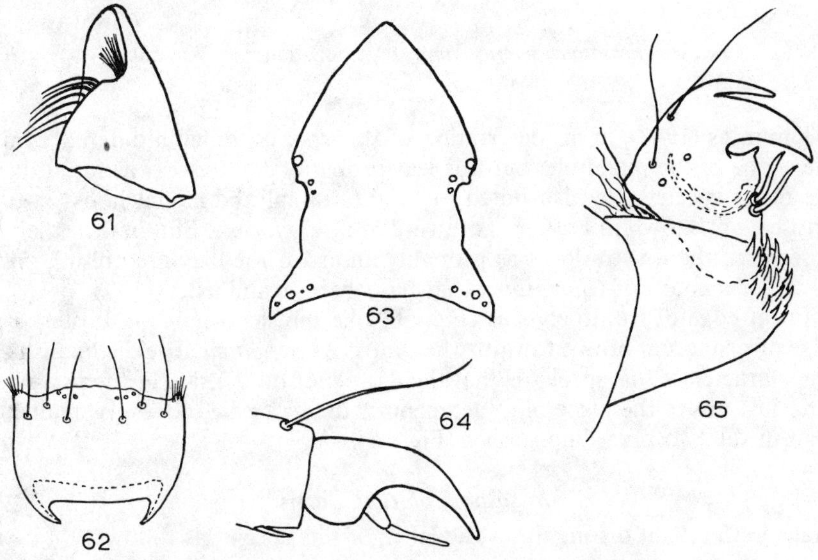

Figs. 61–65. *Glossosoma intermedium* Klap., larva. 61. Mandible; 62. Labrum; 63. Clypeus; 64. Tarsal claw; 65. Terminal abdominal hook. Kimmins.

base and more triangular than those of *Glossosoma conformis*. The labrum and clypeus are also similar (Figs. 62–63), but Lestage does not figure the bristles or their sockets on the latter. The tarsal claw and the terminal abdominal hook (Figs. 64–65) are of the *Glossosoma* pattern. On the apical segments of the latter are two stout setae which are not figured by Lestage.

Mackereth, who obtained larvae of this species from Hayes Water inflow, Cumberland, the River Troutbeck, Westmorland, as well as from the original locality, makes the following comments:

Larvae and cases about the same size as those of the two preceding species.

Fig. 66. *Glossosoma intermedium*; stone pupation case. Kimmins.

Kimmins (1943), in his description of this species, noted a difference in the shape of the mandible, but this was probably due to wear as found in the other species. He also noted, on the terminal abdominal hook, two bristles, not shown in Lestage's drawing of *G. conformis*. But, in fact, their absence in the illustration was probably an error, for they are present on all the specimens of *Glossosoma* that have been examined.

Front edge of fronto-clypeus (Fig. 58) like that of *G. boltoni*. Posterior edge of pronotum brown towards the centre, as in *boltoni*. The distinguishing character of this species is that the dark spot on the side of the pronotum, just above the place of attachment of the first pair of legs, is smaller than in the two preceding species (Fig. 77).

Glossosoma boltoni Curtis

Mackereth (1956) distinguishes the larva of this species as follows:

Larvae and cases about the same size as those of *G. conformis*. Both Ulmer and Rousseau distinguished *boltoni* from *conformis* on pigmentation of head and pronotum, *boltoni* being yellowish-brown and *conformis* dark brown. However, intensity of pigmentation was found to vary a good deal in the specimens examined and not to be a reliable means of separating the species. A better distinction is that the band along the hind margin of the pronotum is brown towards the centre in *boltoni* (Fig. 78) but black along the whole length in *conformis* (Fig. 76). Also the front edge of the frontoclypeus is angled in the middle in *boltoni* (Fig. 59) and uniformly rounded in *conformis* (Fig. 60).

Agapetus ochripes Curtis

Material of this species was collected by Mackereth from the River Wharfe in Mid-Yorkshire and from the River Wey in Surrey. She states that: This species can be distinguished from *A. delicatulus* by the dot on the side of the pronotum being pale and inconspicuous (Fig. 83) and from *A. fuscipes* by the sclerites on the metanotum being small and round.

Agapetus delicatulus McLachlan

Material was obtained by Mackereth from Cumberland, Westmorland, Mid-Yorkshire and Lanark, and she makes the following comments:

This species differs from the other species in the genus in three main characters. Fore margin of frontoclypeus evenly rounded; this distinction is clear on mounted cast skins. The marking on the side of the pronotum just over the point of insertion of the fore-leg dark and conspicuous, as shown in Fig. 84. This character is best seen if the pronotum is removed and squashed flat. Sclerites on the metanotum small and rounded.

Agapetus fuscipes Curtis

The specimens from which this description was made were collected from a small stream at Cranham, Gloucestershire, and compared with larvae collected from a stream at Sutton Park, Warwickshire. Both streams are shallow, swiftly flowing, and have a stony bottom. In both localities the larvae were present in considerable numbers, many thousands of the small cases coating the stones and rocks on the stream bed. Many adults were subsequently reared from larvae collected at each of the localities.

Case (Figs. 67 and 68). Of stone chips, small pebbles and sand-grains with the dorsal surface hemispherical and the ventral surface flat. An opening occurs at anterior and posterior ends, and in the same plane as the ventral surface. Shortly before pupation the case is fastened to a stone or rock. Length 6–8 mm., width 4–5 mm.

Larva (Figs. 69–70). Eruciform, length 6.0 mm., width 1.3 mm. Fusiform curved abdomen, gills absent.

Head (Fig. 71). Orthocentrous, dark brown, oval, widest at aboral end. Eyes set fairly near to oral end and situated in pale areas surrounded by large bristles. Aboral end of head free from bristles. Clypeus wide with faint marks at aboral end, much darker colour at oral end. Antennae (Fig. 72) small, consisting of basal lobe with two finger-like projections. *Mouthparts*. Labrum (Fig. 73) heavily sclerotized except anterior margin

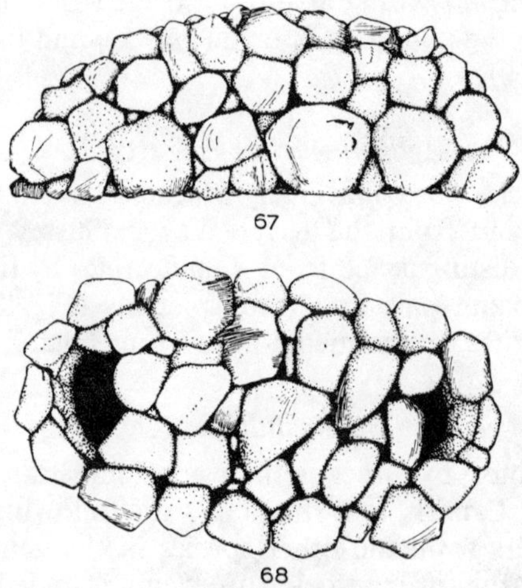

Figs. 67–68. *Agapetus fuscipes*. 67. Larval case, lateral; 68. Larval case, ventral.

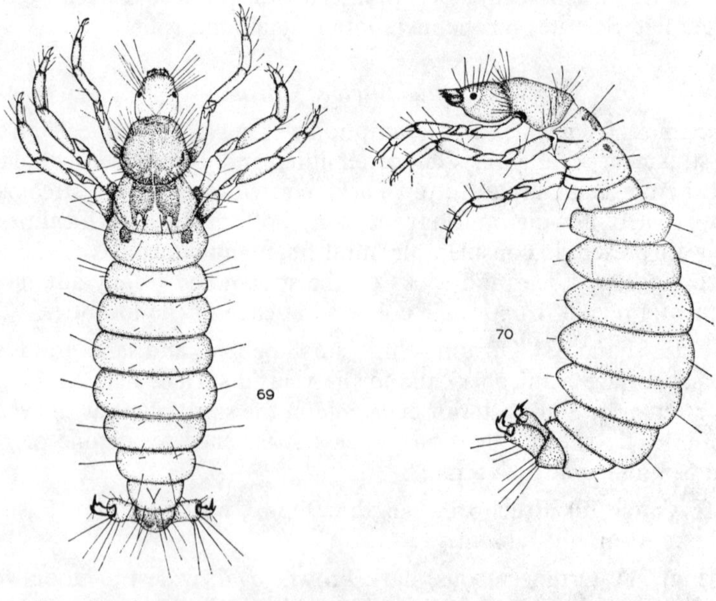

Figs. 69–70. *Agapetus fuscipes*. 69. Larva, dorsal; 70. Larva, lateral.

which is very hairy. Base of labrum with black, bow-shaped sclerite. Mandibles (Fig. 74) with inside edge produced to a slim blade at upper end. Approximately seven long bristles, with secondary hairs, arise from the inside edge. Labium small, segments of maxillary palp indistinguishable, maxillary lobe hairy, labium hairy.

Thorax. Pronotum completely sclerotized, dark brown. Two transverse rows of bristles, a few light spots along posterior row. Median suture. Mesonotum partially sclerotized, two distinct dark brown sclerotized patches have two small projections extending from posterior edge. Metanotum with pair of small, distinct, sclerotized patches.

Legs. Approximately equal. Tarsal claws blunt with a basal spine and a dark mark at the point of insertion in the tarsal segment.

Abdomen. Curved ventrally, widest at about the fifth segment. Lateral line absent. Anal hooks with a single auxiliary hook, extended laterally by a basal lobe with a bulbous projection at the proximal end. Lobes and ninth segment darker in colour.

Mackereth collected larvae of *A. fuscipes* from Cumberland, Mid-Yorkshire, Lanark and Westmorland, and makes the following comments:

Larvae from 4 to 7 mm. long when fully grown. Stone cases about the same length.

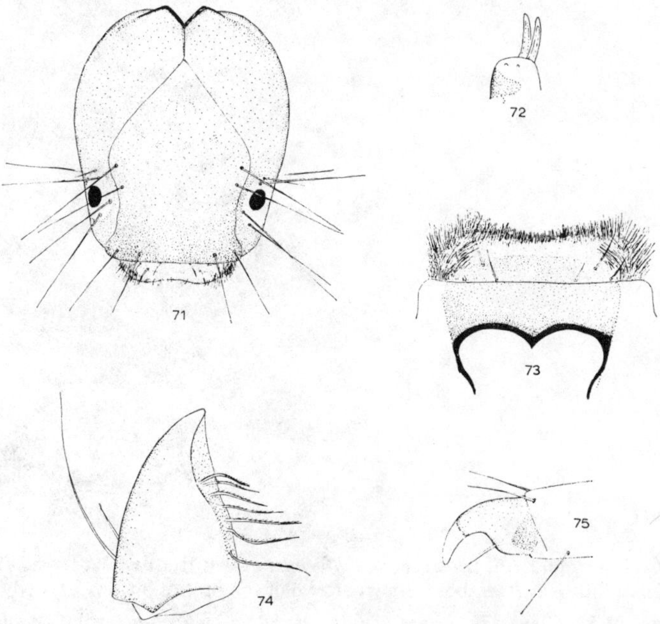

Figs. 71–75. *Agapetus fuscipes*. 71. Head; 72. Antenna; 73. Labrum; 74. Mandible; 75. Tarsal claw of mesothoracic leg.

Head varies in colour in different specimens from light to dark brown. Anterior edge of fronto-clypeus concave and slightly V-shaped.

Pronotum (Fig. 82) bordered anteriorly with a double row of long black setae, and there is a dark band along the posterior border. A mark on the side of the pronotum is barely visible. Sclerotized patches present on meso-

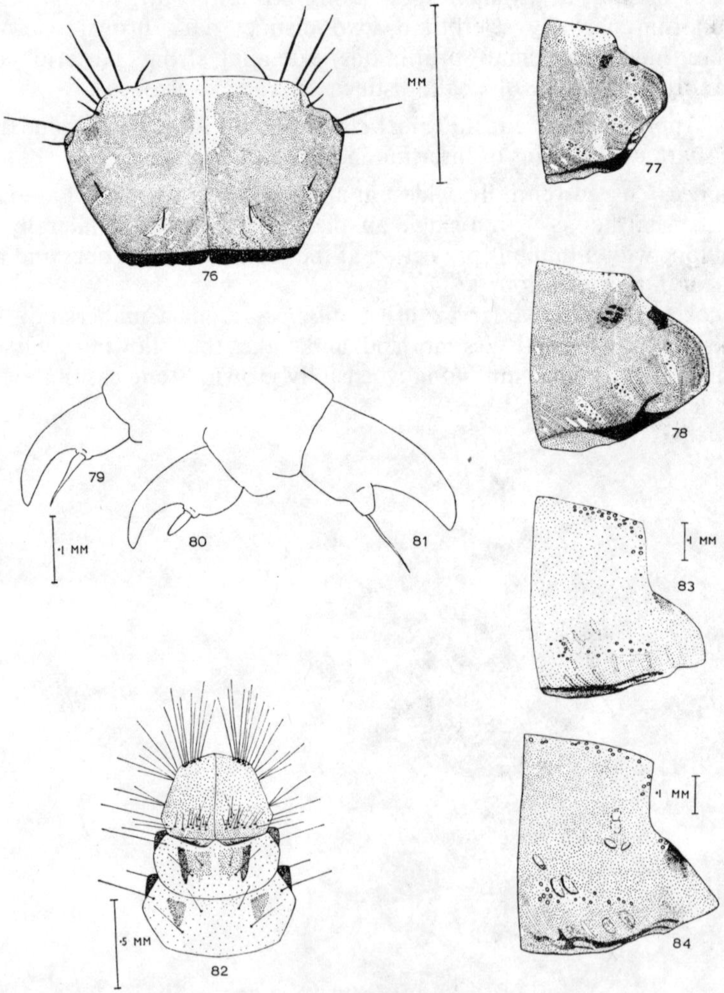

Figs. 76–78. Pronota. 76. *Glossosoma conformis*, drawn from whole larva; 77. *G. intermedium*; 78. *G. boltoni*, both drawn after half the pronotum was pressed out flat.

Figs. 79–81. Claws. 79. *Agapetus fuscipes*; 80. *A. delicatulus*; 81. *G. conformis*.

Figs. 82–84. 82. Dorsal view of thorax of *Agapetus fuscipes*; 83. Pronotum of *A. ochripes*; 84. Pronotum of *A. delicatulus*. Mackereth.

and metanotum, as shown in Fig. 82, these patches being very conspicuous on fully grown larvae.

Legs robust and all of about equal length. Each claw bears, on its inner surface, a single seta arising from a very short footstalk (Fig. 79). Abdomen creamy-yellow; each segment with numerous long hairs ventrally and dorsally. Anal claw short and robust, and has, attached to the outer surface, two and occasionally three auxiliary spines (Figs. 76–84).

Notes on Occurrence and Flight-period

All the records of *G. conformis* are from streams and small rivers, with adults emerging from mid-June to late August. *G. boltoni*, to judge from the published records, is found mainly in larger rivers, with adults emerging from mid-April until late September. *G. intermedium* is known from three localities only. In two of these, larvae have been found over-wintering in their pupal cases, but the time of emergence was not discovered. In the third adults emerged in March. *A. fuscipes* is widespread and larvae have been recorded from streams, rivers and lakes, with adults emerging from May to August. *A. ochripes* has been recorded only from rivers, with adult records from late May to early August. *A. delicatulus* larvae have been collected from both streams and rivers. The adults emerge in August. There are some southern records of adults in July.

Key to Larvae of Glossosomatidae

The following key to the fully-grown larvae is that due to Mackereth:

1. Larvae 8 to 10 mm. in length. Single row of long black setae along the anterior edge of the pronotum (Fig. 76). No sclerotized patches on the meso- or metanotum. Seta on inner surface of claw arising from a footstalk (Fig. 81). Anal claw with a single auxiliary spine attached to the outer surface *Glossosoma* 2

— Larvae 4 to 7 mm. in length. Double row of long black setae along anterior edge of pronotum (Fig. 82). Sclerotized patches present on both meso- and metanotum. Seta on inner surface of claw arising from a very short footstalk (Figs. 79, 80). Anal claw generally with 2 or 3 auxiliary spines attached to the outer surface *Agapetus* 4

2. Spot on side of pronotum small (Fig. 77) *G. intermedium*

— Spot on side of pronotum large (Fig. 78) 3

3. Posterior edge of pronotum entirely black (Fig. 76). Front margin of fronto-clypeus uniformly rounded (Fig. 76) *G. conformis*

— Posterior edge of pronotum brown towards centre (Fig. 78). Front margin of fronto-clypeus angled in the middle (Fig. 59) *G. boltoni*

4. Sclerites on metanotum larger (Fig. 82); setae on claws long (Fig. 79) *A. fuscipes*

— Sclerites on metanotum smaller; setae on claws short (Fig. 80) 5

5. Spot on side of pronotum dark and conspicuous (Fig. 84) *A. delicatulus*

— Spot on side of pronotum light brown and smaller (Fig. 83) *A. ochripes*

HICKIN, N. E. 1943. *Proc. R. ent. Soc. Lond.* (A) **18**, 78–80.

KIMMINS, D. E. 1943. A List of the TRICHOPTERA (Caddis Flies) of the Lake District, with distributional and seasonal data. *J. Soc. Br. Ent.* **2** (4), 153.

KIMMINS, D. E. 1943. A Note on the Caddis Fly, *Mystrophora intermedia* Klapálek (TRICHOPTERA). *Proc. R. ent. Soc. Lond.* (A) **18**, 96–8.

KIMMINS, D. E. 1953. Entomological Results from the Swedish Expedition 1934 to Burma and British India: TRICHOPTERA. *Ark. Zool.* **6**, 167–83.

KIMMINS, D. E. 1965. *Entomologist's Gaz.* **16**, 147–61.

KLOET, G. S. & HINCKS, W. D. 1944. Nomenclatorial Notes on Two Generic Names in the TRICHOPTERA. *Entomologist* **77**, 97.

LESTAGE, J. A., *in* ROUSSEAU, E. 1921. *Les Larves et Nymphes Aquatiques des Insectes d'Europe*, 419–430.

MACKERETH, J. C. 1954. Taxonomy of the Larvae of the British Species of the Genus *Rhyacophila* (TRICHOPTERA). *Proc. R. ent. Soc. Lond.* (A) **29**, 147–52.

MACKERETH, J. C. 1956. *Proc. R. ent. Soc. Lond.* (A) **31**, 167–72.

MORTON, K. J. 1925. *Mystrophora intermedia* Klap., new to the British fauna, and *Apatania muliebris* McLach., two species of TRICHOPTERA from the Lake District. *Entomologist's mon. Mag.* **61**, 130–1.

NEBOISS, A. 1963. The TRICHOPTERA Types of Species described by J. Curtis. *Beitr. Ent.* **13** (5–6), 582–635.

NIELSEN, A. 1942. Uber die Entwicklung und Biologie der TRICHOPTEREN. *Arch. Hydrobiol.* (Suppl.) **17**, 358–74.

SILFVENIUS, A. J. 1905–6. *Acta Soc. Fauna Flora fenn.* **27** (6), 157.

ULMER, G. 1903. *Metamorph. Trichopt.* 130.

ULMER, G. 1909. *Die Süsswasserfauna Deutschlands* **5–6**, 221–2.

8

HYDROPTILIDAE

Adults of the family HYDROPTILIDAE are minute insects with very hairy wings. Some of the hairs are thick and erect and the wing fringes are very long, those of the hind-wing being longer than the greatest width of the hind-wing. The wings are long and narrow with acute apices and the hind-wing is always narrower than the anterior. The length of the anterior wing of several species is no more than 2 mm. In many species it is only 3 mm., whilst in the largest species it is but 5 mm. The antennae are short and stout and are shorter in the female than in the male. Ocelli are present only in some genera. The maxillary palpi of both sexes are five-segmented; the first two segments are very short whilst the terminal one is simple. The neuration of the wings of these very small insects is often difficult to make out but the genitalia are distinctive; the student is advised to have Kimmins's 1958 paper in front of him when identifying adults of the genus *Oxyethira.*

These very small caddis flies are extremely active and are often observed running with great agility on stone walls and bridges near the water in which the immature stages are spent. They are often present in great numbers and they are attracted to light. Our knowledge of the biology and larval taxonomy of this caddis fly family is mainly due to the work of Nielsen in Denmark. Only two detailed descriptions of larvae of British origin have been made but Nielsen has made a very exact study of a species from each of five important genera, and this forms the basis of this chapter.

Hanna (1961) described the last-instar larva of *Hydroptila sparsa* Curtis and Macdonald (1950) described the last-instar larva of *Oxyethira simplex* Ris.

A little is known of the larvae, in varying degrees of detail, of a few other species, but of the known twenty-nine species of the HYDROPTILIDAE in the British fauna, the larvae of fifteen are unknown. It therefore seemed impractical to construct a key for separation when we are in complete ignorance of half of them.

Eight genera containing in all twenty-nine species have been found in Britain, as in the following check-list. *Ithytrichia clavata* was added to the British list in 1939 and does not appear in Mosely's handbook.

Kimmins's (1966) revised check-list made a number of name alterations which are here included with the Kloet & Hincks' 1964 names in parentheses for convenience.

AGRAYLEA

1. *multipunctata* Curtis
2. *sexmaculata* Curtis (*pallidula* McLachlan)

ALLOTRICHIA

3. *pallicornis* (Eaton)

HYDROPTILA

4. *angulata* Mosely
5. *cornuta* Mosely
6. *tineoides* Dalman (*femoralis* (Eaton))
7. *forcipata* (Eaton)
8. *lotensis* Mosely
9. *vectis* Curtis (*maclachlani* Klapálek)
10. *occulta* (Eaton)
11. *pulchricornis* Pictet (*pulchricornis* (Eaton))
12. *simulans* Mosely
13. *sparsa* Curtis
14. *sylvestris* Morton
15. *tigurina* Ris.

ITHYTRICHIA

16. *clavata* Morton
17. *lamellaris* Eaton

ORTHOTRICHIA

18. *angustella* (McLachlan)
19. *costalis* (Curtis) (*tetensii* Kolbe)
20. *tragetti* Mosely

OXYETHIRA

21. *flavicornis* (Pictet) (*costalis* Curtis)
22. *distinctella* McLachlan
23. *falcata* Morton
24. *frici* Klapálek
25. *sagittifera* Ris.
26. *simplex* Ris.
27. *tristella* Klapálek
28. *mirabilis* (Morton) (formerly in OXYTRICHIA)

TRICHOLEIOCHITON

29. *fagesi* (Guinard). In Mosely's handbook this species appears as *Leiochiton felina* Ris. and *fagesii* in Kloet & Hinks' (1964)

Distribution

Agraylea multipunctata. Lakes, ponds, large rivers. Local (M.E.M.); Berks. (M.I.C. *et al.*); Yorks. (P.F.H.).

Agraylea sexmaculata. Very local. Abundant at Mottisfort lake, Hants. R. Kennet and R. Lamborne, Berks., Wicken Fen, Cambridgeshire (M.E.M.); Berks. (M.I.C. *et al.*); Yorks. (P.F.H.).

Allotrichia pallicornis. Local. Inhabits running water (M.E.M.); Berks. (M.I.C. *et al.*).

Hydroptila sparsa. Very widely distributed and abundant. Frequents running water (M.E.M.); Northumberland (G.N.R.); Berks. (M.I.C. *et al.*).

Hydroptila simulans. Widely distributed. Often found in company with *sparsa*, frequenting running water (M.E.M.).

Hydroptila cornuta. Very local. Inhabiting running water and by no means abundant (M.E.M.).

Hydroptila angulata. Local. Abundant along the banks of the River Eden at Carlisle (M.E.M.); Northumberland (G.N.P.); Yorks. (P.F.H.).

Hydroptila sylvestris. Very local. Loch Morlich, Glen More and Aviemore Inverness-shire (M.E.M.).

Hydroptila occulta. Widely distributed and abundant. Frequents running water (M.E.M.); Berks. (M.I.C. *et al.*).

Hydroptila tineoides. Widely distributed and abundant. Lakes and running water (M.E.M.); Berks. (M.I.C. *et al.*); Yorks. (P.F.H.).

Hydroptila pulchricornis. Local. Lakes (M.E.M.); Berks. (M.I.C. *et al.*).

Hydroptila forcipata. Very widely distributed and abundant (M.E.M.); Northumberland (G.N.P.); Yorks. (P.F.H.).

Hydroptila vectis. Very widely distributed and abundant (M.E.M.); Northumberland (G.N.P.); Yorks. (P.F.H.).

Hydroptila tigurina. Local. Northern Scotland (M.E.M.).

Ithytrichia lamellaris. Widely distributed and abundant (M.E.M.); Northumberland (G.N.P.).

Ithytrichia clavata. No information.

Orthotrichia angustella. Rather local. Running water (M.E.M.).

Orthotrichia tragetti. Only known from a private lake at Awbridge Danes, Romsey, Hants. (M.E.M.).

Orthotrichia costalis. Widely distributed, ponds and lakes (M.E.M.), Berks. (M.I.C. *et al.*).

Oxyethira flavicornis. Widely distributed and abundant on most lakes and large ponds (M.E.M.). Rivers, lakes, Devon, Somerset, Dorset, Hants., Surrey, Oxon., Norfolk, Cambs., Beds., Northants, Warwicks, Brecknock, Merioneth, Caernarvon, Westmorland and Lake Lancs., Perth, Kerry, Galway, Dublin, Westmeath, Mayo, Donegal. An abundant species (D.E.K.), Berks. (M.I.C. *et al.*), Yorks. (P.F.H.).

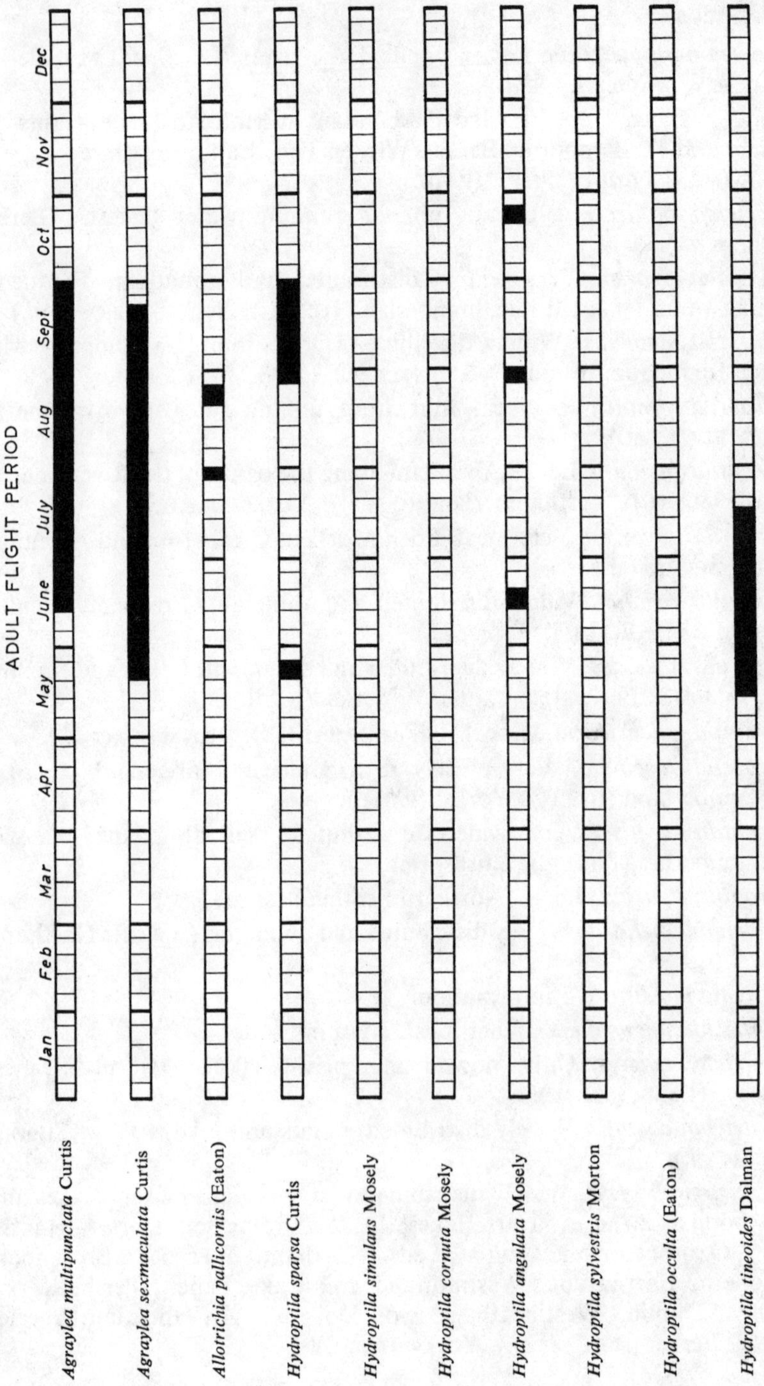
ADULT FLIGHT PERIOD
Jan
Feb
Mar
Apr
May
June
July
Aug
Sept
Oct
Nov
Dec
Agraylea multipunctata Curtis
Agraylea sexmaculata Curtis
Allotrichia pallicornis (Eaton)
Hydroptila sparsa Curtis
Hydroptila simulans Mosely
Hydroptila cornuta Mosely
Hydroptila angulata Mosely
Hydroptila sylvestris Morton
Hydroptila occulta (Eaton)
Hydroptila tineoides Dalman

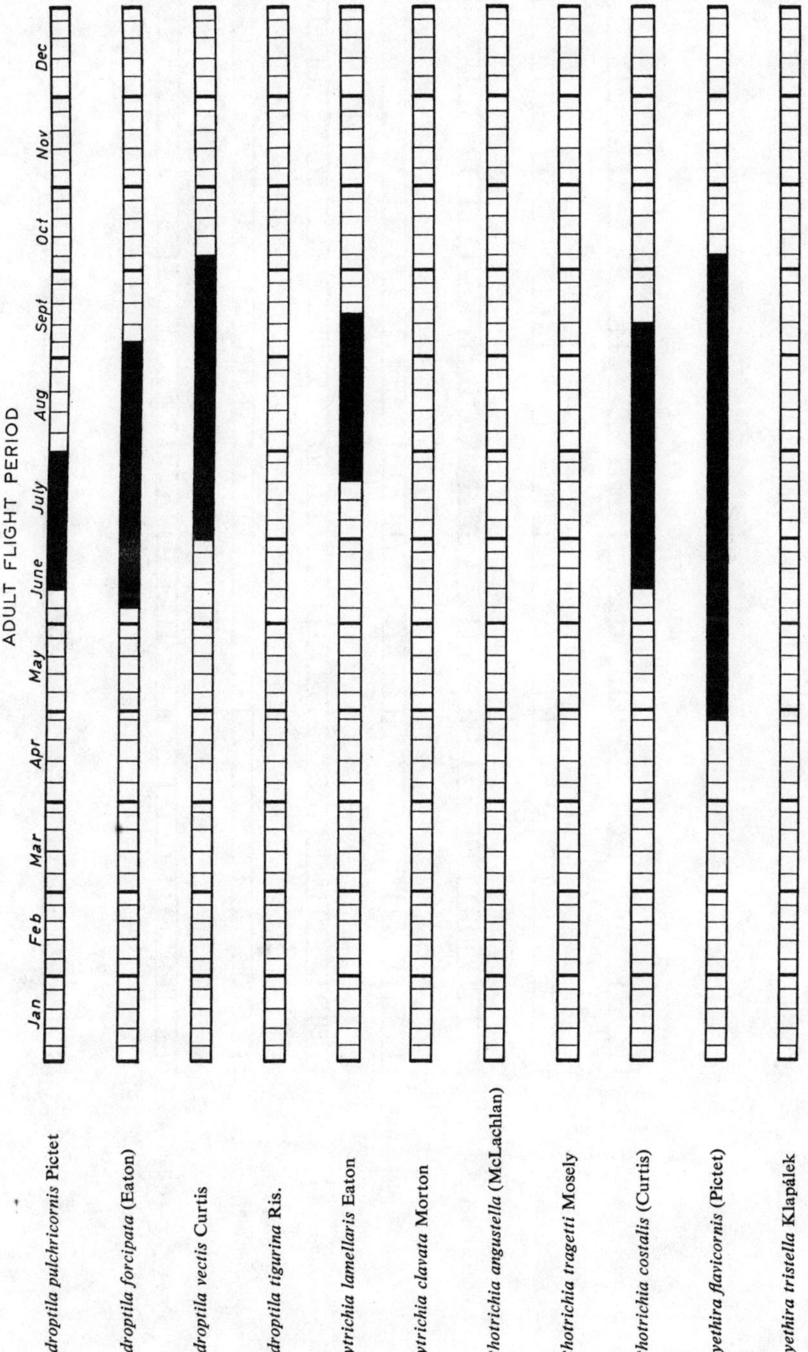
ADULT FLIGHT PERIOD
Jan
Feb
Mar
Apr
May
June
July
Aug
Sept
Oct
Nov
Dec
Hydroptila pulchricornis Pictet
Hydroptila forcipata (Eaton)
Hydroptila vectis Curtis
Hydroptila tigurina Ris.
Ithytrichia lamellaris Eaton
Ithytrichia clavata Morton
Orthotrichia angustella (McLachlan)
Orthotrichia tragetti Mosely
Orthotrichia costalis (Curtis)
Oxyethira flavicornis (Pictet)
Oxyethira tristella Klapálek

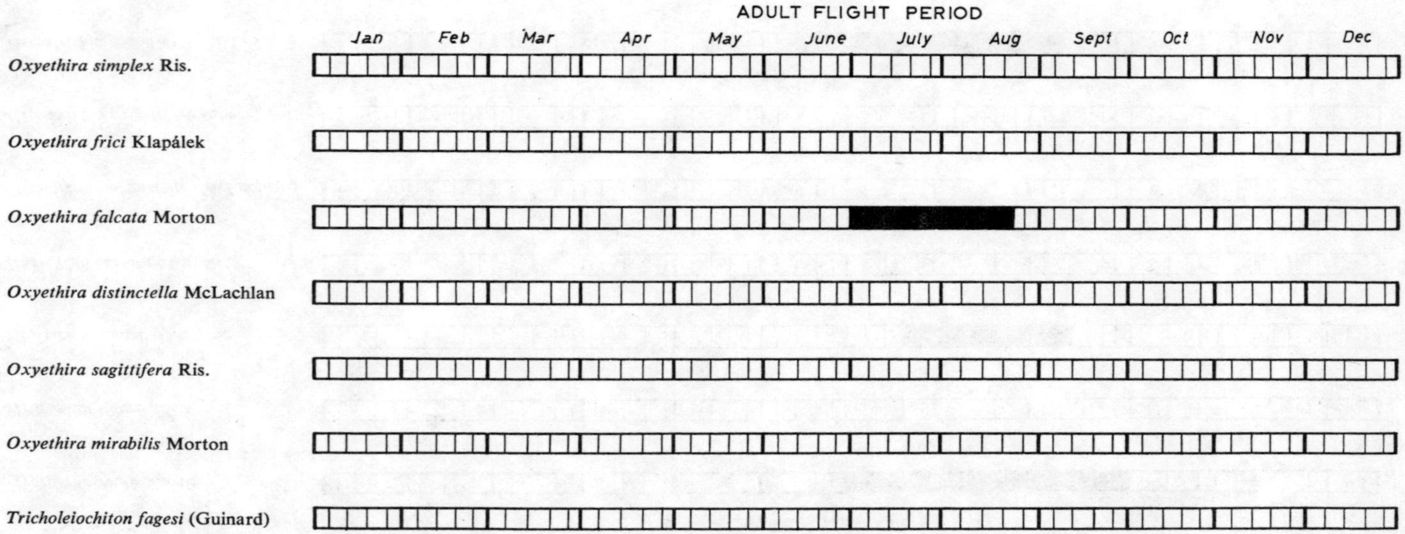

HYDROPTILIDAE. *Adult Flight Periods taken from published records of individual dates.*

Oxyethira tristella. A local species not very abundant where it occurs (M.E.M.). Rivers, lakes, Dorset, Hants., Oxon., Wigtownshire, Inverness, Galway. On slow stream, Moore's R., near Tricketts Cross, Hants-Dorset border (D.E.K.).

Oxyethira simplex. Fairly widely distributed and abundant (M.E.M.). Lakes, Hants, Oxon., Merioneth, Wigtownshire, Perthshire, Inverness, Galway (D.E.K.).

Oxyethira frici. Rather local and not very abundant (M.E.M.); Rivers, Somerset, Hereford, Brecknock, Merioneth, Caernarvon, Westmorland and Lake Lancs., Perth, Inverness, Galway (D.E.K.).

Oxyethira falcata. Widely distributed and abundant. Spring-fed ditch in Hants. (M.E.M.), Rivers, lakes, Devon, Wilts., Dorset, Hants., Sussex, Kent, Bucks., Cambs., Pembroke, Merioneth, Yorks., Wigtownshire, Lanark, Perth, Galway, Mayo (D.E.K.); Berks (M.I.C. *et al.*); Yorks. (P.F.H.).

Oxyethira distinctella. Lake at Awbridge Danes, Hants. (M.E.M.).

Oxyethira sagittifera. Very local, perhaps, more common in Ireland (M.E.M.). Ponds, slow reaches of rivers. Westmorland and Lake Lancs., Perthshire, Galway (D.E.K.).

Oxyethira mirabilis. Little information.

Tricholeiochiton fagesi. Little information.

General Description of Larvae

The first four instars of larval HYDROPTILIDAE differ so widely from the fifth instar that Nielsen (1948) has suggested that hypermetamorphosis is exhibited. Most accounts of the larval stage of species in this family have considered the last instar only, when a case has been constructed. Nielsen, however, pointed out the great differences in appearance between the young larvae (first four instars) and the older larvae (fifth instar). Especially is this so with regard to the prothorax and the anal prolegs (claspers). In addition, however, the habit of the larva usually changes abruptly at this stage. In the young larvae (Figs. 85–87) the prothorax is about as broad as, or is only a little narrower than, the head, and at this point the body tapers rather gradually posteriorly.

The abdomen, which is approximately circular in cross-section, is seen to be slender and the intersegmental grooves are deep, especially when viewed from above. The abdominal setae are much more developed than in last-instar larvae, accentuated perhaps by the slenderness of abdomen. Except in *Oxyethira*, the body of the young larva is covered by hair-like spinules, giving, in the first-instar larvae, a hairy appearance. In *Oxyethira*, however, these hair-like spinules are restricted to the sclerotized areas.

On the other hand, in fifth-instar larvae (Figs. 88–92), the prothorax is as broad as, or is a little broader than, the head, and the meso- and metathorax are equally as broad or may increase in width a little posteriorly. The first abdominal segment is a little wider than the metathorax and a

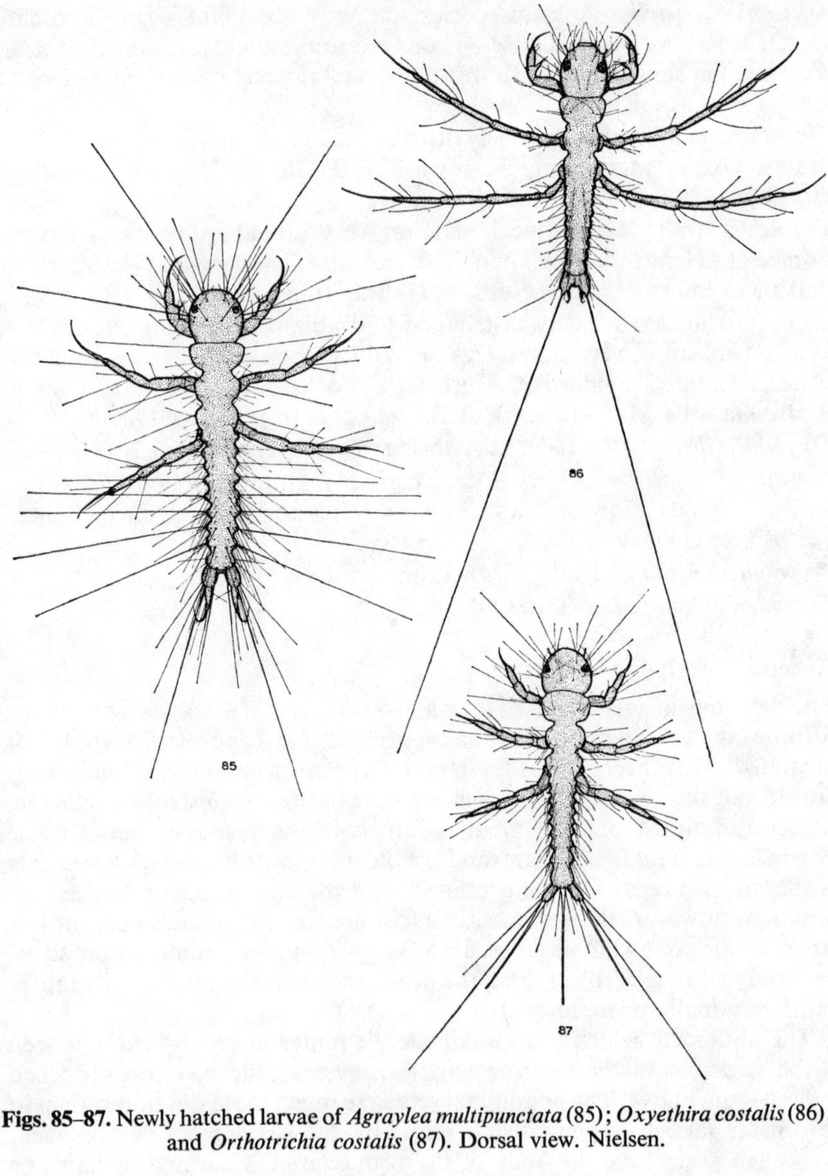
85
86
87

Figs. 85–87. Newly hatched larvae of *Agraylea multipunctata* (85); *Oxyethira costalis* (86); and *Orthotrichia costalis* (87). Dorsal view. Nielsen.

special characteristic of these older larvae is the large distended abdomen in which there is a marked separation of the first segment from the second and the eighth from the ninth.

The abdominal distention, however, takes place soon after the fifth instar. Immediately at ecdysis the abdomen is still slender. The fully grown

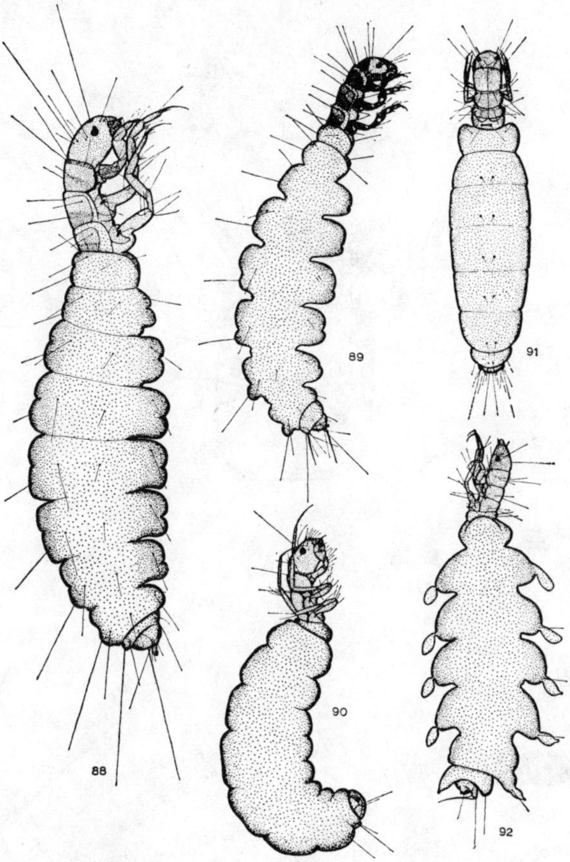

Figs. 88–92. Full-grown larvae of *Agraylea multipunctata* (88); *Hydroptila tineoides* (89); *Oxyethira flavicornis* (90); *Orthotrichia costalis* (91); *Ithytrichia lamellaris* (92). 91 in dorsal, the other larvae in lateral view. Nielsen.

larva has a comparatively large abdomen, usually laterally compressed and deepest (or widest) at segments four to six (Figs. 88–92).

The mouthparts are not prominent and in those species examined are modified for piercing and sucking the cells of filamentous algae, or (as in *Ithytrichia*) they are modified for scooping up and taking in solid food material such as diatoms.

The legs are, in some cases, approximately equal in length and in others the middle and posterior legs are longer. Gills are absent in *Agraylea* and *Oxyethira*, but in *Hydroptila* and *Orthotrichia* three long, filiform, caudal gills are present, and in *Ithytrichia* a single gill is found. In *Ithytrichia* there are ventral and dorsal adominal evaginations which were thought to serve

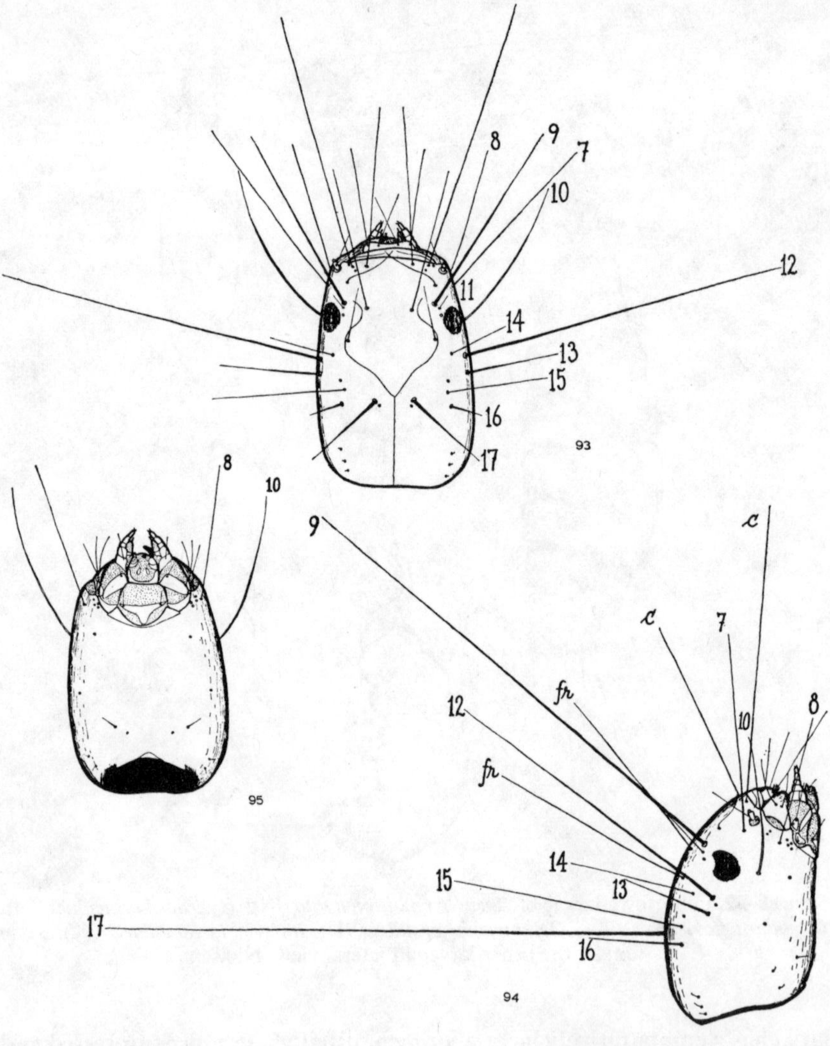

Figs. 93–95. *Agraylea multipunctata.* Head in dorsal (93); lateral (right) (94); ventral view (95). *fr.:* intermediate and posterior latero-marginal setæ of fronto-clypeus. Membranes dotted. Nielsen.

as gills, but Nielsen is not of this opinion. Lateral lines are absent. Small sclerotized patches are sometimes present on the abdominal tergites. In the fifth instar larva the anal appendages are small, apparently of one segment, which is furnished with a sclerotized patch with two posterior hairs and a lateral hair. The terminal claws are short, strong and arched. In the case of larvae in instars one to four the anal appendages are more slender.

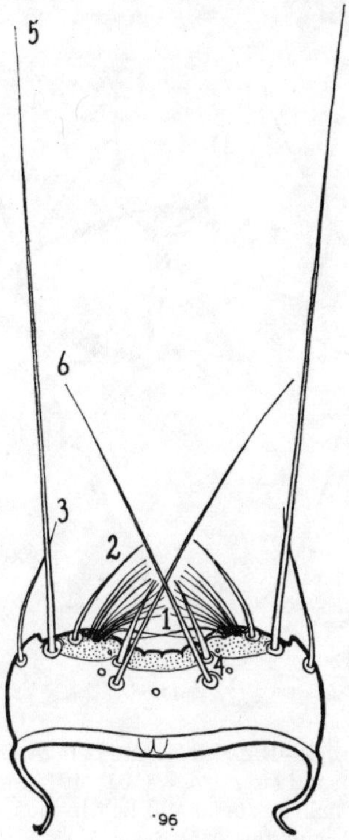

96

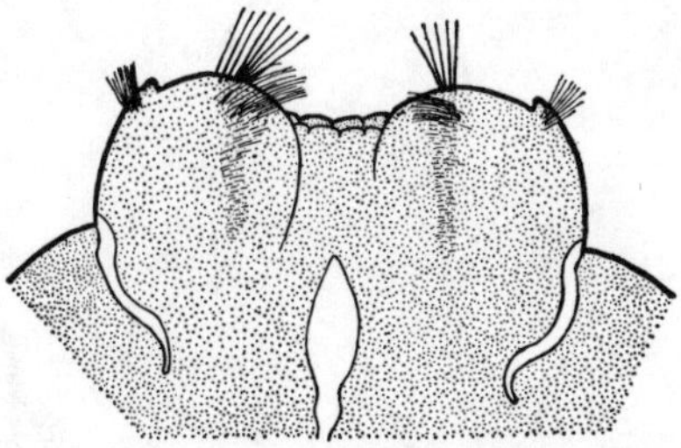

97

Figs. 96–97. *Agraylea multipunctata.* Labrum in dorsal (96) and ventral view (97; + anteclypeus p.p.). Membranes dotted. Nielsen.

In the HYDROPTILIDAE the young larvae, which are quite naked, crawl about the aquatic vegetation or swim feebly, case-building being confined to the fifth instar.

The cases are of characteristic shape, often being vase or seed-shaped, and consist of an inner lining of silk to which sand-grains and detritus may adhere, but also may be of silk-like secretion only. They may be mobile or fixed to the substrate by means of silken threads.

Detailed Descriptions of the Larvae

Agraylea multipunctata Curtis

No description of the larva of this species based on British material is available. The following account is based on Nielsen's detailed study. In the young larva, long bristles occur on the head and abdomen and the anal claws are exceptionally long. With the exception of the sclerites of the legs the body is completely covered with long hair-like spinules of the same

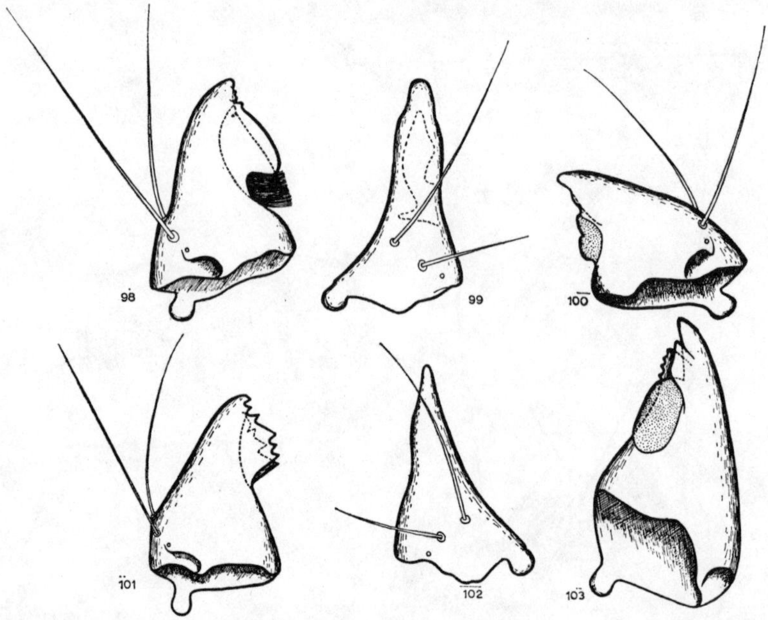

Figs. 98–103. *Agraylea multipunctata.* Mandibles 98–102. 5th instar. Left m. in dorsal (98) and lateral view (99), right m. in dorsal (100) and lateral view (102). 101. 2nd instar; left m. in dorsal view. 103. 4th instar; right m. half in dorsal and half in median view. Membrane dotted. Nielsen.

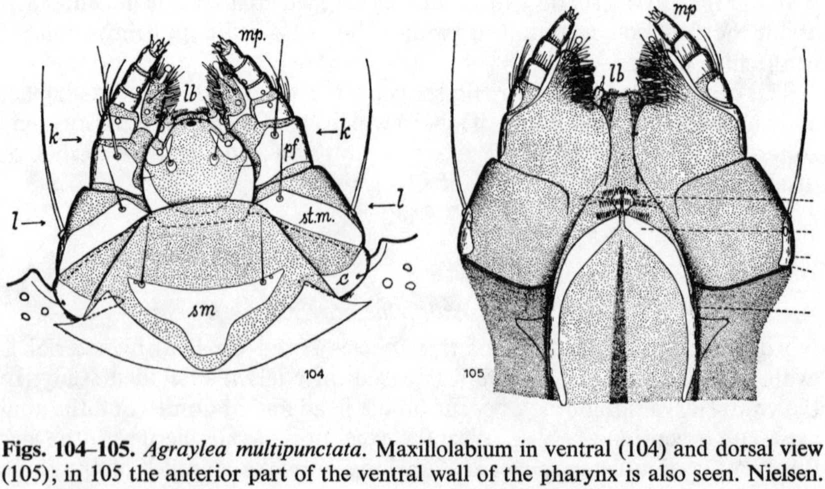

Figs. 104–105. *Agraylea multipunctata.* Maxillolabium in ventral (104) and dorsal view (105); in 105 the anterior part of the ventral wall of the pharynx is also seen. Nielsen.

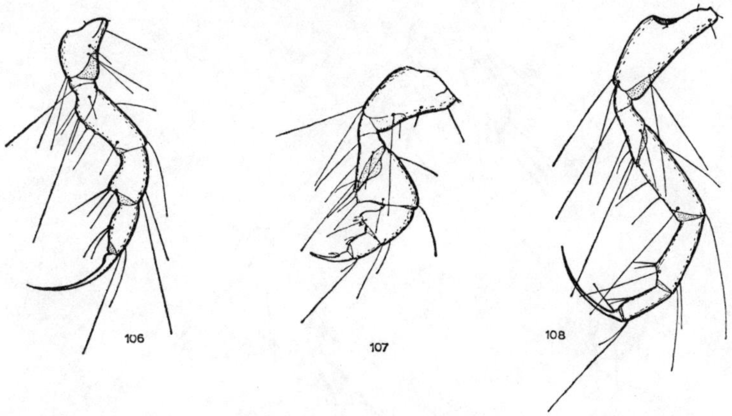

Figs. 106–108. *Agraylea multipunctata.* 106. 1st instar: right fore leg in posterior view. 107, 108. 5th instar; right fore (107) and middle leg (108) in posterior view. On the fore leg the proximal front seta is showing through. Membranes dotted. Nielsen.

size in all four instars. At the fourth instar the head is straw-coloured with a more or less distinct brownish spot on the anterior part of the frons. The pronotum is light brown whilst the meso- and metanota are pale yellow in front and brownish-yellow behind.

The full-grown fifth instar larva, the membraneous parts of which are deep green in colour, measures from 2.7 to 5.1 mm. in length. The first abdominal segment is much broader than the thorax and the width then increases to segment four, when it decreases gradually to segment eight. Segment nine is much smaller. The middle segments are somewhat compressed. The lateral margins of the pronotum are brownish-black, the

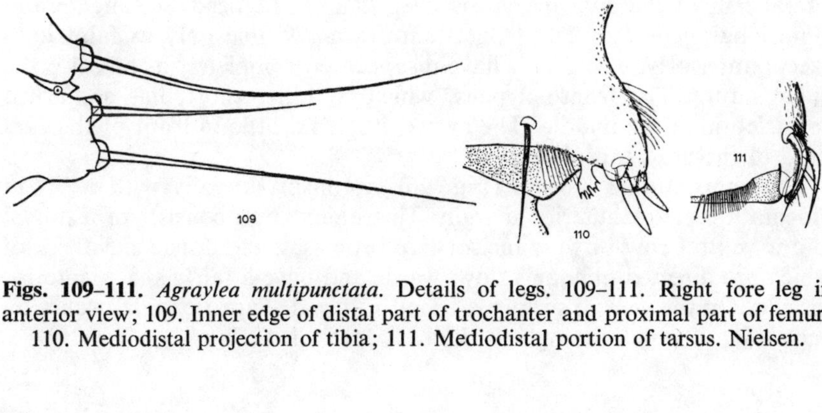

Figs. 109–111. *Agraylea multipunctata.* Details of legs. 109–111. Right fore leg in anterior view; 109. Inner edge of distal part of trochanter and proximal part of femur; 110. Mediodistal projection of tibia; 111. Mediodistal portion of tarsus. Nielsen.

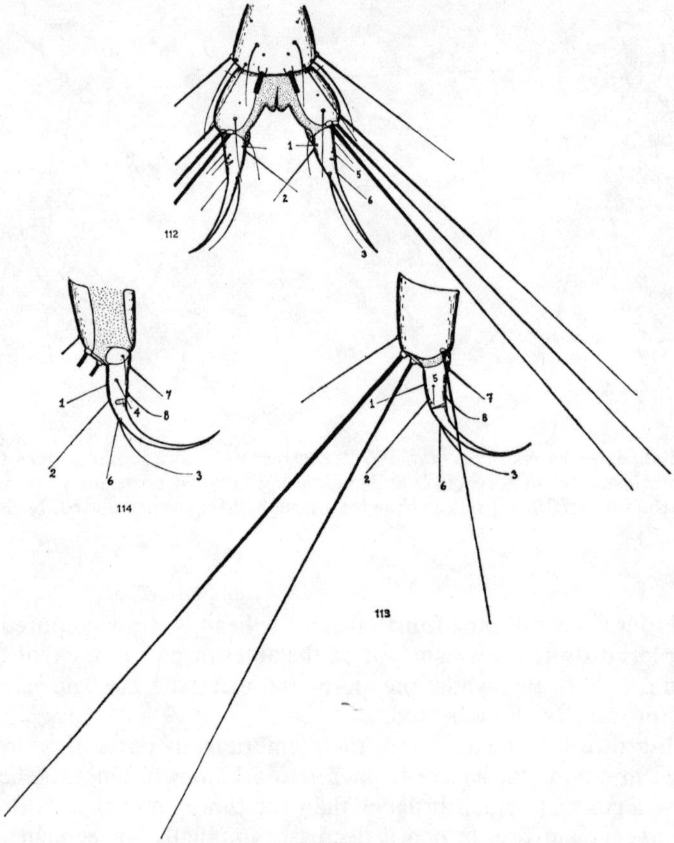

Figs. 112–114. *Agraylea multipunctata*; 4th instar. 112. Segment IX and anal prolegs in dorsal view, 113. Right anal proleg in lateral view; 114. Left one in median view. Membranes dotted. Nielsen.

lateral half of the anal margin is deep black. The head is elongate and almost parallel-sided. The frontal sutures are visible only as faint lines except anteriorly, where they have disappeared completely as has also the gular suture. The fronto-clypeus, which is rather short, has a marked constriction at the middle. The eye is situated a little in front of the oral third of the head capsule.

The short, broad labrum (Fig. 96) is convex laterally with a broad indentation at the anterior margin. This indentation consists of a dorsal and a ventral row of four incisor-like processes, the dorsal members of which are directed obliquely downwards and have a fairly sharp anterior margin. On the ventral surface a broad soft pad-shaped papilla occurs on each side, of which the right is the best developed (Fig. 97).

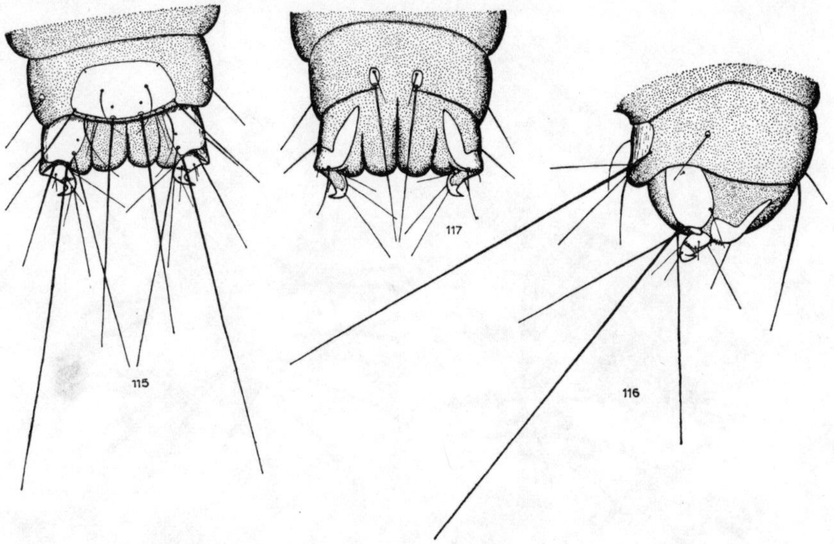

Figs. 115–117. *Agraylea multipunctata.* Segment IX and anal prolegs in dorsal (115), lateral (right, 116), and ventral view (117). Membranes dotted. Nielsen.

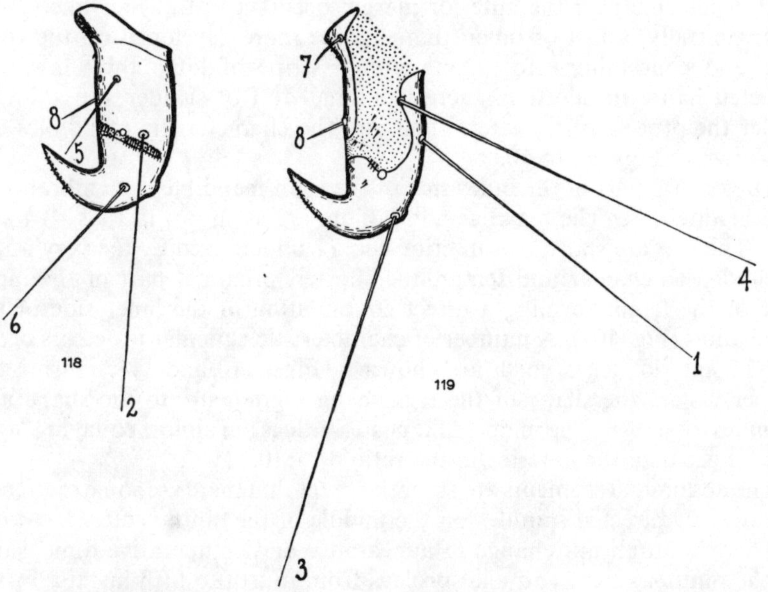

Figs. 118–119. *Agraylea multipunctata.* Left anal claw in lateral view (118), and right one in median view (119). Membrane dotted. Nielsen.

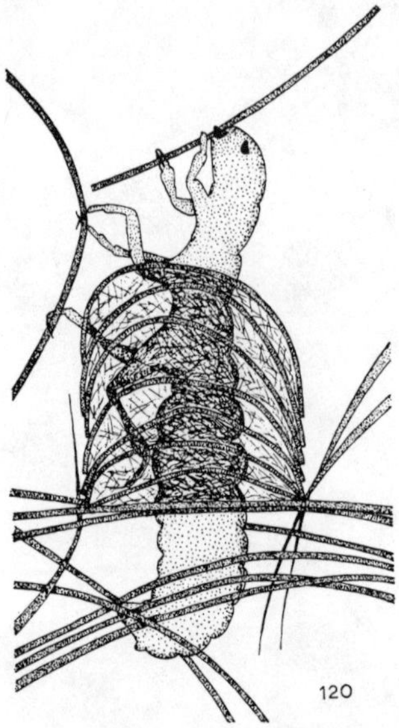

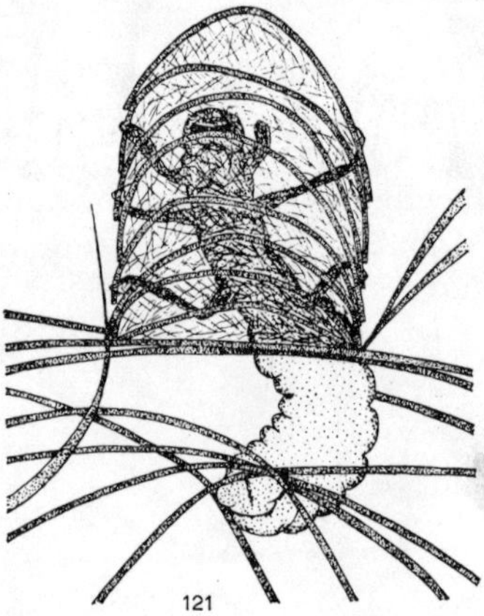

Figs. 120–121. *Agraylea* larva at work. 120. 'Chewing' an algal fragment; 121. Spinning the internal lining. Sketches drawn after photographs. Nielsen.

On each side near the anterior margin occurs a tuft of hair-like spines and, ventrally, a tuft of much thinner ones more developed on the right side and continuing into a longitudinal stripe of long, thin, inwardly directed hairs. In addition, there is a small tuft of slender stiff spinules under the process of the lateral margin. The characteristic mandibles are as shown in Figs. 98 to 103.

The central part of the inner side of the right mandible is membraneous and bladder-like. The maxillae and labium are as shown in Figs. 104 and 105. The legs are short, the anterior being characterized by the very much expanded trochanter and femur, the shorter, proximal part of the inner side of the femur forming a direct continuation of the inner side of the trochanter (Fig. 107). A number of characteristic cuticular processes occur on the anterior legs, which are shown in Figs. 110 and 111. During the earlier instars the shape of the legs changes gradually to the shape and complexity shown. The middle and posterior legs are almost equal in length but longer than the anterior in the ratio of 6:10.

The abdominal segments are smooth except that there are some scattered, backwardly directed spinules on the middle of the ninth venter.

There is an abrupt change in appearance and comparative dimensions of the ninth segment and anal prolegs from fourth to fifth instars. In the young larva, the basal segment of the anal proleg is short and cylindrical, whilst the claw is long and slender and is longer than the basal segment.

The claw is evenly curved and tapers to a fine point (Figs. 113, 114). In the fifth instar the pair of anal prolegs together are hemispherical. The claws are much shorter than in the younger larva (but it should be noted that the claw is still much longer and more slender than is the case in other genera of this family (Figs. 118, 119).

The claw consists of a straight basal part with a strongly recurved apex, and at the point of commencement of the curvature on the outer side there is a small bulge.

Gills are absent in this species.

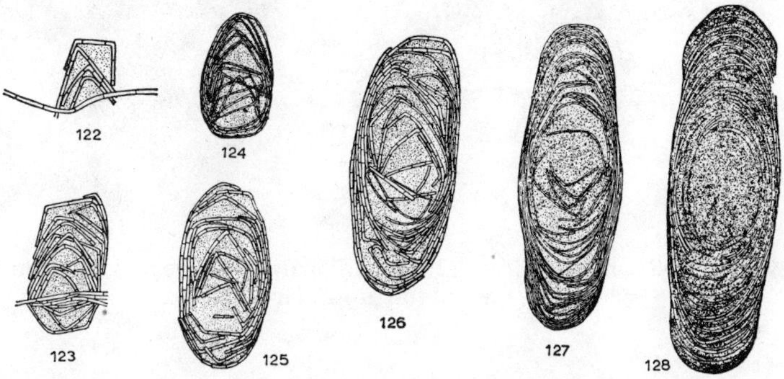

Figs. 122–128. *Agraylea multipunctata.* 122. Early stage of construction of larval case; only one side is represented. 123–128. Development of the case.

The case of *Agraylea multipunctata* consists of two symmetrical halves, very much sagitally compressed and with rounded dorsal and ventral edges. At each end there is a slightly gaping slit-shaped opening bordered by nearly semicircular valves (Fig. 128). The foundation of the case is laid by the larva spinning together a number of algal threads to form a roomy frame around its body. The larva then cuts off pieces of algal filament and spins them on to the frame, bending them around into arch-like shapes until a pair of valves have been built loosely attached together to the surrounding algal filaments with silk threads. The finished case resembles that of *Hydroptila* but the rather open framework of algal filaments takes the place of sand-grains. The case is enlarged by the larva slitting open both dorsal and ventral edges and adding algal filaments along the margins. These latter are regularly and neatly placed.

In the final stages of case-construction, silk is spun on to its inner side and is especially thick along the edges. The algal filaments are worn off, leaving fine concentric depressions around the perimeter of the sides, the central parts are smooth, corresponding to the areas of less firmly attached algal threads, which are rapidly worn off.

The case attains a length of 7.6 mm.

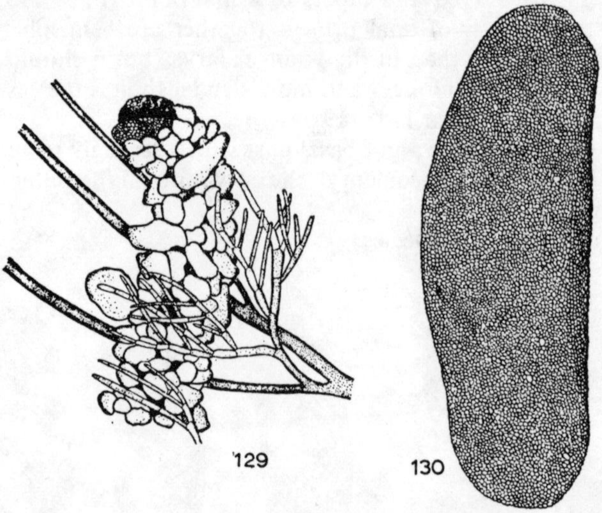

Figs. 129–130. *Hydroptila tineoides*. 129. Provisional larval abode with rudiments of valves. 130. Case of full-grown larva. Nielsen.

Agraylea sexmaculata Curtis

This very local species which appears as *pallidula* McLachlan in Kloet & Hincks (1964) and Mosely, has been collected from a lake in Hampshire and from the River Lamborne in Berkshire, and also from Wicken Fen. A description based on British materials is not available. The larva may be separated from that of *A. multipunctata* by the transverse row of four dark spots on the dorsal surface of the head, isolated clear spots distinct, thoracic segments yellowish, two dark brown spots on each side of the posterior part of the pronotum. On the posterior part of the meso- and metanotum a dark brown mark, and on the anterior margin on each side two brown marks which surround the bases of the long hairs.

Allotrichia pallicornis (Eaton)

Immature stages unknown.

Hydroptila angulata Mosely

Immature stages unknown.

Hydroptila cornuta Mosely

Immature stages unknown.

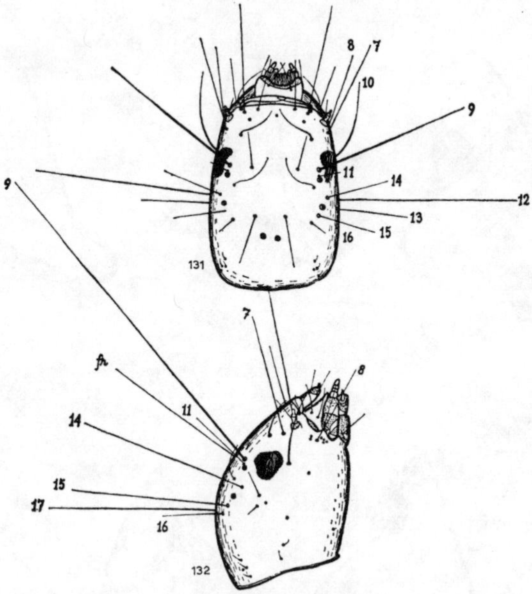

Figs. 131–132. *Hydroptila tineoides*. Head in dorsal (131) and lateral (right) view (132). *fr*: intermediate latero-marginal seta on fronto-clypeus. Membranes dotted. Nielsen.

Hydroptila tineoides Dalman

This widely distributed and abundant species which appears as *femoralis* Eaton in Kloet & Hincks (1964) and Mosely is found in lakes as well as in running water. No description based on British material is available and the following account is based on Nielsen's very detailed study.

The young larva much resembles that of *Agraylea* but is easily distinguished by the presence of caudal tracheal gills. The length of the fifth

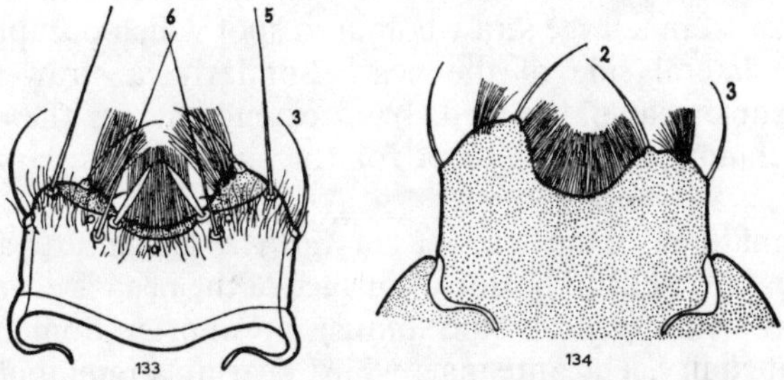

Figs. 133–134. *Hydroptila tineoides*. Labrum in dorsal (133) and ventral view (134; +anteclypeus p.p.). Membranes dotted. Nielsen.

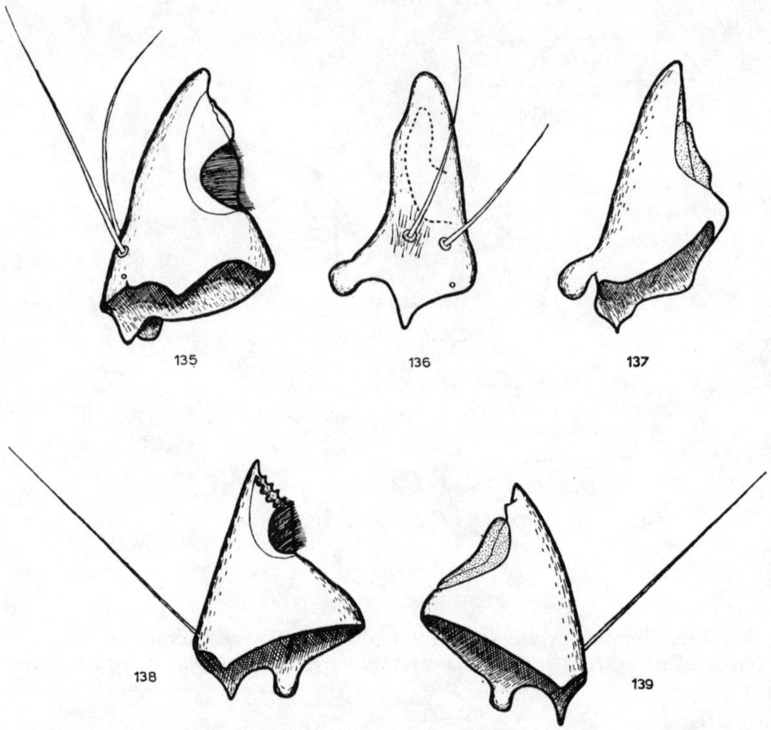

Figs. 135–139. *Hydroptila tineoides.* Left mandible in dorsal (135) and lateral view (136); right one, half in ventral, and half in median view (137; note the edges). 138, 139. *Hydroptila* sp.; 4th instar; left (138) and right mandible (139) seen from the dorsal side and somewhat from the median side. Membrane dotted. Nielsen.

instar larva varies from 1.8 to 3.7 mm. and is similar to the larvae of other species in this genus in being strongly pigmented. The head, thoracic nota and pleura, and the legs, are of a very dark brown colour. The forelegs are, however, a little lighter and the ventral side of the head much lighter. The eyes are situated in a large straw-coloured spot which occupies the greater part of the lateral side of the head. Similarly, a straw-coloured, but frequently somewhat obliterated, band extends across the dorsal side of the head behind the eyes. Except for the sclerites, the larva is green in colour.

The general shape of the head is as in *Agraylea* and sutures are apparently lacking (Figs. 131, 132). The dorsal surface of the head is covered with very long spinules whilst ventrally the spinules are shorter. Sometimes the head appears quite hairy. The antennae, whilst bearing a resemblance to that of *Agraylea*, has the distal joint relatively longer, and at the base of the sensilla there are a few slender, stiff spinules. The labrum (Figs. 133, 134) is

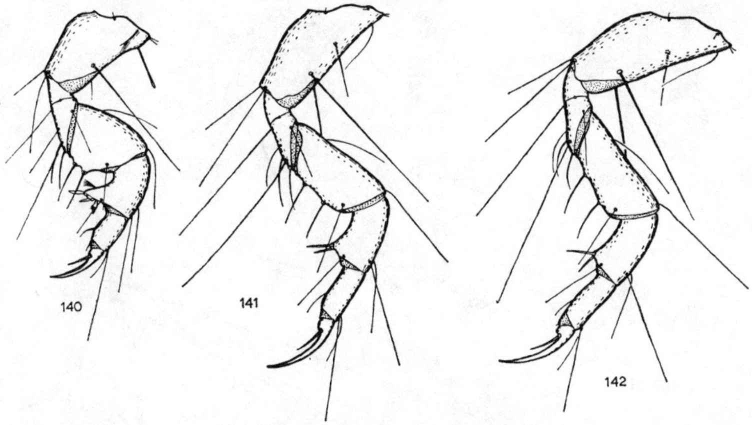

Figs. 140–142. *Hydroptila tineoides.* Right fore (140), middle (141), and hind leg (142) in posterior view. On the middle and the hind leg the proximal front seta is showing through. Membranes dotted. Nielsen.

larger and longer than in *Agraylea* and the lateral margins are almost straight, the anterior margin being very deeply indented. The lobes formed by the indentation are tongue-shaped and the right one is the largest.

The mandibles (Figs. 135–139) again bear a close resemblance to those of *Agraylea* but in the left mandible the indentation at the base of the superior edge is large and parabolic and teeth are practically absent. Both edges of the right mandible are straight. The maxillae and labium resemble those of *Agraylea* but the dorsal side of the stipes is covered with very small but comparatively stout spinules.

When seen from above, the pro- and mesothorax taper strongly posteriorly whilst the metathorax is parallel-sided. Posteriorly, on the pronotum there is a distinct transverse groove. There are four long black bristles on each side of the anterior margin of the pronotum (three in *Agraylea*). The legs also are like those of *Agraylea* except in the following respects. The middle and hind legs are shorter and more robust whilst in the anterior legs the femur is not quite so broad. There are some differences in the cuticular processes of the tibial projection.

The anal prolegs of the young larva are rather shorter than those of *Agraylea*. In the fifth instar larva the claw (Fig. 145) is shorter and thicker than in *Agraylea* and the apex is very much more recurved and the claw bears three dorsal hooks, the median one being rather more slender. Three caudal tracheal gills are present.

The case (Figs. 129, 130) consists of two halves each formed by a single layer of fine sand-grains lined internally with silk. The dorsal and ventral edges are spun together whilst at each end there is a slit bounded by semi-circular valves. The case is enlarged by the larva slitting it open along the

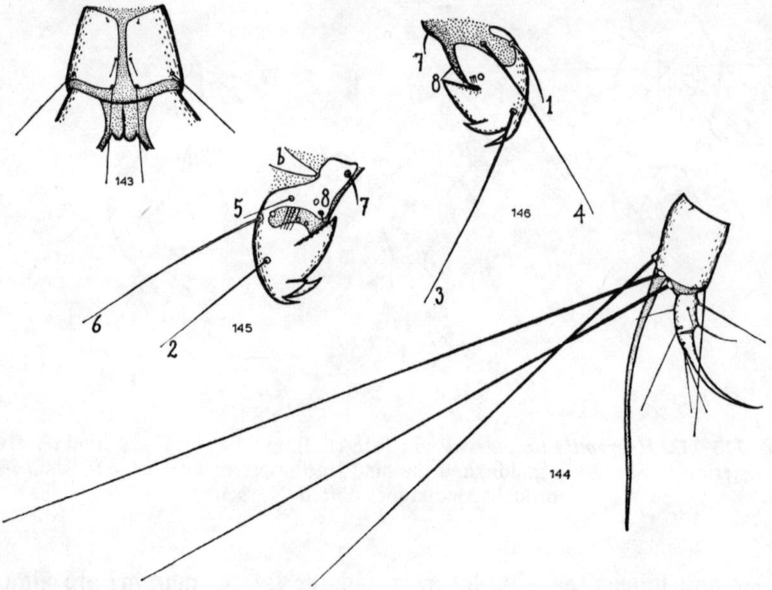

Figs. 143–146. 143, 144. *Hydroptila* sp.: 4th instar; 143. Segment IX and base of anal prolegs in ventral view; 144. Right anal proleg in lateral view; 145, 146. *Hydroptila tineoides*; 5th instar; 145. Right anal claw, half in lateral and half in ventral view; 146. Do. in median view. Membranes dotted. Nielsen.

ventral edge, adding sand-grains, then spinning the two halves together again. At a later stage enlargement takes place entirely at the slit-like ends. The case attains a length of 4.8 mm. and a width of 2.4 mm.

Hydroptila forcipata (Eaton)

Immature stages unknown.

Hydroptila lotensis Mosely

In Britain, this species is known only from the River Wye, Herefordshire. Immature stages unknown.

Hydroptila vectis Curtis

Mosely states that this species is very widely distributed and abundant. No description based on British material is available and the descriptions by Ulmer and Silfvenius are scarcely adequate. This species appears in Kloet & Hincks (1964) as *mclachlani* and Mosely as *maclachlani* Klapálek.

Hydroptila occulta (Eaton)

Immature stages unknown.

Hydroptila pulchricornis Pictet

This local species frequents lakes. No description based on British material is available and the descriptions made by continental authors have not been sufficiently detailed. The author of this species is given in Kloet & Hincks (1964) as Eaton.

Hydroptila simulans Mosely

Immature stages unknown.

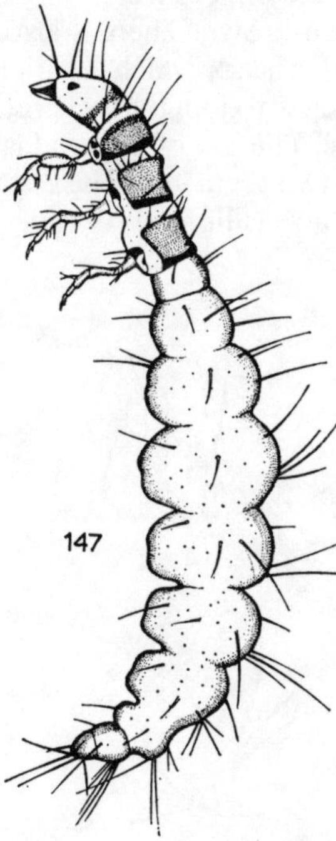
147

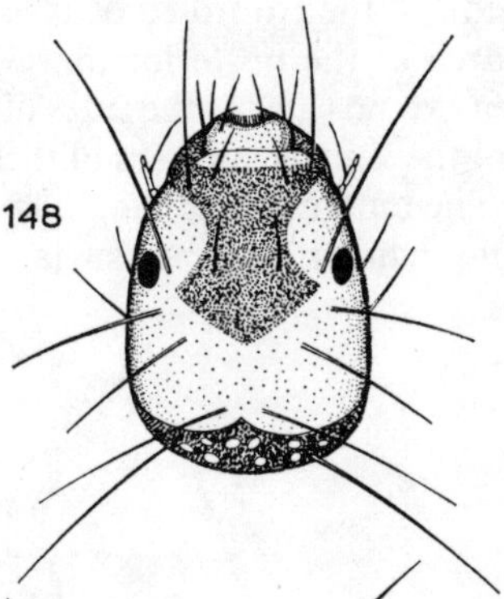
148

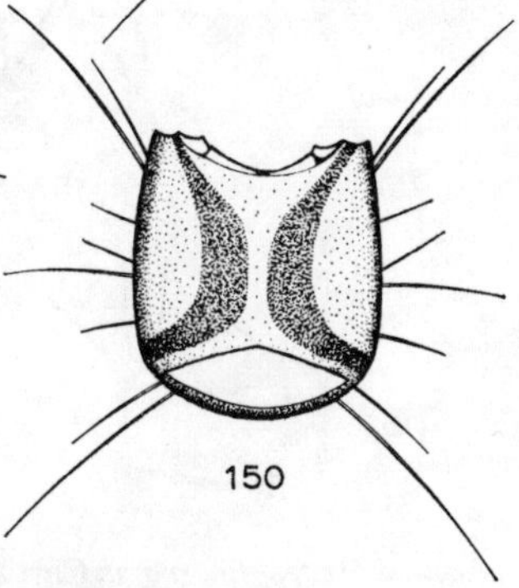
150

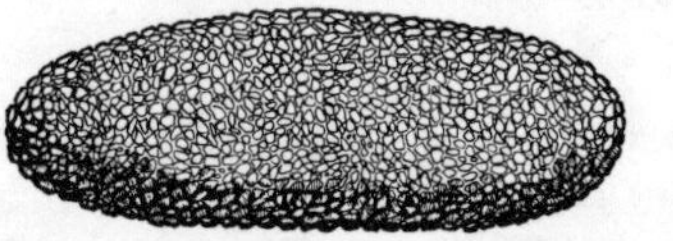
149

Fig. 147. Larva of *Hydroptila sparsa* Curtis (side view).
Fig. 148. Larva of *Hydroptila sparsa* Curtis. Head (dorsal view).
Figs. 149–150. Larva of *Hydroptila sparsa* Curtis. 149. Larval case; 150. Head (ventral view). Hanna.

Hydroptila sparsa Curtis

Hanna gave the following description of the fifth instar larva from specimens collected from a brook near Reading, Berkshire.

Case. The cases (Fig. 149) are up to 4.5 mm. long and 1.6 mm. wide and are made of very fine sand-grains. The dorsal surface of the case is convex, whereas the ventral surface is almost flat. In some of the cases the dorsal surface has a small hump in the middle. The case is sagittally compressed and is composed of two symmetrical halves. There is a slit at each end and each slit is surrounded by a semicircular valve.

Larvae. The larva is campodeiform (Fig. 147). The larvae are 3.6 mm. long and 0.7 mm. wide. The thorax and the abdomen are compressed laterally.

Head. The head is prognathous and has no distinct sutures (Figs. 148, 150). The region of the fronto-clypeus is greyish-brown. There is also a greyish-brown area at the posterior margin of the head, which has lighter spots. This area extends across the sides of the genae to join the two greyish-brown bands on the ventral surfaces of the genae. The rest of the head is creamy in colour. The antennae are long and have two segments. The distal segment has a long hair on its inner surface and a sensillum at its end.

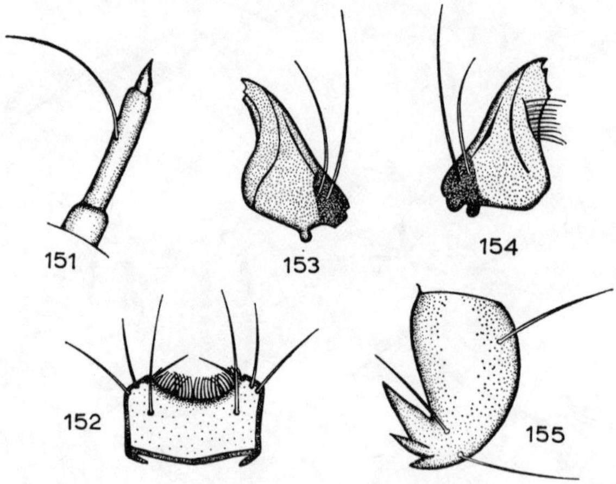

Figs. 151–155. Larva of *Hydroptila sparsa* Curtis. 151. Antenna; 152. Labrum; 153. Right mandible; 154. Left mandible; 155. Anal claw. Hanna.

Labrum. The labrum (Fig. 152) is pale yellow in colour. The anterior margin has a large concavity, which has a hairy brush. On each side of the concavity there are four long setae. The edges of the labrum are highly sclerotized.

Mandibles. The mandibles (Figs. 153, 154) are asymmetrical. The right mandible has three teeth and has no brush on its inner surface. The left mandible has four teeth and there is a hairy brush on its inner surface. Both mandibles have two long setae on their outer surfaces near the base.

Maxilla. The cardo is small and has a single seta. The distal margin of the stipes has two setae. The maxillary palp has five segments, of which the second segment has a few hairs. The lacinia has a few sensilla and a group of hairs.

Thorax. The pronotum is entirely sclerotized, has a median longitudinal suture and a transverse greyish-brown band on its anterior third. The posterolateral and posterior margins of the pronotum are dark brown. There is a greyish-brown band on both sides of the posterior two-thirds of the median suture. The anterior margin of the pronotum is lighter in colour than the rest of the pronotum. The mesonotum and the metanotum are sclerotized except for their margins, have median longitudinal sutures and are lighter in colour than the pronotum. Their posterior margins are dark brown and the anterolateral corners have a dark-brown spot and an oblique greyish-brown band.

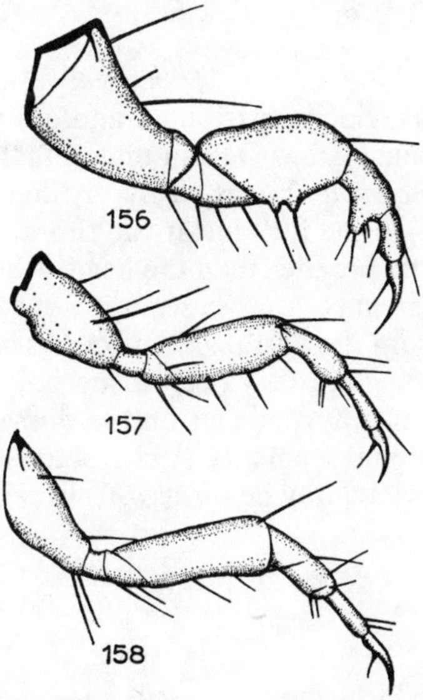

Figs. 156–158. Larva of *Hydroptila sparsa* Curtis. 156. Prothoracic leg; 157. Mesothoracic leg; 158. Metathoracic leg. Hanna.

Legs (Figs. 156–158). The prothoracic leg is slightly shorter than the mesothoracic and metathoracic legs, which are equal in length. The femur of the prothoracic leg is noticeably flattened and has two projections on which there are two thick setae. The tibiae of all the legs have two spurs and those on the prothoracic tibia are situated on a long projection.

Abdomen. The first abdominal segment is small and has no protuberances. The fourth abdominal segment is the largest, after which the segments become smaller towards the posterior end. The lateral line, the abdominal gills and the anal sclerite are absent. The eighth tergum has six long setae and a few short scattered setae at its posterior margin, while the ninth tergum has four long setae, four medium setae and two short setae. The anal claw (Fig. 155) has two auxiliary claws at its base.

Hydroptila sylvestris Morton

Immature stages unknown.

Hydroptila tigurina Ris.

Immature stages unknown.

Ithytrichia clavata Morton

Immature stages unknown.

Ithytrichia lamellaris Eaton

This description of this widely distributed and abundant species is taken from Nielsen's study made from Danish material. The young larva much resembles that of *Orthotrichia* but is easily distinguished by possessing a single caudal gill only. In the fifth instar the thoracic and first abdominal segments are only a little broader than the head. Abdominal segments one and nine are almost circular in cross-section but the other segments are compressed. Viewed from the dorsal side strictures can hardly be seen, but viewed from the side they are very deep. Especially characteristic are the unpaired, egg-shaped, soft appendages on the dorsal and ventral surfaces of several of the abdominal segments. A characteristic of this larva also is that the thorax may very readily be turned 90° in relation to the abdomen.

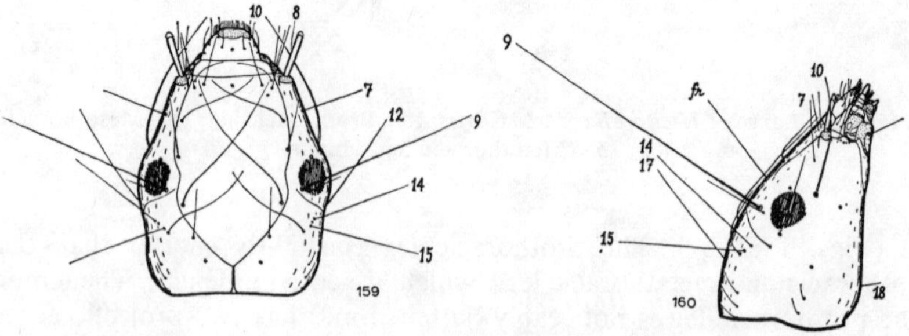

Figs. 159–160. *Ithytrichia lamellaris.* Head in dorsal (159) and lateral (right) view (160). *fr*: intermediate lateromarginal seta of fronto-clypeus. Membranes dotted. Nielsen.

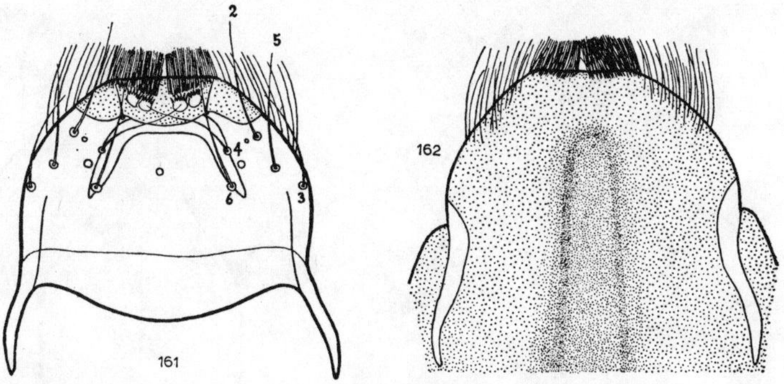

Figs. 161–162. *Ithytrichia lamellaris.* Labrum in dorsal (161) and ventral view (162; +anteclypeus p.p.). Membranes dotted. Nielsen.

The thorax always occupies this position when the larva crawls about and when at rest also the thorax is most frequently somewhat turned. The length at the fifth instar varies from 1.05 to 3.0 mm. The sclerites are brownish yellow with the mandibles a glassy light-brown in colour. When living, the membraneous parts are yellowish.

At the fifth instar the head is moderately broad and rather flattened. Viewed dorsally it is broadest in the middle where there is a prominent and characteristic bulge in the eye region (Fig. 159). Viewed from the side (Fig. 160) the head tapers strongly. The frontal suture is very short and the fronto-clypeus accordingly very long. The anterior corners of the latter are almost right-angled. The gula is entirely undivided and the anterior margin is so concave that it becomes V-shaped, with the arms extending forward on each side of the submental sclerite (Fig. 167).

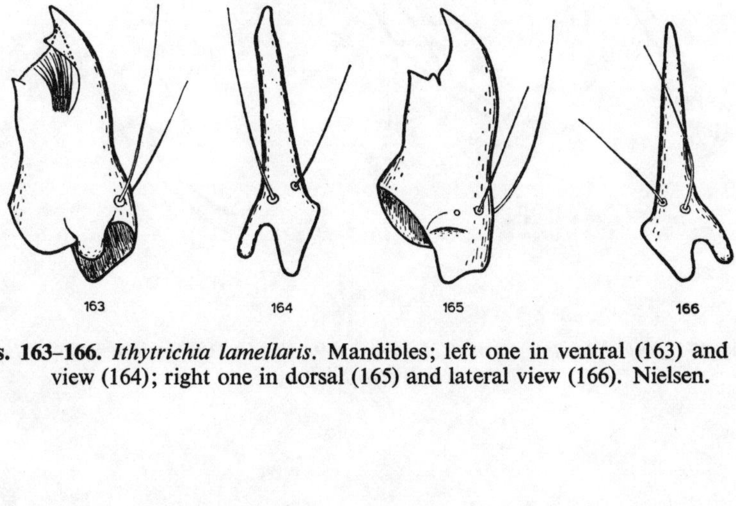

Figs. 163–166. *Ithytrichia lamellaris.* Mandibles; left one in ventral (163) and lateral view (164); right one in dorsal (165) and lateral view (166). Nielsen.

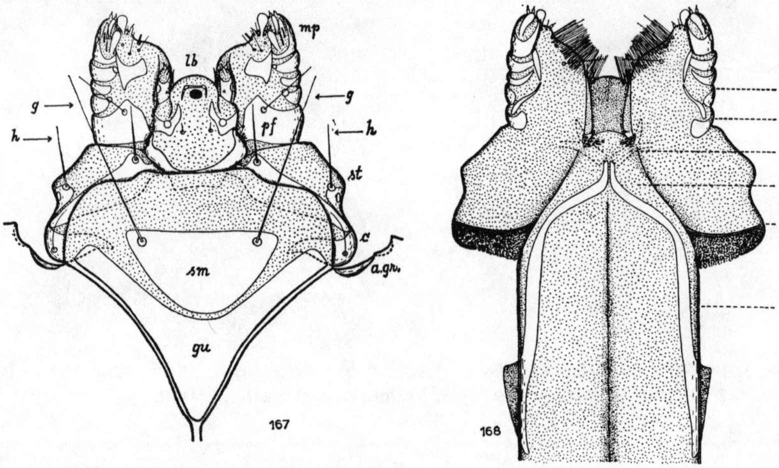

Figs. 167–168. *Ithytrichia lamellaris*. Maxillolabium in ventral (167; +gula) and dorsal view (168; +ventral wall of pharynx). *a.gr*: 'articular groove' for cardo, *c*: cardo, *gu*: gula, *lb*: labial lobe, *mp*: maxillary palp, *pf*: palpifer, *sm*: submental sclerite, *st*: stipes. Membranes dotted. Nielsen.

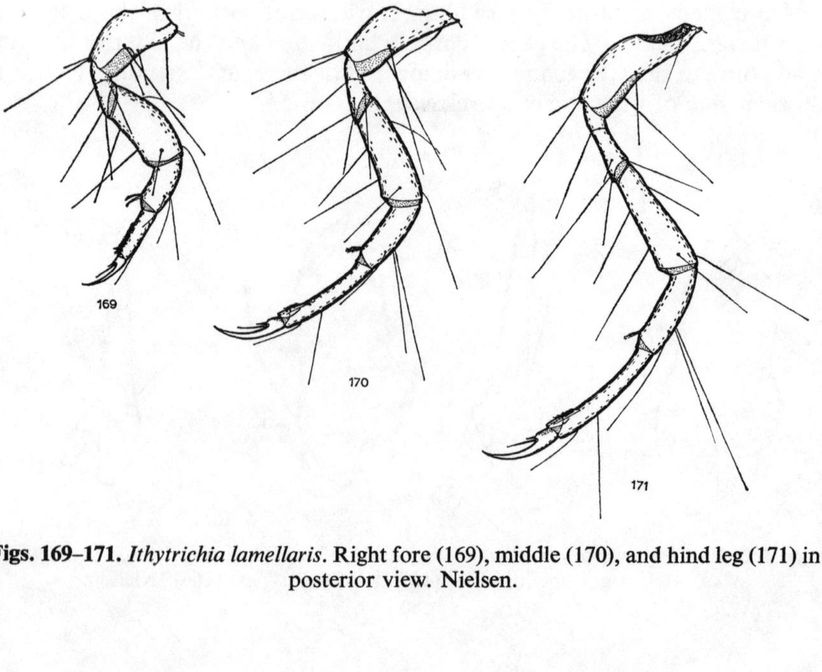

Figs. 169–171. *Ithytrichia lamellaris*. Right fore (169), middle (170), and hind leg (171) in posterior view. Nielsen.

Labrum (Figs. 161–162). This is rather long and narrow and broadest posteriorly. Anteriorly, it narrows with curved sides to a short, abruptly cut-off anterior margin. The anterior margin of the sclerotized part shows three indentations, and an inversely U-shaped internal thickening is noticeable anteriorly. On the ventral side there is a broad longitudinal groove. On the dorsal side, in the membraneous area, there are on each side a pair of colourless plates each equipped with about six forward-directed spinules.

Mandibles (Figs. 163–166). Generally resemble those of *Orthotrichia* but are even more flattened. On the left mandible the short inferior edge is very concave, its posterior corner projecting like a pointed tooth; the serrated superior edge is bounded proximally by a smaller conical tooth. The inferior edge of the right mandible is lacking. Again the maxillolabium

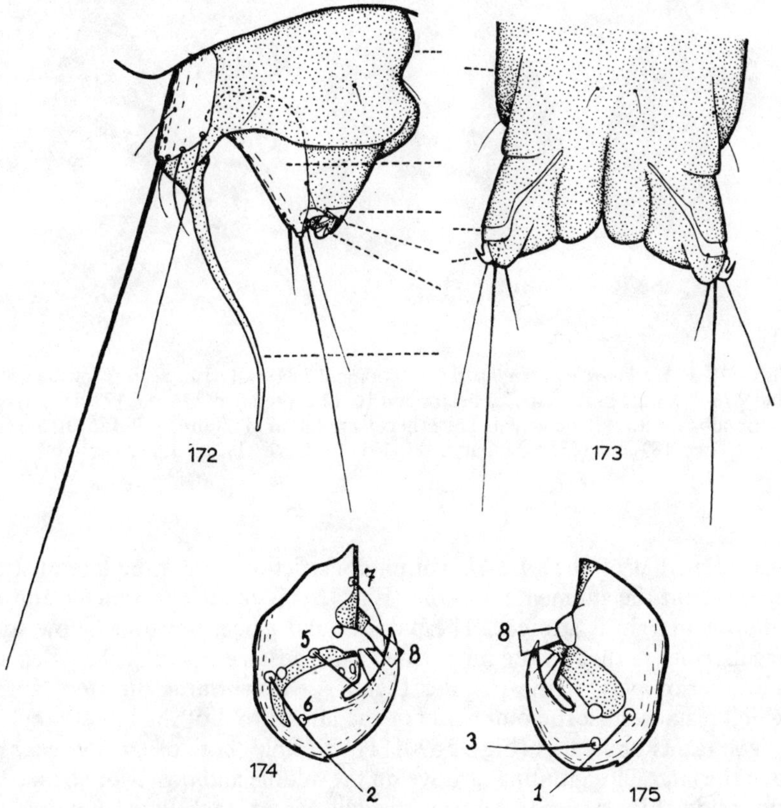

Figs. 172–175. *Ithytrichia lamellaris*. Segment IX and anal prolegs in lateral (172; right) and ventral view (173). Right anal claw in lateral (174) and median view (175). Membranes dotted. Nielsen.

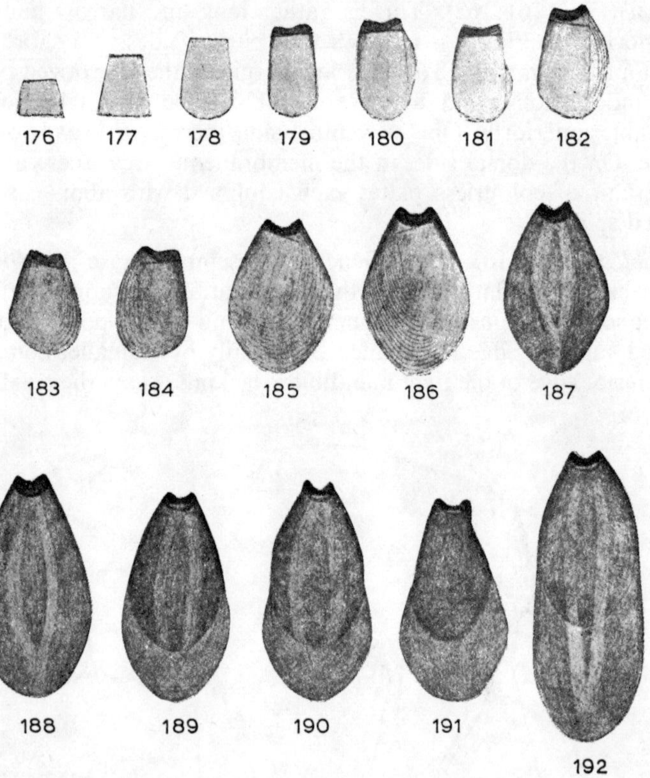

Figs. 176–192. *Ithytrichia lamellaris*. Development of larval case. Seen from the underside. 176–178 were built in the laboratory on 10–12 September; 176 and 177 still attached to the substratum, 178 detached. The others were found in nature; 179–186: 10 September, 187 and 188: 5 January; 189–191: 29 April; 192: 1 June. Nielsen.

resembles that of *Orthotrichia* with a constriction across the lateral stipes sclerite, but the submental sclerite (Fig. 167) is relatively broader and the anterior margin is straight. The palpifer and galea, however, show some peculiarities in that on the inner side of the posterior part of the galea and the anterior part of the palpifer there is a downwards directed lamella which rests against the outer side of the lateral fold of the labial lobe.

The rather short legs (Figs. 169–171) resemble those of *Orthotrichia* but lack the lateral longitudinal groove on the middle and posterior coxae. The inner side of the anterior femur is straight or very slightly concave and the claws are less robust.

In the fifth instar larva the second abdominal segment is furnished with lateral humps and as in *Orthotrichia*, segment nine projects dorsally over the anal prolegs. The dorsal abdominal appendages, already referred to, are developed before the ventral ones. The claw (Figs. 174–175) of the anal prolegs in the fifth instar is very small, with the apex curved much more to the front. There is a single rudimentary dorsal hook. Paired gills are absent but a single gill is situated at the posterior margin of the ninth abdominal dorsum; it hangs down between the distal ends of the anal prolegs. There are six short conical and pointed anal gills.

The case of the full-grown larva (Fig. 192) is strongly sagitally compressed, and its shape has been described as like a pumpkin seed. It is built entirely of silk. At the anterior end there is a hole approximately circular through which the larva stretches its thorax. At the posterior end there is a broad slit bordered by parabolic valves, the ends of the slit lying almost at the broadest point of the case. The slit is open only in the centre. As already mentioned the larva carries its case at right-angles to the line of the thorax, which produces a certain asymmetry in the two sides of the case. On the under-side the anterior margin is deeply indented whilst the upper-side is only slightly indented and projects beyond the under-side.

The length of the case attains 3.65 mm.

Orthotrichia angustella (McLachlan)

Immature stages unknown.

Orthotrichia costalis (Curtis)

This widely distributed species listed in Kloet & Hincks (1964) as *tetensii* Kolbe and in Mosely likewise inhabits ponds and lakes but a description based on British material was not available. The following description is based on Nielsen's work. The young larva is robust in appearance and the three long caudal tracheal gills, which are silvery in colour in live specimens, are characteristic. The whole of the body excepting only the ventral

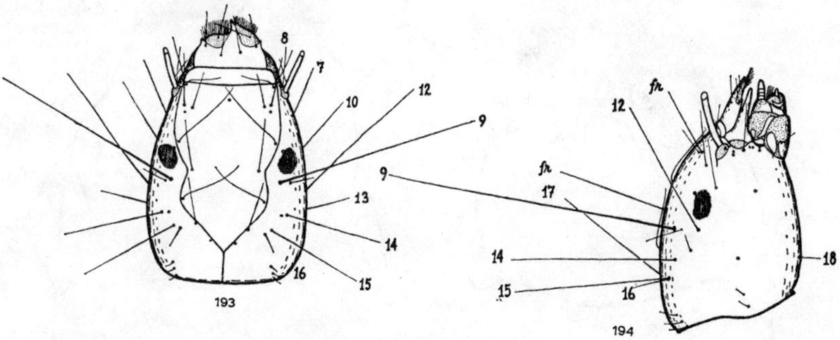

Figs. 193–194. *Orthotrichia costalis*. The head in dorsal (193) and lateral (right) view (194). *fr*: setæ of fronto-clypeus. Membranes dotted. Nielsen.

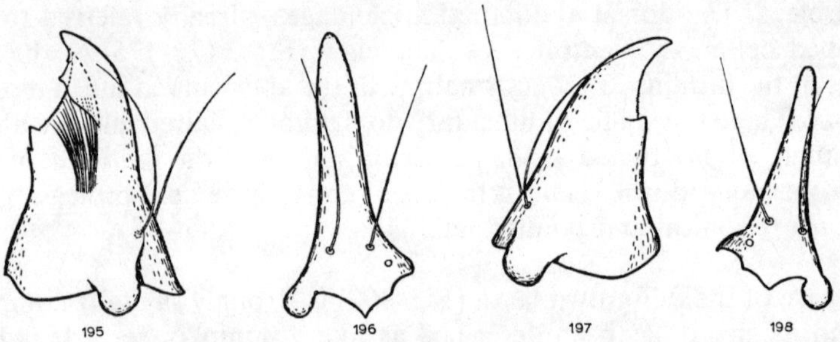

Figs. 195–198. *Orthotrichia costalis*. Mandibles. Left one in ventral (195) and lateral view (196); right one in ventral (197) and lateral view (198). Nielsen.

surface of the head, the tarsal and anal claws, is clothed with hair-like spinules. They are almost of equal length in all instars. They give the first instar larva a shaggy appearance but are not very conspicuous in the fourth instar. When newly hatched the prothorax is as broad as the head whilst the meso- and metathorax are narrower than the head. The first abdominal segment is appreciably narrower than the metathorax.

In the fifth instar larva the prothorax is a little broader than the head whence the breadth increases gradually but only slightly to the first abdominal segment. The latter segment is somewhat overlapped by the succeeding one and in lateral view is partially concealed by the large, forward-directed lateral humps of this segment. Segments three to eight are together barrel-shaped and are only separated by faint strictures, but segment eight is separated by a well-marked stricture from segment seven. In contrast to the other genera the abdomen is not compressed but fairly

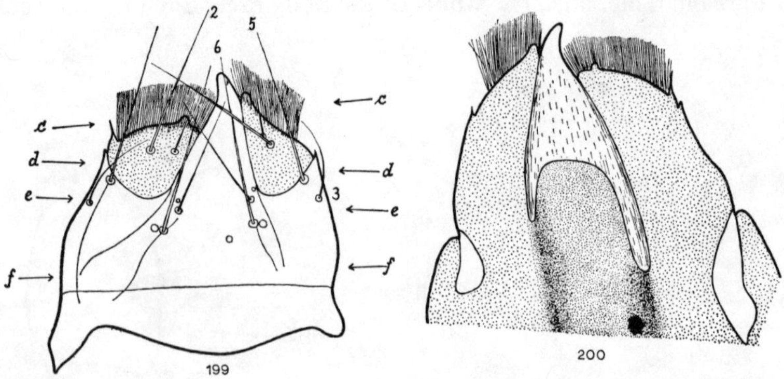

Figs. 199–200. *Orthotrichia costalis*. Labrum in dorsal (199) and ventral view (200). Nielsen.

well flattened. The length of the fifth instar varies from 1.1 to 2.65 mm. The sclerites are generally brownish-yellow. The lateral and posterior margins of the pronotum and the posterior margin of the mesonotum have a narrow brown band. The posterior one-third of the lateral margin of the meso- and metanota, and the whole posterior margin of the latter, have a broad black belt. The anterior corners of the meso- and metanota are dark

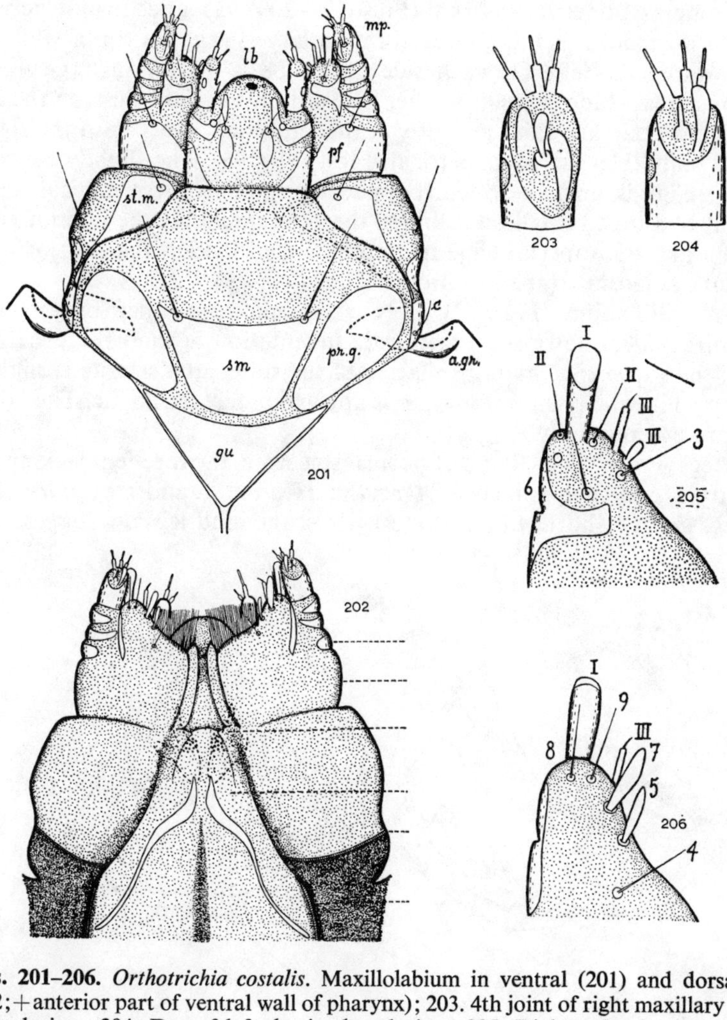

Figs. 201–206. *Orthotrichia costalis.* Maxillolabium in ventral (201) and dorsal view (202; + anterior part of ventral wall of pharynx); 203. 4th joint of right maxillary palp in ventral view; 204. Do. of left do. in dorsal view; 205. Right galea in ventral view; 206. Left do. in dorsal view. *a.gr*: 'articular groove' for cardo, *c*: cardo, *gu*: postgula, *lb*: labial lobe, *mp*: maxillary palp, *pf*: palpifer, *pr.g*: pregula, *sm*: submental sclerite, *st.m*: median stipes-sclerite. Membranes dotted. Nielsen.

brown. The head is not so narrow nor is it nearly so deep as in *Agraylea*, *Hydroptila* and *Oxyethira*. It is widest at about the aboral one-third, strongly narrowing anteriorly (Fig. 193). The fronto-clypeus is fairly long with the anterior corners almost right-angled. The postgula is triangular, rounded aborally and with a concave anterior margin (Fig. 201). The pregula is completely separated from the postgula and forms a pair of elongated egg-shaped sclerites situated on each side of the submental sclerite.

The long asymmetrical labrum (Figs. 199–200) has a peculiar structure, the anterior margin being produced into a large beak, the tip of which is bent towards the left. On each side of the beak there is a large membraneous area which projects farther forward on the right than on the left side. The tormae are stout and broad, the left much larger than the right. The flat mandibles are asymmetrical (Figs. 195–198). The slightly concave superior edge is finely serrated, bounded proximally by a conical tooth adjacent to a much smaller tooth. On the right mandible the inferior edge is absent and the superior edge resembles that of the left but there is only one tooth adjacent to the serrations.

The maxillolabium (Figs. 201, 202), when seen from either the dorsal or the ventral side, shows a characteristic indentation around the proximal part of the stipes. The anterior margin of the submental sclerite is slightly concave whilst the posterior margin is approximately parabolic. The labial palps are exceptionally long.

The legs (Figs. 207–209) are of normal structure, the fore-legs lacking the extensive broadening shown in *Agraylea*, *Hydroptila* and *Oxyethira*. The fore-legs are the shortest and stoutest whilst the hind-legs are longest and

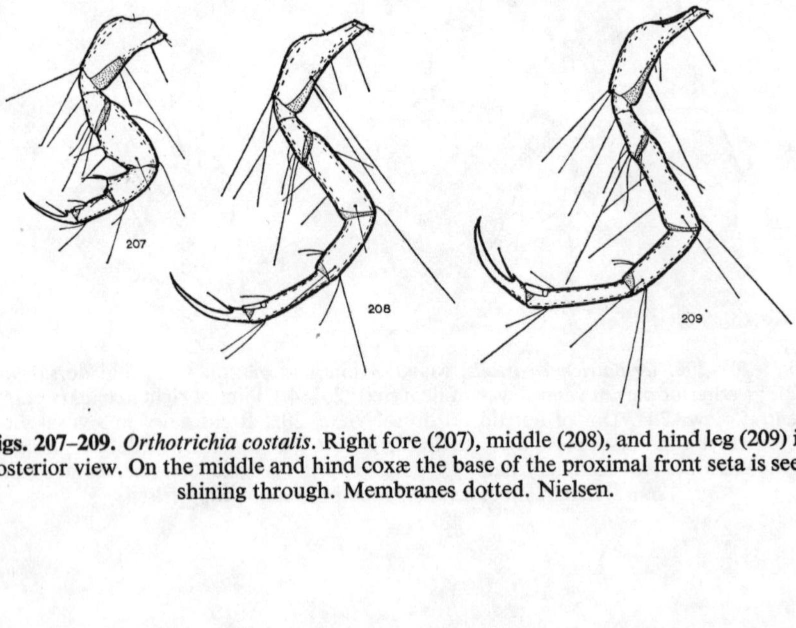

Figs. 207–209. *Orthotrichia costalis*. Right fore (207), middle (208), and hind leg (209) in posterior view. On the middle and hind coxæ the base of the proximal front seta is seen shining through. Membranes dotted. Nielsen.

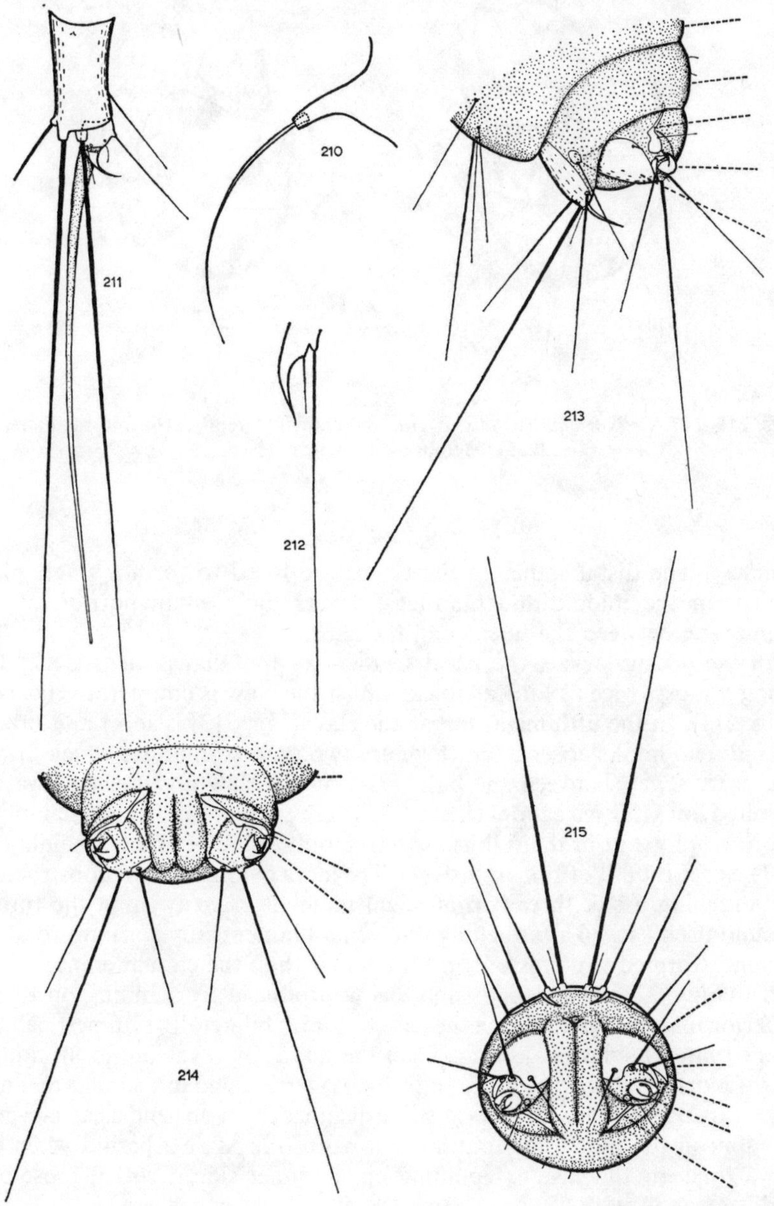

Figs. 210–215. *Orthotrichia costalis.* 210, 211. 4th instar; 210. Medioanal seta from dorsum V; 211. Right anal proleg in lateral view; 212. 1st instar; outline of right anal claw in lateral view; 213–215. 5th instar; segment IX and anal prolegs in lateral (right; 213), ventral (214), and posterior view (215). Membranes dotted. Nielsen.

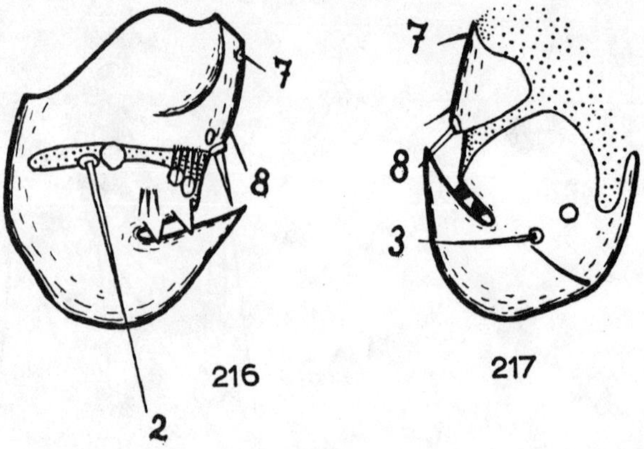

Figs. 216–217. *Orthotrichia costalis.* Right anal claw in lateral (216) and median view (217). Membrane dotted. Nielsen.

thinnest. The distal spines of the tarsus are fused to form a single plate which, in the middle and hind-legs, covers the median portion of the membrane between the tarsus and the claw.

In the young larvae, the basal segment of the anal prolegs is slender, being almost twice as long as thick, whilst the claw is comparatively short (Fig. 211). In the fifth instar larvae the claw (Fig. 213) is short and broad, with dorsal hooks absent, but there are two pointed nodules which bridge the narrow gap between the basal part of the claw and the backwardly directed apex. Three caudal tracheal gills are present but at the final moult they atrophy and in the full-grown larva only small remnants remain.

The case is built of silk exclusively. The larva commences its construction by attaching a silk thread to an algal filament, then twisting the thread around itself as well as attaching it to other filaments. By constant rotation a ring is formed which is only a little larger than the circumference of the larva (Figs. 223–224), lengthening being produced by spinning on to the anterior margin. After a time the case becomes bilaterally symmetrical, the arched lateral sides projecting. Then the larva, by reversing itself, adds a fairly short section at the other end which is terminated in a similar manner. The starting ring thus appears at some distance from one end as an elevated portion on the outer side, but is soon worn off. After a period when the larva thickens the case by spinning on the inner side it cuts it loose and walks away in it.

The finished case (Figs. 222, A, B and C) is a delicate, light-greyish, cylindrical tube. Dorsal and ventral indentations are observed at its ends while the sides are produced into flat curves. The larva then alters the shape, whereby a dorsal and ventral keel, as well as two keels on either side, are produced. The subsequent growth of the case takes place by the larva slitting it open along the ventral edge and spinning a narrow zone along the margin. It becomes so wide at the centre that the larva cannot be extracted.

The case attains a length of 3.65 mm.

Orthotrichia tragetti Mosely

Immature stages unknown.

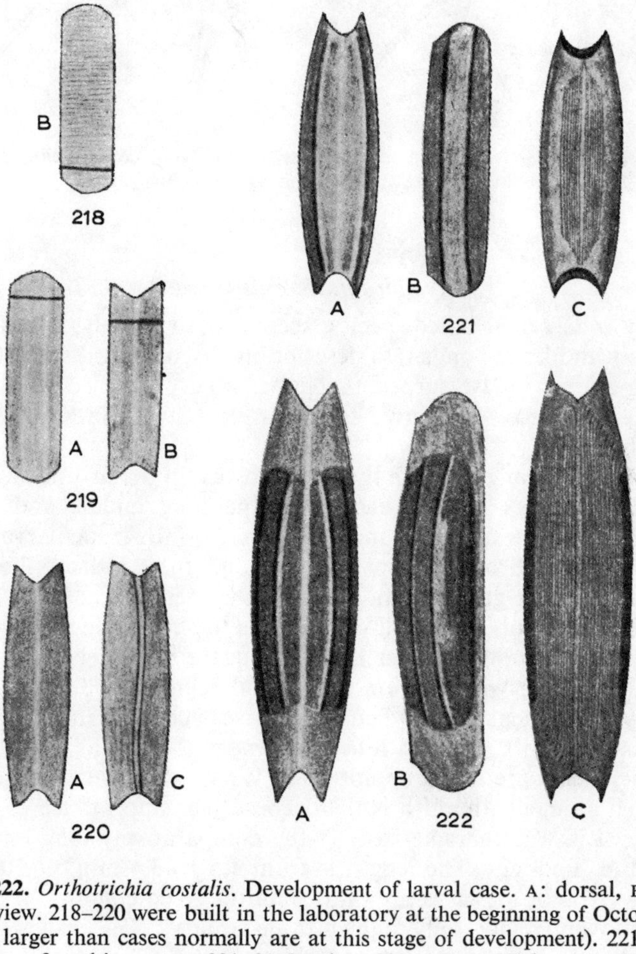

Figs. 218–222. *Orthotrichia costalis*. Development of larval case. A: dorsal, B: lateral, C: ventral view. 218–220 were built in the laboratory at the beginning of October; (220 very likely larger than cases normally are at this stage of development). 221 and 222 found in nature; 221. 21 October; 222. 5 June. Nielsen.

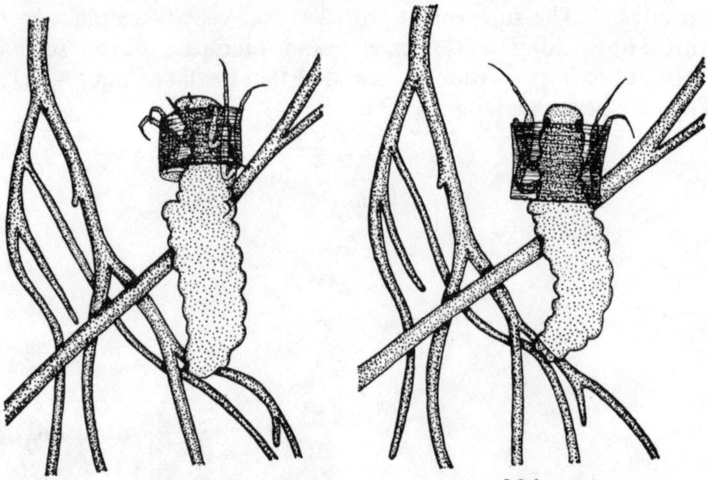

223 224

Figs. 223–224. *Orthotrichia* larva at work on the incipient construction of the case. Sketches drawn after photographs. Nielsen.

Oxyethira flavicornis (Pictet)

This very widely distributed species, according to Mosely, is abundant on most lakes and large ponds. No description based on British material was available and again we turned to Nielsen's very detailed study for this account. This species is referred to as *costalis* (Curtis) in Kloet & Hincks (1964) and Mosely.

The extremely long setae on the anal prolegs (in the first instar one, and in the later instars two on each) and the long middle and hind-legs especially characterize the young larva. In the fifth instar larva the prothorax is a little broader than the head and the width of the thoracic segments increases gradually posteriorly, the metathorax being 1.3 times as broad as the head. In the full-grown larva the first abdominal segment is appreciably broader than the metathorax but the second segment shows an abrupt increase in width. From the second segment the width increases gradually to segment seven when it decreases suddenly in segments eight and nine. Segments one and nine are almost as deep as broad, but the middle segments are much compressed. Whilst the seventh is the widest abdominal segment, the fifth is the deepest. The appearance of the fifth-instar larva is still characterized by the comparatively long and slender middle and hind-legs. The length is from 1.5 to 3.4 mm. and the membraneous parts of the larva are of a beautiful green colour.

The following relates only to the fifth-instar larva. The head is even more elongated and deeper than in *Agraylea* and *Hydroptila* and is a little deeper

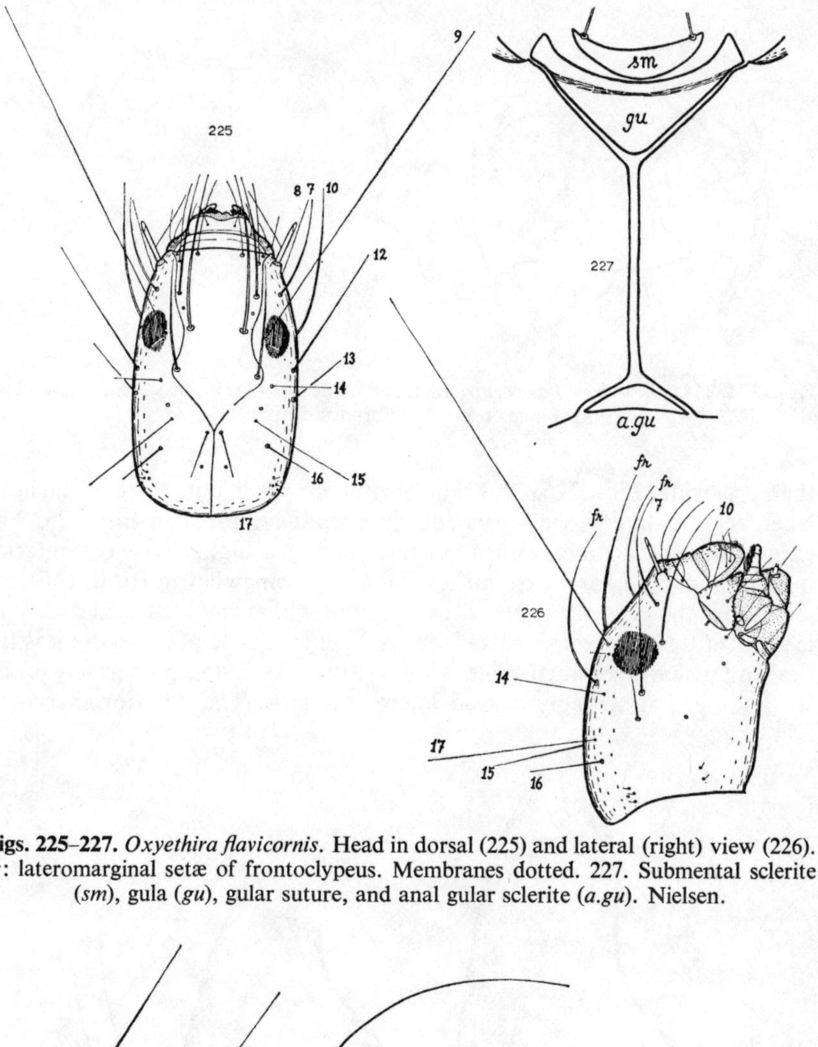

Figs. 225–227. *Oxyethira flavicornis*. Head in dorsal (225) and lateral (right) view (226). *fr*: lateromarginal setæ of frontoclypeus. Membranes dotted. 227. Submental sclerite (*sm*), gula (*gu*), gular suture, and anal gular sclerite (*a.gu*). Nielsen.

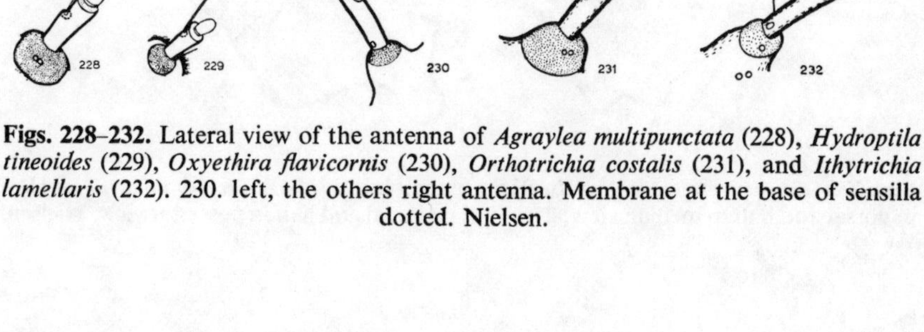

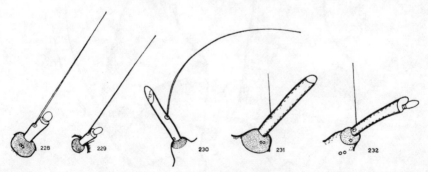

Figs. 228–232. Lateral view of the antenna of *Agraylea multipunctata* (228), *Hydroptila tineoides* (229), *Oxyethira flavicornis* (230), *Orthotrichia costalis* (231), and *Ithytrichia lamellaris* (232). 230. left, the others right antenna. Membrane at the base of sensilla dotted. Nielsen.

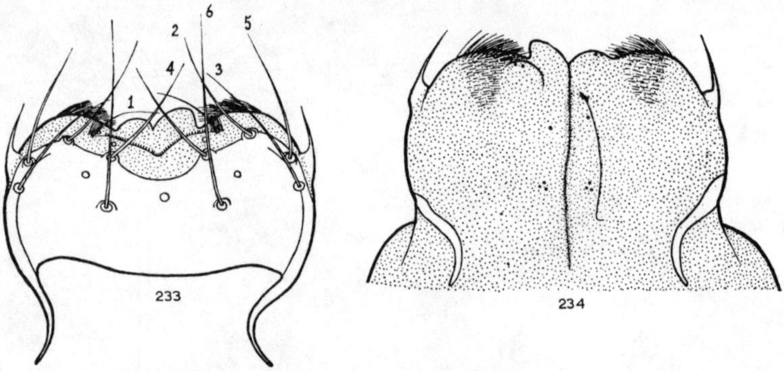

Figs. 233–234. *Oxyethira flavicornis.* Labrum in dorsal (233) and ventral view (234; +anteclypeus p.p.). Membranes dotted. Nielsen.

than it is wide (Figs. 225, 226). The abrupt change in curvature of the head when seen in lateral view particularly should be noted in Fig. 226. The gular suture is distinct with a pregula having a highly concave anterior margin and the lateral extremities extending somewhat in front. (Nielsen states that the postgula is broadly triangular with a rounded anal end. This is apparently incorrect; if this is the anal gular sclerite of his figure it is the oral end which is rounded.) The labrum (Figs. 233, 234) is moderately broad with somewhat strongly curved lateral margins. The anterior margin is

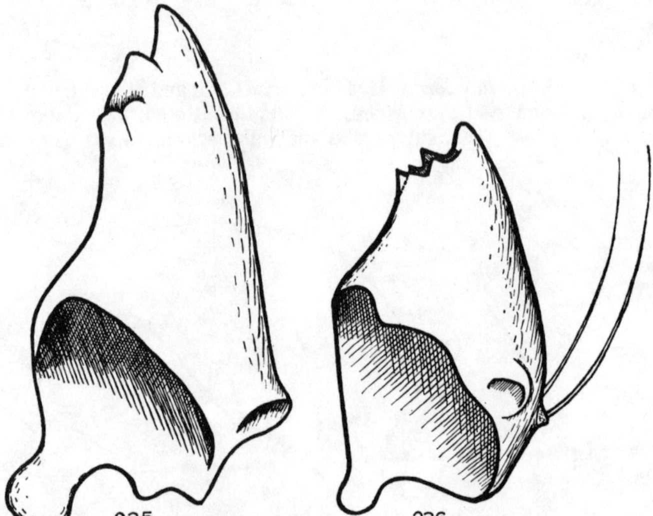

Figs. 235–236. *Oxyethira flavicornis.* Right mandible; 235. 5th; 236. 4th instar. 235. Half in dorsal and half in median view; 236. Half in dorsal and half in posterior view. Nielsen.

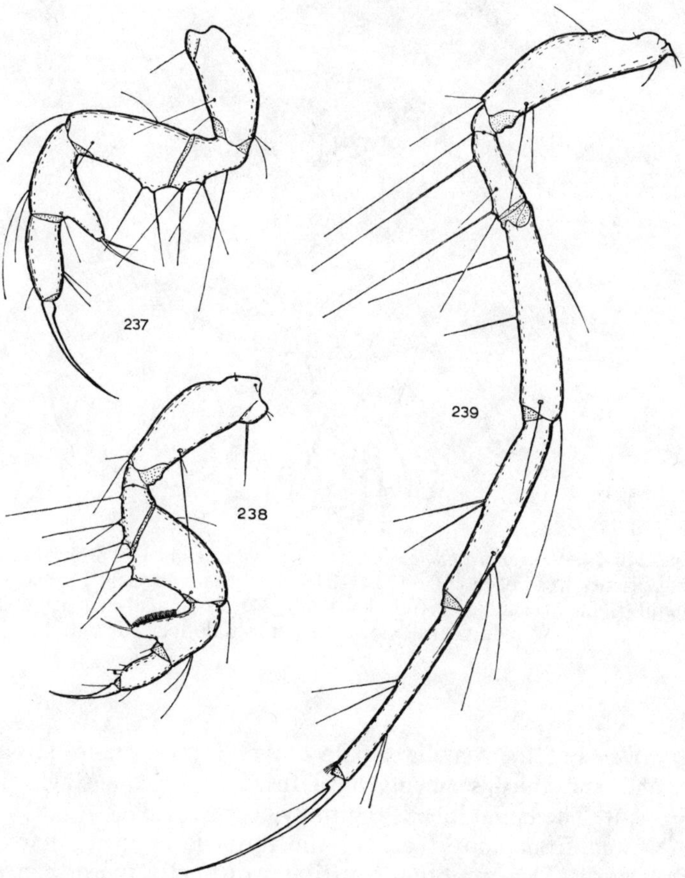

Figs. 237–239. *Oxyethira flavicornis.* 237. 1st instar; left fore leg in posterior view; 238–239. 5th instar. Right fore (238) and middle leg (239) in posterior view. Membranes dotted. Nielsen.

deeply indented but somewhat obscured by the fore-ends of two soft ventral longitudinal folds which adjoin in the middle line. The right fold appears to be rather more developed. The lateral margin of the labrum on each side bears a slender, strongly tapering spine which is very broad at the base and very pale in colour. They are situated somewhat forward of the middle line. The left mandible is very similar to that of *Hydroptila* but the edges are distinctly, though bluntly, toothed (Figs. 235, 236). The edges of the right mandible are scarcely as much reduced as in *Agraylea* and *Hydroptila*. The membraneous part of the inner side, as found in these last-named genera, is absent in *Oxyethira*. The maxillolabium is like that of

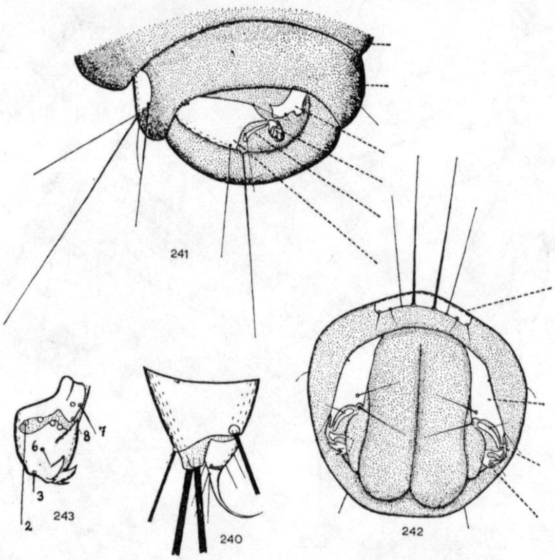

Figs. 240–243. *Oxyethira flavicornis*. 240. 4th instar; right anal proleg in lateral view; the long bristles have been cut off; 241–243. 5th instar. Segment IX and anal prolegs in lateral (right; 241) and posterior view (242); 243. Right anal claw half in lateral and half in ventral view. Membranes dotted. Nielsen.

Agraylea but the maxillary palp curves inwards more strongly and the second and third segments have fused to form a single, long, curved segment. The labial lobe is sagittally compressed.

Viewed from above each of the thoracic segments narrows strongly posteriorly. The pronotum constitutes distinctly more than a semi-circle and the anterior 'corners' are completely rounded. The lateral margins of meso- and metanota are concave. The middle and hind-legs are more than twice as long as the forelegs and the hind legs are only slightly longer than the middle legs. The femur and especially the trochanter of the forelegs are rather less expanded than in *Agraylea* and the claw is slender. The claw is even more slender in the middle and hind-legs. The first abdominal segment is slightly telescoped into the second segment and similarly the ninth segment is somewhat inserted into the eighth, especially dorsally. The anal prolegs of the young larva are different in shape from those of *Agraylea* and *Hydroptila*, the base being short and conical and thicker than long. The claw is the same length as the base, very slender and strongly curved. With the last moult the length of the claw is reduced by half and bears a very great resemblance to that of *Hydroptila*, but a minor dorsal hook is situated above each of the two large dorsal hooks, thus making a total of four hooks on each claw (Fig. 243). Above the hooks there is present a small protuberance similar to that occurring in *Agraylea*. Gills are absent, but Nielsen, on one occasion, found a gill arising from an anal proleg which was clearly an aberration.

Case Construction. The larva secures a lump of fluffy detritus and forms a ring-like belt with it, and with a few silk threads it is attached to an algal filament. The belt is then lined with silk threads until the case begins to grow out of the belt. It now becomes highly sagitally compressed. When the spun case attains the length of the abdomen the larva turns about and the belt first spun becomes the anterior end. The algal debris soon becomes detached. The case is thus entirely composed of silk and is milk-bottle shaped. The rim of the anterior end as well as the thickness and length is added to from time to time. The colour alters from light grey to yellowish-brown and it becomes horny in texture. The length of the case attains up to 4–8 mm. and the width 1–6 mm.

Oxyethira distinctella McLachlan

Immature stages unknown.

Oxyethira falcata Morton

Immature stages unknown.

Oxyethira frici Klapálek

The larva of this local and not very abundant species has been described by Silfvenius and Ulmer but not in sufficient detail to enable it to be separated from other species.

Oxyethira sagittifera Ris.

The larva of this very local species which is thought to be rather more common in Ireland has been described by Silfvenius.

Oxyethira simplex Ris.

Macdonald found larvae of this species in Loch Lomond, Scotland, at 3 m. on a substratum of sand and gravel with *Isoetes*. His description is as follows:

Larval Case. The case (Fig. 244) is 3.5 mm. long and a little over 1 mm. at its broadest part. In lateral view it is flask-shaped, broad posteriorly and tapering anteriorly. The whole case, except the neck region, is strongly laterally compressed. It is entirely composed of silk secreted by the larva and is open at both ends.

Larva. The larva (Fig. 245) is campodeiform, with the head, pro-, meso- and metanotum sclerotized. The abdomen, white in colour, is relatively very large and is slightly curved downwards posteriorly. Like the case, it is

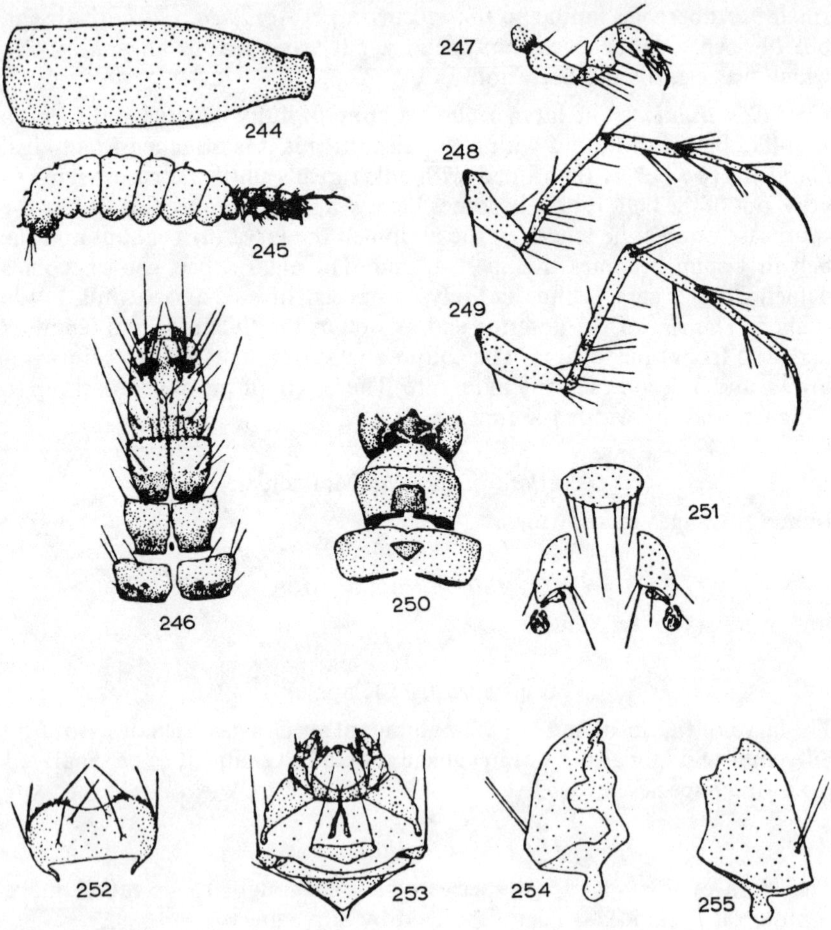

Figs. 244–255. *Oxyethira simplex* Ris. 244. Case; 245. Larva; 246. Head and thorax; 250. Thorax, sternal sclerites; 252. Labrum; 253. Labium, maxillae and gular sclerite; 254. Left mandible (dorsal view); 255. Right mandible (dorsal view); 247. Right prothoracic leg; 248. Right mesothoracic leg; 249. Right metathoracic leg; 251. Anal plate and claws. Macdonald.

flattened laterally, with the exception of the head and thorax. The length of the larva is about 3 mm. and, viewed laterally, it is broadest at abdominal segments four and five, where it is a little less than 1 mm.

Head. The head (Fig. 246) is roughly elliptical in shape, and pale yellow in colour, with the posterior third brownish. There are also two brown patches on the clypeus between the eyes. The latter are very big, with the cellular elements visible through the lightly pigmented integument. The chaetotaxy is as shown in the figure. The antennae arise from the anterior margin of the parietals and consist of a bulbous basal segment and a long stout distal segment. Midway along the latter there is a conspicuous little knob, from which arises a very long seta, and distally a little clear cap. The labrum (Fig. 252) is transverse, with a median anterior concavity. Each lateral margin bears three spine-like setae, the posterior two lying close together. There is also a brush of marginal hairs. Near the centre are two other pairs of spine-like setae, which, like those on the margin, are elevated on small tubercles. The mandibles are asymmetrical. The upper side of the left mandible (Fig. 254) has three teeth, the posterior two small and blunt, the remaining one large and acute. Proximally there is a deep indentation. The lower side has three small rounded teeth. The right mandible (Fig. 255) is obtuse with only three blunt teeth dorsally. Ventrally there is one blunt tooth. Each mandible has two setae arising from the outer surface. The maxilla (Fig. 253) has a four-segmented palp, the distal segment bearing a large papilla on its inner surface. The maxillary lobe is obtuse, and lies close to the palp and bears large numbers of sense organs and hairs on its inner side. The labium is broad and blunt and the labial palps bear small papillae. The gular sclerite (Fig. 253) is transverse and triangular and does not divide the genae.

Thorax. The pro-, meso- and metanotum (Fig. 246) are sclerotized and become progressively broader. The posterior half and lateral margins of each notum are light brown in colour, merging into the pale yellow of the anterior regions. The posterior margin of each segment is black and heavily sclerotized. The metanotum has a few dark spots in the posterior region. Each notum is divided by a longitudinal suture into two sclerites. On the pronotum these sclerites run together anteriorly, but separate posteriorly; on the meso- and metanotum the sclerites are clearly separate, and in the former there is a small elliptical sclerite lying between them. The thoracic segments also have a number of small sternal plates (Fig. 250). The prosternum has a diamond-shaped sclerite supported by two others, which are triangular in shape. The mesosternal plate is quadrangular, almost square, and that of the metasternum is small and triangular, lying nearer the anterior margin of the segment. The meso- and metasternal plates also have two supporting sclerites. The prothoracic leg (Fig. 247) is short, with a relatively broad femur and tibia, the latter with a large dilation, projecting distally, on its under surface. Two setae arise from the tip of the dilation. Along the lower border of the tibia, extending to the extremity of the dilation, there is a series of spinules. Another series is present on the tarsus. The femur also is dilated on its lower proximal surface and two setae arise from this protuberance. The trochanter bears one small and three long setae on its lower surface. The meso- and metathoracic legs (Figs. 248, 249) resemble each other in many respects, though the latter leg is slightly longer (ratio of legs, 6:12:13). The tarsal claw of each leg bears a spine at its base.

Abdomen. The abdomen (Fig. 245) is very large and is flattened laterally. Each segment bears a few small setae dorsally and on the 9th segment there is a dorsal anal plate (Fig. 251), elliptical in shape and bearing four pairs of setae on its posterior margin. There are no external gills nor is there a lateral line. The anal claws (Fig. 251) are very small, each consisting of a median claw and a few smaller supporting ones.

Oxyethira tristella Klapálek

A description of the larva of this local species based on British material is not available.

Oxyethira mirabilis Morton

Immature stages unknown.

Tricholeiochiton fagesi (Guinard)

This species, spelt '*fagesii*', was included in the genus *Leiochiton* by Mosely. The larval case is elliptical, narrower at each end than in the centre.

HANNA, H. M. 1961. The Larva of *Hydroptila sparsa* Curtis (TRICHOPTERA, HYDROPTILIDAE). *Entomologist's Gaz.* **12**, 69–75.
KLAPÁLEK, F. 1890. *Sitzungsb. Böhm. Gesellsch.*, 204–8.
LESTAGE, J. A. *in* ROUSSEAU, E. 1921. *Les Larves et Nymphes Aquatiques des Insectes d'Europe*, 444.
MACDONALD, W. W. 1950. *Proc. R. ent. Soc. Lond.* (A) **25**, 25–6.
NIELSEN, A. 1948. Postembryonic Development and Biology of the HYDROPTILIDAE. *K. danske Vidensk. Selsk. Skr.* **5** (1), 1–200.
SILFVENIUS, A. J. 1904. *Acta Soc. Fauna Flora fenn.* **26** (6), 11–33.
ULMER, G. 1903. *Metamorph. Trichopt.*, 132–5.
ULMER, G. 1909. *Die Süsswasserfauna Deutschlands* **5–6**, 224–6.

9

PHILOPOTAMIDAE

Adults of the family PHILOPOTAMIDAE may be identified by being small to moderate in size, having an anterior wing length from 6 mm. to 11 mm. The tibial spur formula is either 2, 4, 4 or 1, 4, 4, and ocelli are always present. The maxillary palpi are five-segmented and the terminal segment is multi-articulated and flexible. It is also generally much longer than all the other segments together, whilst the first segment is always short.

There is great variation in wing colour. Whilst in *Philopotamus montanus* the wing bears a reticulated pattern of brown and gold, in *Wormaldia subnigra* it is smoky black with a slight iridescence. The yellow longitudinal markings of the wings of *Chimarra marginata* are distinctive also.

Five species only in three genera are known from Britain as in the following check-list. *Wormaldia mediana* was reinstated by Kimmins as a valid species in 1953 after being rejected by Mosely.

PHILOPOTAMUS
1. *montanus* (Donovan)

WORMALDIA
2. *occipitalis* (Pictet)
3. *mediana* McLachlan
4. *subnigra* McLachlan

CHIMARRA
5. *marginata* (L.)

Distribution

Philopotamus montanus. Scotland (M.E.M.); Westmorland (G.B.R.); Yorks. (P.F.H.). Common in rapid hillstreams in the west of the British Isles (D.E.K.).

Wormaldia mediana. This species has obviously a much wider distribution than indicated here, but as Mosely considered it as only a variety of *occipitalis*, it has probably not been separated in collections. Cornwall, Westmorland, Perthshire (D.E.K.).

Wormaldia occipitalis. Local, springs, torrents and waterfalls; fairly abundant where it occurs (M.E.M.); Westmorland (G.B.R.); Northumberland (G.N.P.). Not uncommon, particularly in the west and north of Britain and in western Ireland (D.E.K.).

Wormaldia subnigra. More local than *occipitalis* but found in similar localities (M.E.M.); Westmorland (G.B.R.); Yorks. (P.F.H.); Surrey, west and north-west England, Wales, Scotland and Ireland (D.E.K.).

Chimarra marginata. Local. Waters where there are moss-covered boulders. Symond's Yat, Glos. (A.F.P.); Cumberland, Westmorland (G.B.R.); Northumberland (G.N.P.); Locally common in rocky and fast streams and rivers (D.E.K.).

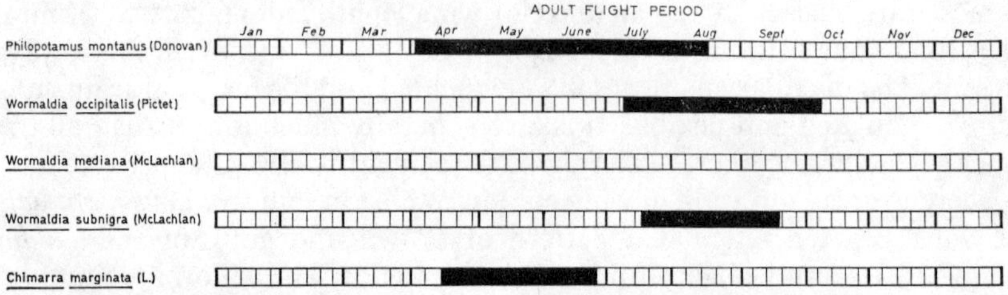

PHILOPOTAMIDAE. *Adult Flight Periods taken from published records of individual dates.*

General Description of Larvae

Campodeiform, slender, compressed dorso-ventrally, generally equal in breadth throughout its length except that the head, prothorax and last abdominal segment are a little narrower; size varying from 6–7 mm. (*Wormaldia*) to 22 mm. (*Philopotamus*).

Oval elongated head; narrow, elongated clypeus, narrowing after the centre, subacuminate at the top; rudimentary antennae; robust prominent mouthparts; labrum not sclerotized, membraneous, soft, pale, very retractile, much broader than it is long, dilated in front with anterior angles prominent. The anterior border ciliated, emarginated at the centre.

Mandibles of the 'scissor' type and curved rather like a bird's beak, the upper edge less developed than the lower, both strongly denticulated, internal brush present.

Maxillae slender; maxillary lobe obtusely conical, finely ciliated on the internal edge; labial lobe not longer than the mandibles, obtusely conical, rudimentary labial palps, depressed; hypostome absent.

Only pronotum is sclerotized, trapezoid sclerotized plate is narrow behind, the posterior angles spiniform and with long prolongation at the rear, as in the POLYCENTROPIDAE and PSYCHOMYIIDAE. Meso-metanotum and abdominal segments membraneous. No prosternal horn.

Legs unequal; two spurs on each tibia; the internal border of the anterior tarsi have a series of short oblique hairs; pubescence very sparse; in front of the supporting plates of the anterior legs is an obtuse appendage; tarsal claws are small, slightly curved, with two sub-basal unequal teeth, the first short and strong, the second setiform.

Abdominal gills and lateral line absent; five anal gills. (Both Ulmer and Lestage give the number of anal gills as four which is clearly not correct.)

Anal appendages well developed, of two cylindrical parts, the first long, the second short, curved underneath and furnished on top with two long terminal hairs; simple, robust, curved claws, with dorsal hooks or ventral spines.

Head and pronotum sometimes pale yellow or pure yellow (*Wormaldia*) sometimes yellowish-brown, with a reddish tinge (*Philopotamus*); posterior border of the pronotum marginated shining black; meso-metanotum and abdomen whitish or yellowish; legs of the same colour as the head; the two posterior pairs often paler.

No larval case is constructed, the larvae living 'free' in nets of a silk-like secretion. The nets are tube-like with an open end attached to a stone and the closed end hanging freely but usually held horizontally by the current.

The nets are about 5 mm. in width and 35 mm. in length. Silt-like material such as particles of organic material—diatoms—are sieved out of the water by the net and are regularly swept up by the characteristic brush-like labrum and taken into the mouth. Larger quarry such as insects are doubtless dealt with efficiently by bringing the sharp mandibles into play.

Several authors have drawn attention to the environment which the larvae of most, if not all, species of this family prefer. This consists of fast-flowing water over moss-covered rocks or boulders. Although streams at high altitude are the usual habitat, fast-flowing water at low altitude may be populated if the water is at a low temperature.

Detailed Description of Larvae

Philopotamus montanus (Donovan)

The present description is taken from specimens collected in Lily bed stream running into the west side of Lake Windermere. This stream cascades over a bed of rocks and is very well aerated. The larva constructs silken nets and tunnels on the underside of the pieces of rock in the torrential part of the stream.

The larva is campodeiform and has the head and prothorax a light chestnut-brown colour. Only the prothorax is sclerotized. The abdomen is white and lateral abdominal gills are absent. In size the larvae go up to 22.0 mm. long and 2.6 mm. broad, but maximum size is variable.

Head. The head is long, the genae running parallel for some considerable distance. It is prognathous. There are no dark spots, the colour being a uniform light chestnut-brown, except at the anterior margins of genae and clypeus, which are more heavily sclerotized and very much darker in colour (almost black). The anterior part of the clypeus is attenuate (Fig. 256). The antennae are rudimentary.

Mouthparts. The labrum is quite distinctive (Fig. 257). It is unsclerotized and extends a considerable distance forwards. Right and left lobes form brushes with straight anterior margins which are transverse. The brushes

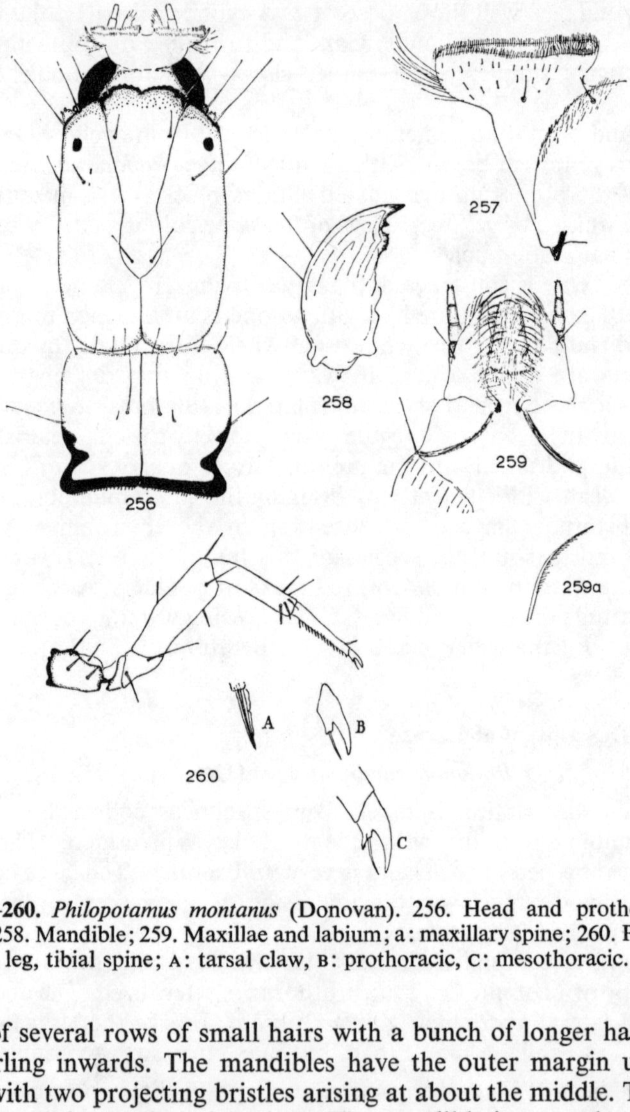

Figs. 256–260. *Philopotamus montanus* (Donovan). 256. Head and prothorax; 257. Labrum; 258. Mandible; 259. Maxillae and labium; a: maxillary spine; 260. Prothoracic leg, tibial spine; A: tarsal claw, B: prothoracic, C: mesothoracic.

consist of several rows of small hairs with a bunch of longer hairs at the sides curling inwards. The mandibles have the outer margin uniformly convex with two projecting bristles arising at about the middle. The outer tooth is much larger than the others. The mandible is more than twice as long as broad (Fig. 258). The maxillae have four-segmented palps with a bristle towards the base (Fig. 259). The maxillae are beset with a number of thin sickle-shaped bristles all bearing secondary hairs (Fig. 259, A). All these bristles are directed inwards towards the labium. The latter is very small, rather hairy, and the labial palps are much reduced.

Thorax. Only the prothorax is sclerotized. It is light chestnut-brown in colour, except the posterior margin, which is more heavily sclerotized and is much darker in colour. A row of bristles is set along the anterior margin of the prothorax and another row runs transversely across the centre.

Legs. The tarsal claws of all the legs have a spine set upon a protuberance. The protuberance is elongated beyond the base of the spine, this being shown better in the meso- and metathoracic legs than in the prothoracic. The tarsal segment is fringed with small spines along its entire length in the prothoracic leg (Fig. 260), but these are restricted to the distal end of the segment in the meso- and metathoracic legs. A small group of spines on the distal end of the tibia are somewhat flattened.

Abdomen. The abdomen, which is white in colour, is devoid of gills except a small group just dorsal to the anus. These are apparently eversible. The abdominal claws are two-segmented, the proximal segment being slightly concave anteriorly and beset with three spines. The claw has a small group of hairs on the convex side.

Wormaldia mediana McLachlan

The larva of this species is unknown.

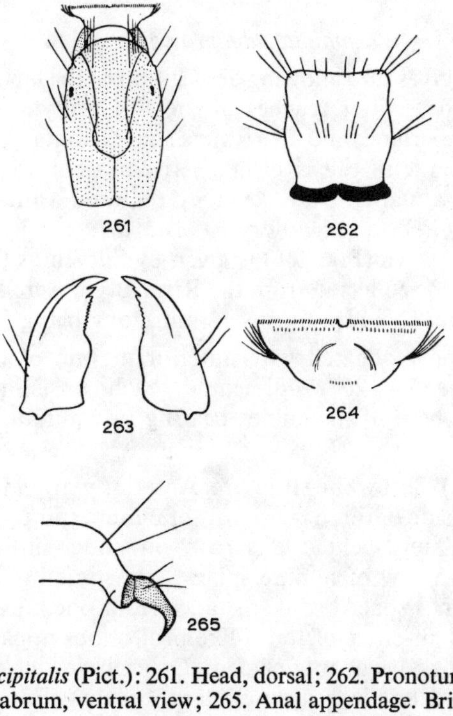

Figs. 261–265. *W. occipitalis* (Pict.): 261. Head, dorsal; 262. Pronotum; 263. Mandibles; 264. Labrum, ventral view; 265. Anal appendage. Brindle.

Wormaldia occipitalis (Pict.)

Brindle (1960) has described the larva of this species from specimens obtained from a small woodland stream at Witherslack, Westmorland, as follows:

Size: 10–12 mm. in length, head elongated, narrow, yellow (Fig. 261), labrum soft, whitish, anterior margin with median notch and with a dense row of setae; a second, less dense row of setae occur near to anterior margin ventrally. Longer setae occur on either side of labrum (Fig. 264).

The median area of the ventral surface of the labrum is more or less smooth with a row of very small teeth-like projections posteriorly and with two curved thickened borders anteriorly (Fig. 264). This latter agrees with the description and figure in Nielsen (1942). Mandibles yellowish-brown, with two setae on external edge, internal edge with apical teeth and serrated on apical half. Pronotum sclerotized, yellow, with black posterior border, and chaetotaxy as Fig. 262. Lateral margins bordered finely with black.

Legs yellowish, with fine setae apically on segments, and with short strong yellow setae elsewhere on coxae, femora and tibiae. Tarsi of meso- and metathorax with fine spines distally.

Abdomen whitish or yellow, without abdominal gills, but with five anal gills. Anal appendages long, ending in strong curved claws (Fig. 265).

Wormaldia subnigra McLachlan

Philipson (1953) gives the following description of the larva from specimens collected from the swifter reaches of the River Blyth near Stannington Bridge, Northumberland, and also from a small tributary, a little more than a foot wide, which joins this river in a series of small cascades. They were reared in stirred aquaria. However, they required a much higher rate of stirring than required for *Hydropsyche instabilis*.

The size of the larva (Fig. 266) is given by Silfvenius (1903) as 6.7 mm. long; the specimens collected from the River Blyth were larger than this—about 9 mm. long, some specimens 12 mm. long being found.

Head. Yellow and elongated with anterior margin of clypeus markedly convex. Antennae (Fig. 267) rudimentary, each consisting of a small light area just behind base of mandibles bearing two minute spines and three fine hairs.

Labrum (Fig. 271). Soft white structure. Anterior margin with small median notch. On either side of this a dense row of minute spines, their ends turned inwards. Immediately behind this row, on under-surface of labrum, a second, less dense row of minute spines, and some little distance behind this a row of spines spaced some distance from one another. In centre of under-surface a half-circle of tooth-like projections apparently formed on projecting anterior edge of ante-clypeus. On each side labrum bears a group of longer incurved bristles forming lateral brushes. Entire labrum can be withdrawn into buccal cavity. Mandibles (Fig. 272) have convex outer surfaces bearing two bristles; internal surfaces grooved. Grooves bordered by a number of teeth; lower edge of each finely serrate proximally. Maxillary palp four-segmented and somewhat incurved. Maxillary lobe bearing a number of fine bristles on internal edge and a small curved bristle on external edge (Fig. 273).

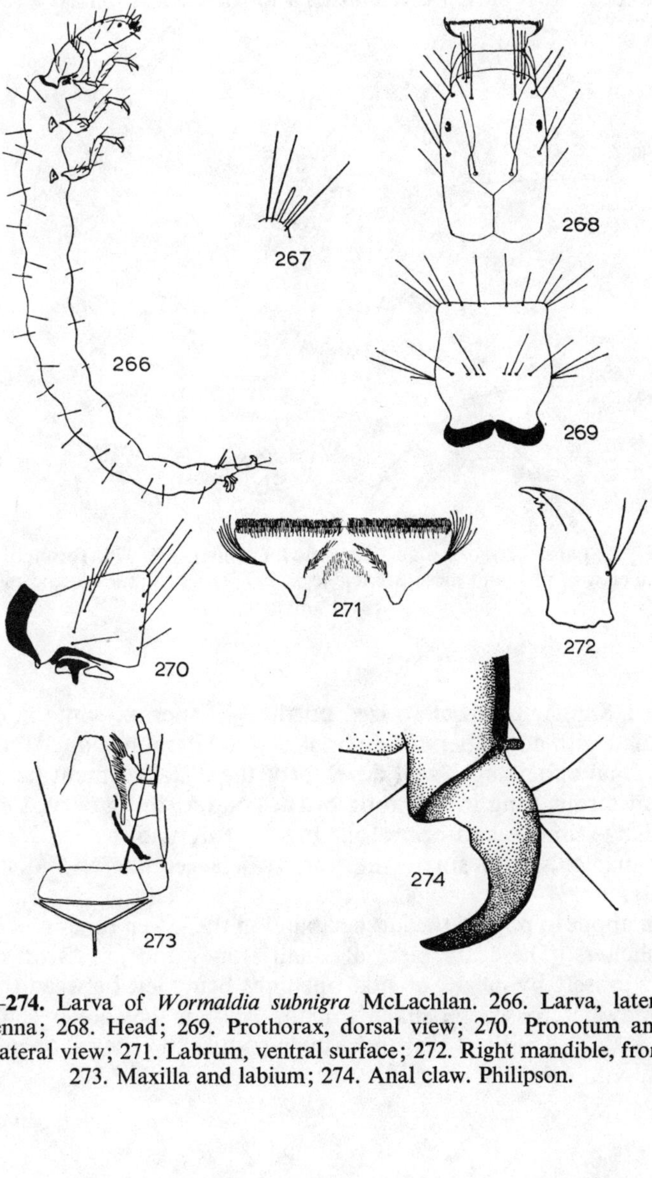

Figs. 266–274. Larva of *Wormaldia subnigra* McLachlan. 266. Larva, lateral view; 267. Antenna; 268. Head; 269. Prothorax, dorsal view; 270. Pronotum and lateral sclerites, lateral view; 271. Labrum, ventral surface; 272. Right mandible, from above. 273. Maxilla and labium; 274. Anal claw. Philipson.

Pronotum (Fig. 269, 270). Sclerotized, yellow in colour, with a broad black posterior border. A number of bristles along anterior margin and a transverse row across the centre.

Legs (Fig. 277–279). Yellow with a few bristles and a number of stronger curved yellow bristles on coxae, femora and tibiae. Tarsi with fringe of fine spines, restricted to distal portion in tarsi of meso- and metathoracic legs. Prothoracic tarsus has in addition a number of pectinate bristles. Claws (Fig. 275, 276) well developed, curved and provided with a small claw-like projection on inner side. Between this and the claw is placed a long spine.

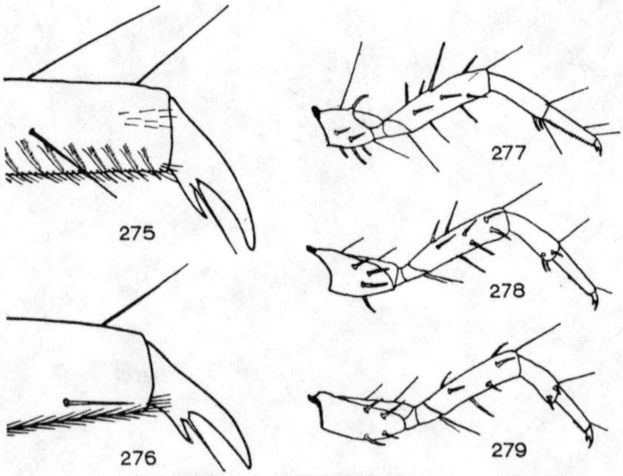

Figs. 275–279. Larva of *Wormaldia subnigra* McLachlan. 275–276. Terminal portion of tarsus and claw of pro- and mesothoracic legs; 277–279. Pro-, meso-, and metathoracic legs. Philipson.

Abdomen. Similar to unsclerotized portions of thorax; white in colour. It is provided with a number of fine bristles. No lateral line and tracheal gills absent. Anal appendages well developed, the distal segment being sclerotized and terminating in two long bristles and a strong curved claw (Fig. 274). This is simple, and bears four bristles externally.

Five anal gills, one small anterior, two larger lateral and two small posterior.

When about to pupate the larvae abandon the silken tubes and construct pupal shelters. These are built of small stones and grains of sand held together loosely by means of silk, openings being left between the stones, which may be crossed by silken threads. Within the shelter a fine silken cocoon is spun. This is of very loose open texture; no special openings were detected.

Chimarra marginata (L.)

No description of the larva of this species based on British material is available. The following information is taken from Marlier's account (1943).

In general form the larva (Fig. 281) is rather like that of *Wormaldia occipitalis* except that it is a little larger, being 7 mm. in length and from 1.0 to 1.2 mm. in breadth. The head is ovoid, amber yellow in colour, a little deeper in shade than that of *Wormaldia* but not quite so deep as that of *Philopotamus*. The anterior margin of the clypeus, which is yellowish-brown in colour, is indented asymmetrically towards the middle. The left lip of the indentation is much more prominent that the right. As in *Wormaldia*, the eyes are situated slightly behind the mandible bases. The thick, strong, mandibles (Fig. 283), which are brown in colour, are asym-

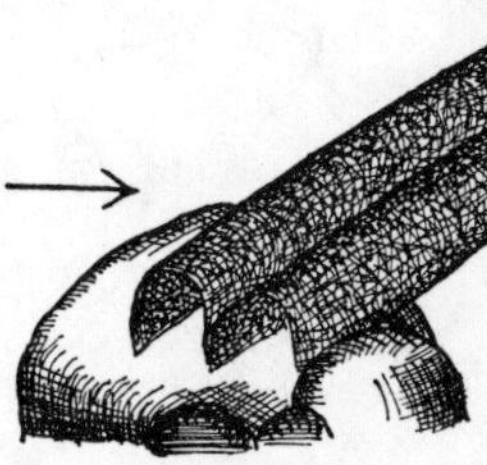
280

Fig. 280. Two silken tubes spun by *W. subnigra* larvae in a stirred aquarium. The arrow indicates the direction of the current. Philipson.

metric and curved. Their internal edges are finely crenulate and they terminate, the right with a blunt apex followed with a small needle-like tooth, the left with a needle-like tooth followed by two smaller ones. The anterior edge of the labrum (Fig. 282) is extended as a pair of long lateral processes which can twist ventrally under the central part (this is the position occupied when the organ is retracted). It is completely bordered by a fringe of serrated hairs. At the anterior angles towards the ventral face, the 'lateral brushes' are much less marked than is the case in *Wormaldia* and *Philopotamus*. The rounded median sclerite is but little developed. The rather large-based hypostome is limited by pronounced sutures and is triangular whilst, in the case of *Wormaldia*, it is partially joined towards the rear to the pleura. As in several other species of the PHILOPOTAMIDAE the occipital margin of the head is little indented and is furnished with a black mark.

The pronotum, which is the only sclerotized thoracic segment, is amber yellow only slightly paler than the head. The posterior margin is raised and saddle-like toward the rear but to a lesser extent than is the case in *Wormaldia*. A large black band covers the posterior edge and the sides of the pronotum as far as the articulation of the epimeron where it terminates.

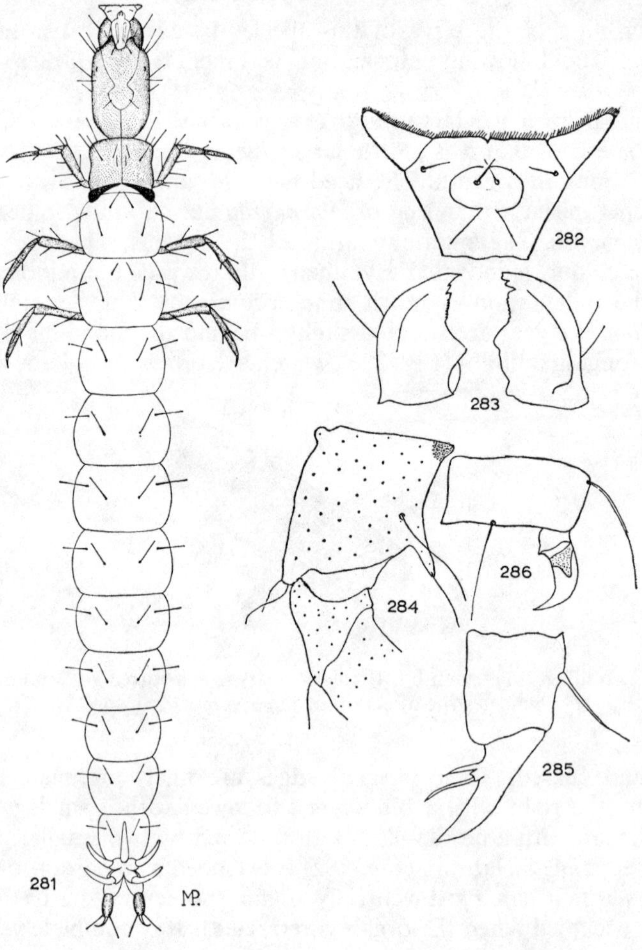

Figs. 281–286. Larva of *Chimarra marginata*. 281. Larva; 282. Labrum; 283. Mandibles; 284. Right anterior coxa; 285. Tarsal claw of anterior leg; 286. Anal claw. Marlier.

The pale-yellow legs are subequal, the anterior pair being the most prominent. The coxa (Fig. 284) is pale and carries on the anterior face a long conical process surmounted by a black hair, and a similar hair is to be found near its base. The structure of the tarsal claws (Fig. 285) is similar to that found generally in the PHILOPOTAMIDAE.

In the case of young larvae, the abdomen (as also the meso- and metathoracic segments) is soft and white but when nearing fully-grown it acquires a yellowish colour which is, in fact, adipose tissue. The bristles of the abdominal segments are black and clearly visible. External abdominal gills are absent but there are five anal gills. The anal appendages are simple and the claws (Fig. 286) are a little more strongly curved than in *Wormaldia*. Pupation takes place in a case resembling that of other PHILOPOTAMIDAE.

Key to Known Larvae of the PHILOPOTAMIDAE

The first couplet is constructed from Marlier's description and the remainder is due to Brindle.

1. Lateral extremities of front edge of labium bending forward and separated by pronounced suture and without tuft of hairs on outer edge *Chimarra marginata*
— Front edge of labium straight, suture not pronounced but with tuft of hairs on outer edge 2
2. Larger, up to 24 mm. in length, head broader in comparison to length, reddish or dark reddish; setae on head mostly short; fronto-clypeus narrow, its anterior margin emarginate and blackish, without long setae (Fig. 256); disc of pronotum with one pair of long setae *Philopotamus montanus*
— Smaller, up to 12 mm. in length, head narrower, yellow; setae on head long; fronto-clypeus broader, its anterior margin convex, not blackish, with long setae; disc of pronotum with two pairs of long setae 3
3. Ventral surface of labrum with a median patch of teeth-like projections (Fig. 271); anterior border of pronotum with long setae (Fig. 269) *Wormaldia subnigra*
— Ventral surface of labrum smooth, without a median patch of teeth-like projections (Fig. 264); anterior border of pronotum with short setae (Fig. 262) *Wormaldia occipitalis*

BRINDLE, A. 1960. The Larva of *Wormaldia occipitalis* (Pict.) (TRICHOPTERA, PHILOPOTAMIDAE). *Entomologist's Rec. J. Var.* **72**, 144–7.

HICKIN, N. E. 1942. *Proc. R. ent. Soc. Lond.* (A) **17**, 16–17.

KIMMINS, D. E. 1953. A Key to the European Species of *Wormaldia* (TROCHOPTERA, PHILOPOTAMIDAE), with descriptions of two new sub-species. *Annals and Magazine of Natural History*. Ser. 12, **VI**, 801–8.

KIMMINS, D. E. 1965. Keys to the British Species of RHYACOPHILIDAE and PHILOPOTAMIDAE. *Entomologist's Gaz.* **16**, 147–61.

LESTAGE, J. A., *in* ROUSSEAU, E. 1921. *Les Larves et Nymphes Aquatiques des Insectes d'Europe*.

MARLIER, G. 1943. Les Metamorphoses de *Chimarra marginata* (TRICHOPTERA, PHILOPOTAMIDAE). *Bull. Mus. Hist. nat. Belg.* **19** (51), 1–8.

NIELSEN, A. 1942. Über die Entwicklung und Biologie der Trichopteren. *Arch. Hydrobiol. Suppl.* **17**, 374.

PHILIPSON, G. N. 1953. The Larva and Pupa of *Wormaldia subnigra* (McLachlan) (TRICHOPTERA, PHILOPOTAMIDAE). *Proc. R. ent. Soc. Lond.* (A) **28** (4–6), 57–62.

SILFVENIUS, A. J. 1903. Über die Metamorphose einiger Hydropsychiden. *Acta Soc. Fauna Flora fenn.* **25** (5), 24 pp.

SILFVENIUS, A. J. 1905. Beiträge zur Metamorphose der Trichopteren. *Acta Soc. Fauna Flora fenn.* **27** (6), 168 pp.

SILFVENIUS, A. J. 1905–6. *Acta Soc. Fauna Flora fenn.* **27** (6), 119, fig. 28, a–c.

ULMER, G. 1903. *Metamorph. Trichopt.*, 117, figs. 10, 21, 35, 71, 80.

ULMER, G. 1903. Über die Metamorphose der Trichopteren. *Abh. naturw. Hamburg* **18**, 1–154.

ULMER, G. 1909. *Trichoptera, Die Süsswasserfauna Deutschlands* **5–6**.

WESENBERG-LUND, C. J. 1943. *Biologie der Süsswasserinsekten*.

10

POLYCENTROPIDAE

The adults of species in the family POLYCENTROPIDAE may be identified by the following characters. In size they are moderately small, the length of the anterior wing being only 6 mm. in *Holocentropus dubius* but in several species it reaches 12–13 mm. The antennae are stout, in length they are either as long as or are shorter than the wings. The maxillary palpi are five-segmented and whilst the first two segments are short the terminal segment is multi-articulated and flexible (using Kimmins's terms). The anterior wings are somewhat ovate and moderately broad and covered with dense pubescence. The discoidal and median cells are always present. The tibial spur formula is 3, 4, 4. A pair of lateral filaments arise from the fifth abdominal segment.

Thirteen species in five genera constitute the British POLYCENTROPIDAE as in the following check-list. It will be noted that *Cyrnus insolutus* McLachlan has been added to the list since the publication of Mosely's handbook.

NEURECLIPSIS
1. *bimaculata* (L.)

PLECTROCNEMIA
2. *brevis* McLachlan
3. *conspersa* (Curtis)
4. *geniculata* McLachlan

POLYCENTROPUS
5. *flavomaculatus* (Pictet)
6. *kingi* McLachlan
7. *irroratus* (multiguttatus (Curtis))

HOLOCENTROPUS
8. *dubius* (Rambur)
9. *picicornis* (Stephens)
10. *stagnalis* (Albarda)

CYRNUS
11. *flavidus* McLachlan
12. *insolutus* McLachlan
13. *trimaculatus* (Curtis)

Distribution

Neureclipsis bimaculata. Large rivers, lakes and canals. Generally abundant where a river enters or leaves a lake (M.E.M.); Berks. (M.I.C. *et al.*); Yorks. (P.F.H.).

Plectrocnemia brevis. Much scarcer than *conspersa* or *geniculata.* Taken in numbers at Seaton, Devon (M.E.M.).

Plectrocnemia conspersa. Widely distributed but lies concealed in crevices and bark of trees, during the day (M.E.M.); Northumberland (G.N.P.); Cardigan (J.R.E.J.); Yorks. (P.F.H.).

Plectrocnemia geniculata. Local, far less frequently taken than *conspersa* (M.E.M.).

Polycentropus flavomaculatus. One of the most widely distributed and abundant of all the TRICHOPTERA (M.E.M.); Northumberland (G.N.P.); Berks. (M.I.C. *et al.*); Cardigan (J.R.E.J.); Yorks. (P.F.H.); Glos. (A.F.P.).

Polycentropus kingi. Local. Abundant in Galway (M.E.M.).

Polycentropus irroratus. Far more local than *flavomaculatus.* Usually taken singly, abundant in Galway (M.E.M.); Northumberland (G.N.P.); Oxon. (L.W.G.).

Holocentropus dubius. Local. Ponds and lakes (M.E.M.); Berks. (M.I.C. *et al.*).

Holocentropus picicornis. Widely distributed and abundant. Lakes, ponds, ditches (M.E.M.; Berks. (M.I.C. *et al.*).

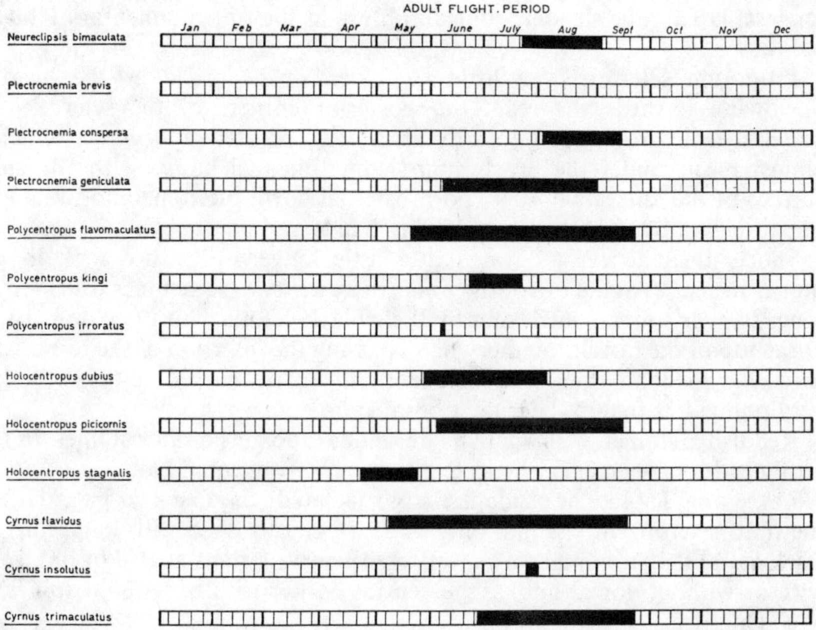

POLYCENTROPIDAE. *Adult Flight Periods taken from published records of individual dates.*

Holocentropus stagnalis. Somerset (Shapwick Heath) and Worcs. (Rubery) (M.E.M.); Hunts. (J.C.).

Cyrnus flavidus. Rather local, lakes (M.E.M.); Berks. (M.I.C. *et al*.).

Cyrnus insolutus. Blelham Tarn. Lake Lancs on rocks at water's edge where shaded by trees (D.E.K.).

Cyrnus trimaculatus. Very widely distributed and very abundant, lakes, ponds, slow-running rivers (M.E.M.); Northumberland (G.N.P.); Berks. (M.I.C. *et al*.); Yorks. (P.F.H.).

General Description of the Larvae

Campodeiform; body a little flattened dorso-ventrally, of large or average size. Fairly large, flat, narrow head, long (*Holocentropus*, *Polycentropus*) or broadly oval (*Plectrocnemia*), having a well-defined marking. No hypostome. Broad clypeus, a little sinuous near the anterior border; eyes inserted on large pale patches; rudimentary antennae, represented by a pale tubercle having three pale hairs and one with three small sensorial spikes. Sclerotized, transverse labrum, more or less hollowed in the middle, hairy in front, with the anterior border, the dorsal surface and the sides having some long hairs. Mandibles of the 'scissor' type, the two edges three-toothed, except the upper edge of the right mandibles, which have only one or two teeth; no brush on the internal edge of the right mandibles; a hairy brush on that of the left; dorsal hairs always present. Maxillary lobe conical; maxillary palps slender, of four parts, the third part being the longest; labial lobe slender, conical, obtuse at the tip; rudimentary labial palps.

Pronotum sclerotized, a little narrower, or as broad as the head, narrowing towards the rear; the posterior corners of the sclerotized patches each prolonged ventrally in a long pointed appendage which almost meets under the prosternum; some unequal hairs on the dorsal surface of the anterior border; no prosternal horn; meso-metanotum and one to eight abdominal tergites having bands and pale spots.

Short, unequal legs—the median a little longer—furnished with long numerous hairs; at the top of the tibiae there are two (sometimes four) very long, slender spurs, contrasting by their black colour with the pale hairs; at the top of the tarsals, feathery hairs passing the insertion of the terminal claw which is long, thin, broad, slender and slightly curved on the second, very pointed at the tip, bearing a basal spur.

Reddish abdomen with whitish lateral lines; no gills or lateral lines, but, on the sides of segments one to eight, numerous unequal hairs.

Large anal leg-like appendages, tri-articulated: the first two parts soft, the third sclerotized; the first only or all three furnished with long, fairly thick hairs; long terminal claws, very sharp curved, right-angled or obtuse, with or without dorsal hooks, the ventral border with or without spines; five anal gills.

No larval case. Silk-like net simple or connected with a narrow tube which may be curved.

Larvae of POLYCENTROPIDAE do not construct transportable cases but spin nets of silk-like material which may be attached to any suitable support such as stones, submerged branches or pond-weed. In shape, they have been described as being like swallows' nests, but are often much more irregular. They occur in many types of water, both flowing and static, and this determines, to a great extent, the particular location of the net. Edington gives the following features which distinguish larvae of POLYCENTROPIDAE from larvae of the other families which do not construct transportable cases:

(a) the mesonotum and metanotum are membraneous (in the HYDROPSYCHIDAE and in *Ecnomus tenellus* (Ramb.) (PSYCHOMYIIDAE) the mesonotum and metanotum are sclerotized);

(b) the labrum is sclerotized (in the PHILOPOTAMIDAE the labrum is membranous, with a brush-like anterior border);

(c) the labium does not project forward beyond the maxillae (in the PSYCHOMYIIDAE the labium usually extends beyond the maxillae and is long and pointed);

(d) lateral abdominal gill tufts are absent (in the British species of *Rhyacophila* (Mackereth, 1954) lateral abdominal gill tufts are present).

In general, larvae of this family may be distinguished by the broad head patterned with dark-coloured spots, the fairly deep intersegmental constrictions of the pinkish or reddish abdomen, and the long, often widely diverging, anal appendages.

In addition to a number of descriptions of separate species by recent authors (Hickin, Macdonald & Brindle), Edington (1964) has given an account of the larval characteristics of this family, excluding only one rare species (*Plectrocnemia brevis*).

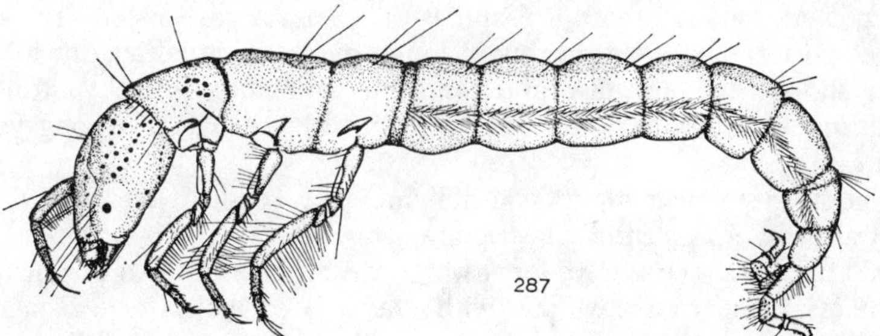

Fig. 287. Larva of *Holocentropus picicornis*, showing large, broad, head; sclerotized pronotum and slender, widely diverging anal appendages.

Detailed Descriptions of Larvae

Neureclipsis bimaculata (L.)

The net made by this species has been described several times. It is said to be trumpet-shaped with the mouth facing the current and the narrow end looped to one side. Net-spinning and behaviour has been described by Brickenstein (1955).

Brindle (1960) has described the final-instar larva from specimens collected in North Wales. The nets were found beneath stones near the edge of the outflow from a lake. Mosely has drawn attention to the habit of this species of preferring this habitat.

Special note: As Edington's figures (1964) illustrate his key I have included them for comparison. This has resulted in a certain amount of duplication of illustrations but it was thought that this was the better arrangement, as far as identification is concerned.

Brindle's description follows:

Larva (final instar). Length 18 mm., breadth 2 mm. Head (Fig. 288) broad, rounded, yellow, slightly darkened along fronto-clypeal and occipital sutures, and with brown spots arranged on the posterior part of the head and on the fronto-clypeus; eyes blackish, rather angular, in whitish areas; chaetotaxy of head as Fig. 288; labrum (Fig. 295) dark yellow with two very long setae laterally and with six shorter setae arranged in a row towards the anterior margin; anterior and lateral margins with short setae, the former also with two long setae directed towards the mid-line; mandibles (Fig. 294) yellowish red; asymmetrical, rather narrow, left mandible with a brush of setae dorsally, and with six sharp teeth (two dorsal and four ventral), the apical one the largest; right mandible without a brush of setae, and with five sharp teeth, the apical one the largest; both mandibles with one pair of long setae on external edge. Pronotum (Fig. 289) yellow with brown spots posteriorly and laterally, and with a well-marked median longitudinal suture, posterior border darker and actual posterior margin black; chaetotaxy as Fig. 289.

Legs (Figs. 291–293) relatively slender, yellow, anterior legs shorter than the posterior pairs; anterior femur with a ventral row of long sub-equal setae; posterior legs with a row of long setae of unequal length; all tarsi with short setae apically, anterior tarsi with short setae ventrally in addition; tarsal claws long, slender, yellowish, each with a long yellow seta.

Abdomen reddish along mid-dorsum, pale reddish laterally, whitish ventrally; intersegmental constrictions very deep; no gills or lateral line. Anal appendages (Fig. 290) long, widely diverging, with two obvious long segments, the basal one whitish, with setae only on the extreme distal edge, the distal segment yellow, blackish proximally, covered with long setae; anal claw long, curved, yellowish, with fine serrations ventrally from near base to mid-length; three sub-dorsal setae.

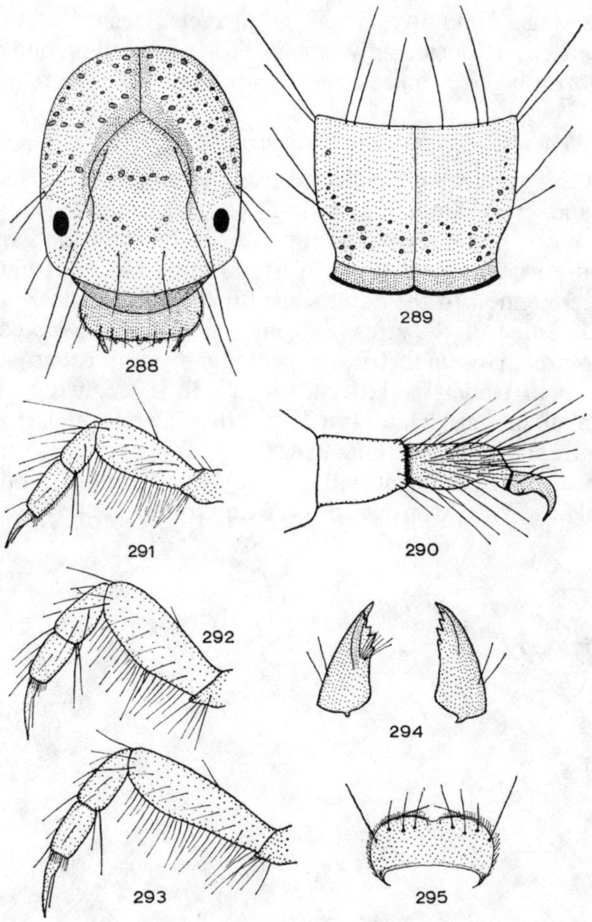

Figs. 288–295. Larva of *Neureclipsis bimaculata* (L.). 288. Head, dorsal; 289. Pronotum, dorsal; 290. Anal appendage; 291. Anterior leg; 292, 293. Posterior legs; 294. Mandibles, dorsal; 295. Labrum, dorsal. Brindle.

Edington (1964) points out that the ninth abdominal segment bears a pair of stout spines on the ventral side (Fig. 354). They project backwards and arise from sclerotized plates. In addition, this author considers the curvature of the anal claw to be of importance in identifying larvae of this family. In this species the claw bears numerous stout small spines on the concave edge but dorsal spines are absent. The black spots on the fronto-clypeus usually form a straight line.

Plectrocnemia brevis McLachlan

Immature stages unknown. There is no recent record of this species in Britain. Indeed, it appears only to have been collected on one occasion and that in Devon in 1903, and a recent search has failed to find it.

Plectrocnemia conspersa (Curtis)

Larvae of this species were collected from a fast stream at South Nutfield, Surrey, and also from Painswick, Gloucestershire; Dowles Brook, Bewdley, Worcestershire; Wilderhope, Shropshire; Sutton Park, Warwickshire, and several localities in Westmorland. Larvae from the Surrey locality were reared to the adult stage for the species determination to be confirmed. Some of the streams from which larvae were collected were nothing more than small trickles running over or round stones or flat pieces of rock. Under the latter a silken web is secreted by the larva for snaring small animals. This consists of a tubular central part and a funnel-shaped entrance and sometimes exit. If the larva is made to suffer a major disturbance of its habitat it will swim by undulating its abdomen in the vertical plane. When transferred to an aquarium, the larva, after a survey

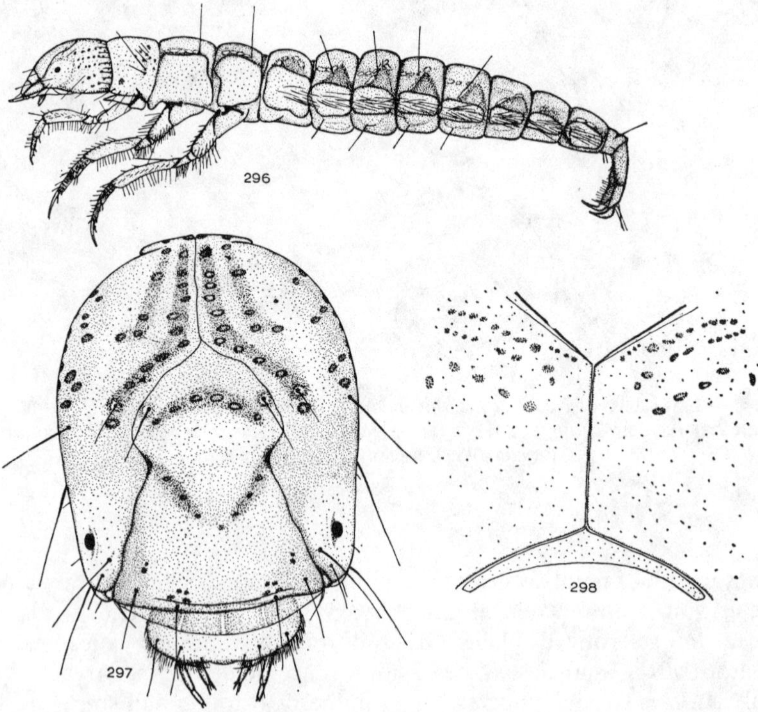

Figs. 296–298. *Plectrocnemia conspersa.* 296. Larva (lateral view); 297. Head; 298. Gular sclerite.

of suitable sites, rapidly constructs a new shelter and snare. It was often found that the larva would desert, for no apparent reason, its newly constructed shelter and build another one. It can run rapidly around its web, occasionally doubling back, and all its movements are precise. From time to time the abdomen is undulated in the vertical plane with the effect of maintaining a current of water over the body for oxygenation. Although the larvae are taken from running water it is easy to rear them through to the pupal stage in an aquarium, but a high proportion die as pupae at the swimming stage immediately before emergence as adults.

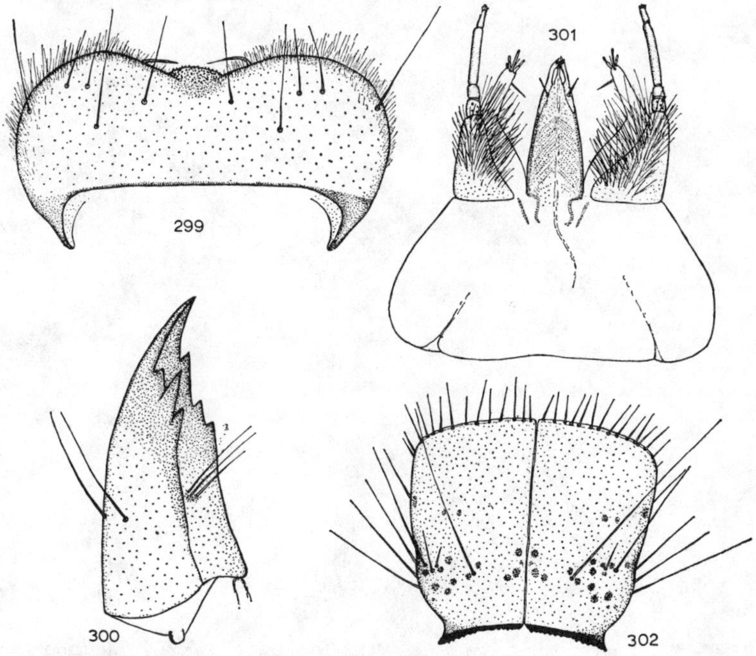

Figs. 299–302. *Plectrocnemia conspersa*. 299. Labrum; 300. Left mandible; 301. Maxillae and labium; 302. Pronotum.

Larva (Fig. 296). Campodeiform, without a case. Up to 22 mm. in length and 3.5 mm. in width. Widest at about the third and fourth abdominal segments. Deep intersegmentation of abdomen. Colour greyish-brown to reddish-brown with a lighter band running longitudinally along dorsal surface of abdomen. A pattern of light spots and dark areas is present.

Head (Fig. 297). Procentrous, yellowish-brown to reddish-brown. Prominent ridges for insertion of head into prothorax. Oral end of clypeus long, narrowing more or less gradually towards aboral end. Light mark in centre of aboral end of clypeus, towards middle of clypeus a 'V'-shaped line of fainter dark marks. These marks are the 'fourteen spots' of previous authors; an additional spot may, however, be present in the centre of the oral group. A fold runs parallel with the anterior margin of the clypeus with a group of four dark spots at each end. Preclypeus faintly divided into four by three longitudinal pale marks. On the genae a row of dark marks adjacent to each side of median suture; these diverge anteriorly and lie adjacent to aboral end of clypeus. A fainter row of marks runs outside each of the rows. Antennae rudimentary, eyes near anterior end of head. Gular sclerite (Fig. 298) very narrow, anterior margin concave. Postgenae contiguous for considerable distance.

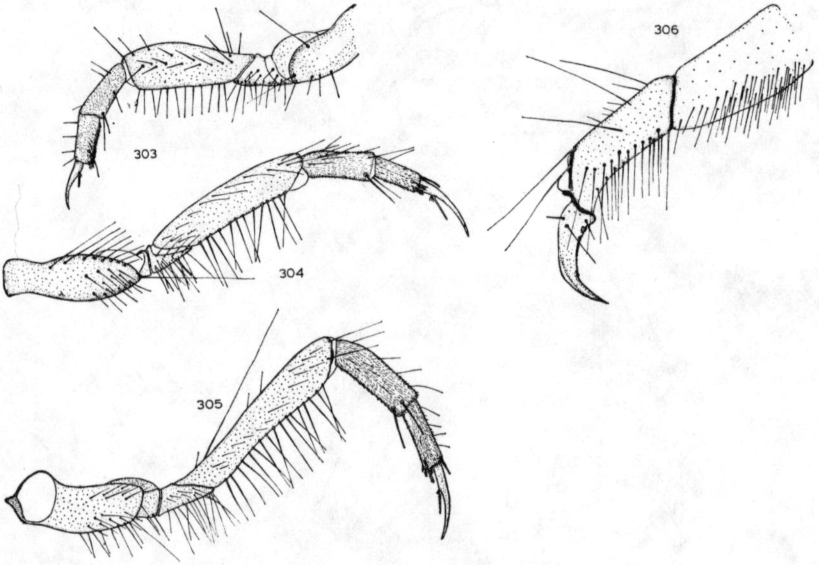

Figs. 303–306. *Plectrocnemia conspersa*. 303. Prothoracic leg; 304. Mesothoracic leg; 305. Metathoracic leg; 306. Anal appendage.

Mouthparts. Anterior margin of labrum (Fig. 299) hairy, concave with a cushion-like protuberance in the centre. Mandibles acute, strongly toothed on inside edge. Left mandible (Fig. 300) has 'brush' of about three bristles on inner edge. This is absent in right mandible. Maxilla (Fig. 301) hairy at base, maxillary palp four-segmented, third segment long. Labium acute, partially sclerotized on outer margins, well-defined spinneret at apex consisting of a central organ with a palp-like organ on each side.

Thorax (Fig. 302). Only the pronotum sclerotized with median longitudinal suture. Slightly lighter in colour than head. Anterior margin with equally spaced bristles. Transverse furrow near posterior margin extending to pleural region. Posterior margin heavily sclerotized, black in colour. A pattern of dark marks present on the posterior third of the pronotum. Prosternal horn absent.

Legs. Prothoracic legs (Fig. 303) shortest. Metathoracic legs (Fig. 305) slightly longer than mesothoracic (Fig. 304). Tibia and tarsus of all legs covered with short bristles. All claws long and slender with bristle on inner edge, with a few fine hairs at the point of insertion of the bristles. Ventral edge of femur of all legs furnished with a row of equally spaced bristles. Anterior projection of sclerite supporting coxa of anterior legs triangular and furnished with sense organs. A few branched bristles at the distal extremity of the tarsus of meso- and metathoracic legs. Two black spines on tibia of all legs on distal ventral edge.

Abdomen (Fig. 296). Flattened dorsi-ventrally, with fringe of long fine hairs on segments one to eight in pleural region. Five small processes in middle of posterior margin on dorsal side of ninth segment are eversible gills. Anal appendages (Fig. 306) long, each of two segments and a long claw. Distal extremity of second segment oblique with a number of fine hairs on inner margin. Hairs on first segment fine and more numerous. Anal claw obtuse with row of extremely fine spicules on the proximal two-thirds of the inner edge.

Edington considers that this species is the typical polycentropid of small streams preferring head-streams and being replaced by *Polycentropus flavomaculatus* in the lower reaches of rivers. However, he also found it in still water in three high tarns.

Note on Varieties of Larval Coloration

Many species of Trichoptera show a range of coloration in the larvae due to variation in the intensity of black pigmentation. A single larva collected from South Nutfield, Surrey, in 1946, which was referable to *Plectrocnemia conspersa* showed variation from the normal coloration of quite a distinct type. The meso- and metapleura of each side were entirely devoid of pigmentation, being creamy white in colour. The remainder of the larva was of normal pinky-brown coloration. This gave the effect of four distinct pale areas when the larva was viewed from above.

Plectrocnemia geniculata McLachlan

Larvae of this species have been collected by Edington from Northumberland, Lancashire and Devon and their specific characteristics noted. Larvae of *Plectrocnemia* are the largest of the polycentropid larvae. The pattern of the head capsule of *P. geniculata* is shown in Fig. 309, but it differs a little from *P. conspersa*. Whilst Ulmer states that in *P. geniculata* the head and pronotum are paler than in the latter species, which may be generally true, some larvae of *P. conspersa* are also light in colour.

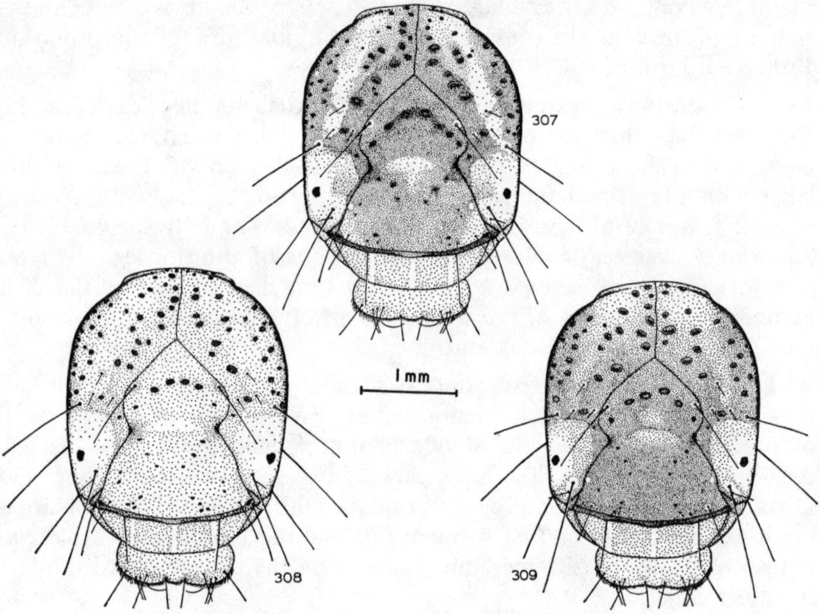

Figs. 307–308. *Plectrocnemia conspersa.*
Fig. 309. *Plectrocnemia geniculata.* Edington.

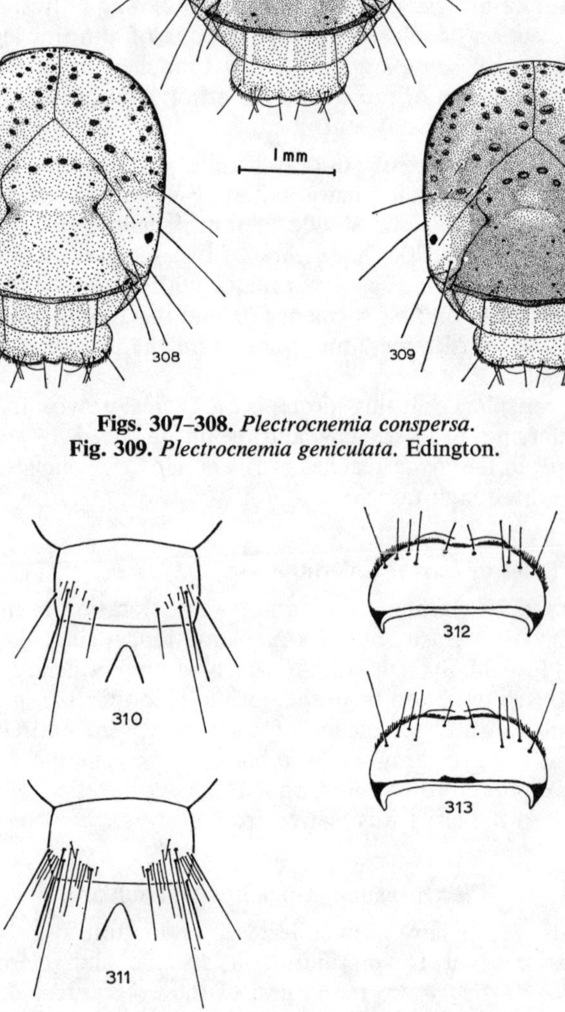

Figs. 310–313. 310. Ventral surface of ninth abdominal segment of *Plectrocnemia geniculata*; 311. Ventral surface of ninth abdominal segment of *Plectrocnemia conspersa*; 312. Labrum (dorsal surface) of *Plectrocnemia geniculata*; 313. Labrum (dorsal surface) of *Plectrocnemia conspersa.* Edington.

Edington (1964) states that the two species are most easily separated by the arrangement of bristles on the ventral side of the last (ninth) abdominal segment. Although in both species four pairs of primary bristles are present, the distinction lies in the number and length of the secondary bristles. Whereas in *P. geniculata* there are only about six pairs of these, and most of them are short (Fig. 310), in *P. conspersa* there are about twelve pairs, and most of them are long (Fig. 311). There is a small distinguishing feature in the labrum. In the case of *P. geniculata* the thin dark line at the posterior margin is uniform (Fig. 312), but in *P. conspersa* there is an elongated expanded area in the median position (Fig. 313) not shown in Hickin's figure. In fifth instar larvae the coloration of the head is much less variable in *P. geniculata* than in *P. conspersa*. In some individuals of the latter species the head is dull yellow with slightly paler areas around the eyes, and at the constriction of the clypeus (the apotome of Edington). In others the head is brown with a yellow band on the posterior part of each parietal. These bands are located about mid-way between the lateral margins of the head and the coronal and frontal sutures. The area around the eyes remains yellow and there is a lighter area at the constriction of the clypeus (apotome). Edington goes on to state that the pigmentation pattern of *P. geniculata* appears to be less variable and in specimens where there are areas of different pigment intensities these are generally not sharply defined.

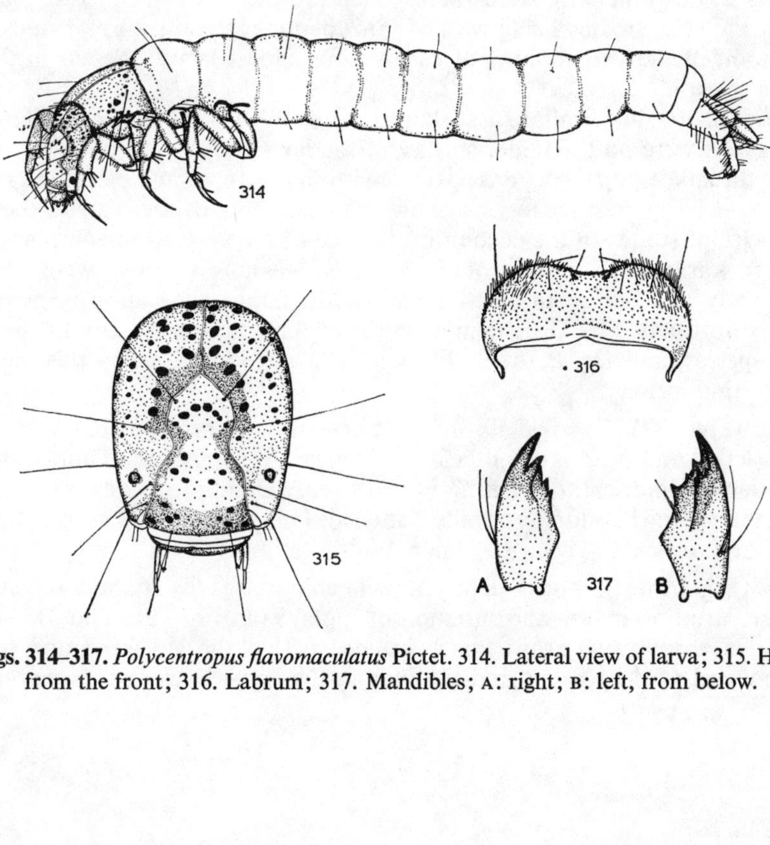

Figs. 314–317. *Polycentropus flavomaculatus* Pictet. 314. Lateral view of larva; 315. Head from the front; 316. Labrum; 317. Mandibles; A: right; B: left, from below.

The appearance of the spots on the lateral margins of the head between the eyes and the occipital foramen, differs in the two species. Whereas in *P. geniculata* the rather indistinct spots are only faintly outlined and the centres are pale, in *P. conspersa* the distinct spots possess a dark outline and tend to have a darker centre.

Polycentropus flavomaculatus Pictet

This most abundant caddis fly is found almost everywhere where moving fresh water is found—not only in streams and rivers but also on the shores of lakes where there is some wave action.

On 31 July, and again on 6 August 1951, I found larvae and pupae of this species exceedingly abundant, together with some hatched adults, in the River Breamish about a mile east of Ingram in Northumberland.

The River Breamish was the scene in the late summer of 1949 of very extensive flooding, the violence of which caused the river-bed to be scoured out by large boulders, destroying all vegetation in the river and washing away all silt. At the present time the bed of the river is composed only of these boulders, the river flowing swiftly over them. The bed of the wide flood channel is similarly composed of boulders, with more slowly flowing subsidiary streams running off the main river and running into it again. These side streams were margined with *Mimulus* and other early colonizing plants in this otherwise sterile area. Small crustaceans and other animals from the side streams being washed into the main stream were obviously the food of the vast number of carnivorous caddis larvae present in the main stream.

The small silk swallow-nest-shaped nets of the *Polycentropus flavomaculatus* were on the under surface of the large stones and boulders, as were the silken pupal cocoons covered with small stones and pieces of grit. It would be interesting to learn how the larva obtains and carries these particles of stones to the cocoon in the face of the very strong current of water. Larvae and pupae of *Rhyacophila dorsalis* (Curtis) were also extremely abundant, but case-bearing caddis larvae were almost absent. I have to thank Mr. D. E. Kimmins for confirming the identity of these two species, and Dr. E. A. R. Ennion, who introduced me to this most interesting locality.

Larva (Fig. 314). Campodeiform, head probably procentrous but larva is very active and head is often held in orthocentrous position. Thorax and abdomen cylindrical, decreasing in width only slightly towards posterior end and somewhat dorsoventrally flattened. Length 14 mm. This is rather larger than the size given by Ulmer. Width 2 mm.

Head (Fig. 315). Elliptical, light yellow in colour, heavily spotted in black (some variation in size and situation of spots) as in Figs. 315 and 319; in addition a yellowish-brown mark extends within the genae around the aboral end of the clypeus and also within the clypeus around its margin

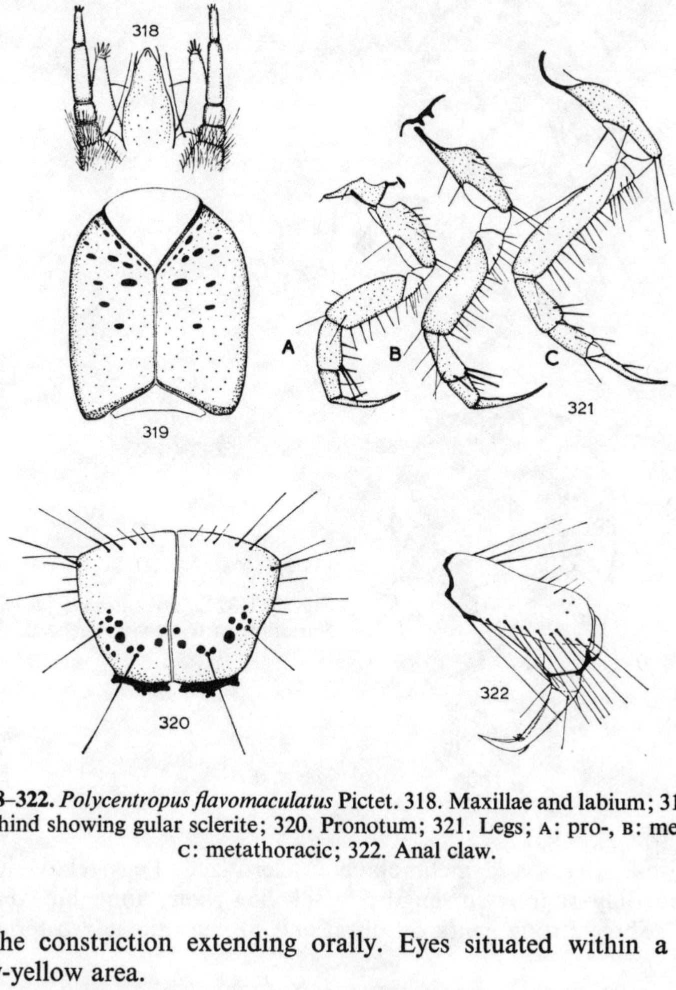

Figs. 318–322. *Polycentropus flavomaculatus* Pictet. 318. Maxillae and labium; 319. Head from behind showing gular sclerite; 320. Pronotum; 321. Legs; A: pro-, B: meso-, and C: metathoracic; 322. Anal claw.

from the constriction extending orally. Eyes situated within a lighter creamy-yellow area.

Gular sclerite. Transverse, genal suture long.

Antennae. Rudimentary.

Labrum (Fig. 316) transverse, sclerotized.

Mandible. Only left mandible has internal brush of hairs (Fig. 317).

Maxilla (Fig. 318). Palp long, four-segmented, third segment much longer than the others. Tips of sense organs at end of galea just reach end of third segment of palp.

Labium (Fig. 318). Conical, palps absent.

Thorax. Pronotum sclerotized, wider anteriorly, light yellow with W-shaped pattern of black spots as in Fig. 320; posterior margin black, heavily

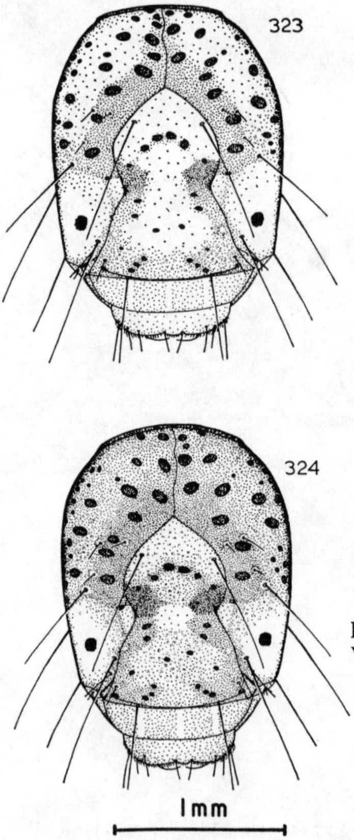

Figs. 323–324. *Polycentropus flavomaculatus*. Variation in front view of head. Edington.

sclerotized. Meso- and metanota not sclerotized. Legs relatively short, differing only slightly in length; tarsal claws very long but very little curved. Three strong spurs on distal end of each tibia. Prosternal horn absent.

Abdomen. Colour variable on dorsal surface—reddish brown, buff, yellowish brown or greenish are common. Whitish on ventral surface. Gills absent, lateral line absent. Abdominal segments distinct but not deeply intersected. Anal appendages each of two long segments, with the terminal claw (Fig. 322) at right angles to them. The claw is slightly scoop-shaped, with two rows of minute spicules on the inner face terminating distally at a group of somewhat larger spicules. Auxiliary claw present as a sharp bristle, closely adpressed, can be seen only with difficulty. Blunt projection on inner face of tip of claw absent.

Edington 1964, has recently studied the larvae of this species. He points out that the anal claw is right-angled. The illustration of my 1952 paper (Fig. 322) does not show quite this character. In addition reference is made

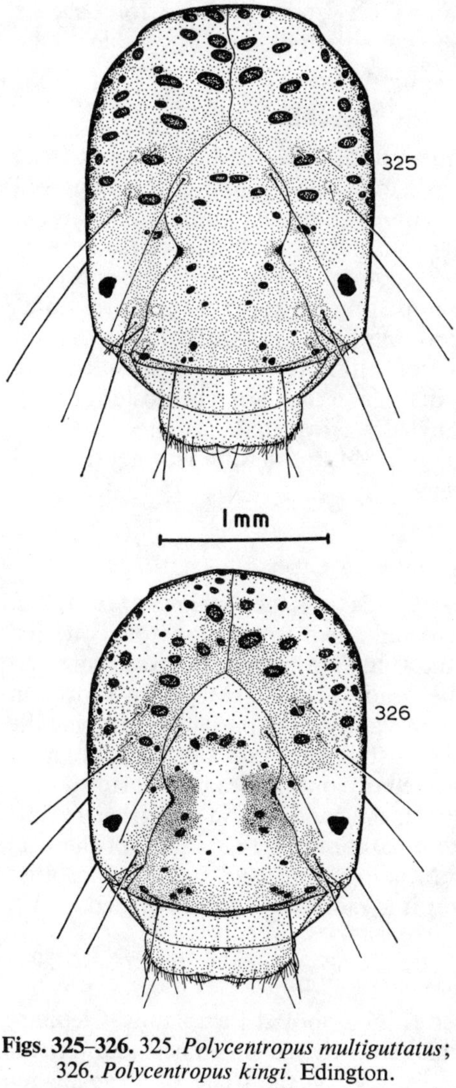

Figs. 325–326. 325. *Polycentropus multiguttatus*; 326. *Polycentropus kingi*. Edington.

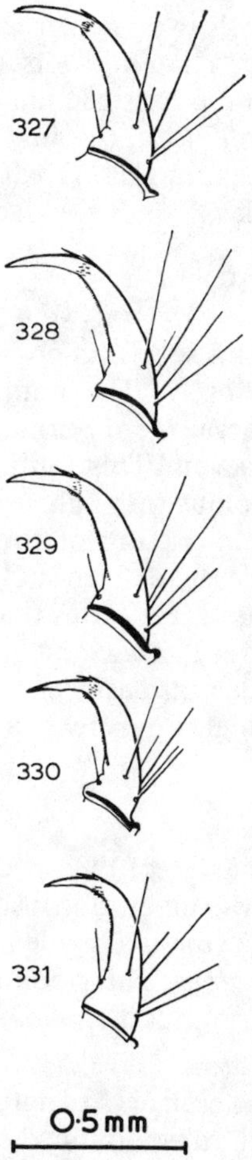

Figs. 327–331. Anal claws, last instar larvae, lateral views of *Plectrocnemia conspersa* (327); *Polycentropus multiguttatus* (328); *Polycentropus kingi* (329); *Polycentropus flavomaculatus* (330); *Holocentropus picicornis* (331). Edington.

to an auxiliary claw situated on the anal claw, whereas Edington believes that there are two such 'dorsal' spines although they are liable to be broken off.

Polycentropus kingi McLachlan

The larva of this species was unknown until collected by Edington from the River Rothay at Ambleside, Westmorland, where a few occurred amongst numerous examples of *P. flavomaculatus*. According to Edington (1964) the head markings of *P. kingi*, and the last-named species, are very similar. In the posterior region of the apotome (clypeus) of both species is an arc of spots (muscle insertions) consisting typically of four large spots with two smaller and separate lateral ones. Whereas in *P. flavomaculatus* these lateral spots are usually contained within the darkly pigmented area at the construction of the apotome, in *P. kingi* they often lie posterior to it. In *P. kingi* the anal claw is obtuse-angled whereas in *P. flavomaculatus* it is right-angled. Great care must be taken to ensure that, in viewing the anal claws, their side-faces are at right-angles.

Polycentropus irroratus Curtis

This species is given as *multiguttatus* (Curtis) in Kloet & Hincks (1964) and Mosely. The immature stages were unknown until Edington collected larvae from Stirlingshire, Northumberland, Westmorland, Lancashire and Devon. This author states that the head (Fig. 325) is yellow-brown in colour with lighter areas around the eyes and on the parietals alongside the constriction of the apotome (clypeus). The spots on the head are very clearly defined and there are no marked discontinuities of pigment on the head. He states that this contrasts with the condition in *P. flavomaculatus* and *P. kingi* and affords a means of separation. The curvature of the anal claw also assists in this regard, as in *P. multiguttatus* and *P. kingi* it is obtuse angled, whereas in *P. flavomaculatus* it is more nearly right-angled.

Holocentropus dubius Ramb.

Larvae of this species were collected from a pool at Barnslands, Cleobury Mortimer, Shropshire. The larvae had constructed silken nets and retreats on the underside of the floating leaves and on the stems of *Potamogeton natans*. This pool has a margin of trees but has very little vegetation apart from the *Potamogeton* already referred to.

Larva (Fig. 332). Campodeiform, head procentrous, only prothorax sclerotized. Length 12–14 mm., width 2.0–2.5 mm., widest at about the third abdominal segment. Abdomen flattened dorso-ventrally, tapering distally with deep intersegmentation.

Head. Light golden-brown, elliptical with genae parallel. Prominent transverse ridge for insertion of neck muscles. Irregular and variable pattern

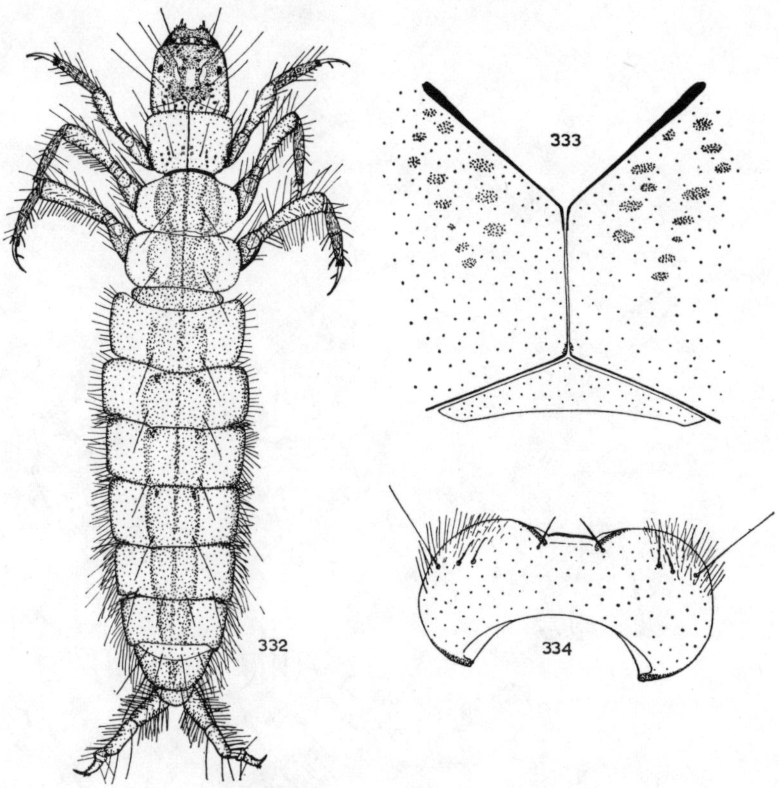

Figs. 332–334. *Holocentropus dubius*. 332. Larva (dorsal view); 333. Gular sclerite; 334. Labrum.

of dark marks at aboral end of head. Clypeus greyish-brown with a medium light golden-brown rectangular mark at the aboral end and a pair of triangular light marks fusing with two light marks on the genae. Except for a patch near the aboral end of the clypeus, the region of the genae adjacent to the clypeus is greyish-brown. The anteclypeus is large and well defined. It is divided into four dark areas by three longitudinal light marks. Antennae very small.

Mouthparts. Labrum (Fig. 334) with concave posterior margin and hemispherical lateral margins. Central region of anterior margin concave with flat cushion-like protuberance.

Mandibles (Fig. 335). Acute, inner face concave. Thin black line almost encircling base.

Maxillae (Fig. 336). Maxillary palp of four segments, third segment long, a small sclerotized patch at base of palp and one on each side adjacent to

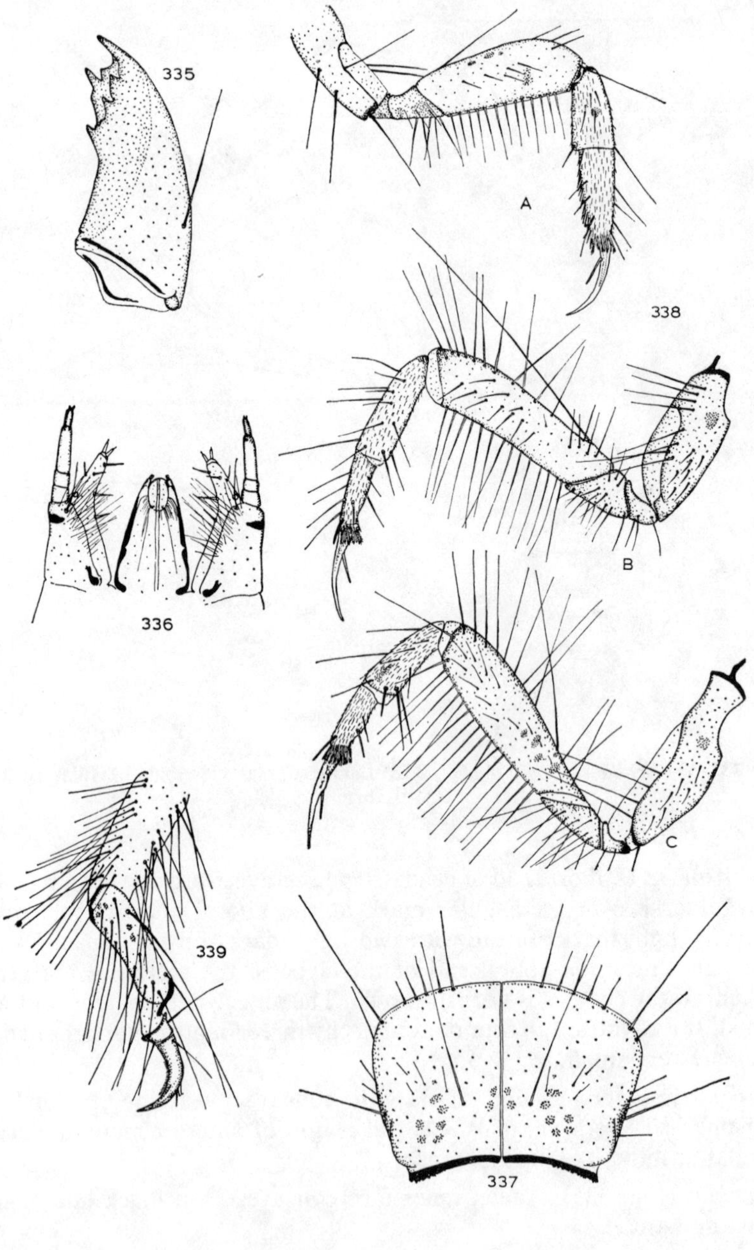

Figs. 335–339. *Holocentropus dubius*. 335. Mandible; 336. Maxilla and labium; 337. Prothorax; 338. Legs; A: prothoracic; B: mesothoracic; C: metathoracic; 339. Anal claw.

insertion of base of labium into mentum. Tip of maxillary palp slightly darker in colour. Mala extends approximately to the centre of the third (long) segment of the maxillary palp. Four spines on the mala are directed inwards and one small one outwards.

Labium. Conical spinneret well defined and terminates in an oval organ situated between the small labial palps.

Gular sclerite (Fig. 333). Narrow, transverse, does not divide the genae.

Thorax. Prothorax (Fig. 337) sclerotized with longitudinal median suture. Posterior margin heavily sclerotized and black in colour. Bristles situated on anterior margin but not in central region near the suture. One long bristle on each side in pleural region. A few dark marks in posterior half of prothorax with bristles of unequal length extending transversely, a dark patch on each side. Meso- and metathorax elliptical, lighter at the sides, metathorax slightly wider than mesothorax. Prosternal horn absent.

Legs (Fig. 338, a, b and c). In each case tibial and tarsal segments covered with fine hairs. Claw long, with base encircled by group of branched bristles. Claw spine on mesothoracic legs approximately half-way along claw. In each case ventral edge of femur with equally spaced bristles.

Abdomen. First segment narrow and thin, segments two to eight darker in the centre on the dorsal surface, and with fairly long hairs in pleural region. Posterior margin of seventh segment with fold of skin partly overlapping segment eight. Anal claw (Fig. 339) long, with fringe of spinules on concave edge and three spines (one large and two small) on the dorsal convex edge. Claw strongly curved. The two long supporting segments each covered with many long black hairs.

Edington states that the intensity of pigmentation on the head (Fig. 340) is much less than in the case of *H. picicornis* and *H. stagnalis*, and its distribution is different. There is no continuous light stripe on the apotome but there is a yellow rectangle which extends backwards from the constriction. There are light areas on the parietals alongside the constriction and these are continuous with two further light areas on the apotome lying anterior to the first pair. The remainder of the apotome and adjacent parietal area is lightly pigmented. There is no marked pigmentation around the coronal suture. These head markings serve to separate *H. dubius* quite easily from the other two species in the genus.

Holocentropus picicornis (Stephens)

Edington (1964) collected larvae of this species from still water in Northumberland, Lancashire and Hertford, where they were often associated with *Lemna trisulca* L. This author states that the head of *H. picicornis* (Fig. 341) and *H. stagnalis* (Fig. 342) larvae bears dark bands of pigment originating from along the coronal suture. They enclose the extreme posterior apex of

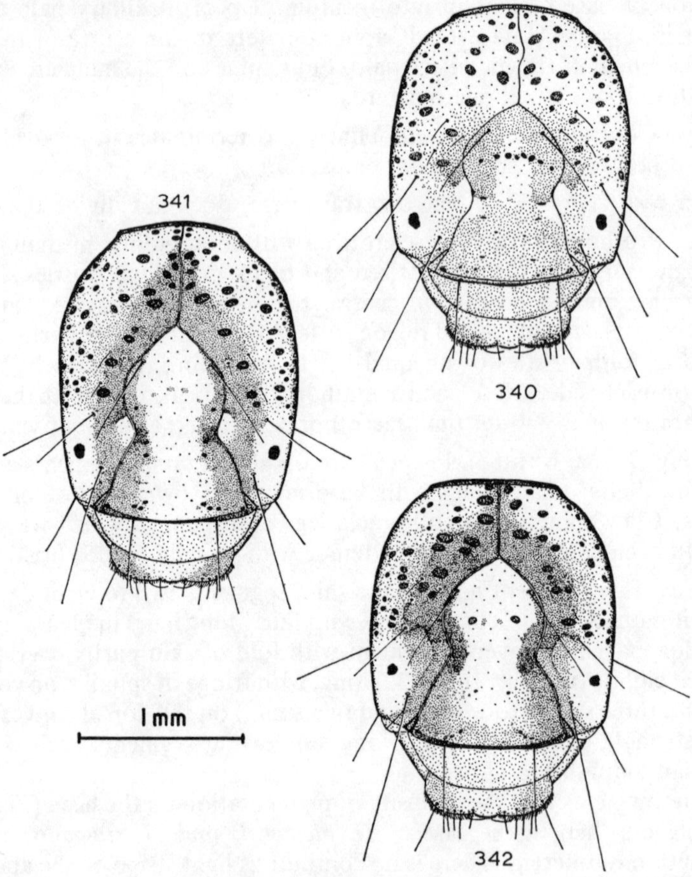

Fig. 340. *Holocentropus dubius.*
Fig. 341. *Holocentropus picicornis.*
Fig. 342. *Holocentropus stagnalis.*
Edington.

the apotome (clypeus) and then pass forward, spreading across the lateral margins of the apotome and the adjacent regions of the parietals. A yellow longitudinal stripe remains on the apotome. Another useful character serving to identify this species from *H. stagnalis* is the presence of distinct light patches on the parietals alongside the constriction of the apotome. They are absent from fifth instar larvae of *H. stagnalis*. In addition, the head of *H. picicornis* is narrow and elongated whilst that of *H. stagnalis* is broader and shorter. The nets of this species are often to be found on the underside of the leaves of *Nymphaea*, *Potamogeton natans*, and *Lemna trisulca*, as mentioned above.

Holocentropus stagnalis Albarda

Larvae of this species were collected by Edington from small ponds in Northumberland. He did not consider that the character given both by Silfvenius and Ulmer for the separation of this species from *H. picicornis* was satisfactory. This concerned the width of the brown band in the region of the apotome. On the other hand, a good separation was found in that the anterior region of the apotome was light-coloured in *H. picicornis*, but dark-coloured in *H. stagnalis*. In addition, in *H. picicornis* the light stripe continues to the anterior margin of the apotome but in *H. stagnalis* the light stripe is invaded by pigment at its anterior end (Fig. 342).

When this species was known only from Shapwick Heath in Somerset I accidentally reared a large number of adults in an aquarium in which were larvae of *Oligotricha ruficrus*. The water-weed for the aquarium was collected from Broadmoor Wood on the outskirts of Birmingham in Worcestershire. This series is in the B.M. collection.

Cyrnus flavidus McLachlan

Macdonald (1950) collected larvae of this species from Loch Lomond, Scotland, and his description is given below.

Larva. The larva is campodeiform and has no case. The head and prothorax are sclerotized, the rest of the body is soft. The abdomen is slightly flattened dorso-ventrally. In a fully-grown larva the length is about 14 mm. and the width 2 mm. It is widest at the third and fourth abdominal segments.

Head. The head (Fig. 343) is prognathous, with the genae running parallel for most of its length. It is golden yellow in colour, with light-brown spots, which are not very distinct. The width of the clypeus at its anterior margin is almost as great as its length. Near the middle of this region there is a clear patch with a transverse row of about eight spots posteriorly, and two pairs antero-laterally. At the apex of the clypeus there is another clear area, triangular in shape and bounded on two sides by the converging frontal sutures. Near the anterior margin of the clypeus are two short oblique series of spots. On the parietals, at the level of the central clear patch of the clypeus, there is a fusiform light area, bounded on the inner side by the frontal suture. The posterior region of the parietals and genae have a large number of irregularly placed brownish spots. The chaetotaxy is as shown in the figure. The antennae arise on the anterior margin of the parietals and each is reduced to a small plate bearing three or four small setae and one or two sensory papillae. The labrum (Fig. 344) has an anterior median emargination with a flat base, from each side of which arises a seta. Each antero-lateral region bears four setae and a brush of marginal hairs. The mandibles are acute and asymmetrical. The left mandible (Fig. 346) has four teeth on the upper side and three on the lower, two of the latter being very small. The right mandible (Fig. 347) has three teeth on both the upper and lower sides, and again two of the lower teeth are very small. Only the left mandible has an inner brush of three or four setae, but both have two setae arising from the outer margin. The maxilla (Fig. 345) has a five-segmented palp, the third segment being very long. The basal segment bears large numbers of fine hairs and a strong seta on the inner distal margin. The maxillary lobe bears two papillae at its tip, and one or two spines on its inner margin. The labium (Fig. 345) has an acute ligula with the palps lying close to it, and a broad mentum bearing two setae. The gular sclerite (Fig. 345) is narrow, transverse and triangular and does not divide the genae.

Thorax. The pronotum (Fig. 343) is golden yellow in colour, except along the posterior margin, which is dark brown. There is a transverse furrow posteriorly, extending to the pleural regions. Scattered over the surface are a number of light-brown spots. There is no prosternal horn. The meso- and metanotum are soft and do not bear any sclerites. The legs are all strongly built, lengthening progressively posteriorly. The prothoracic leg (Fig. 348) has a regular series of setae on the under surfaces of the trochanter and femur. From the upper borders of the tibia and tarsus arise large numbers of setae, and on the lower border of the latter segment there is a series of small spinules. The mesothoracic (Fig. 349) and metathoracic (Fig. 350) legs resemble each other in many points, each having a series of fine setae on the lower margin of the trochanter and a large number of strong setae on the upper and lower surfaces of the femur. The tibia and tarsus of both have an upper and lower series of setae, and the most distal seta of the lower tarsal series is plumose. On each tarsus also there is a small spine on the upper margin at the distal end. The tarsal claw of all three legs has a ventral spine near its base, around which is a series of fine hairs and several small plumose setae.

Abdomen. The abdomen is slightly flattened dorso-ventrally, with clearly defined segments. Each segment has a faint pattern dorsally and there is a fringe of long fine hairs laterally on segments one to eight. On the dorsal side of the ninth segment are five small eversible gills. The anal claw (Figs. 351, 352) is long and strongly curved, with two fringes of spinules on the inner margin extending to the angle of the claw, where they meet a group of small spines. Distally the inner edge is barbed, forming four blunt teeth pointing towards the base. On the outer side, at the angle of the claw, there is a strong projecting spine. From the proximal region of the claw arise a number of long setae. The two supporting segments, which with the claw form the anal appendage, are densely fringed with long fine setae.

Edington's larvae were obtained from various tarns and lakes in the Lake District. He found that the statement of Silfvenius that the larvae of this species can be separated from those of the other two species by the indistinctness of the spots on the pronotum did not apply to British

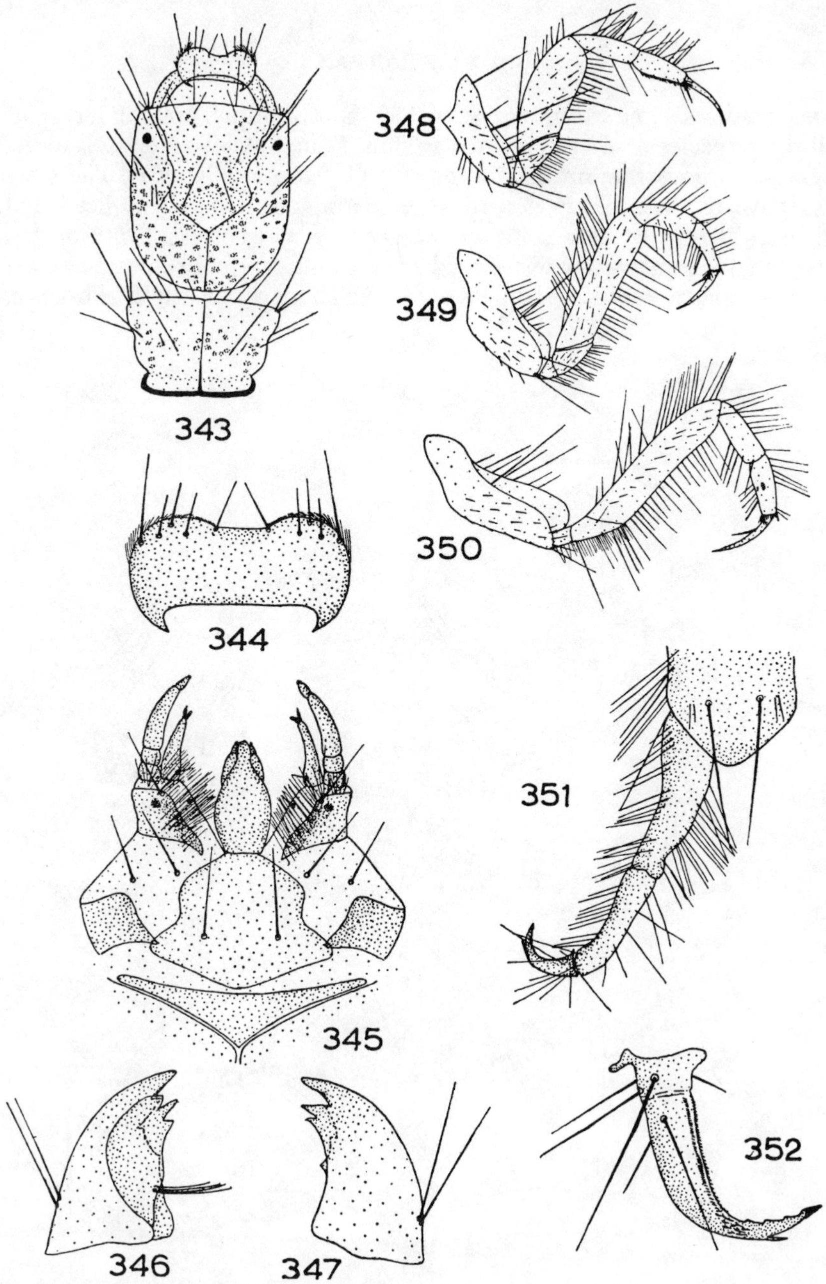

Figs. 343–352. *Cyrnus flavidus* McLachlan. 343. Head and prothorax; 344. Labrum; 345. Labium, maxillae and gular sclerite; 346. Left mandible (dorsal view—dotted line represents ventral teeth); 347. Right mandible (dorsal view—dotted line represents ventral teeth); 348. Right prothoracic leg; 349. Right mesothoracic leg; 350. Right metathoracic leg; 351. Anal segment and appendage (dorsal view); 352. Anal claw (ventral view). Macdonald.

specimens, as some can be found with distinct spotting. Having larvae of all three species of *Cyrnus* in front of him, Edington calls attention to the light areas occurring on the apotome of *C. flavidus* (Fig. 364). The most extensive of these occupies a roughly central position on the head and includes within its area some of the spots in the posterior row on the apotome. Symmetrically distributed around this central area, and also on the apotome, are four round light areas, which mark the origin of bristles.

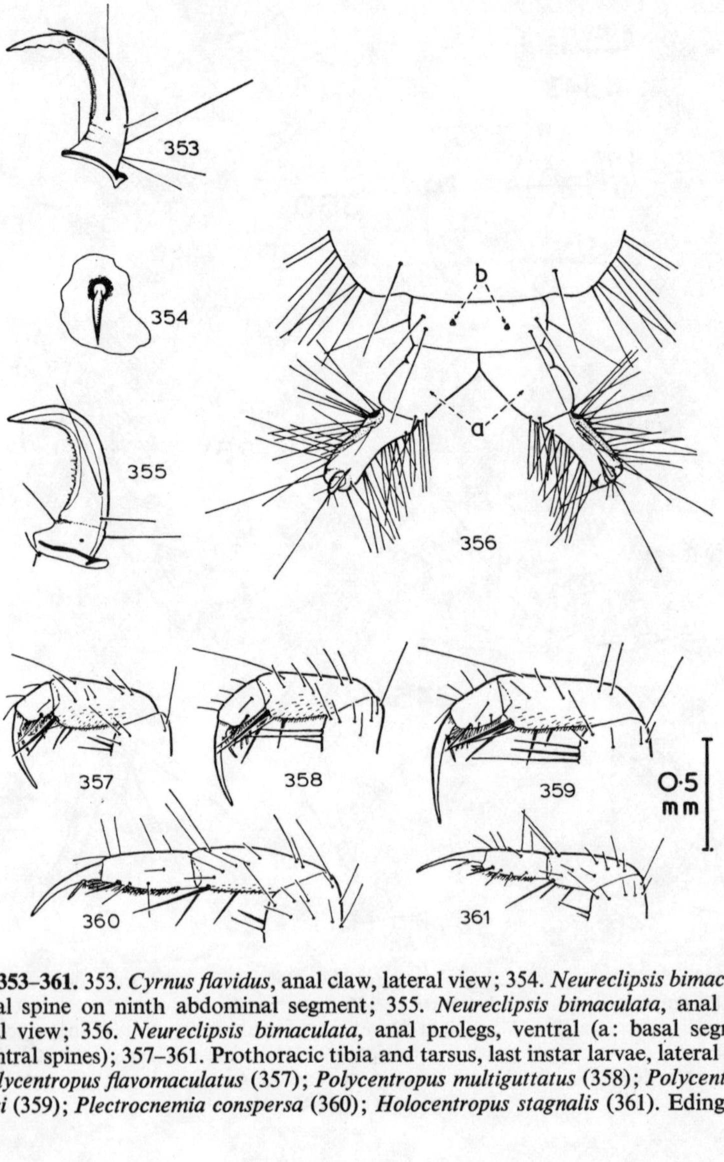

Figs. 353–361. 353. *Cyrnus flavidus*, anal claw, lateral view; 354. *Neureclipsis bimaculata*, ventral spine on ninth abdominal segment; 355. *Neureclipsis bimaculata*, anal claw, lateral view; 356. *Neureclipsis bimaculata*, anal prolegs, ventral (a: basal segment; b: ventral spines); 357–361. Prothoracic tibia and tarsus, last instar larvae, lateral views of *Polycentropus flavomaculatus* (357); *Polycentropus multiguttatus* (358); *Polycentropus kingi* (359); *Plectrocnemia conspersa* (360); *Holocentropus stagnalis* (361). Edington.

At the posterior apex of the apotome a light area is typically, but not invariably present. On the parietals alongside the constriction of the apotome is a pair of distinct light areas.

Edington points out that the head capsules of *C. flavidus* and *H. dubius* are somewhat similar but can be separated as follows. (*H. dubius*, of course, lacks the four blunt teeth on the anal claw which are possessed by *C. flavidus*.) In the case of *H. dubius* there are light areas on the parietals on either side of the constriction of the apotome. Continuous with these and lying anterior to them are two further light areas. The second light areas lie on the apotome. In *C. flavidus*, on the other hand, in this anterior position there are only two round light bristle bases and these are not continuous with the light areas on the parietals.

Cyrnus insolutus McLachlan

Edington states that this species still occurs at Blelham Tarn in the Lake District where it was discovered by Kimmins in 1942, and he obtained his larvae from this still unique locality. Edington found that the distribution

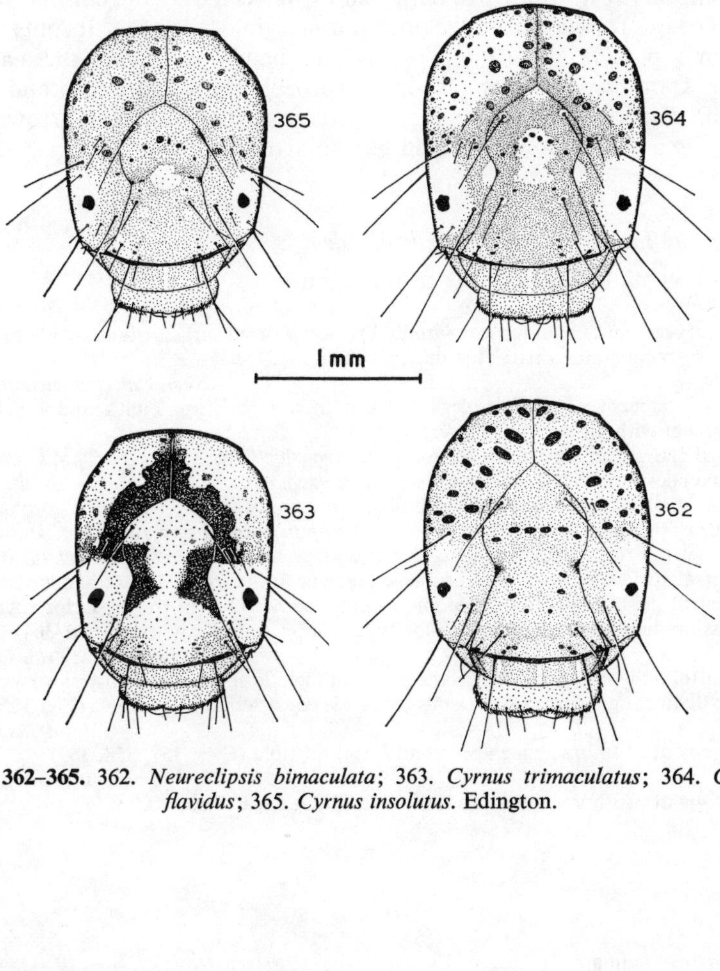

Figs. 362–365. 362. *Neureclipsis bimaculata*; 363. *Cyrnus trimaculatus*; 364. *Cyrnus flavidus*; 365. *Cyrnus insolutus*. Edington.

of the light areas on the head contrasted with the pattern as shown by *C. flavidus*. Thus, in *C. insolutus* (Fig. 365), the central light area is small and does not reach the posterior row of spots on the apotome. There is no light area at the posterior apex of the apotome nor are there distinct light areas on the parietals alongside the constriction of the apotome.

Cyrnus trimaculatus Curtis

Larvae of all three British species of *Cyrnus* have been collected by Edington (1964) who examined them for separable characters, checking with the descriptions of Silfvenius from Finnish material.

Larvae of *C. trimaculatus* were found in a number of tarns, lakes and streams in the Lake District.

The larvae of all species of *Cyrnus* are furnished with four blunt teeth on the concave margin of the anal claw (Fig. 353). This can be seen even in the case of early instar larvae. The larva of *C. trimaculatus* is most distinctive on account of the presence of dark bands on the front of the head capsule (Fig. 363). The dark band is bissected by the coronal suture which passes forward from the posterior margin of the head. It splits at the apotome, passing along the parietals. The bands pass inside the margins of the apotome, however, as they approach the constriction and then become dispersed. The pigmented area is thus shaped like an arrowhead, enclosing a pale area also shaped like an arrowhead.

Key to the Larvae of the Polycentropidae

The following key is that due to Edington

1. Basal segment of anal prolegs without bristles. Ventral surface of ninth abdominal segment bears a pair of stout spines (Figs. 354, 356)
   Neureclipsis bimaculata (L.)

— Basal segment of anal prolegs bears numerous bristles. Ninth abdominal segment without ventral spines 2

2. Anal claws bear four blunt teeth on concave edge (Fig. 353) *Cyrnus* spp. 3

— Anal claws without four blunt teeth on concave edge 5

3. Dorsal surface of head with dark pigmented bands. These lie on the coronal suture, surround the posterior apex of the apotome and pass on to the lateral margins of the apotome at its constriction (Fig. 363) *C. trimaculatus* (Curt.)

— Dorsal surface of head does not have pigment in the form of dark bands 4

4. Central light area on the apotome includes some spots in the posterior row. Distinct light area on parietals alongside constriction of apotome (Fig. 364)
   C. flavidus McL.

— Central light area on the apotome does not include spots in the posterior row. No distinct light areas on parietals alongside constriction of apotome (Fig. 365)
   C. insolutus McL.

5. Tarsus of prothoracic leg about half length of tibia (Figs. 357, 358, 359)
   Polycentropus spp. 6

— Tarsus of prothoracic leg about same length as tibia (Figs. 360, 361) 8

6. Dorsal surface of head without marked discontinuities of pigment (Fig. 325) *P. irroratus* (Curt.)
— Dorsal surface of head with marked discontinuities of pigment (Figs. 326, 324, 323) 7
7. Anal claw obtuse-angled (Fig. 329) *P. kingi* McL.
— Anal claw right-angled (Fig. 330) *P. flavomaculatus* (Pict.)
8. Anal claw obtuse-angled (Fig. 327) *Plectrocnemia* spp. 9
— Anal claw right-angled (Fig. 331) *Holocentropus* spp. 10
9. Bristles on ventral side of last abdominal segment numerous (Fig. 311). Median pigment mark on posterior margin of labrum (Fig. 313) *P. conspersa* (Curt.)
— Bristles on ventral side of last abdominal segment sparse (Fig. 310). Median pigment mark on posterior margin of labrum absent (Fig. 312) *P. geniculata* McL.
10. Dorsal surface of head with dark bands of pigmentation. Pale median stripe on apotome (Figs. 342, 341) 11
— Dorsal surface of head without distinct dark bands of pigmentation. No continuous median stripe on apotome (Fig. 340) *H. dubius* (Ramb.)
11. Head broad and short. Pale median stripe on apotome clouded with pigment at anterior end. No light areas alongside constriction of apotome (Fig. 342) *H. stagnalis* (Alb.)
— Head narrow and elongate. Pale median stripe on apotome clear of pigment at its anterior end. Light areas alongside constriction of apotome (Fig. 341) *H. picicornis* (Steph.)

BRICKENSTEIN, C. 1955. Über die Netzbau der Larve von *Neureclipsis bimaculata* (L.). *Abh. bayer. Akad.* **69**, 1–44.

BRINDLE, A. 1960. *Ent. Rec.* **72**, 244–5.

EDINGTON, J. M. 1964. Taxonomy of British Polycentropid. Larvae (TRICHOPTERA). *Proc. zoo. Soc. Lond.* **143** (2), 281–300.

EDINGTON, J. M. 1964. *Proc. zool. Soc. Lond.* **143** (2), 281–300.

EDINGTON, J. M. 1965. *Mitt. int. Verein. theor. angew. Limnol.* **13**, 40–8.

HICKIN, N. E. 1943. *Proc. R. ent. Soc. Lond.* (A) **18**, 19–21.

HICKIN, N. E. 1947. *Proc. R. ent. Soc. Lond.* (A) **22**, 114–17.

KIMMINS, D. E. 1942. *Cyrnus insolutus* McL. (TRICHOPTERA) new to Britain, *Entomologist* **75**, 66–8.

LESTAGE, J. A., *in* ROUSSEAU, E. 1921. *Les Larves et Nymphes Aquatiques des Insectes d'Europe*, 487–96.

MACDONALD, W. W. 1950. *Proc. R. ent. Soc. Lond.* (A), **25**, 19–28.

SILFVENIUS, A. J. 1904. *Acta Soc. Fauna Flora fenn.* **26**, 3, 12.

SILFVENIUS, A. J. 1904. *Acta Soc. Fauna Flora fenn.* **24**, 7, 10.

SILFVENIUS, A. J. 1905–6. *Acta Soc. Fauna Flora fenn.* **27**, 124, 131–2.

SOMME, S. 1934. Varfluelarver som angriper ørnetyngel. *Norsk. Jaeg. og Fisk. Foren. Tidsskr.* **63**, 70.

ULMER, G. 1903. *Metamorph. Trichopt.*, 119.

ULMER, G. 1909. *Die Süsswasserfauna Deutschlands* **5–6**, 229–30.

WESENBERG-LUND, C. 1911. *Int. Revue ges. Hydrobiol. Hydrogr., Biolog. Suppl.*, ser. 111, 10.

WESENBERG-LUND, C. 1911. *Int. Revue ges. Hydrobiol. Hydrogr., Biolog. Suppl.*, ser. 111, 15.

WESENBERG-LUND, C. 1911. *Int. Revue ges. Hydrobiol. Hydrogr. Suppl.* **3**, 1–63.

WESENBERG-LUND, C. 1943. *Biologie der Süsswasserinsekten*, 195.

11

PSYCHOMYIIDAE

Species of five genera are represented in Britain as in the following check-list. *Tinodes pallidula* has been added to the British list since the publication of Mosely's handbook.

PSYCHOMYIIDAE

ECNOMUS
1. *tenellus* (Rambur)

TINODES
2. *assimilis* McLachlan
3. *maclachlani* Kimmins
4. *dives* (Pictet)
5. *maculicornis* (Pictet)
6. *pallidulus* McLachlan
7. *rostocki* McLachlan
8. *unicolor* (Pictet)
9. *waeneri* (L.)

LYPE
10. *phaeopa* (Stephens)
11. *reducta* (Hagen)

METALYPE
12. *fragilis* (Pictet)

PSYCHOMYIA
13. *pusilla* (Fabricius)

Distribution

Ecnomus tenellus. Inhabits lakes and ponds, rather local (M.E.M.); Lake District (G.B.R.); Berks. (M.I.C.); River Brue, Edington Burtle, Somerset (J.C.).

Tinodes assimilis. More widely distributed than *T. maclachlani*, neighbourhood of waterfalls and rock-springs (M.E.M.); Berks. (M.I.C.); Port Erin, Isle of Man and Tramore, Co. Waterford, Ireland (N.E.H.); North-East Lancs. (A.B.).

Tinodes maclachlani. Local (M.E.M.); Lake District (G.B.R.); Yorks. (J.M.B. and H.W.); Derbyshire on wet rock surfaces (J.M.B.).

Tinodes dives. Rather local, streams, springs, waterfalls and running ditches in alpine or sub-alpine districts (M.E.M.); Lake District (G.B.R.); Yorks. (J.M.B. and H.W.); Derbyshire (J.M.B.); Malham Tarn, Yorks. (P.F.H.); North-East Lancs. (A.B.).

Tinodes maculicornis. There appear to be no recent records of this species.

Tinodes pallidulus. Known only from two localities in Surrey (N.E.H.).

Tinodes rostocki. Hindhead, Surrey (M.E.M.); Near Cheadle, Staffs. (J.E.); probably this species, Malham Tarn, Yorks. (P.F.H.).

Tinodes unicolor. Local, abundant in Yorkshire (M.E.M.); Berks. (M.I.C.); Yorks. (J.M.B. and H.W.); Bristol, Glos. (N.E.H.); North-East Lancs. (A.B.).

Tinodes waeneri. Widely distributed and abundant, lakes and rivers (M.E.M.); Lake District (G.B.R.); Berks. (M.I.C.); Northumberland (G.N.); Yorks. (J.M.B. and H.W.); Derbyshire (J.M.B.); North-East Lancs. (A.B.).

Lype phaeopa. Very widely distributed and abundant (M.E.M.); Lakes Windermere and Coniston (N.E.H.); Berks. (M.I.C.); Northumberland (G.N.P.); Yorks. (J.M.B. and H.W.); Derbyshire (J.M.B.); Malham Tarn, Yorks. (P.F.H.); near Stroud, Glos., and Bletchingly, Surrey (N.E.H.).

Lype reducta. Only British examples seen by Mosely were from Somersetshire and Shropshire (M.E.M.); Berks. (M.I.C.); Oxfordshire and Herefordshire (L.W.G.); East Bergholt, Suffolk (N.E.H.).

Metalype fragilis. Rather local, abundant on the Hampshire Test and on Lough Corrib, Galway (M.E.M.); Durleigh, Somerset (J.C.); Malham Tarn, Yorks. (J.M.B., H.W. and P.F.H.).

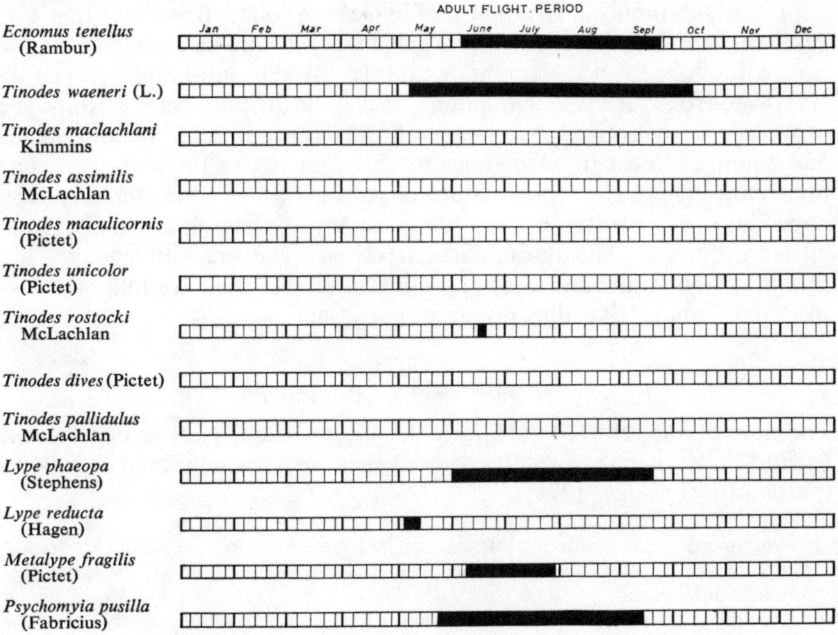

PSYCHOMYIIDAE. *Adult Flight Period taken from published records of individual dates.*

Psychomyia pusilla. One of the most widely distributed of the caddis-flies. Inhabits rivers and streams and on large rivers swarms in hundreds of thousands (M.E.M.); Lake District (G.B.R.); Berks. (M.I.C.); Northumberland (G.N.P.); Symonds Yat, Glos. (A.F.P.); Yorks. (J.M.B. and H.W.); Derbyshire (J.M.B.); North-East Lancs. (A.B.).

General Description of the Larvae

The campodeiform larvae of this family are relatively small, being only from 8 to 11 mm. in length. They neither construct transportable cases nor nets, but inhabit fixed tunnels made of silk-like material which are attached along the whole of their length to stones, rock, submerged wood or other substrate. The tunnel is usually several times the length of the inhabiting larva, being from 30 to 75 mm. in length. The tunnels often follow the course of cracks or fissures in rock or wood and sometimes become partially covered with silt of mineral and vegetable origin, making detection very difficult. Sometimes the tunnel entrance may be above water-level (*Tinodes pallidulus*, *Tinodes unicolor*), or they may be constructed in a thin water film on wet rocks (*Tinodes assimilis*). The larvae of *Ecnomus tenellus* are stated to construct their tunnels in or on freshwater sponges. The larvae of *Lype phaeopa* may bite out grooves in the soft outer layer of submerged wood. A separate and more substantial barrel-shaped cocoon is constructed of silk, to which adhere particles of extraneous material. This is formed either within the silken tube near the entrance or entirely free from it.

In the sub-family *Ecnominae*, of which the sole British example is *Ecnomus tenellus*, the pro-, meso- and metanota are sclerotized and the premental lobe of the labium is simple. In the sub-family PSYCHOMYIINAE, to which the remaining species in Britain belong, only the pronotum is sclerotized, the meso- and metanotum being membraneous and coloured in a similar manner to the segments of the abdomen. The premental lobe of the labium is prolonged in front as a slender very long appendage. The prosternal horn is absent as are abdominal gills but anal gills are present. The labrum is sclerotized. The anal appendages are prominent but short and deep. Lestage states that they are long and leg-like, and Brindle, that they are very short.

Ecnomus tenellus (Rambur)

The larva of one genus of the family, *Metalype*, Klapálek, with one species, *fragilis*, Pictet, is unknown. The other genera may be separated as follows (partly after Lestage, 1921):

1. Pro-, meso-, and metanota sclerotized; in tubes on stones in static water, reported to be often associated with fresh-water sponges, in which the larval tunnels may occur *Ecnomus*
— Only the pronotum sclerotized, the meso- and metanota membraneous 2

2. Head yellow, almost unicolorous, except for a darker area on fronto-clypeus; pronotum yellow darkened on posterior border and slightly on disc; median and hind tarsal claws with two sub-equal spines; anal claw with five prominent spines ventrally *Psychomyia*

— Head not yellow and almost unicolorous but with extensive dark markings; pronotum uniformly dark or with four subequal longitudinal yellow patches; median and hind tarsal claws with one spine; anal claw without prominent spines ventrally 3

3. Head yellowish with a very broad transverse daık parallel-sided band, the borders of which are straight and clearly defined (the band is usually interrupted at the fronto-clypeal sutures); anal claw without ventral teeth; in tubes on submerged wood in static or slowly moving water *Lype*

— Head yellow or brownish without such a band but with an irregularly bordered dark area; anal claw with five or six small ventral teeth; in tubes usually on submerged or wet rock or stones, sometimes on wood, in faster running water as a rule, though *T. waeneni* often occurs in static water *Tinodes*

Detailed Descriptions of the Larvae

Ecnomus tenellus (Rambur)

No description based on British material is available. The campodeiform larva of this local species is 8 mm. in length and 1.2 mm. in width. The Labial lobe is conical and very slender with the labial palps long and of two segments. The pro-, meso- and metanotum entirely sclerotized, the nota slightly larger in front than behind.

The larva lives in galleries made of secretion sometimes associated with the freshwater sponge *Spongilla*. They are found only in still water in ponds, canals, etc.

Tinodes assimilis McLachlan

The adult fly of this species has, for a long time, been associated with waterfalls and rock springs. Its larva was, however, unknown. I am, therefore, very grateful to Dr. E. J. Popham, who kindly gave me some larvae of this species which he had collected from Braddon Head, Port Erin, Isle of Man, on 24 March 1951. Mr. W. D. Hincks had originally collected an adult from this locality on 23 March 1950, which was identified by Mr. Henry Whitehead. Then, on 12 June 1952, I collected a number of the larvae of *Tinodes assimilis* McL. on about four square feet of wet rock surrounded by pavements and brick walls in the centre of Tramore, Co. Waterford, Ireland.

The long fixed galleries are very similar to those already described for the species of the family PSYCHOMYIIDAE. They are about 5 to 6 cm. in length and 3 to 4 mm. wide, composed of silk to which detritus is attached, and occupy irregularities in the rock surfaces over which a thin film of water is running. The pupal cocoon is built within the larval gallery. Popham (1951) has made a study of the fauna associated with thin films of water running over vertical rock surfaces—the *Fauna Hygropetrica*—in which, in certain conditions, *Tinodes assimilis* larvae play an important part.

Larva. Campodeiform, head procentrous. Head and pronotum dark greyish-brown, rest of body dark greyish-brown gradually turning to light chestnut-brown towards the hinder end. Length 8 mm., width 1.5 mm. The larva closely resembles that of *Tinodes pallidulus* McLachlan, and in this description only points of difference between them will be discussed.

Head (Fig. 366). Dark greyish-brown except for patch around eyes, light yellowish-brown. Four pairs of small light marks on the clypeus and genae (the central constriction of the clypeus), as in Fig. 366. On ventral surface of head the dark greyish colour reaches anterior margin of genae, except for patch in region of eyes. The fold at the anterior margin of genae on the ventral surface much wider than in *Tinodes pallidulus*. Labrum golden brown. Teeth of mandibles sharper than in *Tinodes pallidulus* (Fig. 368). Antennae somewhat larger and antennal bristle more prominent than in *Tinodes pallidulus*. Pronotum slightly darker than the head.

Legs. Prothoracic legs slightly larger and more robust than meso- and metathoracic legs.

Tinodes maclachlani Kimmins

This species is referred to as aureola (Zetterstedt) in Kloet & Hincks (1964) and in Mosely. No description based on British material is available. It is a

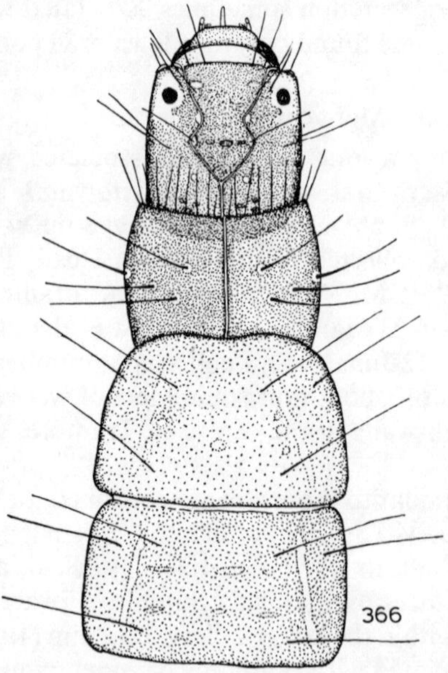

366

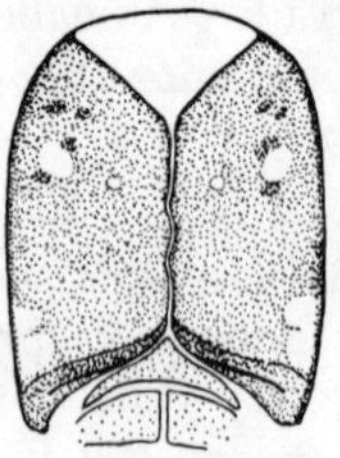

367

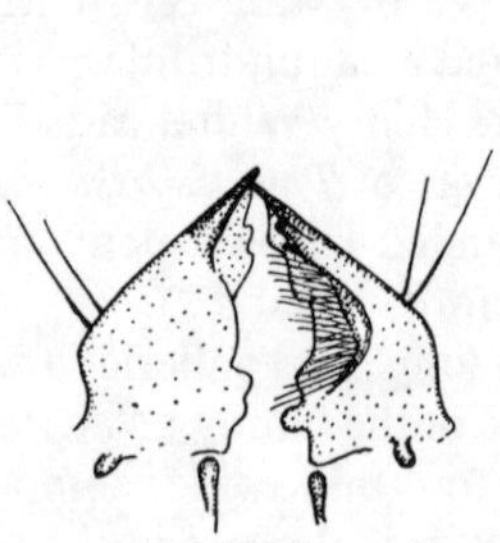

368

Figs. 366–368. *Tinodes assimilis* McLachlan; 366. Head and thorax of larva; 367. Head of larva from beneath; 368. Mandibles of larva from beneath.

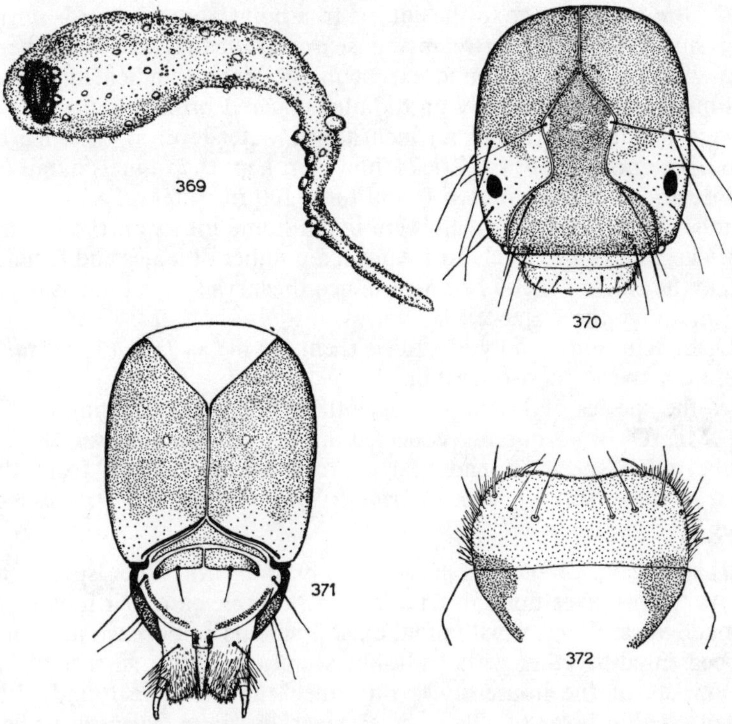

Figs. 369–372. *Tinodes pallidulus.* 369. Tunnel of larva (with pupal case within) after emergence of adult; 370. Head of larva from above; 371. Head of larva from below; 372. Labrum.

local species and the descriptions given in the literature cited below are insufficiently detailed to enable separation to be made from the other species in the genus.

Tinodes dives (Pictet)

Immature stages unknown.

Tinodes maculicornis (Pictet)

Immature stages unknown.

Tinodes pallidulus McLachlan

On 26 June 1949 I collected a batch of larvae from the stream which I have previously called the Nutfield stream, Surrey, at the point where it emerges from a culvert under the railway embankment. At this point the stream runs due north and is the parish boundary between Bletchingley and Nutfield. It is about a mile due south of Bletchingley village. The larvae were collected from the entrance to the culvert to a point about 20 yards north of this point. Although the stream was searched on this and subsequent occasions, larvae were not found except in the locality indicated above. Larval tunnels were found only on old alga-covered broken house-bricks which were in all cases about an inch above water-level at the time of collection. The porosity of the bricks, however, kept the tunnels damp on the outside and they were indeed found to be full of water.

A number of the larval tunnels were taken home intact on the broken house-bricks, and from 17 July to 1 August a number of males and females emerged in the jars in which I had maintained the larvae—the tunnels being about ½ in. above the water-level.

Mr. D. E. Kimmins kindly identified them for me as *Tinodes pallidulus* McLachlan, a species new to Britain.

Besides the species of Trichoptera mentioned here as occurring in this locality, Mr. D. E. Kimmins collected adults of *Micropterna sequax* McLachlan (this species I had reared from larvae in 1948 from the Outwood stream, Surrey), *Cyrnus trimaculatus* Curtis, and *Athripsodes bilineatus* L.

Larval Tunnel (Fig. 369). This larva in common with other species of PSYCHOMYIIDAE does not construct a transportable case, but inhabits a tunnel made of sand-grains with mud, algal filaments and occasional small pebbles cemented together with a silk-like secretion. Lower surface of the tunnel consists of the house-brick (on which tunnels were found), but coated with a thin layer of silk. Tunnel generally sinuous, posterior end tapering, the anterior end spatulate. Often difficult to observe when occupying concavities in the brick. Length very variable, often the current of the stream washes away the posterior part of the tunnel, leaving only the anterior spatulate part. Width of anterior end up to 6 mm.

Larva. Campodeiform, head procentrous. Head and pronotum dark brown, remainder of body light yellowish-green with large area on meso- and metanotum dark and with dark streak down centre of remainder of body—brownish to dark green. Length 8 mm., width 1.4 mm. Head and prothorax and posterior two abdominal segments narrower than remainder of body. The hinder four segments tapering.

Head (Figs. 370 and 371). Oval. Genae brown but region around eyes pale cream; ventral region also brown. Clypeus dark brown. Antennae very small.

Labrum (Fig. 372). Excision on anterior margin shallow. Light yellowish brown with dark brown areas in the hinder part of the lateral margins.

Mandibles (Figs. 373a and 373b). Asymmetrical, both mandibles with two long bristles on outer edge, left mandible only with extensive brush of fine hairs on inner edge. Both mandibles have auxiliary teeth at base of inner edge.

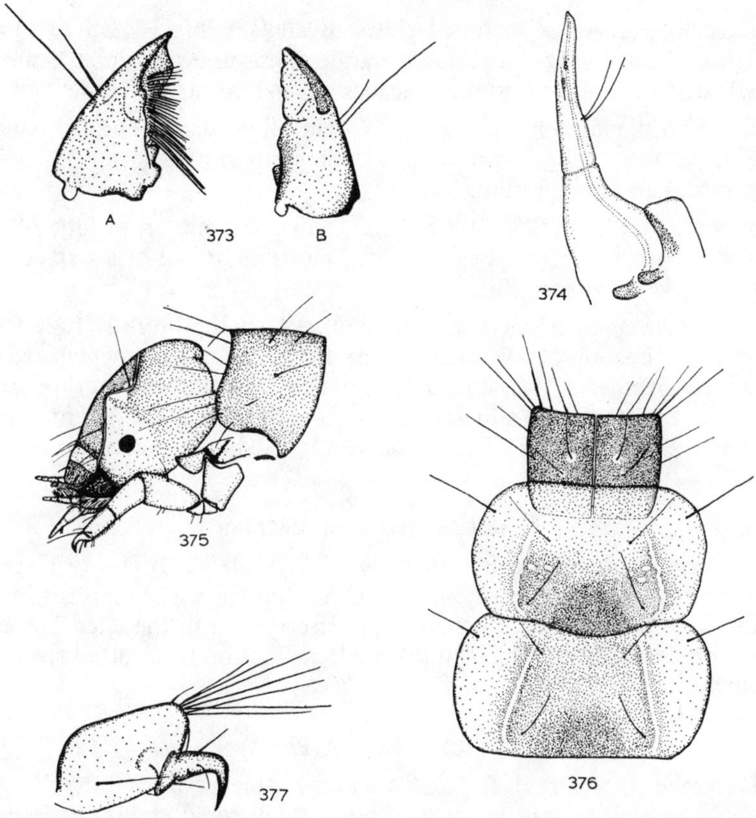

Figs. 373–377. *Tinodes pallidulus*. 373. A: left mandible; B: right mandible; 374. Labium, lateral view; 375. Head and prothorax, lateral view; 376. Thoracic nota; 377. Anal claw.

Maxillae. Palp four-segmented, third segment longer. Mala hairy, yellowish brown.

Labium (Fig. 374). Long and acicular, palps absent but pair of fine hairs on ventral surface, lateral sense organs situated between the hairs and the tip, and a transverse groove approximately half-way along the labium are well shown in this species.

Thorax (Fig. 376). Prothorax sclerotized, dark brown, but the four bristles in the centre of the pronotum arise from small light-coloured areas, the anterior pair larger than the posterior pair. Prosternal horn absent, but posterior margin of prosternum produced forward in the middle to form a large blunt sclerotized spine as in *Tinodes waeneri*. Meso- and metathorax not sclerotized, but with variable pattern of light-coloured streaks and spots formed by presence and absence of minute spicules. Prothoracic legs slightly more heavily sclerotized and more robust than meso- and metathoracic legs. Bristles on tarsal claws of all legs in addition to spines. Fringe of hairs along entire hind margin of prothoracic tarsi. Anterior sclerite on trochantin of prothoracic legs not so large as in *Tinodes waeneri*.

Abdomen. Intersegmentation deep. Lateral line absent, branchial gills absent, but five anal gills present. Variable pattern of light marks formed as in meso- and metanotum.

Anal claw (Fig. 377). Sclerotized ring at posterior margin of lobe of anal claw well shown in cleared specimens. The series of five or six spicules at base of claw not discernible.

Note on Formation of Cocoon (Fig. 369). When the larva is fully fed it constructs a cocoon of silk and small particles of sand, etc., which are well cemented together. This is constructed within the enlarged anterior part of the larval tunnel, and is adherent to the base of the tunnel. After the pupa has emerged the cocoon still remains firmly fixed in the larval tunnel.

Tinodes rostocki McLachlan

No description based on British material is available. It is a rare species hitherto collected only from Surrey and Staffordshire and searches for the larvae have failed to reveal them. The descriptions in the cited literature below are insufficiently detailed to enable separation from other species in the genus to be made.

Tinodes unicolor (Pictet)

In December 1950 Dr. H. E. Hinton sent me some larvae of a Psychomyiid caddis fly which he had collected from a small rapid stream at Failand, near Bristol, which is in a limestone region.

On examination I found these larvae different from the other Psychomyiid species in my collection and arrangements were accordingly made for me to rear some during 1951. Later Dr. Hinton sent me a number of living larvae in their fixed galleries on pieces of lime-encrusted rock from the bottom of the stream. The pieces of rock were wrapped in damp tissue paper and all wedged tightly into a strong wooden box. The larvae appeared to suffer little from their two-day journey to my home in Surrey, and quickly settled down in a large bowl into which the pieces of rock had been placed. Many of the galleries were purposely left above the water-level in the bowl, as Dr. Hinton had already observed that many of the galleries at Failand were in such a position, being left above water-level when the stream falls. He did not think they are deliberately constructed out of water. Such conditions also occur in the case of *Tinodes pallidulus* McLachlan, and it also seems certain that in *Tinodes assimilis* McLachlan the film of water in which the galleries lie must sometimes dry up for a time without necessarily being lethal to the larvae.

Two males and eight females emerged and were identified by Mr. D. E.

Kimmins as *Tinodes unicolor* (Pictet). As females of most of this genus cannot at present be identified, this collection of adult caddis of this species is retained at the British Museum as providing associated females.

The larval galleries are constructed of sand-grains held together by a silk-like secretion from the labial glands. Each is open at both ends and the Larvae feed on green algae, which they scrape with their mandibles from the surface of the stones.

The sites of the galleries are determined by the direction of stream flow only in the sense that they are built in depressions, grooves or behind ridges on the surface of the stone or its crust of tufa. Laboratory experiments by Mr. A. M. E. Mortimer clearly show that this is so. Controlled currents of water were directed over stones on which a layer of fine sand had been placed. In time the current removed the sand from all surface areas except

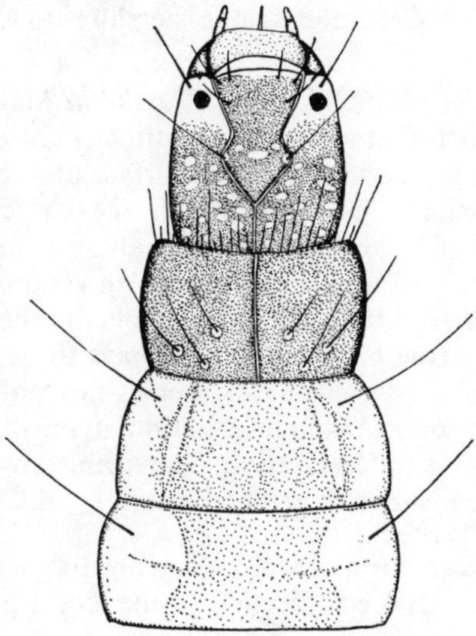

378

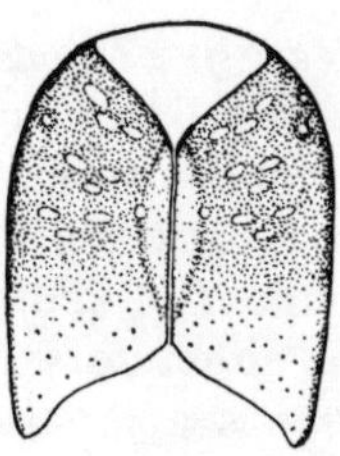

379

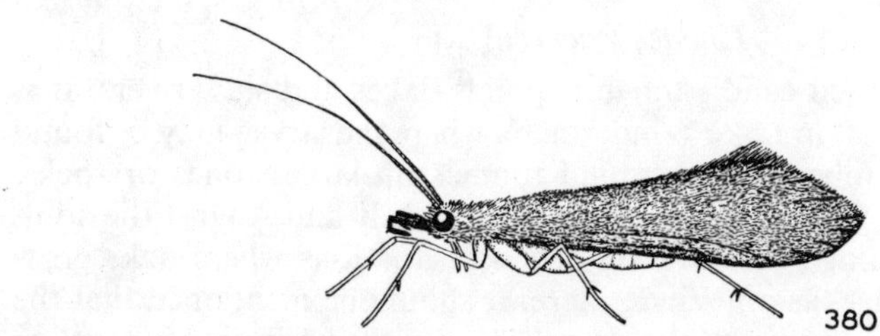

380

Figs. 378–379. *Tinodes unicolor* (Pictet). 378. Head and thoracic segments of larva from above; 379. Head of larva from beneath.

Fig. 380. *Tinodes unicolor* (Pictet). Adult.

depressions or areas immediately behind ridges. Larvae were then placed on the stone-constructed galleries only where sand was present. Such experiments seem to confirm that the direction of the current, by determining the sites of local accumulations of sand, determines the sites where galleries will be constructed. But within the areas sufficiently protected for sand to accumulate, the orientation of the galleries was random with respect to the direction of flow.

The number of galleries on stones with a calcium carbonate crust is much higher than on stones similar in every respect but not encrusted. The reason seems to be that the surface of encrusted stones is much more irregular, and therefore provides a larger number of suitable niches. The tufa on some stones is partly tunnelled, and it is not uncommon to find occupied galleries in these tunnels. Larvae are usually absent on the surface of stones in waterfalls or on very smooth stones.

During 1952 I reared another batch of larvae of this species collected from a small stream near Cranham Common, Gloucestershire—a limestone region.

Larva. Very like *Tinodes pallidulus* McLachlan and *T. assimilis* McLachlan, but has the following characteristic features. About fourteen light-coloured spots on dorsal surface of each gena and on ventral surface about nine on each gena (pattern of spots shown in Figs. 378 and 379). The dark chestnut-brown colour of aboral region of head does not reach anterior margin of genae on either dorsal or ventral surfaces. Genal suture on ventral surface of head has a longitudinal furrow on each side which, in the centre, extends as far as the bristle pits (the bristles appear always to be absent).

There is a wide variation in the larva of this species not only in the ground-colour of the head and pronotum but, in addition, in the patterning of the spots and their intensity of differentiation. They can, however, be identified quite easily using the characters given above.

I am indebted to Dr. H. E. Hinton for allowing me to incorporate Mr. A. M. E. Mortimer's notes on the formation of the larval gallery of this species.

Tinodes waeneri (L.)

This widely distributed caddis inhabits ponds, lakes and slow rivers. It is particularly abundant in Lake Windermere, where the larvae may be found in considerable numbers in their mud tunnels on large stones or rocks, usually 6 feet or so from the edge of the lake. In July and August the adult flies are fond of sitting on oak leaves in those situations where oaks grow near the water's edge, as on Windermere. It should be mentioned that the other seven British species of the genus *Tinodes* are uncommon, local or rare, but *T. waeneri* is the largest.

I am indebted to Dr. T. T. Macan, Mr. G. Thompson and Mr. S. Nield, of the Freshwater Biological Association, for getting live *T. waeneri* larvae from Wray Castle, Windermere, to my home in Surrey during a hot midsummer.

Larva (Fig. 381). Campodeiform, head procentrous, legs small in relation to size of body, larva uniformly cylindrical. When seen from above, the head, prothorax and last abdominal segment less wide than the rest of the body, and a slight tapering from about the fourth abdominal segment is evident when the larva is moving. Length 10 mm. Width 1.2 mm.

Larval Tunnel (Fig. 382). Larvae of *Tinodes waeneri* construct a long thin tunnel of mud and sand particles, vegetable debris, algae and silk. The large quantity of silk that goes to make up the tube gives the tube a characteristic greyish colour. The tubes are up to 30–35 mm. in length and from 1.5–2.0 mm. in diameter. Superficially they resemble the tunnels constructed by the marine worm *Pomatoceros*, as they are often sinuously curved. Wesenberg-Lund (1943, fig. 156B) gives a photograph of a number of these tunnels. The tunnel is fixed to the stone with silk all along its length. When the larva is approaching full growth the anterior end of the tunnel is widened (up to about 6 mm.) and it is in this wider anterior end that the change to pupa is effected. When this has occurred the narrow posterior part of the tube is often washed away by wave action (as on Lake Windermere) or by currents. The larva spins extra silk before pupation, which fairly closely envelops the pupa when the metamorphosis has taken place. It is, however, adherent to the inner wall of the tunnel, so that it does not form a cocoon separate from it (cf. *Tinodes pallidulus* and *Lype reducta*).

Head (Figs. 383 and 384). Oval, long, yellowish or greenish-yellow with clypeus dark greyish-brown, oral end less dark. The areas of the genae near the vertex of the clypeus also dark greyish-brown. Eyes situated in a forward position. Antennae small, bulbous, situated in pit-like foldings of the anterior margin of the genae. The mouthparts are capable of considerable extension; both figures of the head show the mouthparts in the extended position.

Labrum (Fig. 385). Sclerotized, dark brown, darkest at lateral areas of the hind margin, prolongations of hind margin long and black in colour except distal tips. Anterior marginal excision shallow and wide. The whole labrum is covered with fine hairs.

Mandibles (Figs. 386 and 387). Asymmetrical, scoop-shaped. Teeth sharper on left mandible, auxiliary toothed area at base of each mandible on inner edge and outer teeth of the opposite mandibles grind against them. Each mandible has two bristles on the outer edge, but only the left mandible has a brush of hairs on the inner edge.

Maxillae. Maxillary palp four-segmented, the third segment being the longest; mala covered with hairs directed inwards, the inner hairs turning outwards as they meet the long acicular *labium*. Base of labium with

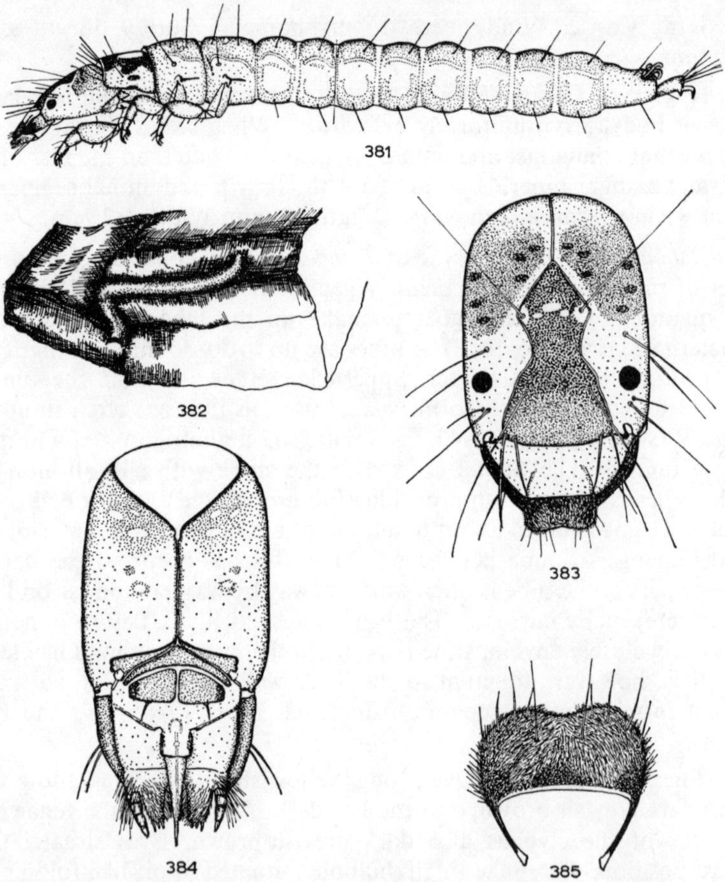

Figs. 381–385. *Tinodes waeneri* (L.). 381. Lateral view of larva; 382. Larval tunnel; 383. Head from above; 384. Head from below; 385. Labrum.

sclerotized collar-like ring. Labial palps absent, but a pair of fine hairs present on ventral surface.

Gular sclerite does not divide the genae, it is bow-shaped; lying just anterior to it is a pair of large sclerites. In the genae just posterior to the gular sclerite is a black, thickened furrow, which, commencing from the oral extremity of the ventral genal suture, is directed outwards until it reaches a point opposite the outer horns of the gular sclerite when it curves forwards to finish on a cube-shaped organ immediately adjacent to the outer horns of the gular sclerite.

Thorax (Fig. 388). Only the prothorax sclerotized. Prosternal horn absent. Prothorax only slightly wider than the head, very dark brown except for four large creamy-yellow areas, the median pair being larger. Anterior margin also creamy yellow. Ventral posterior margin black and produced forwards in the median line to form a robust spine-like projection. Meso- and metathorax larger than prothorax, dorsal surface light chestnut brown with greenish-mauve tinge, which is given by presence of minute spicules, with very variable pattern of white streaks and spots (when the spicules are absent); pleural and ventral regions bright apple green. The amount of

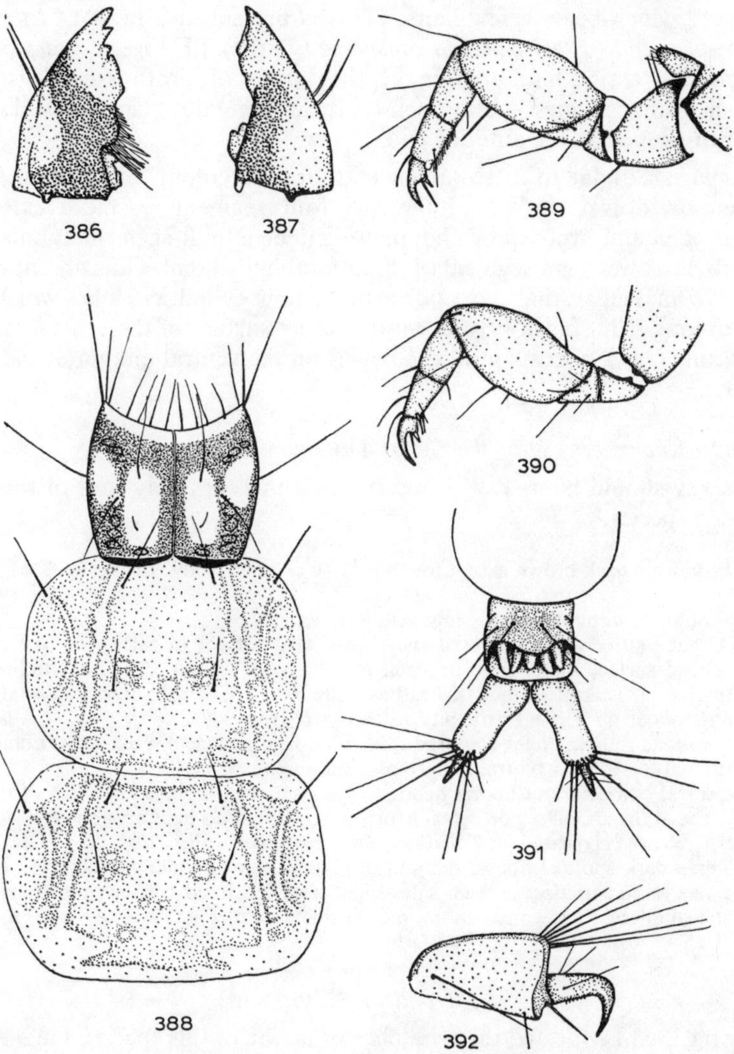

Figs. 386–392. *Tinodes waeneri* (L.). 386. Left mandible; 387. Right mandible; 388. Thoracic nota; 389. Prothoracic leg; 390. Mesothoracic leg; 391. Anal segments from above; 392. Anal claw from the side.

green coloration is subject to wide variation. Some larvae are vivid bluish-green, whilst other larvae are found with the green colour entirely absent.

Legs (Figs. 389, 390). Small in relation to size of body. Prothoracic legs slightly larger than meso- and metathoracic legs. Tarsal claws capable of considerable flexion in relation to tarsus when tarsal spines interlock with tarsal claw. On all tarsal claws a bristle is present on the claw in addition to the spine at the base. A flap-like forward extension of the trochantin is present which bears a few hairs. (This is present also in POLYCENTROPIDAE, such as *Plectrocnemia conspersa* (Curtis).) Fringe of hairs present along the entire hind margin of the tarsus of prothoracic legs only. Trochantins of meso- and metathoracic legs are black, rod-like and stirrup-shaped at the anterior end.

Abdomen. Similar to meso- and metathorax in colour, but area of bright green on dorsal surface of the last four segments is more extensive. Intersegmental grooves of abdomen distinct. Ninth segment tubular and much narrower than segments 1–8. Lateral line absent. Gills absent except for five anal gills. Anal claws borne on distinct cylindrical lobes which bear seven long stout black bristles on the upper surface of the ventral margin. The anal claws bear six minute spicules on the ventral margin of the basal part.

Key to Known Larvae of the Genus Tinodes

This key should be used with reserve as it includes only four of the eight British species.

1. Pronotum dark brown except for four large creamy-yellow areas (Fig. 388) *waeneri*
— Pronotum without large creamy-yellow areas 2
2. About fourteen light-coloured spots on dorsal surface of each gena and on ventral surface about nine on each gena (Figs. 378 and 379). Longitudinal furrow on ventral surface of head extends from the genal suture orally and aborally as far as the bristle pits at the centre (Fig. 379) *unicolor*
— Less than fourteen light-coloured spots on dorsal surface of each gena. Longitudinal furrows on ventral surface of head absent 3
3. Central clear mark at aboral dilation of clypeus. Dark pigmented area of the genae at the aboral region extends forwards to enclose a light area on each side between the clypeal aboral dilation and constriction (Fig. 370) *pallidulus*
4. Three dark spots at aboral dilation of clypeus. Small white spot situated on genae at constriction of each side. Light areas enclosed between the clypeal aboral dilation and constriction on each side absent (Fig. 366) *assimilis*

Lype phaeopa (Stephens)

During 1953 I collected three batches of larvae of this species. On 4 April, when collecting with Mr. James Edwards in the locality of Alton, North Staffordshire, where the very rare Psychomyiid caddis fly *Tinodes rostocki* McLachlan is well established, we found a number of fixed galleries on branches submerged in a fast-running stream near a waterfall. From this material I was able to rear out a male adult on 6 June. The identification was confirmed by Mr. D. E. Kimmins. Like many other species of caddis it emerged during the night or early morning. The long blackish pubescence which clothes the anterior wing has a dark brown sheen when seen from some angles. The underside of the thorax and legs are very pale straw in colour and there is a ring of this colour surrounding the eyes. On comparing these confirmed larvae with a batch of larvae found on submerged branches in a small stream at Cherrington near Stroud, Gloucestershire, when collecting with Mr. Arthur Peacey on 9 April 1953, it was observed that they were identical. Mr. Peacey had collected adults of *Lype phaeopa* at this locality in previous years. Lastly, I have found some larvae in galleries on a submerged branch in a small trickle of water at Bletchingley, Surrey, on 24 May 1953.

Lype phaeopa is one of the commonest of caddis flies. It is therefore amazing that its larva has remained virtually unknown until the present time. When once its fixed galleries can be recognized, almost any stick or branch submerged in water—from the veriest trickle to a lake—will be found to contain them. The final galleries may be up to 60 to 70 mm. in length and up till the present I have found them only on submerged wood. The surface of the wood appears to be bitten out into a groove, the top of the groove being covered with a skin-like secretion to which sand-grains adhere. The larva can move rapidly along its gallery and when the wood supporting the galleries is taken from the water the larvae will often leave them.

The larva of *Lype phaeopa* is almost identical with that of *Lype reducta* (Hagen) in my collection. My specimens of *phaeopa*, however, differ from *reducta* as follows: the dark spots on the lateral surfaces of the genae (the 'cheeks') are quite discrete in *phaeopa* but tend in *reducta* to run together with a greyish suffusion. The ventral surfaces of the genae are also much greyer in *phaeopa* than in *reducta*, although I would now like to make my drawing of these in *reducta* much darker than I actually did. In the same drawing I would now like to make the oral margin of the clypeus black, and the wide oral part grey, but not so dark as the wide, almost black, transverse band.

In *phaeopa*, also, I found that the pronotum is rather darker in colour than in *reducta*, and the white areas around the bristles are slightly larger.

Lestage quotes Ulmer's statement that the larva of *Lype* resembles that of *Tinodes assimilis* McLachlan to an extraordinary degree. I have found that both British species of *Lype* do resemble those species of *Tinodes* which I have collected (*assimilis* McLachlan, *unicolor* (Pictet), *pallidulus* McLachlan), with the exception of *Tinodes waeneri*, which can be easily separated from these species with the naked eye.

Note. It is not certain that A. J. Siltala (1905–6, *Acta Soc. Fauna Flora fenn.* **27** (6), 138) refers to this species. *See* Rousseau (1921), 527.

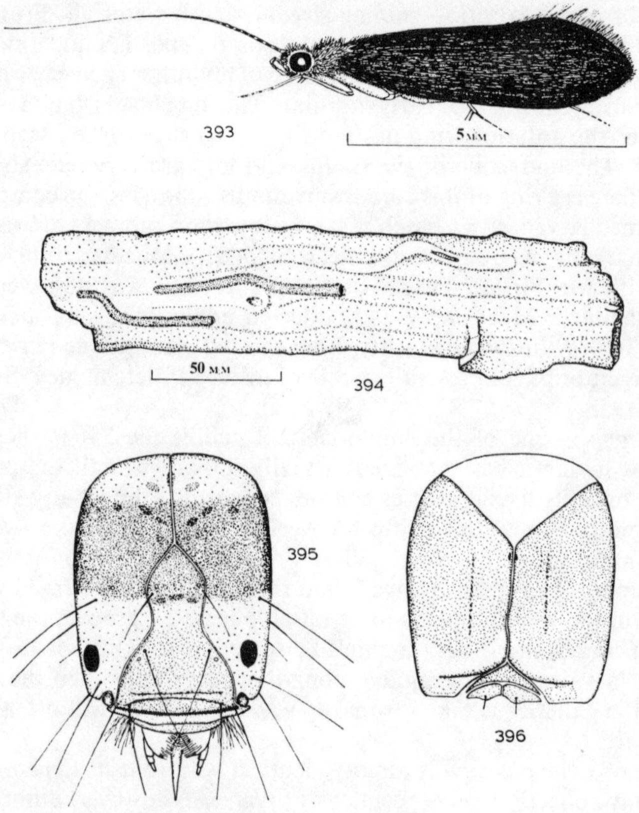

Figs. 393–396. *Lype phaeopa* (Stephens). 393. Adult; 394. Three larval tunnels on piece of submerged wood, tunnel on right has had covering removed to show comparative size of larva; 395. Head of larva from the front; 396. Head of larva from behind, mouthparts omitted.

Lype reducta (Hagen)

Lype reducta is a rare caddis in Britain. One adult male fly was taken by McLachlan in 1884. Since then (in the period 1936–40) it has been taken only by the Reverend Professor L. W. Grensted and his son David at localities in Somerset, Shropshire, Oxfordshire and Herefordshire, mostly near or at rather small, fairly rapid streams. The immature stages were entirely unknown. I was therefore fortunate, when staying with Dr. E. A. R. Ennion at Flatford Mill, Suffolk, on 5 July 1949, to collect a small batch of larvae which subsequently turned out to be this species. They occurred in a small fast stream to the west of St. Martin's Hill, East Bergholt, Suffolk. This stream runs through open rough meadows and skirts some small woods. There is luxuriant marginal vegetation, and other caddis present in the stream are large quantities of *Agapetus fuscipes* with a smaller number of *Limnephilus lunatus* and *Anabolia nervosa*. The larval tunnels of *Lype reducta* were found only on submerged rotting sticks, where the stream ran alongside a copse. No tunnels were found on stones. A bundle of the sticks was collected and placed in the boot of my car. On the following day, when the boot was opened in order to wrap the sticks in wet sacking for the journey to my home, one was found to have emerged. A further two specimens emerged, one three days and the other five days later, the sticks having been placed in a shallow tray of water. Mr. D. E. Kimmins kindly confirmed my determination of species.

Lype phaeopa (Stephens), the only other species in the genus on the British list, is very abundant and widely distributed, and appears to be more generally found near large rivers and still water.

Larval Tunnel (Fig. 397). Usually sinuously curved. Fixed along the whole of its length to submerged rotting sticks. Maximum length of tunnel 75 mm., varying in width from 5.5 mm. at the anterior end to 1 mm. at the

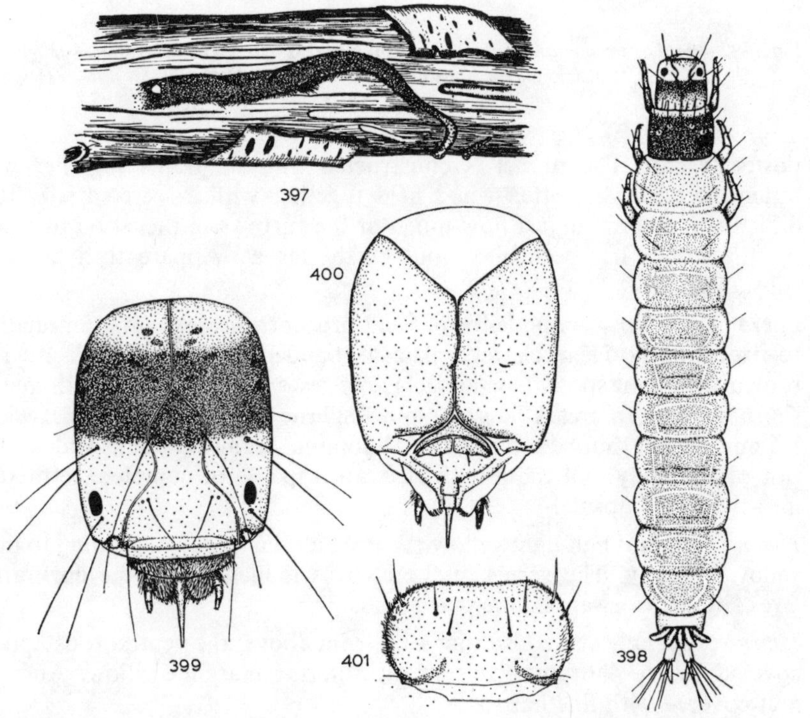

Figs. 397–401. *Lype reducta* (Hagen). 397. Larval tunnel; 398. Larva from above; 399. Head from the front; 400. Head from below; 401. Labrum.

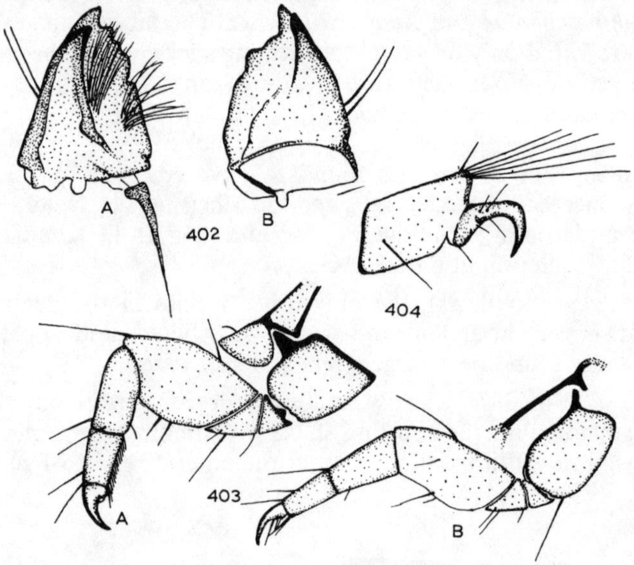

Figs. 402–404. *Lype reducta* Hagen. 402. A: left mandible; B: right mandible; 403. A: prothoracic leg; B: mesothoracic leg; 404. Anal claw.

posterior end. The tunnel is constructed of sand-grains together with bitten-off pieces of rotten wood held together with a secreted silk. It is difficult to determine just how much of the furrows in the wood in which the tunnels lie has been bitten out by the larvae. Apparently it is fairly substantial.

Larva (Fig. 398). Campodeiform, head procentrous, legs small in relation to size of body. Head characteristically banded. Prothorax dark greyish-brown with light spots. Dorsal surface of rest of body pinkish with mauve sheen shading to cream laterally and underneath. Length 11 mm.; width 1.5 mm. Head, thorax and last two abdominal segments not as wide as the rest of the body. All the mouthparts are capable of complete retraction into the head capsule.

Head. Oval, oral half light yellow, aboral half dark brown, shading to light yellow aborally. The vertex of the clypeus is included in the dark area. Eyes situated forwards, antennae small.

Labrum. Strongly curved, so that seen from above, the ventral rods appear covered by the labrum. Excision on anterior margin shallow. Anterior half covered with fine hairs.

Mandibles (Figs. 402a and b). Asymmetrical, blunt and scoop-shaped. Five teeth on left mandible, four on right. Two bristles on outer edge of each mandible, but brush of long hairs on inner face of left mandible only. These hairs arise from two distinct areas. Right mandible somewhat thicker than the left.

Maxillae (Figs. 399 and 400). Palps four-segmented, third segment longest, mala hairy, the hairs being directed inwards.

Labium. Long and acicular, labial palps absent but bearing pair of fine hairs on ventral surface. Base of labium consists of ring-like collar supported by sclerotized rods.

Gular sclerite. Genae not divided by the gular sclerite. Genal suture sinuous. Gular sclerite shaped as in Fig. 400, with lateral horns directed outwards and forwards. The paired sclerites lying anterior to the gular sclerite less deep than in *Tinodes waeneri.*

Thorax. Only prothorax sclerotized. Prosternal horn absent. Prothorax only slightly wider than the head, dark greyish brown with four light creamish-yellow spots, inner pair anterior to outer pair and a long bristle arises from centre of each. Anterior margin creamish in colour, posterior margin black. Meso- and metathorax wider than prothorax, pinkish in colour with mauve sheen, light cream laterally and underneath. The pinkish colour of the dorsal surface appears to be given by minute spicules, which are absent from a pattern of light streaks and spots, which, however, is very variable.

Legs. Relatively small for size of larva. All legs approximately equal in length, but prothoracic legs slightly larger and more robust than meso- and metathoracic legs. Tarsus of prothoracic legs only with fringe of small hairs along entire hind edge. A large peg on the anterior margin of the coxa articulates with the trochantin. Large flap-like anterior extension of the trochantin present. Tarsal claws of all legs bear a bristle in addition to a spine at the base. Both spines and bristles are, however, often lost on the prothoracic claws.

Abdomen. Lateral line absent. Width of segments, coloration and pattern of light streaks and spots similar to meso- and metathorax, but very variable. Eighth segment triangular, ninth segment small and cylindrical, bearing five anal gills. Anal claws simple. I have not been able to find the series of six minute spicules on the ventral margin of the basal part of the claw, as found in *Tinodes waeneri.*

Note on Cocoon Formation. When the larva is fully grown it spins a silken web-like cylindrical cocoon around itself. On to this particles of sand become somewhat loosely attached, but the whole cocoon is situated within the larval tunnel—at the anterior end. Thus when the pupa emerges it has to cut through both the cocoon and the larval tunnel. The exceptionally long curved scythe-like pupal mandibles may be directly connected with this.

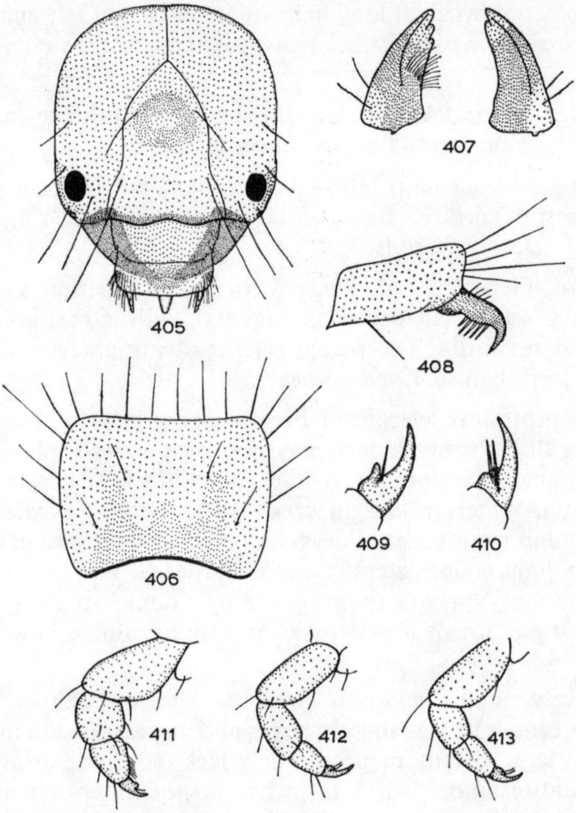

Figs. 405–413. Larva of *Psychomyia pusilla* (F.). 405. Head, dorsal; 406. Pronotum; 407. Mandibles, dorsal; 408. Anal appendage; 409. Tarsal claw, anterior leg; 410. Tarsal claw, posterior legs; 411. Anterior leg; 412 and 413. Posterior legs. Brindle.

Metalype fragilis (Pictet)

Immature stages unknown.

Psychomyia pusilla (Fabr.)

Brindle gives the following account.

The larva of *Psychomyia pusilla*, the only species in this genus, were obtained from silken tubes constructed on the sides of submerged stones in the River Hodder, near Whalley, Lancashire, in May 1960. The stones lay in small backwaters of the river where the bed was silted; they did not occur where the current was at all perceptible. The adults are very common along the river during the summer, flying freely in the day when the weather is hot and humid.

Larva (final instar). Length 8 mm., breadth 1.5 mm. Head (Fig. 405) broad, almost quadrangular, yellowish, unicolorous except for a rather variable darker mark on fronto-clypeus; anterior margin of fronto-clypeus emarginate; darkened; labrum yellow with long setae laterally; labium of usual Psychomyid type with long, pointed premental lobe, both this and the tips of the maxillary palpi projecting below the anterior margin of labrum; antennae reduced, bulbous, near anterior border of head below eyes; eyes prominent, blackish; mandibles (Fig. 407) asymmetrical, blackish or reddish-brown, with reddish or yellowish apical and basal parts, scoop-shaped, each with a pair of setae towards external edge, the left mandible with a dorsal row of setae and with three blunt teeth, right mandible dorsally without setae and with two blunt teeth. Pronotum (Fig. 406) yellow with darker markings on disc and with actual posterior margin black; quadrangular, anterior corners rounded, posterior corners rather produced posteriorly; a series of long setae on anterior margin and on anterior half of lateral margin; two pairs of long setae on disc. Anterior legs (Fig. 411) short, broad, chaetotaxy reduced, tarsi with a row of short setae ventrally, the tarsal claw (Fig. 409) with one spine; posterior legs (Figs. 412, 413) subequal, longer and not so broad as anterior pair, chaetotaxy similar except for the absence of the short ventral setae on tarsi; tarsal claws with two subequal spines (Fig. 410). Abdomen usually reddish dorsally, greenish ventrally, but this was subject to variation in specimens examined; no lateral line or gills. Anal appendages (Fig. 408) short, of typical Psychomyid type with five long dark setae distally and one ventrally; claws strongly curved with four short setae dorsally and five spines ventrally, the proximal one shorter than the others.

BRINDLE, A. 1960. The Larva of *Psychomyia pusilla* (F.). *Entomologist's Rec. J. Var.* **72**, 265–7.

EDWARDS, J. 1952. *Tinodes rostocki* McL. (TRICHOPTERA) in Staffordshire. *J. Soc. Br. Ent.* **4**, 111–12.

HICKIN, N. E. 1941. Some Records of Trichoptera in the Lake District. *Entomologist's mon. Mag.* **77**, 252.

HICKIN, N. E. 1950. *Proc. R. ent. Soc. Lond.* (A) **25**, 67–74, 103–6.

HICKIN, N. E. 1952. *Proc. R. ent. Soc. Lond.* (A) **27**, 89–90.

HICKIN, N. E. 1953. *Proc. R. ent. Soc. Lond.* (A) **28**, 36–8.

HICKIN, N. E. 1953. *Tinodes pallidula* McLachlan (TRICHOPTERA, PSYCHOMYIIDAE). A second British Station. *Entomologist* **86**, 113.

HICKIN, N. E. 1954. *Proc. R. ent. Soc. Lond.* (A) **29**, 93–5.

KIMMINS, D. E. 1938. British Ephemeroptera, Plecoptera and Trichoptera in 1936. *J. Soc. Brit. Ent.* **1**, 202–9.

KIMMINS, D. E. 1943. A List of the Trichoptera (Caddis Flies) of the Lake District with Distributional and Seasonal Data. *J. Soc. Br. Ent.* **2**, 136.

KIMMINS, D. E. 1949. *Tinodes pallidula* McLachlan, an addition to the British list of Trichoptera. *Entomologist* **82**, 269.

LESTAGE, J. A., *in* ROUSSEAU, E. 1921. *Les Larves et Nymphes Aquatiques des Insectes d'Europe.*

McLachlan, R. F. 1878. *Tinodes pallidula* McL. *Rev. Syn. Trich.*, 419, pl. xliv.
McLachlan, R. F. 1884. *Lype reducta* Hagen, an addition to the British Trichoptera. *Entomologist's mon. Mag.* **21**, 113.
Mosely, M. E. 1939. *The British Caddis Flies* (*Trichoptera*). London.
Popham, E. J. 1951. *Proc. R. ent. Soc. Lond.* (C) **16**, 48.
Silfvenius, A. J. 1905–6. *Acta Soc. Fauna Flora fenn.* **27** (6), 135–40.
Tomaszewski, C. 1955. *Ekologia Polska* (A) **111** (4), 85–99.
Ulmer, G. 1903. *Metamorph. Trichopt*, 121–3.
Ulmer, G. 1909. *Die Süsswasserfauna Deutschlands* **5–6**, 231–3.
Wesenberg-Lund, C. 1943. *Biologie der Süsswasserinsekten*, fig. 156B.

12

HYDROPSYCHIDAE

All British representatives of the family HYDROPSYCHIDAE are placed in the sub-family HYDROPSYCHINAE. They are characterized as follows. They are moderately small insects with an anterior wing length of about 10–11 mm. in the genus *Hydropsyche* but in *Diplectrona felix* and *Cheumatopsyche lepida* it is somewhat smaller, being from 7–9 mm. Ocelli are absent. The antennae are either longer or shorter than the wings, but the first segment is short and bulbous. In the genus *Hydropsyche* each segment of the antenna bears a blackened spiral ridge which becomes rather indistinct in the terminal segments. The maxillary palpi have a short basal segment and a long terminal one which is divided into a number of small parts which articulate and are flexible. In the male the inferior appendages are long and slender and are always composed of two segments. Females of the genus *Hydropsyche* have hitherto been difficult to determine but Kimmins's 1957 paper gives a key and a series of illustrations.

In the British list there are ten species placed in three genera as in the following check-list. Since Mosely's handbook, *Hydropsyche saxonica* and *Hydropsyche contubernalis* have been added.

HYDROPSYCHE

1. *angustipennis* (Curtis)
2. *contubernalis* McLachlan
3. *exocellata* Dufour
4. *fulvipes* (Curtis)
5. *guttata* Pictet
6. *instabilis* (Curtis)
7. *pellucidula* (Curtis)
8. *saxonica* McLachlan

CHEUMATOPSYCHE

9. *lepida* (Pictet)

DIPLECTRONA

10. *felix* McLachlan

Distribution

Hydropsyche angustipennis. Widely distributed and very abundant (M.E.M.); Berks. (M.I.C. *et al.*); Northumberland (G.N.P.).

Hydropsyche contubernalis. Berks. (M.I.C. *et al.* McLachlan (1865) records R. Thames, Kew to Richmond (Surrey).

Hydropsyche exocellata. Local. Thames at Weybridge (M.E.M.). McLachlan (1865) records R. Thames, Kew to Richmond (Surrey).

Hydropsyche fulvipes. Local. Dorset, Somerset (M.E.M.); widely distributed in Britain, Dunbartonshire, Lanarkshire, North Lancs., West Yorks., Cheshire, Staffs., Somerset, Perthshire (R.M.B.).

Hydropsyche guttata. Rather local in the British Isles (M.E.M.). McLachlan (1865) records R. Thames, Kew to Richmond (Surrey), but Kimmins doubts identification.

Hydropsyche instabilis. Very widely distributed and abundant (M.E.M.); Berks. (M.I.C. *et al.*); Yorks. (P.F.H.); Cardigan (J.R.E.J.); Northumberland in faster water than *pellucidula* (G.N.P.).

Hydropsyche pellucidula. Distributed throughout the British Isles and abundant (M.E.M.); Symond's Yat, Glos. (A.F.P.); Northumberland (G.N.P.).

Hydropsyche saxonica. Only known from a fast-flowing stream at Headington, Oxford (L.W.G.).

Cheumatopsyche lepida. Fast rivers, rather local (M.E.M.); Symond's Yat, Glos. (A.F.P.); Northumberland, fast-flowing water (G.N.P.).

Diplectrona felix. Widely distributed but rather local, small rapid streams in shady woods (M.E.M.).

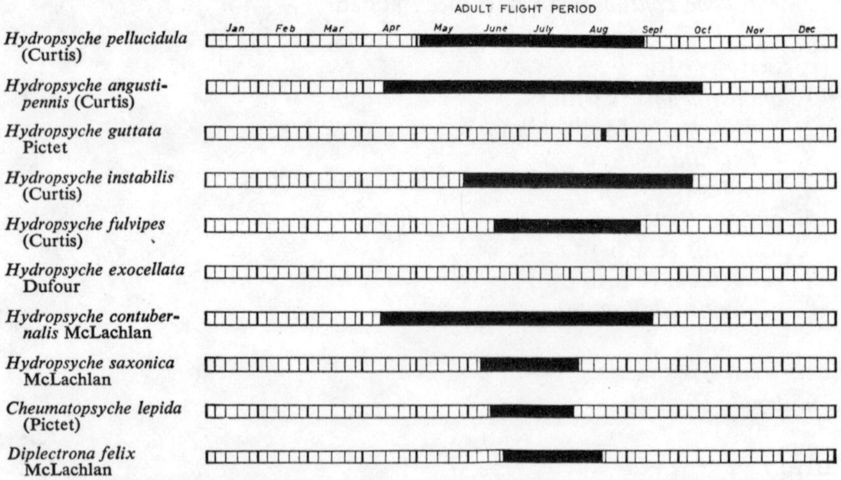

HYDROPSYCHIDAE. *Adult Flight Period taken from published records of individual dates.*

General Description of Larvae

Lestage gives a description of larvae of this family as follows:

Campodeiform, of average size (10–20 mm. long) narrow (1–2 mm. wide), parallel sides, of almost uniform width or gradually narrowing

(*Hydropsyche*), sometimes with abdominal segments successively wider up to the fourth and then diminishing (*Diplectrona*).

Head relatively small, a little narrower than the pronotum, ovoid (top view), flattened dorso-ventrally.

No hypostome. Broad clypeus, the sides slightly sinuous, not narrowing. Rudimentary antennae. Buccal parts slightly prominent: labrum sclerotized, transverse-elliptical, nearly straight in front, slightly angular in the middle of the sides, with short numerous dorsal hairs. In front of each side usually a longer hair; lateral brushes formed of thick, very long hairs, robust mandibles, triangular, with two ridges more or less denticulated; the right without internal brush, both with numerous hairs on the dorsal surface. Narrow maxillary lobe, conical with long sensorial bristles at the tip and long numerous hairs on the internal border. Maxillary palps of five segments. Labial lobe obtusely conical, rudimentary labial palps, of two segments, the second short and bearing three small sensorial bristles.

Pro-, meso- and metanota sclerotized and quadrangular, unequal, generally greyish-brown; that of the pronotum black on the sides, at the anterior corners and posterior corners and generally in the middle of the posterior border; those of the meso- and metanota broadly black at the anterior corners, more narrowly at the posterior corners and on the sides as well as on the middle of the posterior border which gives the appearance of triple-sinus; on each of these patches are numerous hairs and spikes.

Short, robust legs, the anterior ones larger, the posterior ones a little longer; supporting plates of the anterior legs terminated in the form of a fork; femora, tibiae and tarsi furnished with internal yellowish spines; anterior trochanters having long yellowish-brown hairs; coxae and femora with four feathery spines on the broad surface, longer than on the anterior legs; claws of the median and posterior tarsi with a robust basal spine; that of the anterior claws setiform; in addition, on all the legs, particularly on the internal side of the anterior femora, numerous robust black hairs, tibiae without apical spines.

Abdominal segments, clearly defined and covered with short numerous black hairs.

No lateral line, well-developed and much-branched gills in groups inserted on the meso-metasterna and the 3–7 ventral segments, sometimes 3–6 only; in addition 4 (*Hydropsyche*) or 5 (*Diplectrona*) anal gills, on the 3–7 segments, 1–3 delicate conical appendages (gills), moreover, on each segment are dark short thick hairs, more numerous than on the eighth and ninth segments where they are mixed with a few yellow spines.

Long leg-like anal appendages of two parts, the basal part long, sclerotized, with black hairs and having a remarkable group of long, broad black hairs; very short second part, bent underneath in a right-angle and terminated by a simple claw, large and broad, strongly curved, with fine hairs, but ventral spines and dorsal hooks absent.

No larval case, the larvae living free in silken nets which they spin (at least in the *Hydropsyche*), usually under stones and usually to be found in fairly fast running water.

Key to Larvae of British Genera of HYDROPSYCHIDAE

1. Head brown without lighter marks; abdomen gradually broadened to fourth segment and narrowed after; five anal gills *Diplectrona*
— Head with lighter marks; abdomen not broadened; four anal gills 2
2. Mandibles ventrally with proximal tooth largest; right mandible with largest tooth furnished with a dorsal row of setae *Cheumatopsyche*
— Mandibles ventrally with proximal tooth less than the next distal; no dorsal row of setae on right mandible *Hydropsyche*

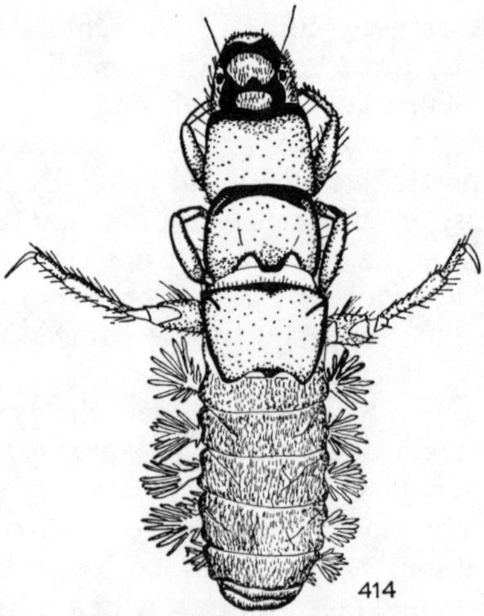

414

Fig. 414. Larva of a Hydropsychid from above, shown in its typical resting position with the abdomen curled beneath.

Hydropsyche angustipennis (Curtis)

Brindle (1960) has described larvae of this species collected from Surrey, as follows:

The nets made by the larvae correspond to those described for *H. instabilis* by Philipson (1953), being rough elongated silken shelters attached to the underside of stones in the stream from which a net extended into the current. The pupae were found inside silken shelters, incorporating small pieces of stone and debris, attached to the underside of larger stones. The shelters were weak and often the pupae were injured when removing the shelters from the stones. Some yellowish deposit occurred in the water which affected the larval coloration even after removal of the gross material. It is evident therefore that in other habitats the coloration of the larvae may

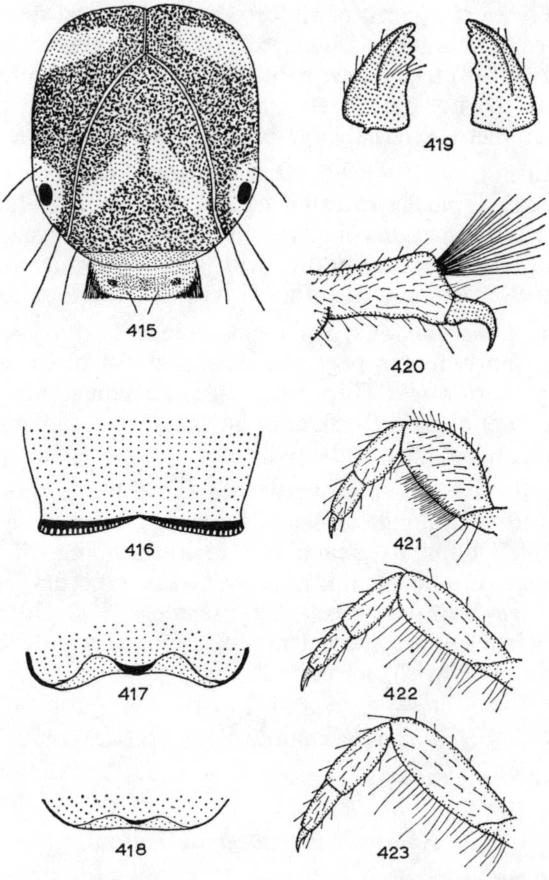

Figs. 415–423. Larva of *Hydropsyche angustipennis*. 415. Head, dorsal; 416. Pronotum, posterior margin; 417. Mesonotum, posterior margin; 418. Metanotum, posterior margin; 419. Mandibles, dorsal; 420. Anal appendage; 421. Anterior leg; 422–423. Posterior legs. Brindle.

be darker than that described below. It was also noted that in a second habitat where the larval nets were constructed on a vertical rock down which water was flowing, the nets were almost semicircular.

Larva (final instar). Length 17 mm., breadth 2.5 mm. Head (Fig. 415) dark yellowish-brown, very rough owing to the covering of small spicules, fronto-clypeus with a Λ-shaped lighter patch, lighter patches on occipital areas. In some larvae the latter patches were only slightly paler than the rest of the head, and young larvae tended to be lighter in colour and generally with the markings larger; eyes surrounded with yellow; antennae indistinct; labrum transverse, darker on disc with dense lateral tufts of blackish fine setae, the points of all forming a horizontal straight edge, disc with scattered black setae and two large setae near anterior border; mandibles (Fig. 419) triangular, reddish-brown, left mandible with dorsal brush of setae and five blunt teeth, right mandible without brush of setae and four blunt teeth, external edge of both with short setae.

Thorax. Each nota with a sclerotized plate, yellowish or greyish brown covered with small spicules and all margined with black laterally. Pronotum with black posterior margin (Fig. 416), mesonotum with broad black curved median line (Fig. 417), metanotum with short black narrow curved line (Fig. 418), both meso- and metathorax with much branched ventral gills.

Legs. All with a covering of sparse black setae, chaetotaxy as Figs. 421–3; anterior legs short, femur very broad with dense black setae ventrally (Fig. 421), posterior legs (Figs. 422, 423) subequal, longer and more slender than anterior pair, the femora having a row of long unequal black setae ventrally, tarsal claws subequal, short, each with one strong seta.

Abdomen. Yellowish-brown covered with short black spines, a pair of much branched ventral gills on segments one to seven, each gill arising as one stalk which bifurcates, each bifurcation sending off two or more branches. Some variation in this pattern, however, occurs. Four anal gills. Anal appendages well developed, distal segment (Fig. 420) lightly sclerotized, bearing black setae, with a dense tuft of long brownish setae on distal edge; anal claw yellowish, sclerotized, sharply curved.

Hydropsyche contubernalis McLachlan

Immature stages unknown.

Hydropsyche exocellata Dufour

Immature stages unknown.

Hydropsyche fulvipes (Curtis)

Immature stages unknown.

Hydropsyche guttata Pictet

Immature stages unknown.

Hydropsyche instabilis (Curtis)

Philipson gives the following description from larvae collected from the River Blyth in Northumberland. This is a small river which shows marked alternation of shallow swift-flowing and deeper slow-flowing reaches. The larvae were found in the swift-flowing waters, their distribution contrasting sharply with that of *H. pellucidula* (Curtis), the larvae of which were found to be generally distributed throughout the river.

Larva (Fig. 424). Campodeiform, about 17 mm. long and 2 mm. broad.

Head (Fig. 425). Yellow with characteristic dark brown or black markings. On clypeus a pattern is marked out consisting of four yellow areas surrounded by black, of which that farthest from the mouth is usually the most distinct; oral and lateral areas may be confluent by absence of the dividing arms of the forwardly directed Y-shaped figure. Eyes surrounded by yellow areas.

Antennae are described for the genus (Lestage, 1921) as being 'rudimentaires, formées d'un petit tubercle pâle inséré prês de la base des mandibles et portant deux soies pâles et deux bâtonnets sensoriels'. In this species no distinct antennal area could be discerned. Many short sclerotized spicules on sclerites of head.

Labrum (Fig. 429). Transversely elliptical and bears well-developed lateral brushes and a number of small bristles on upper surface. Two large bristles inserted anteriorly.

Mandibles (Figs. 430, 431). Roughly triangular in shape, each bearing a narrow groove on external surface in which are situated a number of bristles. Internal surface concave, tip and lower edge of concavity bearing a number of teeth. Left mandible bearing a small internal brush which is absent from the right mandible. Mentum, which is contiguous proximally with the gular plate, bifurcated distally and bearing a number of bristles.

Maxillae and *labium* (Fig. 432). Can be completely withdrawn within limits of distal border of mentum, or protruded beyond limits of labrum. Maxillary lobe long and finger-like, bearing small papillae at tip, numerous hairs on inner surface and two groups of bristles at base. Labial palps rudimentary.

Thorax. Pro-, meso- and metanota greyish-brown in colour and covered with short sclerotized spicules. Pronotum bordered with black posteriorly and laterally. Meso- and metanota have black lateral and anterior borders. On posterior margin of mesonotum (Fig. 426) a curved, M-shaped black mark; within central angle of the M are five to seven dark spots and there are similar marks posteriorly in lateral angles. In middle of posterior edge of metanotum (Fig. 427) a black band separated from the dark margins by a row of small spots. Immediately behind prothoracic legs is a small ventrally placed sclerite, transversely elongate, with a black anterior border. Two small sclerotized areas lie posterior to this.

Legs (Figs. 434–436). Strong and well provided with strong black bristles. Femora, tibiae and tarsi provided internally with yellow spines. Trochanters of prothoracic legs provided with long yellow bristles. Plumose hairs (Fig. 433) occur on coxae and femora of mesothoracic legs and on femora of metathoracic legs. A single plumose hair is placed at about the centre of internal edge of femur of each prothoracic leg. Claws of meso- and metathoracic legs strong and much curved, and each has a strong basal

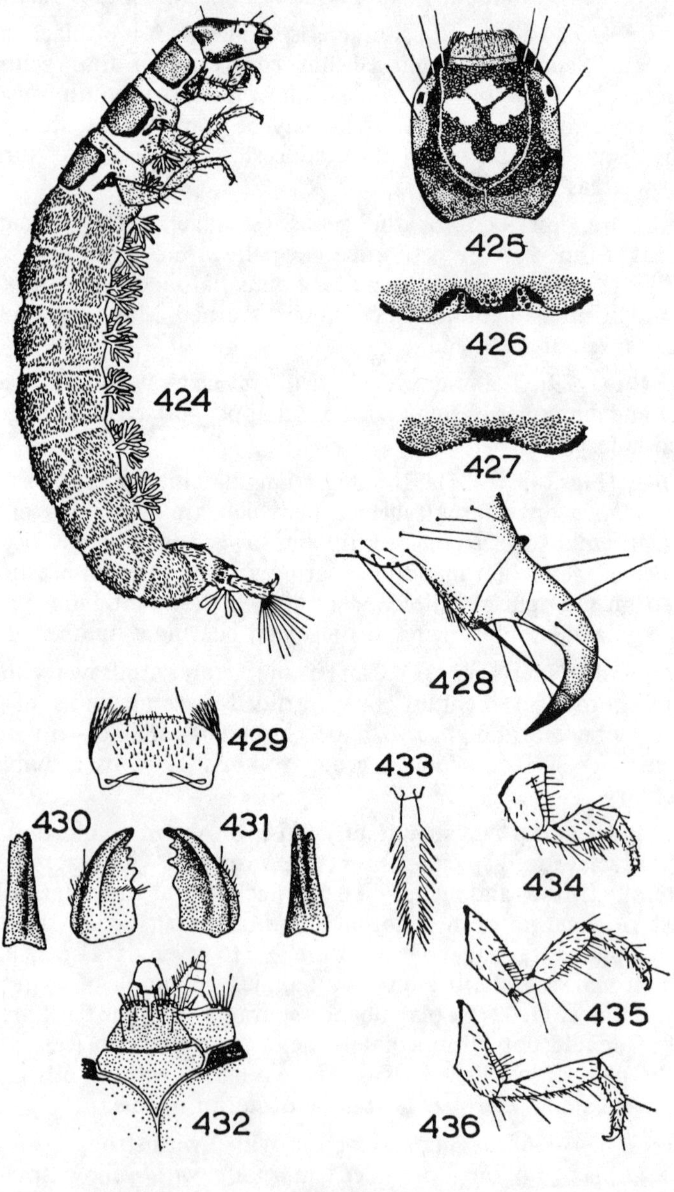

Figs. 424–436. Larva of *Hydropsyche instabilis* (Curtis). 424. Larva, lateral view; 425. Head; 426. Posterior border of mesonotum; 427. Posterior border of metanotum; 428. Anal claw; 429. Labrum; 430. Left mandible, internal surface and from above; 431. Right mandible, from above and internal surface; 432. Maxilla and labium; 433. Plumose hair from leg; 434, 435, and 436. Pro-, meso-, and metathoracic legs. Philipson.

spine. Claws of prothoracic legs less curved and provided with finer basal spines. Prothoracic pleural sclerites bifurcate anteriorly.

Abdomen. Covered with numerous small black spines. No lateral line. Gills well-developed branched structures, each arising from a single stalk; situated ventrally on the meso- and metathorax and abdominal segments one to six inclusive. Gills may occur singly or in pairs, the two basal stalks arising side by side. On each side they are arranged in the following manner: mesothorax bearing a single tuft placed near insertion of coxa. A similarly placed single tuft on metathorax together with a single tuft median to it. Two tufts occur on first abdominal segment; on abdominal segments two to six a double tuft placed laterally and a single median tuft. A number of conical gill-like processes develop laterally on the abdomen in segments three to six when the larva is constructing its shelter preparatory to pupation: these occur on posterior margins of segments, there being one on the third segment, while the fourth, fifth and sixth segments each have three placed one behind the other, the anterior being the smallest, the posterior the largest.

A number of minute yellow backwardly directed spines occur on triangular sclerotized areas of eighth and ninth abdominal segments. Two such areas occur ventrally on eighth abdominal segment, apices of triangles being directed forwards. Along bases of triangles are a number of large backwardly directed bristles. Two similar though larger areas occur on ninth abdominal segment, which also bears two smaller areas on each side.

Anal appendages (Fig. 424) well developed and leg-like, each with a long sclerotized segment bearing numerous small bristles and terminating in a strong curved claw, which (Fig. 428) has a large basal bristle and a number of smaller bristles. Above attachment of claw a number of very long black bristles directed posteriorly.

Four anal gills present.

(Ulmer separated the larvae of *H. instabilis* from the other known species of *Hydropsyche* by the absence of abdominal gills on the seventh abdominal segment in this species and by the markings on head and thorax.)

The larvae build a rough type of elongated shelter attached to the substratum. The shelter is open at both ends and incorporates a few pebbles, pieces of vegetable debris, etc. From one end of the shelter a net is spun stretching out into the current. The net may be large, extending an inch or more from the mouth of the shelter. A small central portion is neatly spun with a small rectangular mesh. Outside this area the spinning is less regular and the mesh larger. The small central portion is kept clean by the activities of the larva, the outer part becoming covered with silt. The larvae have not been observed feeding from this outer area and it is probable that this, when silted up, acts as a funnel directing the stream through the feeding area.

Most frequently, in the aquaria, the larvae would build their retreats horizontally, from which the net would stretch up more or less vertically. The larva would then be suspended, ventral side uppermost, head into the stream (Fig. 437). In a few instances more or less vertical retreats were formed.

When feeding, the larva holds itself in position by the anal appendages and meso- and methathoracic legs, while the prothoracic legs are held close to the under-surface of the head, which is moved rapidly from side to side, covering the central area of the net. Any particles becoming entangled in the net are immediately seized if large enough. The action of seizing is exceedingly rapid. The prey is apparently grasped by the mandibles and the prothoracic legs simultaneously. Copepods and small *Daphnia* were

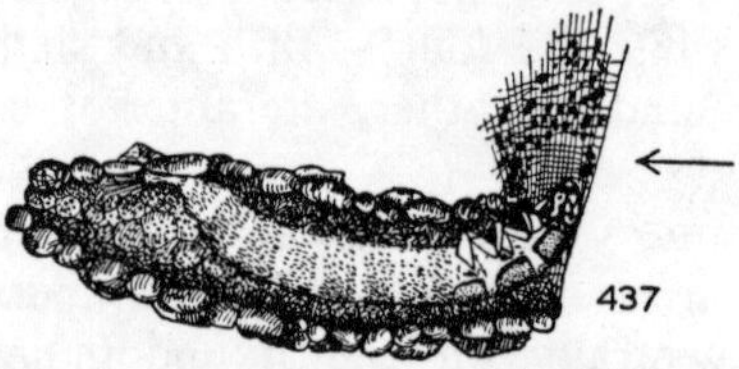

437

Fig. 437. *H. instabilis* larva feeding at its net, as viewed in a stirred aquarium. The shelter of stones and the distal ends of the threads of the net are attached to the glass walls of the aquarium. The arrow indicates the direction of the water current. Philipson.

observed to be taken and a small fresh-water Oligochaete was seized and devoured. Particles which are apparently unsuitable—for example large *Daphnia*, or inedible particles—are removed from the net and pushed out into the current; if this is not possible then they are cemented firmly in a convenient place by means of silk.

In ten freshly collected specimens the alimentary canals were found to contain animal material among which portions of a mayfly nymph and a caddis-fly larva were recognized. In one individual two small Cladocera were found. Filamentous algae, Diatoms and Desmids were present, among which *Closterium* figured largely. Fragments of plant tissue were found in two instances. Slack (1936) states that the 'gut contents consisted of small organisms of an order of size consistent with the method of obtaining food by *Hydropsyche*, i.e. by means of a net collecting minute organisms and debris particles carried to it by the flow of the stream . . . diatoms proved to be dominant'. Lestage considers *Hydropsyche* larvae to be omnivorous, and cites Silfvenius reporting an instance of *Hydropsyche* larvae attacking the wooden piles of a bridge.

In addition to the snatching of particles from the net, the larvae were observed in the aquaria to press the mouthparts against the net and sweep it clean by the side-to-side movements of the head. Lestage states that the long black bristles on the end of the anal appendages are used by *Hydropsyche* larvae to clean the net. I have, in no instance, observed this, the only method being the use of prothoracic legs and mouthparts already described. The larvae have been frequently observed to reverse their position in the retreat for a short period. In this movement the head is first pushed down between the legs, the thorax is then sharply flexed, the larva becoming bent into a V shape. The head end continues to move towards the hind end, which is at the same time drawn up towards it. By a continuation of this movement the body is eventually straightened again. The final straightening movement consists of flicking back the reflected anal appendages, a movement by which the posterior group of bristles is forced against the sides of the retreat. This would serve to remove any foreign particles from the inner wall of the retreat itself and so help to maintain a free passage for the movement of water. The waving of the posterior end of the larva during the rhythmic abdominal movements described below also serves to keep the posterior end of the retreat free of obstruction, and is aided in this function by the presence of the posterior bristles. The function of cleansing the retreat is ascribed by Lestage to the comb of bristles on the prothoracic legs. In view of the use of these legs in capturing food it would appear that the main function of the bristles is to increase the efficiency of the legs as raptorial organs.

Rhythmic abdominal movements are made by the larva, which holds its position in its retreat by means of the claws of the legs and abdominal appendages, between which the abdomen lies slightly curved dorsally. At the commencement of a rhythmic movement the abdomen is moved ventrally. Sometimes the abdominal appendages then lose their grip of the retreat and the abdomen becomes more or less straight. In recovery, with the re-assumption of the curved position of the abdomen, the abdominal appendages again grip the retreat. In still water these movements occur practically continuously, but in flowing water they occur in groups, with periods of quiescence following periods of movement.

It was observed in the laboratory aquarium that prior to pupation the larva generally leaves its old shelter and commences to construct the pupal shelter in a fresh location. The pupal shelter is built of small pebbles loosely joined together so that small gaps are left between them, which may or may not be crossed by thick silken threads. After completing the shelter the larva spins a thin, loosely woven, silken cocoon within it. This is separate from the inner silken wall of the shelter except at the ends, where they are continuous.

Hydropsyche pellucidula (Curtis)

No description of the larva is available based on British material except for Badcock's brief observations. Unlike most members of its family it is found in lakes and ponds.

Hydropsyche saxonica McLachlan

No description of the larva is available based on British material.

Key to known Larvae based on notes by Badcock

1. Double-tufted gills present on the seventh abdominal segment 2
— Double-tufted gills absent on the seventh abdominal segment *Hydropsyche instabilis*
2. Surface of fronto-clypeus smooth and practically glabrous except near anterior margins and sutures 3
— Surface of fronto-clypeus rugose and covered with many fine hairs, many of which are branched as on the genae. Aboral fleck of fronto-clypeus when present, small round blob. Pale oral fleck absent *Hydropsyche angustipennis*
3. The anteriorly bifid mentum either uniformly sclerotized and dark brown in colour or the centre is darker than the periphery. Pale yellow aboral fleck of fronto-clypeus U, V or Y shaped or rarely absent. Pale oral fleck indistinct and variably present *Hydropsyche fulvipes*
— Centre of mentum never darker in colour than the periphery although both parts may be of same pale brown colour. Pale oral fleck of fronto-clypeus strongly marked *Hydropsyche pellucidula*

Cheumatopsyche lepida (Pictet)

This species was formerly placed in the genus *Hydropsyche*. There is no description available based on British material.

Diplectrona felix McLachlan

The larva of this species is generally similar to that of *Hydropsyche* and is of similar habit, inhabiting under stones in rapid streams.

The head is almost as long as wide, furnished with spine-like hairs and with several large frontal hairs. The eyes are situated in yellowish areas on the otherwise brownish head. The labrum is rounded with well-developed lateral brushes of hairs. The asymmetrical mandibles are furnished with several hairs on the dorsal surface. The monotum is rather more rectangular than the meso- and metanota. The metanotum is paler than the nota of the preceding two segments. The median and posterior legs are rather longer than the anterior. The abdomen increases in size up to the fourth segment when it narrows progressively. Five anal gills present. Length about 10 mm., width 1.0 mm.

BADCOCK, R. M. 1955. Widespread Distribution in Britain of our Allegedly rare Caddis, *Hydropsyche fulvipes*, Curtis (TRICH., HYDROPSYCHIDAE). *Entomologist's mon. Mag.* **91**, 30–1.

BRINDLE, A. 1960. The Larva of *Hydropsyche angustipennis* Curtis (TRICHOPTERA: HYDROPSYCHIDAE) *Entomologist's Rec. J. Var.* **72**, 267–70.

GRENSTED, L. W. 1955. The Occurrence of *Hydropsyche saxonica* McLachlan in Britain. *Entomologist's mon. Mag.* **79**, 35–8.

KIMMINS, D. E. 1957. Notes on some British Species of the Genus *Hydropsyche* (TRICHOPTERA) *Entomologist's Gaz.* **8**, 199–210.

LESTAGE, J. A., *in* ROUSSEAU, E. 1921. *Les Larves et Nymphes Aquatiques des Insectes d'Europe*. Brussels.

PHILIPSON, G. N. 1953. A Method of Rearing Trichopterous Larvae Collected from Swift-flowing Waters. *Proc. R. Ent. Soc. Lond.* (A) **28**, 15–16.

PHILIPSON, G. N. 1953. The Larva and Pupa of *Hydropsyche instabilis* Curtis (TRICHOPTERA, HYDROPSYCHIDAE). *Proc. R. ent. Soc. Lond.* (A) **28** (1–3), 17–23.

SILFVENIUS, A. J. 1905–6. Beiträge zur Metamorphose der Trichopteren. *Acta Soc. Fauna Flora fenn.* **27** (6).

SILFVENIUS, A. J. 1906. *Acta Soc. Fauna Flora fenn.* **28** (4), 25.

SLACK, H. D. 1936. The Food of Caddis Fly (Trichoptera) Larvae. *J. Anim. Ecol.* **5**, 105–15.

ULMER, G. 1903. Über die Metamorphose der Trichoptera. *Abh. naturw. Hamburg* **18**, 1–154.

ULMER, G. 1903. *Metamorph. Trichopt.*, 115.

ULMER, G. 1909. Trichoptera. *Die Süsswasserfauna Deutschlands* **5–6**.

WHITEHEAD, H. 1949. Notes on a Net Spinning Caddis Larva, *Hydropsyche angustipennis* (Curtis). *Naturalist*, 89–90.

13

ODONTOCERIDAE

There is only one European species in the family ODONTOCERIDAE. This is *Odontocerum albicorne* (Scopoli). The adult is recognized by its moderate size, the length of the silvery-grey or whitish anterior wing in the male being 13 mm. and the female 18 mm., and the long antennae which are toothed on the inner side. In the male the hind-wing is triangular and bears a long pencil of hairs which is absent from the more rounded hind-wings of the female.

Distribution

Widely distributed where there is running water (M.E.M.).

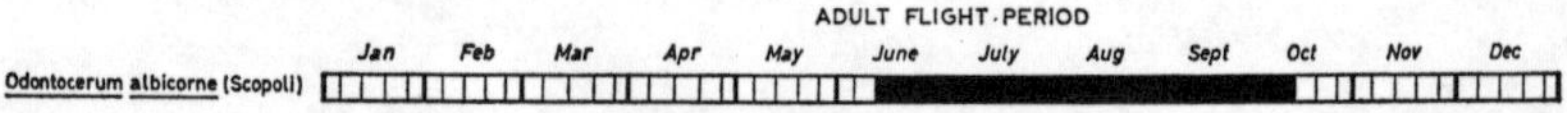

ODONTOCERUM ALBICORNE. *Adult Flight Period taken from published records of individual dates.*

Odontocerum albicorne (Scopoli)

The larva of this species inhabits swiftly running streams with a rocky bottom, the specimens described below having been obtained from Dowles Brook, Bewdley, Worcestershire, a brook which, although practically at sea-level, has many characters of a subalpine stream. It is swiftly running over a rocky bed and there is no phanerogamic vegetation present. The water is cold even in summer, being shaded by dense woodland for many miles of its length.

Odontocerum albicorne is the sole British representative of the family ODONTOCERIDAE.

Case. The larval case is composed of a single layer of large grains of sand, the exterior having a fairly smooth texture. It is slightly curved and tapers towards the posterior end (Fig. 438a). One or several pebbles incompletely block the tail end, held in position by a web of dark brown secretion, and before pupation the head end is blocked also by a small pebble. Size: up to 20 mm. in length and 4 mm. in breadth at the larger end.

Larva. The eruciform larva is cylindrical and is widest at the thorax. The sclerotized parts are light chestnut brown in colour whilst the abdomen is greyish-white. Size: up to 18 mm. in length and 3 mm. in breadth.

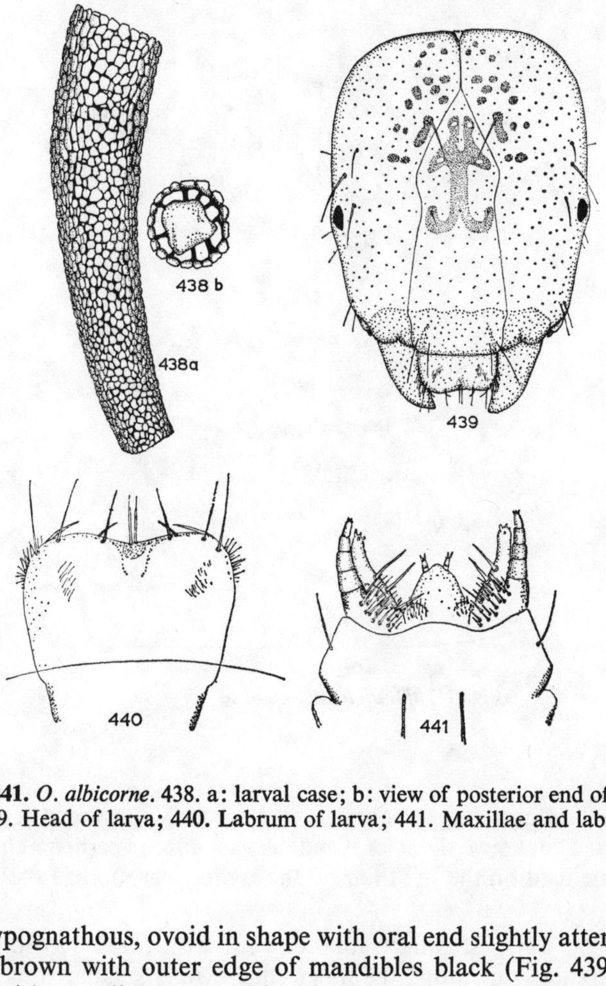

Figs. 438–441. *O. albicorne*. 438. a: larval case; b: view of posterior end of larval case; 439. Head of larva; 440. Labrum of larva; 441. Maxillae and labium.

Head. Hypognathous, ovoid in shape with oral end slightly attenuate, light chestnut brown with outer edge of mandibles black (Fig. 439). Clypeus elongate with two slight concavities along each lateral margin, one smaller near the oral end and the larger, near the middle. At the aboral end of the clypeus is an anchor-shaped mark with the 'eye' divided. A spine is borne at both oral apices of the mark whilst a light coloured spot is situated at each of the remaining four apices. Several groups of dark spots are situated on the frons near the aboral end of the clypeus and near the median frontal suture. A spine projects from a single larger spot lying on each side of the aboral end of the clypeus. A group of three or four spines lies near each eye.

Mouthparts. Labrum (Fig. 440) with anterior margin concave and small median sclerotized depression. Four pairs of bristles and two pairs of inwardly bent spines along anterior margin.

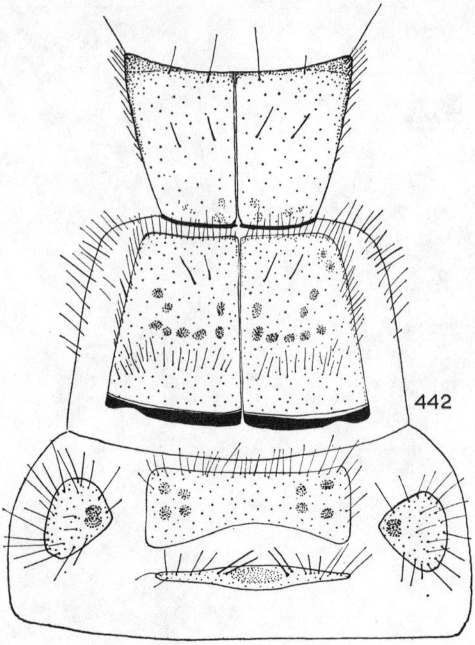

Fig. 442. *O. albicorne*. Thorax of larva.

Mandibles. Wedged-shaped obtuse and slightly asymmetrical. Outer margin folded and heavily sclerotized. A few blunt teeth on the cutting edge more defined on the left than on the right. A group of bristles on the outer edge.

Maxillae (Fig. 441). Maxillary lobe conical with three curved inwardly projecting spines lying along the inner edge. Many smaller spines at base of lobe. Maxillary palp four-segmented with insertion into lobe sclerotized, giving appearance of another segment. (Ulmer gives five segments, Lestage four.) Labium conical, labial palps two-segmented. Two hairy lobes at base of labium.

Thorax (Fig. 442). Pronotum sclerotized, concave anteriorly for insertion of head. Hairy and darker in colour at the sides. Sclerotized portion of mesonotum consists of two large plates reaching laterally almost as far as pleural region. Posterior margin black. A few dark spots run transversely at about the middle, and just posterior to these is a transverse row of hairs. Anterior and outer margins of sclerotized plates hairy. Metanotum has four sclerotized patches. Two are transverse, lying one behind the other. The anterior patch is large, with its posterior margin deeply concave.

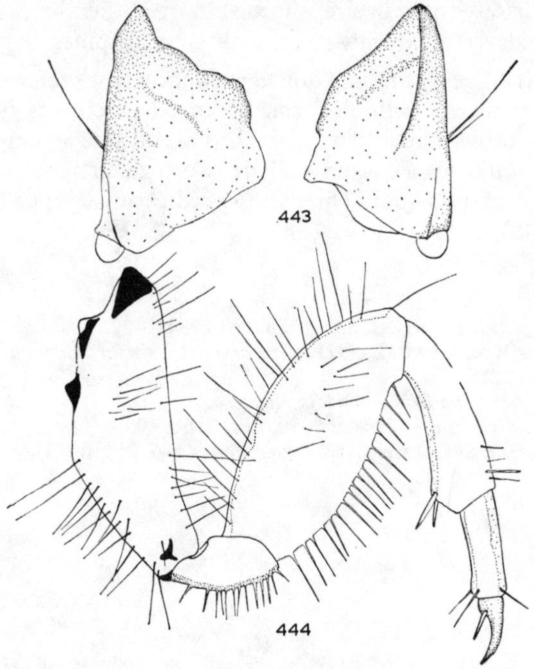

Figs. 443–444. *O. albicorne*. 443. Mandibles of larva; 444. Prothoracic leg of larva.

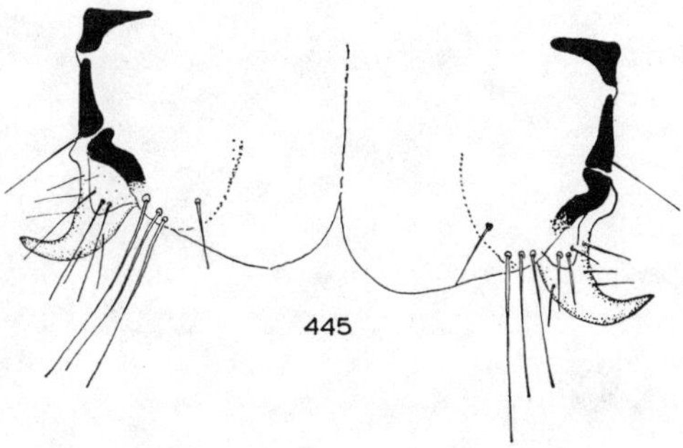

Fig. 445. *O. albicorne*. Anal claws of larva.

Posterior patch narrow with anterior and posterior margins convex and with a large dark mark in the centre. A row of hairs runs transversely along this patch. The remaining pair of sclerites are lateral patches, one in each pleural region. They are hairy and each bears a dark mark in the region nearest the dorsal surface.

Legs. Approximately equal in size, shortest in front. Femur and tibia hairy, tarsus not so hairy. Tarsal claw long, with a basal spine.

Abdomen. Fairly large stump-like dorsal sucker on first segment, but lateral suckers much reduced. Gills filiform, arranged in circlets near anterior margin of each of segments two to seven. Lateral line weakly developed, formed of fine hairs. Anal claws small, of two segments, but supported by three sclerites black in colour. Three long bristles adjacent to the posterior sclerite (Fig. 445).

HICKIN, N. E. 1942. *Proc. R. ent. Soc. Lond.* (A) **17**, 119–22.

LESTAGE, J. A., *in* ROUSSEAU, E. 1921. *Les Larves et Nymphes Aquatiques des Insectes d'Europe*, 592, fig. 202.

MOSELY, M. E. 1939. *The British Caddis Flies*, 143.

ULMER, G. 1903. *Metamorph. Trichopt.*, 99, figs. 7, 61, 100.

ULMER, G. 1909. *Die Süsswasserfauna Deutschlands* **5–6**, 252, fig. 379.

14

PHRYGANEIDAE

The characters by which adults of species in this family are identified are as follows. In size they are moderate to large: whilst a male of *Trichostegia minor* has an anterior wing length of only 10 mm. that of the female of *Phryganea grandis* is 27 mm. Some of our largest caddis flies belong to this family. Wing colour varies to a great degree from the sooty-black of *Oligotricha ruficrus* to the milky colour of *Agrypnetes crassicornis*, and from the black and yellow reticulated pattern of *Oligotricha clathrata* to the cryptic streaked-brownish of *Phryganea striata*. The antennae are stout, about as long as or only slightly shorter than the anterior wing. Ocelli are present. In the male the maxillary palpi consist of four segments whilst in the female they consist of five segments. In both sexes, however, they are generally similar being only slightly pubescent, the basal segment short and all the segments being cylindrical. The wings are rather broad with scanty pubescence in some genera but dense in others. The neuration is distinct and the discoidal cell whilst being closed in both wings is long and narrow in the anterior but short in the posterior. The median cell is not present and the cellula thyridii is very long. Apical forks Nos. 1, 2, 3 and 5 are present in the anterior wing of the male and Nos. 1, 2 and 5 in the posterior. The legs are usually fairly stout and short and the tibial spur formula is 2, 4, 4.

Check-list of PHRYGANEIDAE. Generic names shown on right-hand side are those used in Mosely's handbook.

PHRYGANEIDAE

OLIGOTRICHA
1. *clathrata* (Kolenati)=*Neuronia clathrata*
2. *ruficrus* (Scopoli)=*Neuronia ruficrus*

PHRYGANEA
3. *grandis* L.
4. *obsoleta* McLachlan
5. *striata* L.
6. *varia* Fabricius

TRICHOSTEGIA
7. *minor* (Curtis)=*Nannophryganea minor*

AGRYPNETES

8. *crassicornis* McLachlan

AGRYPNIA

9. *pagetana* Curtis

10. *picta* Kolenati

Distribution

Oligotricha clathrata. Very local. Abundant at Chartley Moss, Staffs. (M.E.M.). Recently collected in Shropshire (A.B.).

Oligotricha ruficrus. Local, found in deep weedy pools often at considerable altitude (M.E.M.); Yorks. (P.F.H.); Berks. (M.I.C. *et al.*); Worcs., Lancs., Sussex (N.E.H.).

Phryganea grandis. Widely distributed; Lakes and slow-running rivers (M.E.M.); Northumberland (G.N.P.); Berks. (M.I.C. *et al.*).

Phryganea obsoleta. Pond and lake species, common in the north and in high altitudes (M.E.M.); Yorks. (P.F.H.); Northumberland (G.N.P.); Westmorland, Radnor (N.E.H.).

Phryganea striata. Widely distributed; low-lying lakes, slow-running rivers, also tarns and lochs at considerable altitudes (M.E.M.); Yorks. (P.F.H.); Northumberland (G.N.P.); Berks. (M.I.C. *et al.*).

Phryganea varia. Ponds and lakes (M.E.M.); Northumberland (G.N.P.); Berks. (M.I.C. *et al.*); Staffs. (N.E.H.).

Trichostegia minor. Retiring in habits, concealed in daytime in bark of trees at edge of ponds and lakes. Sluggish streams (M.E.M.); Cannop Pool, Forest of Dean, Glos. abundant (N.E.H.).

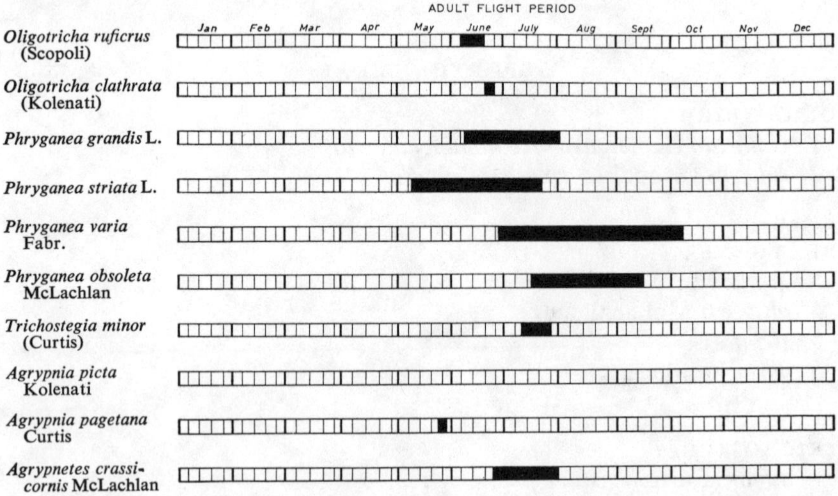

PHRYGANEIDAE. *Adult Flight Period taken from published records of individual dates.*

Agrypnetes crassicornis. Malham Tarn, Yorks. Only known British locality, fairly numerous some years (P.F.H.).

Agrypnia pagetana. Abundant on many lakes and ponds, particularly in the fens. Near Edinburgh, Scotland (M.E.M.); Northumberland (G.N.P.).

Agrypnia picta. Very local. Scarce in collections (M.E.M.).

General Description of Larvae

The sub-eruciform body is cylindrical and of medium or large size, measuring from 15–40 mm. It is widest at the first two abdominal segments. The head, which is inserted downwards, inclined slightly but not forming a right angle with the rest of the body, is pale in colour with dark slightly curved 'furcal' bands. The elongated clypeus has the anterior angles projecting and the lateral margins are indented. The short antennae consist only of a single segment carrying an apical hair. The mouthparts project distinctly, the elliptical labrum being large and concave at the anterior border and rounded laterally and with yellow curved spines on the anterior margin. Dorsally one pair of hairs is situated towards the centre and there are two on each side. The robust asymmetrical mandibles are of the 'scissors' type and, with the exception of *Trichostegia minor*, are without internal brushes. There are two dorsal hairs at the base and the right mandible is generally less well furnished with small teeth. The maxillary lobe is finger-like and the five-segmented maxillary palps are long and curved. The labial lobe is larger, curved and conical and the labial palps are of two segments.

The thoracic segments increase gradually in size posteriorly but only the pronotum is sclerotized, the meso- and metanotum being membraneous. The mesonotum is furnished with lines of distinct spots and lateral partially sclerotized plates. The prosternal horn is strongly developed. The legs are light in colour with dark patterns and are proportionally quite long. The two anterior pairs are sub-equal but the anterior are the more robust. The posterior legs are very long. The terminal spurs of the anterior tibiae are inserted on a distinct protuberance and those of the median tibiae are inserted on a very short tubercle. The terminal spur is generally large and there are long yellow spines on all the trochanters and shorter ones on the tarsi. The long curved claw of the anterior leg is sometimes longer than the tarsus and the basal spine of the anterior and median claws is large, but that of the posterior claw is smaller, spine-like or seta-like. The well-separated abdominal segments are reddish, greenish or whitish, the first two being the widest and the ninth the narrowest. The lateral line is well developed as well as are the protuberances on the first segment. The slender, long black hairs of the lateral line start to diminish at the third segment. The hexagonal partially sclerotized plates on the ninth tergite carry four hairs on the posterior margin. The gills are long, filiform, simple and robust and commence on the ventral surface of the first abdominal segment. The posterior gills of the dorsal series of segments two to seven are much more pubescent than are the others. Anal gills are absent although in the case of *Agrypnetes crassicornis* there is a short branched protuberance on each side of the anal appendages near the hind margin of the tenth sternite. The two segmented anal appendages are well developed and the terminal claws are robust and armed with numerous dorsal crochets of varying shape.

The Larval Case. This is composed of vegetable material applied in a left-turning spiral, although in *Agrypnia picta* it may consist of a single piece of twig. The cases of PHRYGANEIDAE are generally straight and open at both ends. They are often of very large size. The individual segments of the case consist of long rectangles and are usually of the same size in any particular case. Cases of *Oligotricha* cut along the turns of the spiral were found to measure almost 15 cm. when opened out and consisted of from 100 to 150 separate pieces.

Larvae of PHRYGANEIDAE are often carnivorous and cannibalistic when nearly fully grown but *Trichostegia minor*, *Agrypnia pagetana* and *Agrypnetes crassicornis* are known only to be vegetable feeders.

Oligotricha clathrata Kolenati

I do not know anyone who has collected this larva, either here or on the Continent in recent years. The characters used in the key are those given in the works of Struck, Silfvenius and Ulmer.

Detailed Descriptions of Larvae

Oligotricha ruficrus Scopoli

My first experience of this species was in 1937 when I collected three larvae from a pool in Broadmoor Wood, Rubery, Worcestershire (about 650 feet above sea-level). These larvae were successfully reared to the adult stage. Each year since I have taken some adults, always resting on stems of marginal water plants, but although I collected extensively each spring in this locality I failed to obtain more larvae. In September 1943, however, Mr. W. E. China very kindly showed me the pool in Brathay Quarry (about 200 feet above sea-level), near Wray Castle, Lake Windermere, where he and Mr. D. E. Kimmins had taken a series of the adults at the beginning of June 1942. The larvae were quite plentiful and I collected a number. They were walking over the decaying leaves which formed the floor of the pool. Then on 10 October I revisited the Broadmoor Wood Pool and was successful in collecting a dozen larvae. The pool at Broadmoor Wood, Rubery, and this pool at Brathay Quarry have several common characters. Both are surrounded by a dense growth of trees, contain much decaying foliage, and much thick black mud at one side, and in parts are fairly deep. Lestage in Rousseau states that the larvae are crepuscular or nocturnal. My experience leads me to believe that the larvae spend the winter in deeper water, or covered with leaves, as they are plentiful in autumn, where

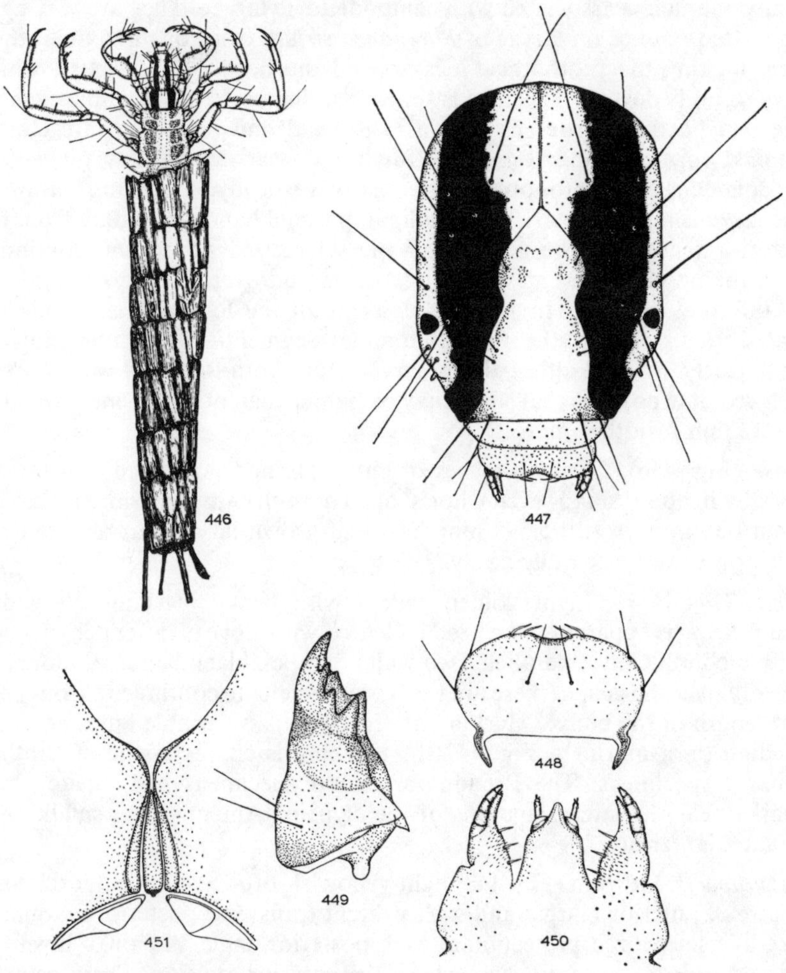

Figs. 446–451. *Oligotricha ruficrus*. 446. Larva and case; 447. Head; 448. Labrum; 449. Mandible (left); 450. Maxillae and labium; 451. Gular sclerite.

they occur, but in spring have disappeared, while at the beginning of June the adults are plentiful again.

This is perhaps a convenient place to note that the neck and prothorax of the adult when freshly emerged are bright yellow in colour, making the insect very striking in appearance.

Larva (Fig. 446). Betten's term 'suberuciform' for larvae of PHRYGANEIDAE is an appropriate one; there is a fairly deep intersegmentation of the abdominal segments. Larvae of *O. ruficrus* watched in an aquarium show many characters associated with campodeiform larvae. They are very agile and often pounce on larvae of *Phryganea striata* twice or three times their size, locking the prothoracic legs around the body of the victim, which they steadily devour. The head is somewhat depressed but not approaching the truly orthocentrous condition. The head and pronotum are closely applied together, and it is doubtful if the head enjoys any movement independent of the pronotum—laterally or vertically. A striking feature of the larva is a pair of dark, longitudinal, parallel bands extending from the anterior margin of the genae over the sclerotized pronotum, continued over the unsclerotized meso- and metanota and over the abdomen almost to the anal segment. In Ulmer's description the longitudinal bands are stated to extend to the first abdominal segment—often much farther posteriorly. So that although my larvae from both localities were heavily marked, it is possible that some may be lighter than others. Length of larva 20–22 mm., width 4 mm.

Case (Fig. 446). Of small pieces of cut roots and pine needles arranged spirally in about six to seven whorls, open at both ends, anterior end larger, length 35 mm., width 5.5–7 mm. The half-grown larvae can reverse their position in the case quite easily.

Head (Fig. 447). Light golden yellow with heavy marking. Very fine sculpturing is apparent if the head is cleared with clove oil after dehydration with alcohol. Genae parallel. Two wide, parallel, black bands run longitudinally over the genae (these are the bands which are continued throughout the length of the body). Marking on the clypeus is variable but there is no median dark mark. In Fig. 447 the marking is characteristic of Brathay Quarry specimens. The Broadmoor Wood specimens had lunate black marks near the lateral margins of the clypeus. Antennae are small. Eyes situated anteriorly.

Mouthparts. Labrum (Fig. 448) light yellowish-brown, elliptical, excised at centre of anterior margin and with straight transverse posterior margin. A peg-like sclerotized projection at each posterior angle. A pair of inwardly curved yellow spines on each side of the anterior excision. Three pairs of black bristles on the dorsal surface, and three pairs of short blade-like yellow spines on the ventral surface with some short hairs. Mandibles strong (Fig. 449) with three subsidiary teeth on the upper cutting edge and two on the lower. Teeth of the right mandible not so prominent as those of the left. No brush of hairs on the inner face but two bristles situated on the outer face. Maxillary palp (Fig. 450) five-segmented and sclerotized but proximal segment incompletely sclerotized on the inner face. Maxillary lobe almost as long as the palp. Three inwardly directed yellow spines on base of maxilla with some short hairs. Labial palps of two segments, the proximal one bulbous, the distal long with some sensory papillae at the tips A sensory pit on the distal segment slightly removed from the tip is quite prominent. Labium conical, spinneret pronounced. Gular sclerite (Fig. 451)

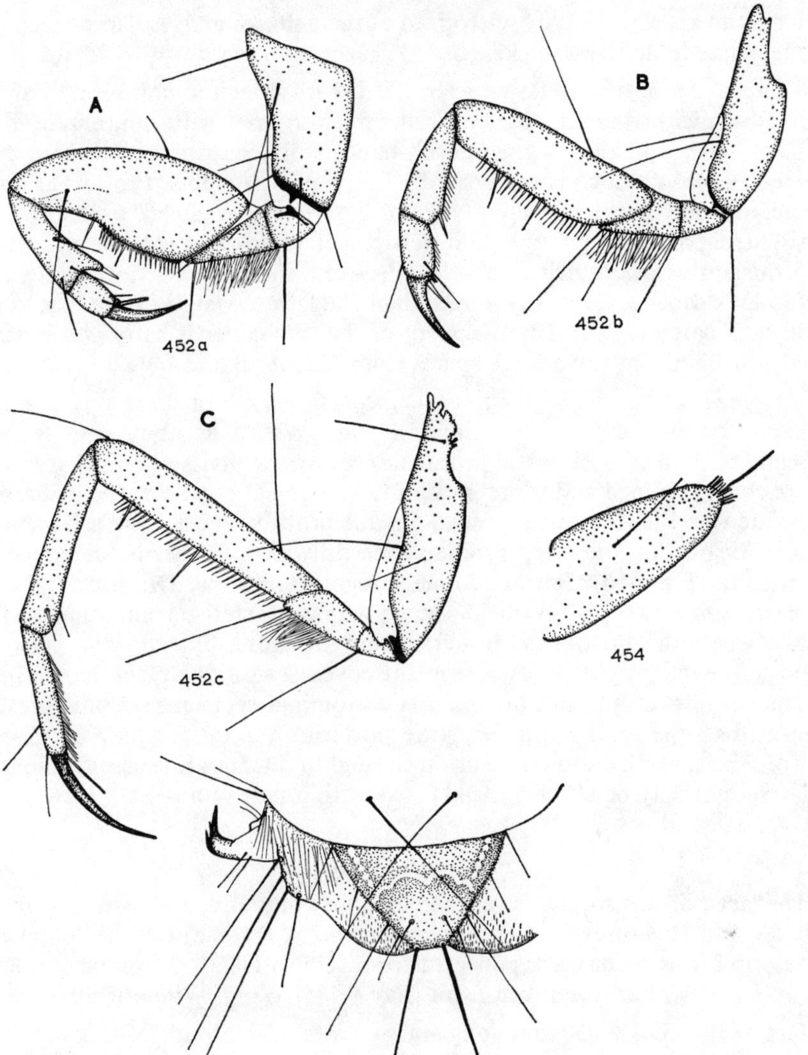

Fig. 452. *Oligotricha ruficrus*. A, prothoracic leg; B, mesothoracic leg; C, metathoracic leg.

Fig. 453. *Oligotricha ruficrus*. Ninth segment of abdomen.

Fig. 454. *Oligotricha ruficrus*. Lateral protuberance on first abdominal segment.

long and narrow, heavily sclerotized at the anterior end. Adjacent areas of the genae folded and thickened.

Thorax. Pronotum yellow, sclerotized with longitudinal black bands. Meso- and metanota unsclerotized, greenish-grey with continuation of black bands. In each segment each band is divided transversely into two areas, each of which is traversed by a row of light spots. Prosternal horn present, not large, apparently bent anteriorly. A small protrusion is situated on the anterior lateral vertices of meso- and metathorax, each furnished with a bunch of bristles. Pro- and mesothoracic legs short, the former deeper and slightly longer than the latter. Metathoracic legs long. In each case only the distal segment of the trochanter is hairy and ventral edge of femur has two large spines (Fig. 452, A, B and C).

Abdomen. Deep intersegmentation, greenish-grey with dark grey parallel bands continued from head and thorax. Widest at about the second segment. Lateral and dorsal protuberances on the first abdominal segment are clearly defined and white in colour. The dorsal protuberance is sharply pointed and is devoid of bristles, lateral protuberances have a bunch of sensory bristles and a long spine at the tip with a further bristle situated laterally (Fig. 454). First abdominal segment narrow. Dorsal margin of ninth segment (Fig. 453) shield-shaped with characteristic pattern and with two very long bristles on posterior margin. Long filiform gills present, posterior lateral gills in each segment covered with fine black hairs. Two ventral gills are present on the first abdominal segment, six gills on the second to the seventh segment (the posterior ventral gill may be absent from the seventh), and two gills on the eighth segment (of these the dorso-lateral gill may be absent). Anal claws with a pair of auxiliary claws.

Phryganea grandis L.

The larva of this species is very large and is commonly met with in ponds, lakes and slow-moving rivers. The specimens from which the following description is made were collected from King's Heath, Birmingham, and compared with larvae taken from Clay Pond, Wray, Windermere.

Case (Fig. 455). 30–50 mm. long and 8–9 mm. wide. Made of fragments of leaves uniformly cut, arranged in a spiral whorl and tapering down to the distal end. When dragging its case along, the larva makes spasmodic jerking movements with the prothoracic legs.

Larva. Eruciform. The head is hypognathous and only the prothorax is sclerotized. With the exception of the legs the rest of the body is white. Length 30–40 mm., width 4–5 mm.

Head (Fig. 456). The head is bright yellow with two dark bands running over it from the anterior margin of the genae to cervical region of the frons, where they meet. These dark bands skirt the clypeus and appear as an inverted 'V'. A single dark band runs down the median line of the clypeus.

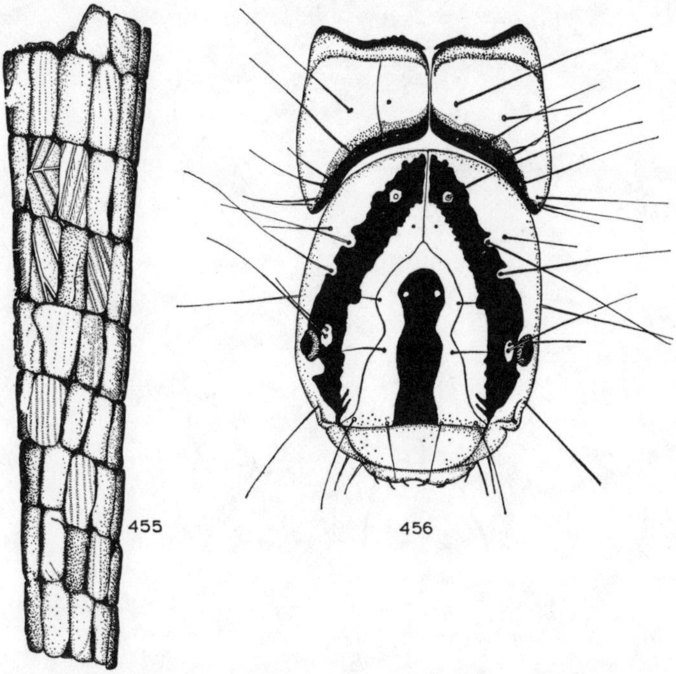

Figs. 455–456. *P. grandis*. 455. Case; 456. Head and prothorax.

Within the 'V'-shaped band, two large bristles emerge from a yellow patch near the eye on each side. A pair of single bristles also emerges from this band posterior to the eyes.

Mouthparts (Fig. 457). The labrum has the median part of the anterior margin concave. Six spines arise from this margin. A pair of spines also occupies a median position, with a spine on each lateral edge. The mandibles are three-toothed and rather long. The left mandible (Fig. 458) has the teeth doubled. From the base of the maxillae arise a large number of blunt spines (Fig. 459). The palp is five-segmented and the mala is palp-like, having a few tubercles at the distal end. The labial palps are distinct and have tubercles on the distal extremities. Two long bow-shaped sclerites are visible within the labium.

Thorax. Only the prothorax is sclerotized. It is bright yellow in colour with a transverse dark band across the anterior margin and another running across the posterior margin. A row of large bristles is set within the anterior band and a second row emerges across the centre of the prothorax transversely. A group of bristles occupies a lateral position on each side of the meso- and metathorax.

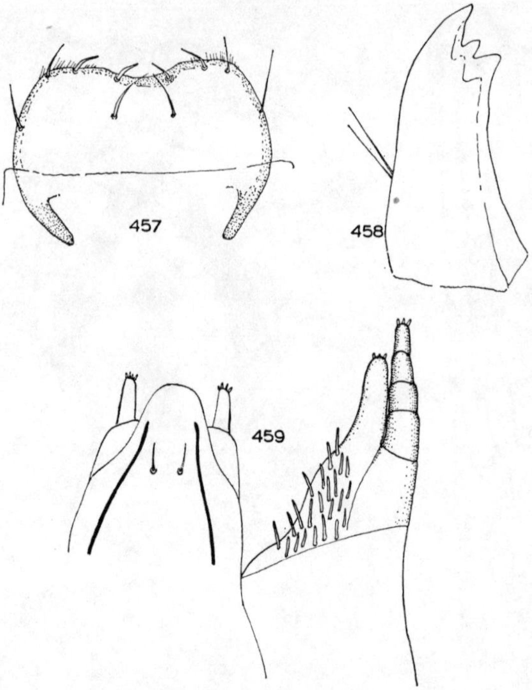

Figs. 457–459. *P. grandis*. 457. Labrum; 458. Mandible; 459. Labium and maxilla.

Legs (Figs. 460–1). A prosternal horn is present, i.e. an unsclerotized finger-like prolongation of the prosternum lying between the femora of the prothoracic legs. Prothoracic legs shortest, mesothoracic slightly longer, whilst the metathoracic legs are considerably longer. In all legs the ventral margin of tibia and the distal part of the trochanter are edged with bristles between which lie small spines. The latter are also present on the tarsus of the meso- and methathoracic legs. In all legs a spur is situated at the base of the tarsal claw. In the metathoracic leg the tarsal claw is very long. On the femur of prothoracic leg are a number of small scales edged with fine hairs.

Abdomen (Fig. 462). On the first abdominal segment is a large median process which is pointed and directed posteriorly and a pair of blunt lateral processes which curve forwards. The lateral processes have a circlet of small setae situated at their extremities. The remainder of the abdominal segments have filamentous gills. The eighth segment has only a single pair of gills—situated at the anterior pleural region of the segment. Anal claws (Fig. 463) with three auxiliary claws.

Phryganea obsoleta McLachlan

I first collected the adults of this species in 1938 at Grisedale Tarn, near Dolly Waggon Pike, Westmorland, at a height of 1,768 feet. I secured also a number of larvae. They were very common, and quite conspicuous as they walked over the submerged sphagnum and shoreweed (*Litorella*

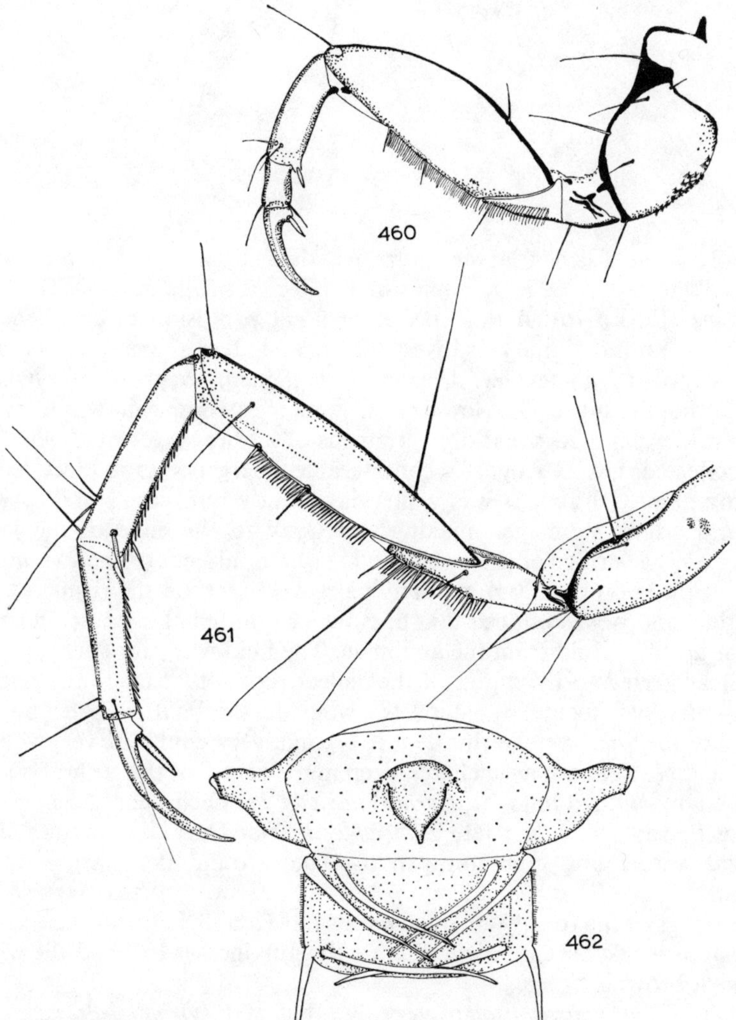

Figs. 460–462. *P. grandis*. 460. Prothoracic leg; 461. Metathoracic leg; 462. Abdominal segments.

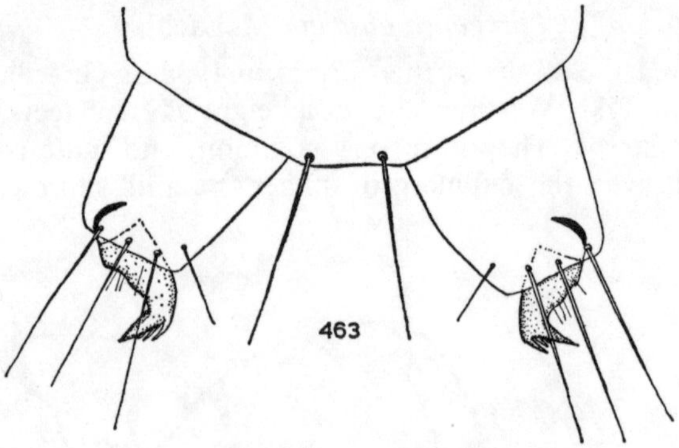

Fig. 463. *P. grandis*. Anal claws.

uniflora). Some further larvae from Scandale Tarn at 1,820 feet were obtained where Mr. D. E. Kimmins had collected adults in 1941. Then on 22 August 1953 I found this species in great profusion in and around Llynhillan, Radnorshire. This lovely lake is 1,243 feet above sea-level, it is nowhere more than a few feet deep and *Litorella uniflora* grows everywhere in it. During the subsequent few days, although a strong wind was blowing in this rather exposed situation, specimens of all life-stages of *P. obsoleta* were collected. On 23 August a considerable emergence took place, commencing about 10 a.m. It was a fine day, windy but warm, and several thousand adults were seen making their way to the edge of the lake, skating on the surface of the water, and very reminiscent of the swimming pupa. From noon to 3.30 p.m. many pairs were seen on the rushes at the lake edge, and it was noticed that apparent virgin females exuded a small bead of greenish fluid from the abdomen. The fluid was very sticky, and if it got transferred to the wings of the collected insects it made the setting operation very difficult, by gluing the wings to the boards. Later on the same day the weather deteriorated, it became very cold and very windy, and the *P. obsoleta* adults sheltered deep in the centre of the rush tussocks, in company with adults of *Cyrnus flavidus* McLach. and *Limnephilus vittatus* F. and a few very late *Triaenodes bicolor* Curt. Egg masses of *P. obsoleta* were found commonly in the lake around the margin on 31 August.

The wings of the freshly emerged adults have a milky appearance, and it is not until the day after emergence that the milkiness is lost and the wings assume a brownish tinge.

Larva of *Phryganea obsoleta* very like that of *P. grandis*, *P. varia* and *P. striata*. Length 25 mm., width 4 mm. Case length 28–35 mm., width 4.5 mm. Formed spirally with cut pieces of shoreweed, leaves or pieces of vegetable debris. Head (Figs. 465 and 467) very like that of *P. striata* L.; examination of Figs. 465 and 466 will, however, show several rather small

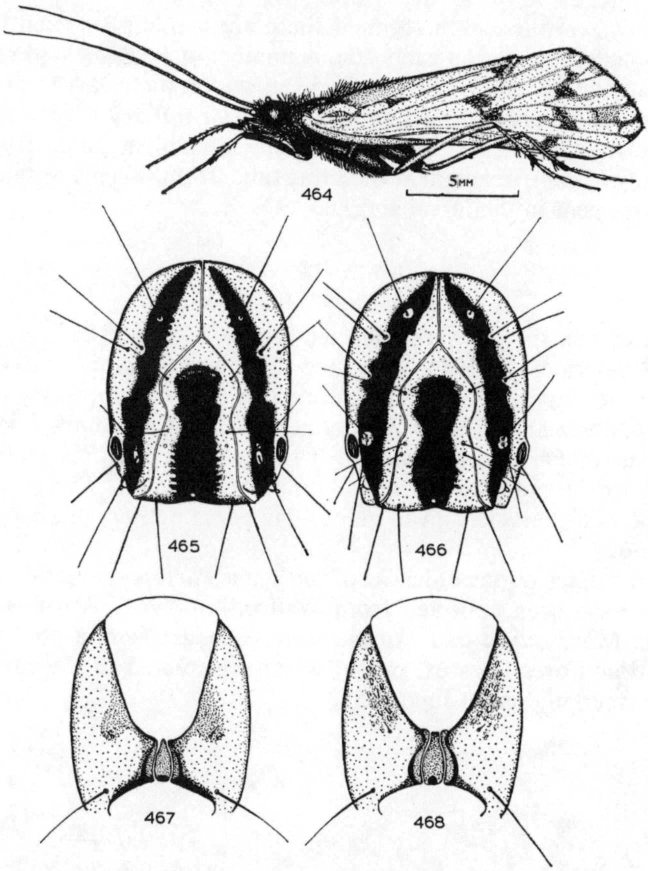

Figs. 464–468. *Phryganea obsoleta*. 464. Adult, lateral view. Head of larva (465), anterior aspect; 466. Ventral aspect; mouthparts omitted. *Phryganea striata*; Head of larva (467), anterior aspect; 468. Ventral aspect, mouthparts omitted.

differences. In the case of *P. striata* the black band in the clypeus narrows almost to a point at the oral margin, but in *P. obsoleta* the black band does not narrow and the oral margin of the clypeus is much darker. From beneath, the oral margin of the genae is black and is continuous, with two wing-like dark marks extending to the aboral margin of the genae. The muscle-spots are barely discernible in *P. obsoleta* but are readily seen in *P. striata*.

Abdominal gills. All larvae of the British species of *Phryganea* have long filiform gills. In *P. obsoleta*, *P. striata* and *P. grandis* there are two gills on each side of the first abdominal segment, then on the second to the seventh segments there are six gills on each side, and in the case of *P. grandis* and

P. obsoleta, on the eighth segment there are two gills on each side; in *P. striata* there are three on each side. A number of *P. obsoleta* were collected from Llynhillan on 23 August 1953. Two of these larvae, kept in an aquarium, measured 23 mm. in length on 13 February 1954, whilst larvae hatched from an egg-mass collected from Llynhillan on 23 August 1953 were only 11 mm. in length at the same time. It thus seems certain that two years are spent in the larval stage.

Phryganea striata L.

Larvae of this species were collected from Clay Pond near Wray Castle, Lake Windermere, in 1945. They are very common in this small pond, and, in addition, larvae have been collected in many other lakes, ponds and sluggish streams near Lake Windermere. When I made my description of the larvae of *Phryganea grandis* L. I included amongst my material some half-grown larvae which I now know to have been *Phryganea striata*. However, no characters given in my 1942 paper on *Phryganea grandis* were erroneous.

For my descriptions of larvae of the genus *Phryganea*, further *Phryganea grandis* have been collected from Welford-on-Avon, Warwickshire, and studied. More larvae of *P. striata* were collected from a pool at Nutley, Ashbourne Forest, Sussex, one of which was reared to the adult stage, a female emerging on 10 June 1953.

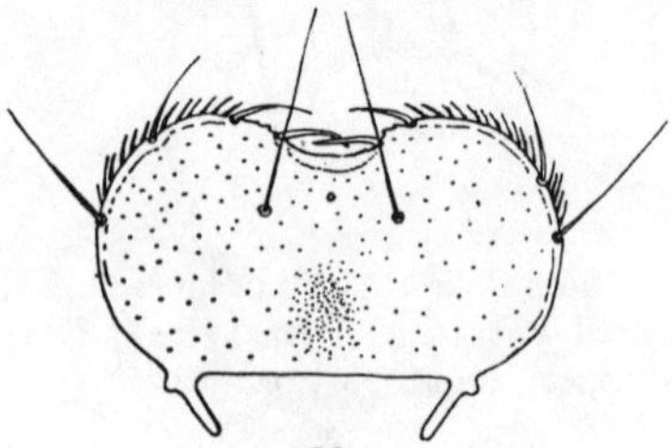

469

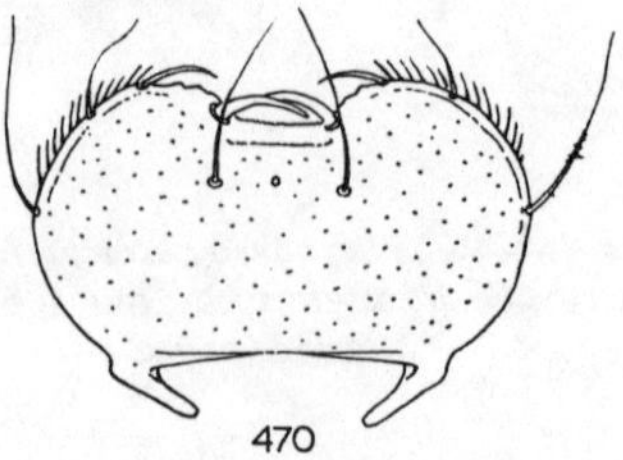

470

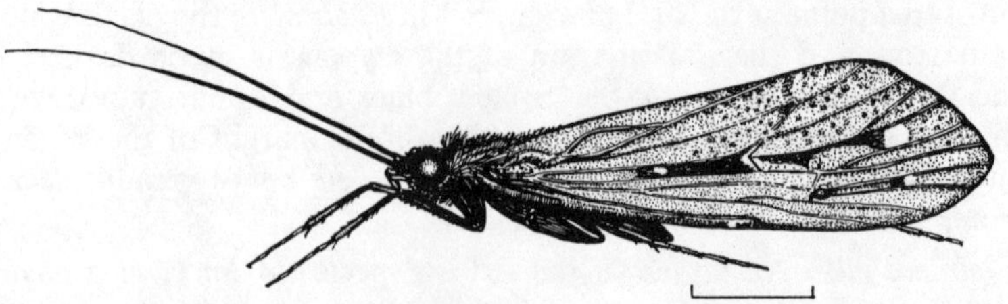

471

5 MM

Figs. 469–471. Labrum of larva of (469) *Phryganea obsoleta*; 470. *Phryganea striata*; 471. *P. striata*. Adult, lateral view.

Phryganea varia Fabricius

During an unsuccessful search for the larvae of *Oligotricha clathrata* Kolenati at Chartley Moss, Staffordshire, in March 1944, a few Phryganeid larvae were collected which ran down in Ulmer's key to *Phryganea varia* Fabricius and, indeed, one specimen was reared to the adult stage, emerging on 7 June, its identity being confirmed by the late Mr. Martin Mosely. The pool at Chartley Moss is well known as the habitat of *O. clathrata*; it is a sphagnum bog with but a few square yards of open water in the centre. The nymphs of the rare dragon-fly *Leucorrhinia dubia* van der Linden were exceedingly common here, several being present in every handful of sphagnum examined. The rare caddis *Rhadicoleptus alpestris* (Kolenati) also occurs at Chartley Moss. Whether the larvae of *O. clathrata* are to be found in the sphagnum bog or in the deep ditches which are present here is not known.

The larva of *Phryganea varia* is a typical Phryganeid larva and, in general form, is similar to that of *P. grandis* L.

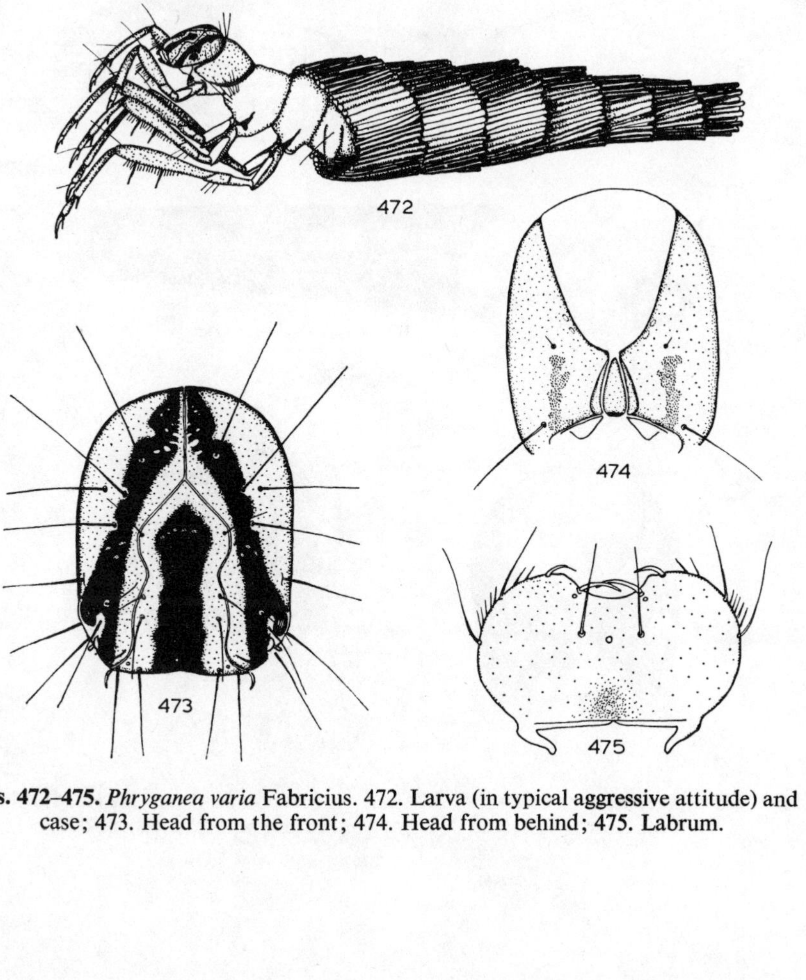

Figs. 472–475. *Phryganea varia* Fabricius. 472. Larva (in typical aggressive attitude) and case; 473. Head from the front; 474. Head from behind; 475. Labrum.

Larva (Fig. 472). 16–22 mm. in length, 4 mm. in width, i.e. only two-thirds the size of *Phryganea grandis* and *Phryganea striata* and the same size as *Phryganea obsoleta*. It is, however, easily distinguished from these other species by the presence of a pair of dark-brown longitudinal marks on the ventral surface of the genae (Fig. 474) and the absence of gills on the eighth abdominal segment.

Ulmer gives the size of the case as up to 55 mm. in length and up to 7 mm. wide. My specimens were 53 mm. long and 5 mm. wide. They are spirally wound and made of cut lengths of sphagnum stalks.

Trichostegia minor (Curtis)

Adults of this uncommon caddis fly were found in profusion by Mr. A. Peacey on 11 July 1953. They were resting in the crevices of the bark of oak-trees surrounding Cannop Pool, Forest of Dean, Gloucestershire. This was the first record for the county. I accompanied Mr. Peacey to Cannop

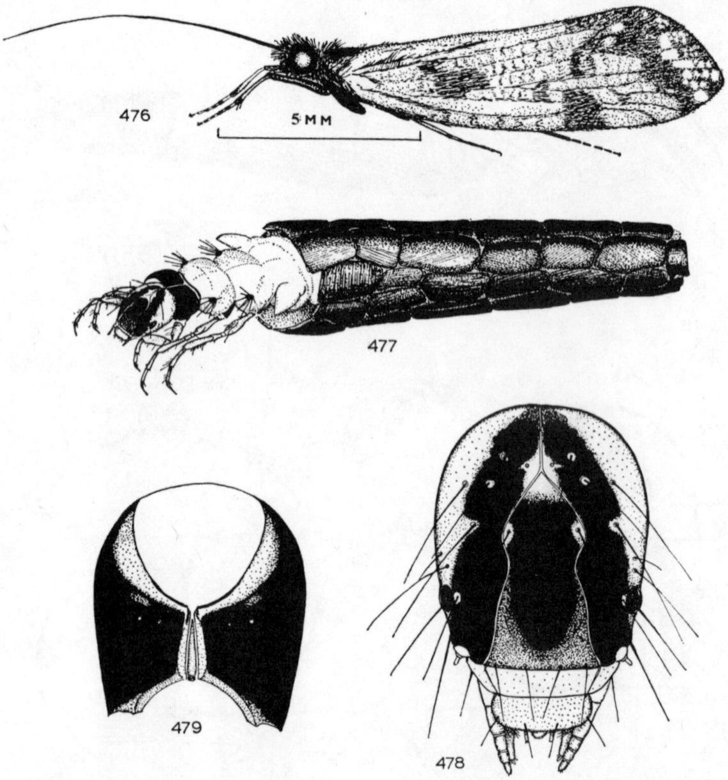

Figs. 476–479. *Trichostegia minor*. 476. Adult; 477. Larva in case; 478. Head of larva from the front; 479. Head of larva from behind, mouthparts omitted.

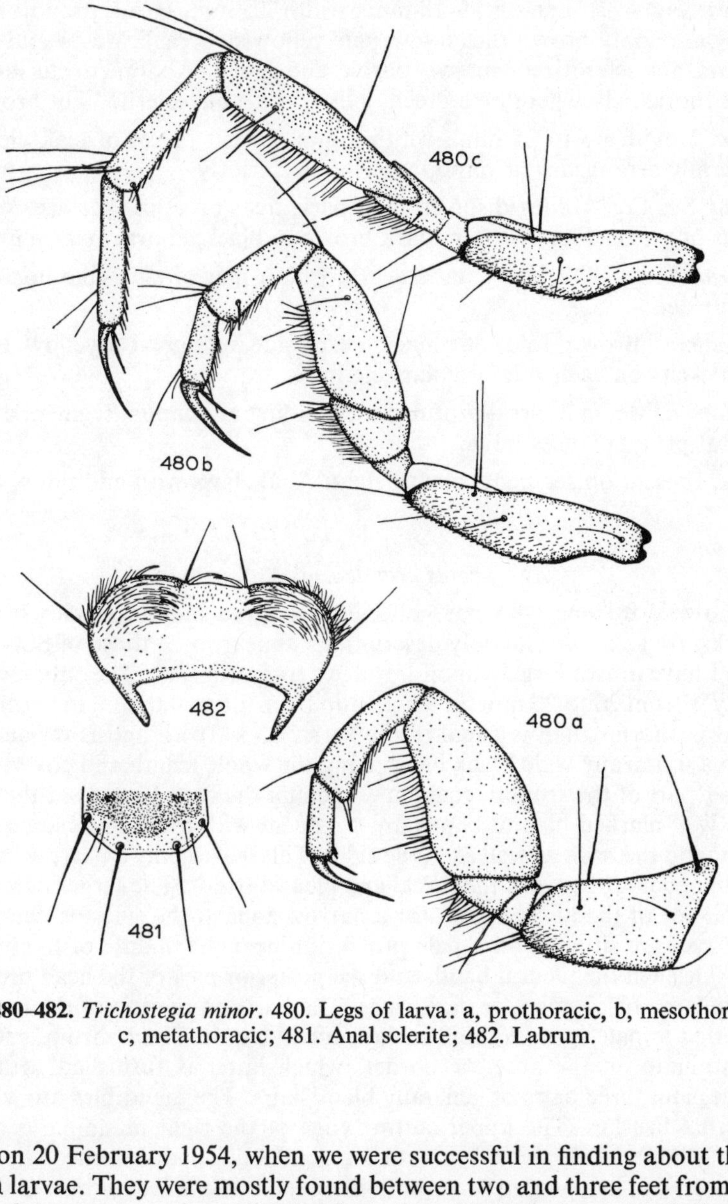

Figs. 480–482. *Trichostegia minor*. 480. Legs of larva: a, prothoracic, b, mesothoracic, c, metathoracic; 481. Anal sclerite; 482. Labrum.

Pool on 20 February 1954, when we were successful in finding about three dozen larvae. They were mostly found between two and three feet from the edge of the water, and at a depth of two feet. There had previously been a period of severe frost. The larvae appeared to be most common at the entrance to the bay-like indentations, and larvae of *Limnephilus rhombicus* (L.), *L. flavicornis* (Fabr.) and *Glyphotaelius pellucidus* (Retz.) were common also.

Larva (Fig. 477). Length 15–18 mm., width 2.5 mm. Head, pronotum and legs dark sooty-brown (head with pale yellowish areas). Meso- and metanotum not sclerotized, mauve–purple and green. Abdomen bluish-green near thorax, elsewhere lime-green, gills white, anal sclerite light brown.

Case. Length up to 25 mm., width 3 mm. Of cut pieces of leaf, etc., not generally arranged in a spiral, tapering posteriorly.

Head. See Fig. 478 for disposition of dark areas on upper surface. Ventral areas of genae (Fig. 479) very dark brown to black, aboral margin greyish.

Labrum (Fig. 482). Hairy underneath, with a pair of claw-like bristles on each side.

Mandibles. Brown, folds of genae on each side light greyish-yellow. Pair of bristle-pits on each side of gular sclerite.

Abdomen. Median dorsal protuberance on first abdominal segment absent, lateral protuberances large.

Gills. Present on segments one to eight. Anal claws with additional claws.

Agrypnetes crassicornis McLachlan

I know of no one who has collected the larva of this species from its Yorkshire locality. The only descriptions appear to be those of Silfvenius, and I have drawn largely upon Lestage's transcriptions. The suberuciform body is from 20 to 25 mm. in length and from 2.5 to 4.0 mm. in width. The head is shorter than is usual in the PHRYGANEIDAE and is ornamented with a uniformly wide black band along the whole length and covering the larger part of the frontal 'ecusson', reaching the furcal bands at the back. The well-marked pleural bands are confluent with the furcal bands at the front and the back as well as at the sides. This leaves only a narrow band of original colour between the furcal and pleural bands. The larger part of the underside of the head, except for a narrow zone at the sides, is dark grey. The eyes are situated on a pale protrusion near the middle of the head at the sides. On the pleural bands and the posterior part of the head there are quite large black spots or dark rings. The hairs on the top of the head are inserted in pale spots within the dark furcal bands. The Labrum is weakly emarginate on the anterior border, which latter is furnished with long spines and three pairs of generally black hairs. The mandibles are without internal brushes. The upper cutting edge of the right mandible is armed with a large tooth and often with a second smaller distal one. The lower cutting edge has one or two stump-like teeth. The left mandible is double toothed at the apex whilst on the upper cutting edges bears three teeth and the lower two or three teeth.

The pronotum is bordered with black, the anterior margin often dark with a zone situated behind the sclerotized 'lisense', darker than the one at the front. On the 'ecusson' there are numerous well-marked large black spots. The prosternum is darker in the middle of the posterior margin and more strongly sclerotized. The legs are reddish-yellow and gradually increase in size posteriorly (1:1.02:1.05:1.22). The posterior femurs are furnished with small pectinations. The spurs of the anterior legs are short and yellow and the upper spur of the anterior tibia is distinct. There is a yellow spur on the distal part of the upperside of the anterior femur, a curved spur yellow, or dark, on the proximal part of the posterior margin of the same femur, whilst in the median and posterior there are black hairs replacing the spurs. The sides of the eighth abdominal segment are prolonged into small rounded lumps abutting the laterial line.

The gills are 2+6+6+6+6+5+5+1=37, those of the dorsal series usually being absent from the anterior margin of the sixth and seventh segments. On the ninth segment there is a sclerotized plate carrying hairs of which the lateral are longer than those in the centre. At the back of the tenth sternite there is, on each side, close to the dual appendages, a small gill-like protuberance.

The straight larval case is from 21 to 30 mm. in length, 5 mm. in width and is open at both ends. It narrows slightly towards the hinder end and is composed of vegetable materials applied in a left-turning spiral. There are from six to twelve turns.

Biological notes on AGRYPNETES CRASSICORNIS

The larval habitat of this species is somewhat of a paradox. It appears to live equally well in sea-water amongst various algae but especially *Fucus*, as in fresh water. In Britain it is known only from Malham Farm, Yorkshire.

Few insects appear to have become adapted to a marine environment but information is accumulating involving an increasing number of species of caddis flies that are able to thrive in such conditions. The larva of *A. crassicornis* feeds mainly on algae and is much less carnivorous in habit than other species in the same family. The larval case varies from 21 to 30 mm. in length and is about 5 mm. in width. It is constructed of thin fragments of wood, algae, portions of *Chara*, roots, twigs, plant fibres and debris of *Mytilus* shells. Most frequently, however, the case is composed exclusively of the debris of decomposed stems of *Fucus*. The latter may alternate with other material but usually the *Fucus* is always situated at the front of the case and is then applied in an haphazard way. The case is built in a left-handed spiral, the case of the fully grown larva showing six to twelve complete turns. Shortly before pupation the case is plugged with algae mixed with sand-grains. At the posterior end there is a grille with numerous perforations. The anterior grille, however, is only very slightly constructed. The pupa reaches the surface by swimming, being aided by the flattened and fringed median tarsi.

The adult female also swims well, both the median and posterior legs being fringed with hairs. The egg-mass is similar to the characteristic circlet-like egg-mass of the PHRYGANEIDAE and is laid under water. It measures from 23 to 31 mm. in length and from 17 to 20 mm. in width, is about 5 to 8 mm. in diameter and is greenish or brownish in colour.

Agrypnia pagetana Curtis

Hanna gives the following description of the larva of this species from specimens collected on aquatic vegetation in the canal at Reddish, near Manchester. Adults emerged on 26 May.

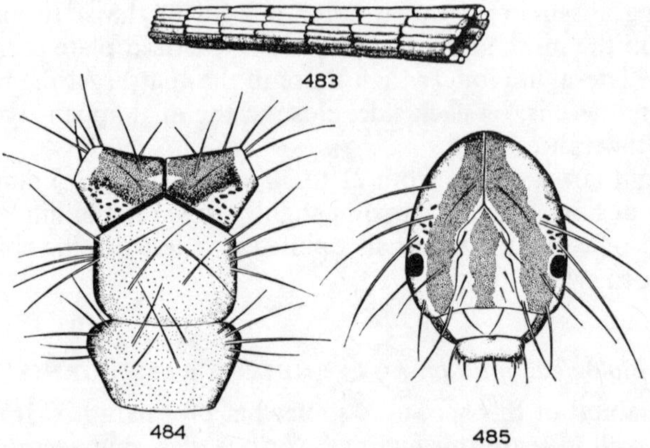

Figs. 483–485. *Agrypnia pagetana* Curtis. 483. Larval case; 484. Thoracic nota from above; 485. Head of larva from the front. Hanna.

Case. The cases (Fig. 483) were up to 35 mm. in length and 5 mm. wide. Ulmer (1909) recorded cases up to 72 mm. in length. The cases were made of pieces of leaves arranged longitudinally. In addition many larvae were found inhabiting empty stems of water-plants. Spiral cases were much less common. All the cases were cylindrical in form.

Larva. The larva is suberuciform. The larvae examined were up to 26 mm. long and 3 mm. wide. Ulmer (1909) recorded larvae up to 24 mm. in length.

Head. The clypeal band (Fig. 485) is dark brown and slightly dilated posteriorly. The dark brown bands on the genae converge to make a V. Each gena also has a dark brown band along the side. Posteriorly the genae bear dark brown spots. The gular sclerite separates the genae completely.

Labrum. The anterior margin of the labrum has a slight concavity on each side of which there are three dark brown bands, a median seta and two marginal setae.

Mandibles. The mandibles are asymmetric and each bears two long lateral setae near the base. The right mandible has four teeth. The left mandible has six teeth.

Maxilla. The cardo is small and devoid of setae. The stipes has an oblique proximal margin while the distal margin has a prolongation. The maxillary palp has five segments. The lacinia bears some sensilla and many hairs along its inner margin.

Thorax. The pronotum, which is entirely sclerotized, is chestnut brown except for its lateral and posterolateral margins, which are golden yellow with dark brown spots. The posterior margins are heavily sclerotized and are almost black, while the lateral margins are dark brown. The mesonotum and metanotum are soft. The prosternal horn is present. The prosternal sclerite is triangular.

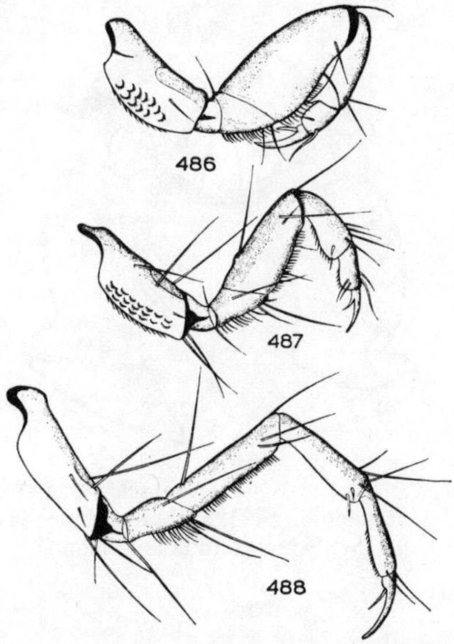

Figs. 486–488. *Agrypnia pagetana* Curtis. 486. Prothoracic leg; 487. Mesothoracic leg; 488. Metathoracic leg. Hanna.

Legs. The prothoracic legs (Fig. 486) are more flattened and slightly longer than the mesothoracic legs. The inner surfaces of the femora, tibiae and tarsi of all the legs carry small spines and there are a few hairs on the segment of the trochanters.

Abdomen. On the first abdominal segment there are three protuberances of which the dorsal one is the longest and devoid of setae, while each of the others bears two setae and some short spines. The gill filaments which are

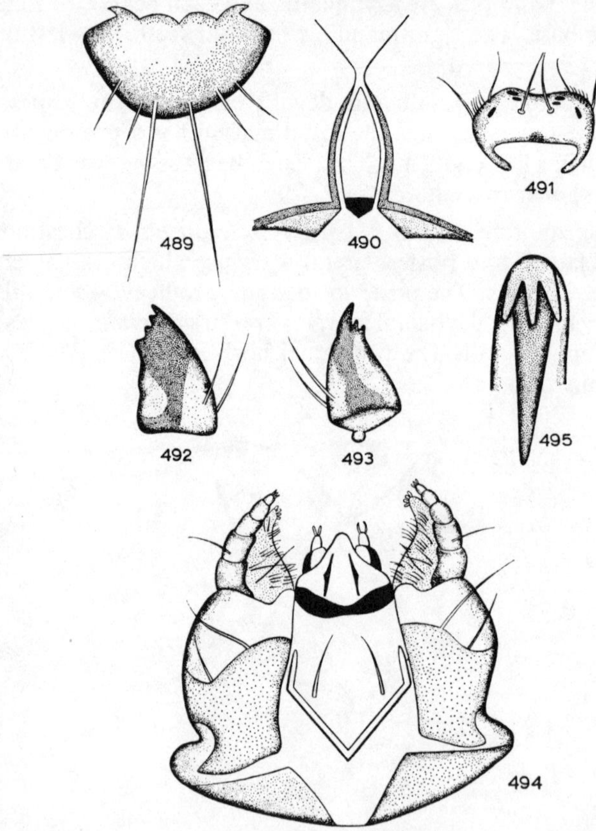

Figs. 489–495. *Agrypnia pagetana* Curtis. 489. Anal sclerite; 490. Gular sclerite, posterior view; 491. Labrum; 492. Left mandible; 493. Right mandible; 494. Maxillae and labium; 495. Anal claw from below. Hanna.

present on segments one to eight are single. The anal sclerite bears two anterior projections, while the posterior margin is markedly convex and possesses two long setae and some short ones. The anal appendages are two segmented and the claw has three auxiliary claws at its base. The lateral line running from segments three to eight is formed of fine hairs.

The clypeal band is straighter in the larva of *Agrypnia* than in the larvae of *Agrypnetes* and of *Phryganea*. In *Oligotricha* there is no clypeal band and the two bands on the anterior surfaces of the genae are parallel. In *Trichostegia* the head, pronotum and the legs are dark brown.

The clypeal band was used by Lestage (1921) for separating *Agrypnia pagetana* Curt. from *A. picta* Kol. In the latter species it is more dilated aborally, reaching the converging bands on the genae.

Agrypnia picta Kolenati

This is a very local species and no descriptions have been made from British larvae. It is said to be found amongst vegetation in lakes and ponds. It was, however, known to Silfvenius and Ulmer whose descriptions were used by Lestage. The larval case may be up to 40 mm. in length and has a spiral formation. All definite specific characteristics are embodied in the key.

Key to the Larvae of PHRYGANEIDAE

This key has been modified from that of Lestage. It should be noted that larvae of the British species all belong to class III in Lestage's key, the mesonotum in every case being entirely unsclerotized.

1. Two blackish sub-parallel bands on the head, the thorax and the first abdominal tergite *Oligotricha ruficrus*
— Thorax without black parallel bands 2
2. Two blackish longitudinal bands on the frons *Oligotricha clathrata*
— A single blackish band on the frons 3
3. Medio-frontal band very broad and almost covering the frons 4
— Medio-frontal band narrow, mandibles without internal brush, left mandible with one tooth 5
4. Mandibles without internal brush; left mandible bidentate; dorsal series of gills generally absent on abdominal segments 6–7. Length 20–25 mm. Spiral case *Agrypnetes crassicornis*
— Mandibles with an internal brush; left mandible bearing only a single tooth; dorsal series of gills present on abdominal segments 6–7, length about 15 mm. Case generally not spiral *Trichostegia minor*
5. Anterior border of the pronotum mainly dark around whole edge. Central area of the pronotum light 6
— Pronotum with, on each side, a dark zone from anterior border at least to the middle of each half of the segment, and occasionally median and lateral band can be seen beneath the dark area 7
— No spots or dark bands on the head; size 30–44 mm. 8
6. Two dark spots almost entirely covering the undersides of the head; dorsal series of gills present only on segments 2–5. Case generally composed of leaf fragments *Phryganea varia*
— Two dark longitudinal parallel bands on the head; dorsal series of gills present on segments 2–8; case composed of wood debris and roots *Phryganea obsoleta*
7. In the centre of the anterior border of the labrum a large pale area of tubercles; eight abdominal segments furnished with on each side at the front a gill filament *Phryganea grandis*
— Pale tuberculated zone at the anterior border of the labrum much smaller; eight abdominal segments with the anterior gill filament generally absent *Phryganea striata*
8. Median band of the clypeus very strongly dilated behind, often covering almost the whole of the posterior part of the clypeus and touching, more or less, the forked bands. All the bands are sprinkled with numerous clear points, dark golden; large sclerotized plate on the prosternum absent, gills $2+6+6+6+6+5+5+1=37$. Spiral case *Agrypnia picta*

— Median band of the clypeus only feebly dilated behind and not reaching the forked bands; several brownish spots on the head; a large transverse sclerotized plate on the prosternum; gills 2+6+6+6+6+4+2=32, case exceptionally spiral, generally formed from a single piece of reed *Agrypnia pagetana*

BRINDLE, A. 1961. The Larval Taxonomy of the British TRICHOPTERA 2. PHRYGANEIDAE. *Entomologist's Rec. J. Var.* **73**, 235.

HANNA, H. M. 1957. Larva of *Agrypnia pagetana* Curtis. *Entomologist's Gaz.* **8**, 110–14.

HICKIN, N. E. 1941. Some Records of Trichoptera in the Lake District. *Entomologist's mon. Mag.* **77**, 130.

HICKIN, N. E. 1942. Larvae of the British Trichoptera. *Phryganea grandis*. *Proc. R. ent. Soc. Lond.* (A) **17**, 134–7.

HICKIN, N. E. 1944. Larvae of the British Trichoptera, *Neuronia ruficrus*. *Proc. R. ent. Soc. Lond.* (A) **19**, 12.

HICKIN, N. E. 1944. Larvae of the British *Trichoptera*. *Trichostegia minor*. *Proc. R. ent. Soc. Lond.* (A) **29**, 174–6.

HICKIN, N. E. 1944. Chartley Moss and the Caddis Fly *Neuronia clathrata* Kol. (Phryganeidae, Trichoptera). *Entomologist* **77**, 20.

HICKIN, N. E. 1953. Larvae of the British Trichoptera. *Phryganea varia* Fabricius. *Proc. R. ent. Soc. Lond.* (A) **28**, 39–40.

HICKIN, N. E. 1955. Larvae of the British Trichoptera *Phryganea striata* and *P. obsoleta*. *Proc. R. ent. Soc. Lond.* (A) **33**, 176–8.

KIMMINS, D. E. 1943. A List of the Trichoptera (Caddis Flies) of the Lake District with Distributional and Seasonal Data. *J. Soc. Br. Ent.* **2**, 136.

LESTAGE, J. A., *in* ROUSSEAU, E. 1921. *Les Larves et Nymphes Aquatiques des Insectes d'Europe*, 546, 572, 580, fig. 196.

MOSELY, M. E. 1939. *The British Caddis Flies*, 27, 28, 30–32.

SILFVENIUS, A. J. 1902. Über die Metamorphose einiger Phryganeiden und Limnephiliden. *Acta Soc. Fauna Flora fenn.* **21** (4), 22–4.

SILFVENIUS, A. J. 1903. Über die Metamorphose einiger Phryganeiden und Limnephiliden II. *Acta Soc. Fauna Flora fenn.* **25** (4), 7.

SILFVENIUS, A. J. 1905–6. *Acta Soc. Fauna Flora fenn.* **27** (2), 5, 12, 13, 15, 16, 19, 26.

SILFVENIUS, A. J. 1908–9. *Acta Soc. Fauna Flora fenn.* **31** (3), 11.

STRUCK, R. 1903. Beiträge zur Kenntnis der Trichopterenlarven. *Mitt. geogr. Ges. Lübeck* **17**, 4.

ULMER, G. 1903. *Metamorph. Trichopt.*, 37–41.

ULMER, G. 1903. Über die Metamorphose der Trichopteren. *Abh. naturw.* Hamburg **18**, 40.

ULMER, G. 1909. *Die Süsswasserfauna Deutschlands* **5–6**, 238, 239, fig. 363, B, 240, 303.

WESENBERG-LUND, C. 1911. *Int. Revue ges. Hydrobiol. Hydrogr.* **4**, 65, pls. 9–10.

15

LIMNEPHILIDAE

This is perhaps the most important family of TRICHOPTERA in the northern hemisphere, yet they are, with but a few exceptions, absent from south of the equator. A few species are known from Chile only.

Limnephilids are essentially insects of temperate regions, they become progressively rarer as more tropical regions are approached.

In the British list there are fifty-six species in the LIMNEPHILIDAE out of a total number of 193 species of all families.

The British species are fairly robust caddis flies and although a few are small in size most are moderately large. The largest species is *Stenophylax permistus* McLachlan, with an anterior wing length of about 25 mm. in the female.

The following characters serve to identify the adults of LIMNEPHILIDAE. The moderately stout antennae are just about as long or are slightly shorter than the wings. The basal segment of the antenna is rather stout and bulbous. Ocelli are never absent. The maxillary palpi, although of similar form in both sexes, the basal segment being short, are three-segmented in the male and five-segmented in the female. The labial palpi are small and the terminal segment is often concave.

The number of tibial spurs serves to identify certain genera but the anterior legs bear only one spur or none. The intermediate legs never bear more than three. Strong spines are often borne on the tibiae and tarsi. The genitalia of the male conform to a fairly constant pattern. The penis is short and the sheaths are fringed or pectinate. A vulvar scale with side lobes and central tongue is present in the female.

The anterior wings are usually parchment-like and except in *Drusus* there is little pubescence and they are always much narrower than the posterior wings. The latter are always less pubescent than the anterior wings and often somewhat iridescent. The sexes have generally similar neuration although the male of certain genera bears a long pencil of hairs in a fold of the hind-wing near the base. The discoidal cell is always closed in both wings with the exception of the hind-wings of *Apatania*, and the medium cell absent whilst the cellula thyridii is long and narrow. Apical forks 1, 2, 3 and 5 are present in both wings.

The genus *Enoicyla* of which one, *pusilla* (Burmeister), of the three European species is found in Britain, is remarkable in that the larvae are terrestrial and the wings of the female are vestigial.

Classification

Since the publication of Mosely's handbook the classification of some of the species of the LIMNEPHILIDAE has undergone extensive revision. The genera in which no less than fifteen species were placed have been changed. In the following check-list, which is that of Kimmins, the generic names appearing on the right-hand side refer to the genera in which Mosely placed the species in his handbook.

Grammotaulius nitidus was omitted from Kloet & Hincks' check list, 2nd edition, 1964, and Kimmins (1966) revised the names of *Limnephilus xanthodes* and *Potamophylax stellatus* and *P. latipennis*.

DICOSMOECINAE
IRONOQUIA
1. *dubia* (Stephens)=*Caborius dubius* (Stephens)

APATANIINAE
APATANIA
2. *auricula* (Forsslund)=*Apatidea fimbriata* Morton nec Pictet
3. *muliebris* McLachlan=*Apatidea muliebris* (McLachlan)
4. *wallengreni* McLachlan

DRUSINAE
DRUSUS
5. *annulatus* Stephens

ECCLISOPTERYX
6. *guttulata* (Pictet)

LIMNEPHILUS
7. *affinis* Curtis
8. *auricula* Curtis
9. *bipunctatus* Curtis
10. *borealis* (Zetterstedt)
11. *centralis* Curtis
12. *coenosus* Curtis=*Asynarchus coenosus* (Curtis)
13. *decipiens* (Kolenati)
14. *elegans* Curtis
15. *extricatus* McLachlan
16. *flavicornis* (Fabricius)
17. *fuscicornis* (Rambur)
18. *fuscinervis* (Zetterstedt)
19. *griseus* (Linnaeus)
20. *hirsutus* (Pictet)
21. *ignavus* McLachlan
22. *incisus* Curtis=*Colpotaulius incisus* (Curtis)

23. *lunatus* Curtis
24. *luridus* Curtis
25. *marmoratus* Curtis
26. *nigriceps* (Zetterstedt)
27. *politus* McLachlan
28. *rhombicus* (Linnaeus)
29. *sparsus* Curtis
30. *stigma* Curtis
31. *subcentralis* (Brauer)
32. *vittatus* (Fabricius)
33. *binotatus* Curtis=*xanthodes* McLachlan

GRAMMOTAULIUS

34. *atomarius* (Fabricius)
35. *nitidus* (Müller)

GLYPHOTAELIUS

36. *pellucidus* (Retzius)

NEMOTAULIUS

37. *punctatolineatus* (Retzius)

ANABOLIA

38. *brevipennis* (Curtis)=*Phacopteryx brevipennis* (Curtis)
39. *nervosa* (Curtis)

RHADICOLEPTUS

40. *alpestris* (Kolenati)=*Stenophylax alpestris* Kolenati

POTAMOPHYLAX

41. *cingulatus* (Stephens)=*Stenophylax latipennis* auct.
42. *rotundipennis* (Brauer)=*Stenophylax rotundipennis* (Brauer)
43. *latipennis* (Curtis)=*Stenophylax stellatus* (Curtis)

HALESUS

44. *digitatus* (Schrank)
45. *radiatus* (Curtis)

MELAMPOPHYLAX

46. *mucoreus* (Hagen)=*Halesus guttatipennis* McLachlan

ENOICYLA

47. *pusilla* (Burmeister)

STENOPHYLAX
48. *lateralis* (Stephens)=*Micropterna lateralis* (Stephens)
49. *permistus* McLachlan
50. *sequax* (McLachlan)=*Micropterna sequax* McLachlan
51. *vibex* (Curtis)

MESOPHYLAX
52. *aspersus* (Rambur)
53. *impunctatus impunctatus* McLachlan
53a. *i. zetlandicus* McLachlan

ALLOGAMUS
54. *auricollis* (Pictet)=*Halesus auricollis* (Pictet)

HYDATOPHYLAX
55. *infumatus* (McLachlan)=*Stenophylax infumatus* McLachlan

CHAETOPTERYX
56. *villosa* (Fabricius)

Distribution

Ironoquia dubia. Rarely seen, Windsor Forest (M.E.M.); Berks. (M.I.C. *et al.*).
Apatania auricula. Cold springs (M.E.M.); Yorks. (P.F.H.).
Apatania muliebris. No male known of this species (M.E.M.); Berks. (M.I.C. *et al.*).
Apatania wallengreni.
Drusus annulatus. Northumberland (G.N.P.); Yorks. (P.F.H.).
Ecclisopteryx guttulata. Very abundant in mountain regions. Fly in clouds round outstanding objects on river banks; running water (M.E.M.); Northumberland (G.N.P.).
Limnephilus affinis. Very widely distributed, still water. Often found far from water (M.E.M.); Berks. (M.I.C. *et al.*); Som. (J.C.).
Limnephilus auricula. Widely distributed, abundant, shelters under leaves of trees sometimes at a distance from water (M.E.M.); Northumberland (G.N.P.).
Limnephilus bipunctatus. Local, still water (M.E.M.); Berks. (M.I.C. *et al.*).
Limnephilus borealis. North and western Scotland, mainly lakes and ponds (M.E.M.).
Limnephilus centralis. Very widely distributed, usually found in considerable numbers, marshy localities (M.E.M.); Berks. (M.I.C. *et al.*); Yorks. (P.F.H.).
Limnephilus coenosus. Rather local, probably cold springs (M.E.M.); Som. (J.C.); Yorks. (P.F.H.).

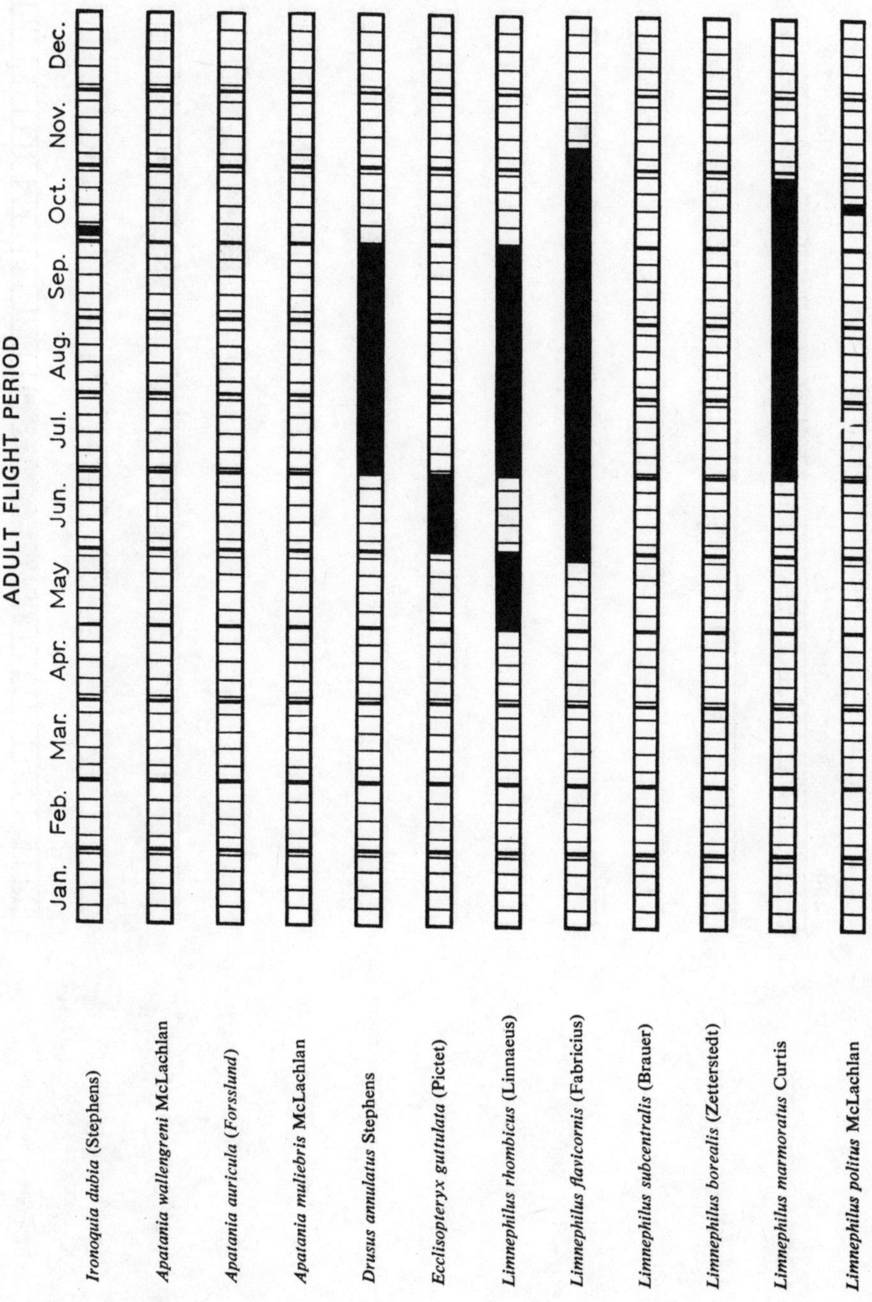
ADULT FLIGHT PERIOD
Jan. Feb. Mar. Apr. May Jun. Jul. Aug. Sep. Oct. Nov. Dec.
Ironoquia dubia (Stephens)
Apatania wallengreni McLachlan
Apatania auricula (Forsslund)
Apatania muliebris McLachlan
Drusus annulatus Stephens
Ecclisopteryx guttulata (Pictet)
Limnephilus rhombicus (Linnaeus)
Limnephilus flavicornis (Fabricius)
Limnephilus subcentralis (Brauer)
Limnephilus borealis (Zetterstedt)
Limnephilus marmoratus Curtis
Limnephilus politus McLachlan

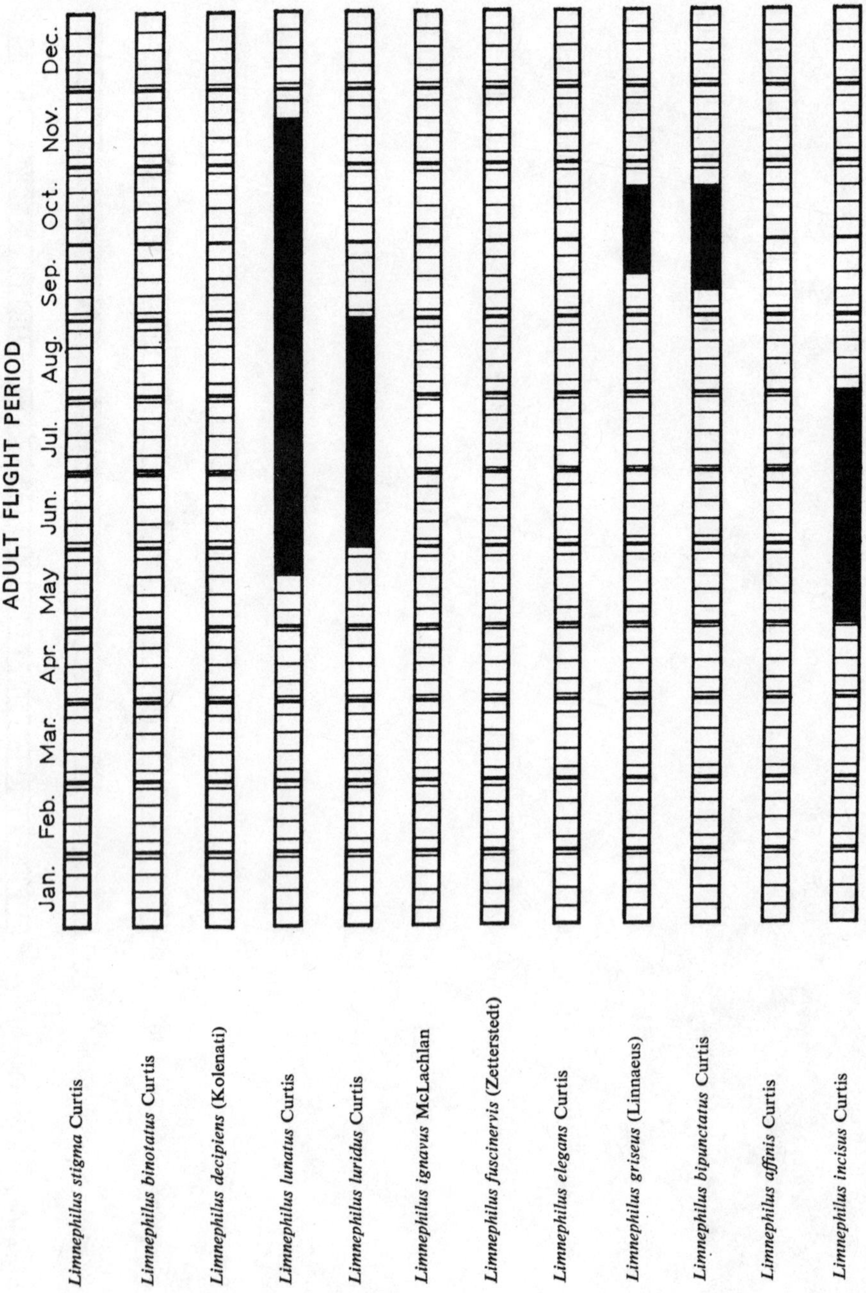
ADULT FLIGHT PERIOD
Jan.
Feb.
Mar.
Apr.
May
Jun.
Jul.
Aug.
Sep.
Oct.
Nov.
Dec.
Limnephilus stigma Curtis
Limnephilus binotatus Curtis
Limnephilus decipiens (Kolenati)
Limnephilus lunatus Curtis
Limnephilus luridus Curtis
Limnephilus ignavus McLachlan
Limnephilus fuscinervis (Zetterstedt)
Limnephilus elegans Curtis
Limnephilus griseus (Linnaeus)
Limnephilus bipunctatus Curtis
Limnephilus affinis Curtis
Limnephilus incisus Curtis

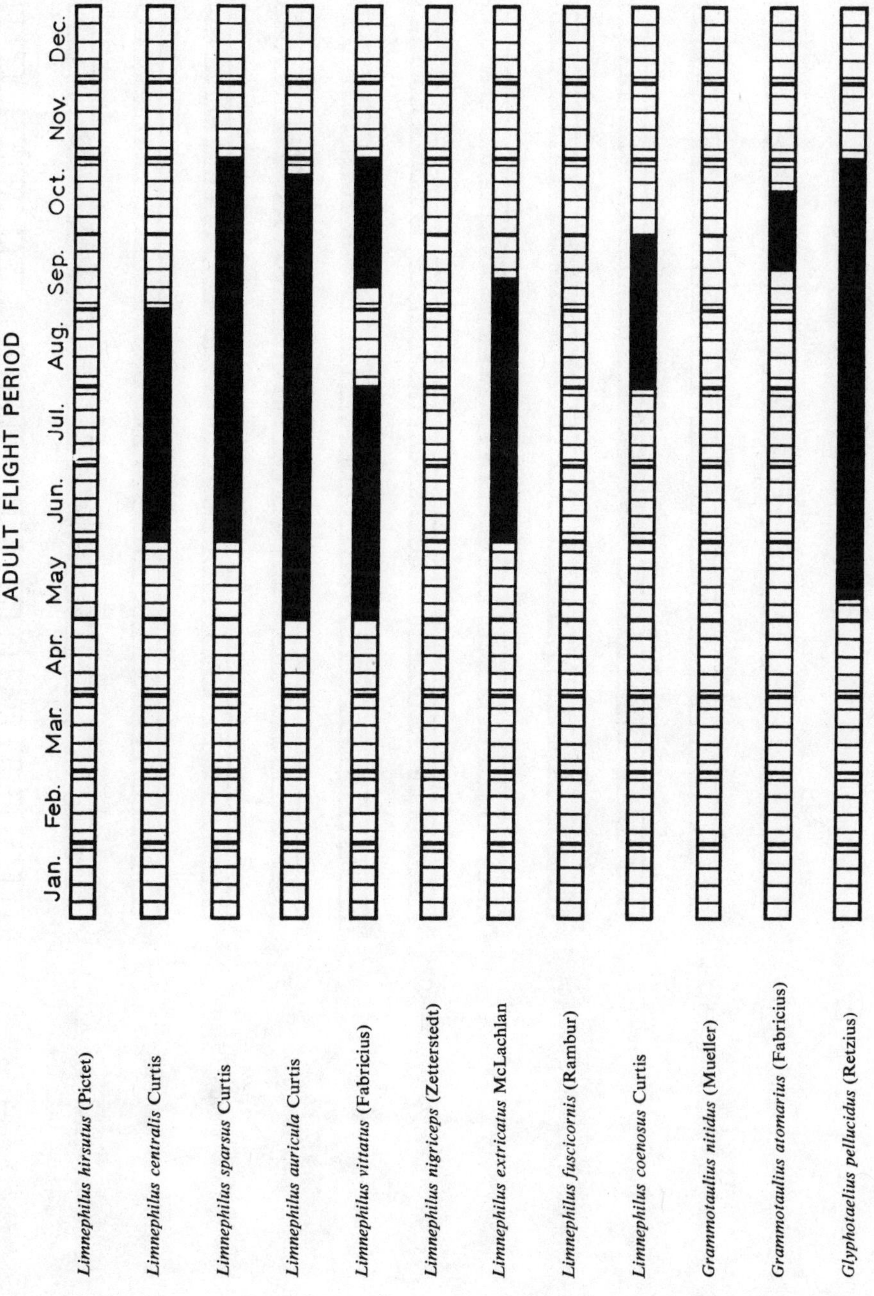
ADULT FLIGHT PERIOD
Jan. Feb. Mar. Apr. May Jun. Jul. Aug. Sep. Oct. Nov. Dec.
Limnephilus hirsutus (Pictet)
Limnephilus centralis Curtis
Limnephilus sparsus Curtis
Limnephilus auricula Curtis
Limnephilus vittatus (Fabricius)
Limnephilus nigriceps (Zetterstedt)
Limnephilus extricatus McLachlan
Limnephilus fuscicornis (Rambur)
Limnephilus coenosus Curtis
Grammotaulius nitidus (Mueller)
Grammotaulius atomarius (Fabricius)
Glyphotaelius pellucidus (Retzius)

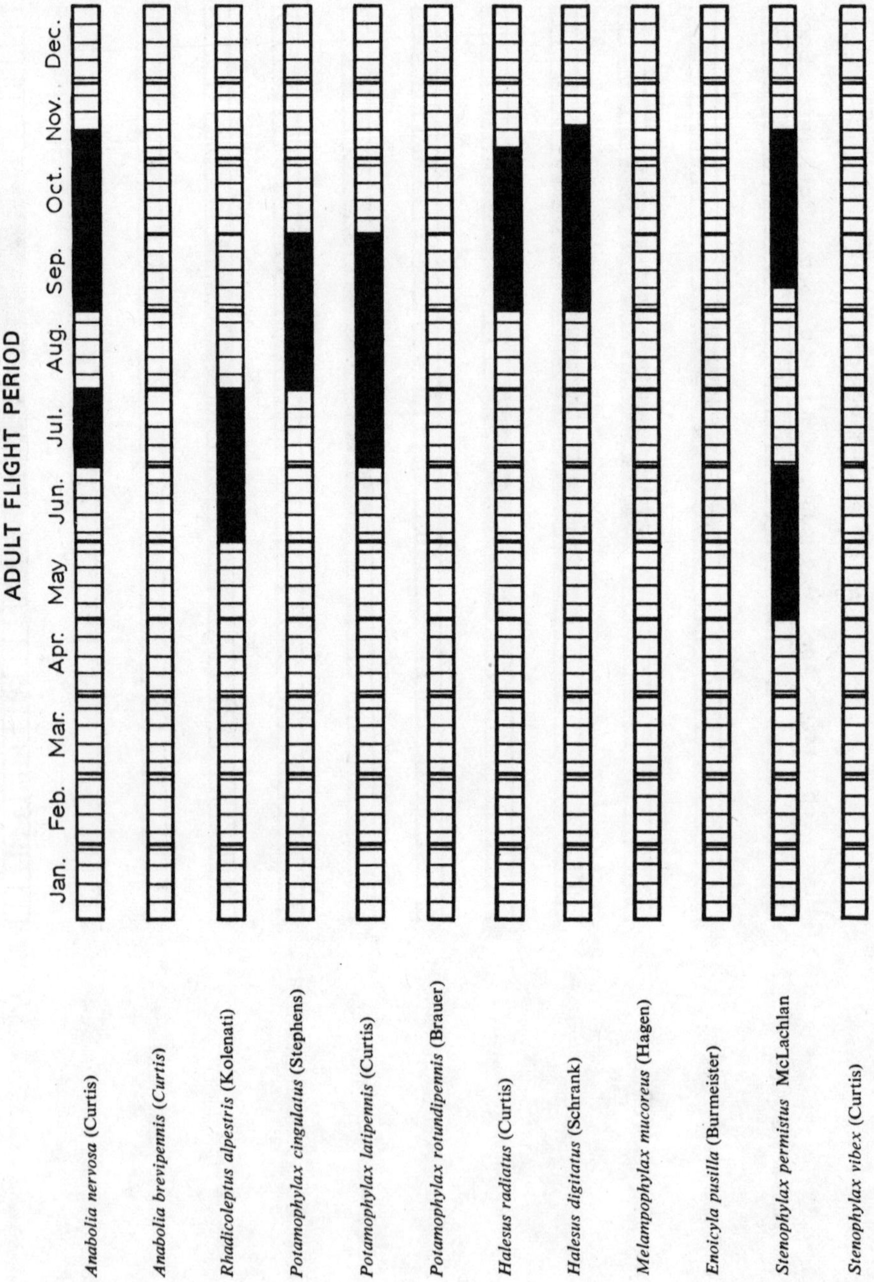
ADULT FLIGHT PERIOD
Jan.
Feb.
Mar.
Apr.
May
Jun.
Jul.
Aug.
Sep.
Oct.
Nov.
Dec.
Anabolia nervosa (Curtis)
Anabolia brevipennis (Curtis)
Rhadicoleptus alpestris (Kolenati)
Potamophylax cingulatus (Stephens)
Potamophylax latipennis (Curtis)
Potamophylax rotundipennis (Brauer)
Halesus radiatus (Curtis)
Halesus digitatus (Schrank)
Melampophylax mucoreus (Hagen)
Enoicyla pusilla (Burmeister)
Stenophylax permistus McLachlan
Stenophylax vibex (Curtis)

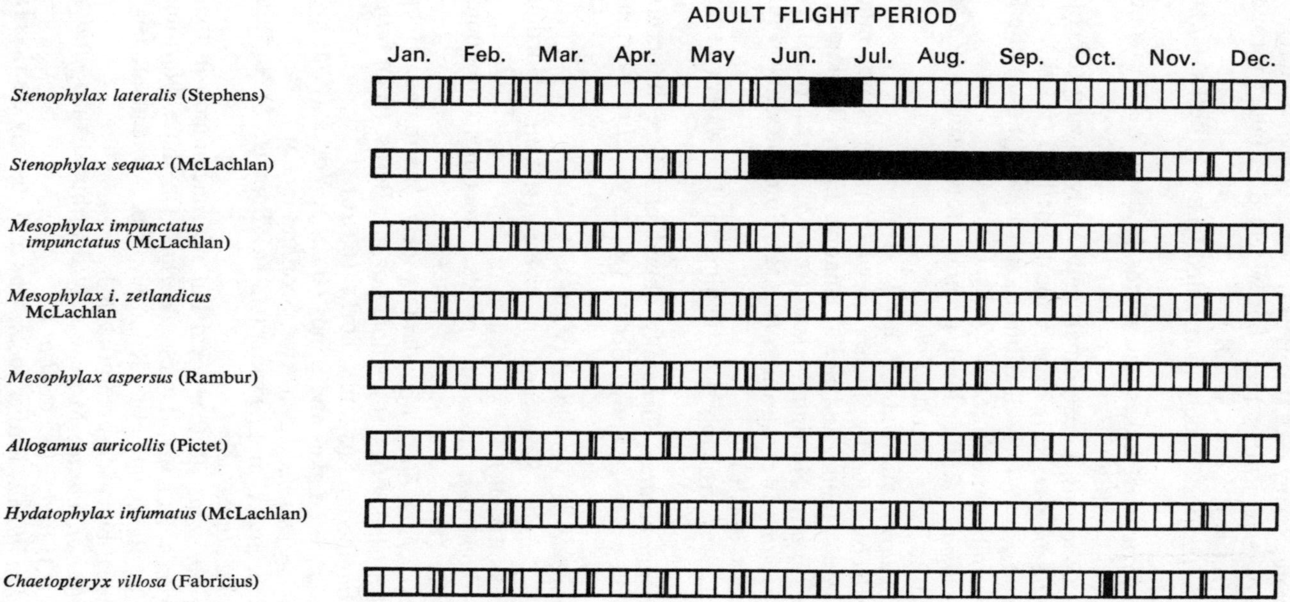

LIMNEPHILIDAE. *Adult Flight Period taken from published records of individual dates.*

Limnephilus decipiens. Rather local, abundant in Hants., and at Wicken Fen, Haslemere, Surrey (M.E.M.); Berks. (M.I.C. *et al.*); Som. (J.C.).

Limnephilus elegans. Rather local but widely distributed throughout the British Isles (M.E.M.); Yorks. (P.F.H.).

Limnephilus extricatus. Widely distributed (M.E.M.); Berks. (M.I.C. *et al.*); Yorks. (P.F.H.).

Limnephilus flavicornis. Lakes and ponds (M.E.M.); Berks. (M.I.C. *et al.*); Northumberland (G.N.P.).

Limnephilus fuscicornis. Local, canals and slowly flowing rivers (M.E.M.)

Limnephilus fuscinervis. Rare or very local. Soctland (M.E.M.).

Limnephilus griseus. Widely distributed, still water (M.E.M.); Berks. (M.I.C. *et al.*); Northumberland (G.N.P.); Yorks. (P.F.H.).

Limnephilus hirsutus. Rather local, generally found singly (M.E.M.); Sligo, Ireland (N.E.H.); Berks. (M.I.C. *et al.*).

Limnephilus ignavus. Still water (M.E.M.).

Limnephilus incisus. Rushy ditches in which there is not much water, Windsor Forest, Hants., and near Llangorse Lake, Brecknock (M.E.M.); Berks. (M.I.C. *et al.*); Northumberland (G.N.P.); Yorks. (P.F.H.).

Limnephilus lunatus. In most waters (M.E.M.); Som. (J.C.); Northumberland (G.N.P.); Yorks. (P.F.H.).

Limnephilus luridus. Probably slow-running, shallow water (M.E.M.); Yorks. (P.F.H.).

Limnephilus marmoratus. Generally abundant where it occurs, almost every type of water (M.E.M.); Berks. (M.I.C. *et al.*); Northumberland (G.N.P.).

Limnephilus nigriceps. Very local (M.E.M.).

Limnephilus politus. Slow-running rivers (M.E.M.); Northumberland (G.N.P.); Yorks. (P.F.H.).

Limnephilus rhombicus. Lakes and ponds, sometimes slow-running rivers (M.E.M.); Berks. (M.I.C. *et al.*); Yorks. (P.F.H.).

Limnephilus sparsus. Very abundant, widely distributed. Hebrides and other Scottish islands (M.E.M.); Glos. (A.F.P.); Berks. (M.I.C. *et al.*); Som. (J.C.); Northumberland (G.N.P.); Yorks. (P.F.H.).

Limnephilus stigma. Lakes and ponds (M.E.M.); Yorks. (P.F.H.).

Limnephilus subcentralis. Loch Awe, Scotland (M.E.M.).

Limnephilus vittatus. Standing water (M.E.M.); Som. (J.C.); Northumberland (G.N.P.); Yorks. (P.F.H.).

Limnephilus binotatus. Rather local but not uncommon in the fens and Norfolk Broads (M.E.M.); Lancs., Yorks., Cheshire, Northumberland, Salop, Staffs., Berks., Bucks., Killarney, Co. Monaghan, Isle of Man. Somewhat local but widespread (L.N.K.).

Grammotaulius atomarius. More widely distributed than *nitidus*; habitat similar (M.E.M.); Berks. (M.I.C. *et al.*).

Grammotaulius nitidus. Rather local, fens and marshy districts (M.E.M.).

Glyphotaelius pellucidus. Widely distributed (M.E.M.); Som. (J.C.).

Nemotaulius punctatolineatus. A single specimen was discovered in Inverness-shire in 1965.

Anabolia brevipennis. Very local, Yorks. (M.E.M.).

Anabolia nervosa. Very widely distributed, running and still water, on some rivers as great clouds (M.E.M.); Berks. (M.I.C. *et al.*).

Rhadicoleptus alpestris. Very local (M.E.M.); Chartley Moss, Staffs. (N.E.H.); Yorks. (P.F.H.).

Potamophylax cingulatus. Local, chiefly in south-west of England (M.E.M.); Northumberland (G.N.P.); Yorks. (P.F.H.).

Potamophylax rotundipennis. Local. Recorded by McLachlan from Scarborough and Carluke, Lanark. Near Oxford by L.W. and A.D.G. (M.E.M.); Northumberland (G.N.P.).

Potamophylax latipennis. Very widely distributed. In great numbers on rivers where they occur (M.E.M.); Som. (J.C.); Northumberland (G.N.P.); Yorks. (P.F.H.).

Halesus digitatus. Hants (M.I.C. *et al.*); Som. (J.C.); Northumberland (G.N.P.); Yorks. (P.F.H.).

Halesus radiatus. Widely distributed. Small examples from parts of Ireland. Galway (M.E.M.); Berks. (M.I.C. *et al.*); Northumberland (G.N.P.); Yorks. (P.F.H.).

Melampophylax mucoreus. Rather local, lower altitudes as a rule than *Allogamus auricollis*. Northumberland (G.N.P.); Yorks. (P.F.H.).

Enoicyla pusilla. Wyre Forest, Worcestershire, at foot of oak trees under dead leaves.

Stenophylax lateralis. Hants. (M.I.C. *et al.*); Som. (J.C.).

Stenophylax permistus. Widely distributed, attracted to light (M.E.M.); Som. (J.C.); Northumberland (G.N.P.); Yorks. (P.F.H.).

Stenophylax sequax. Small, clear brooks (M.E.M.); Berks. (M.I.C. *et al.*); Northumberland (G.N.P.); Yorks. (P.F.H.).

Stenophylax vibex. Less widely distributed than *permistus* (M.E.M.); Som. (J.C.); Yorks. (P.F.H.).

Mesophylax aspersus. By no means common. Often found in caves (M.E.M.). Actually three British specimens only. Ipswich (1895). 1 Winchester (1961) and 1 Portland (1961). Kimmins discusses the possibility of these specimens being immigrants (D.E.K.).

Mesophylax impunctatus. Very rare or local. Scotland (M.E.M.); Yorks. (P.F.H.).

Allogamus auricollis. Widely distributed in alpine regions (M.E.M.); Northumberland (G.N.P.).

Hydatophylax infumatus. Rather local, moorland streams (M.E.M.); Yorks. (P.F.H.).

Chaetopteryx villosa. Running water (M.E.M.); Hants. (M.I.C. *et al.*); Northumberland (G.N.P.); Yorks. (P.F.H.).

General Form of the Larvae

Lestage has given the following description, mainly based on the work of Struck, Siltala and Ulmer. The eruciform larvae are generally large and robust, more or less cylindrical, widest at the metathorax and the first abdominal segment. The head and prothorax almost equal in size, the ninth abdominal segment the narrowest. Head strongly inclined downwards and generally shorter than in the PHRYGANEIDAE. Eyes usually situated on a transparent prominence. Antennae are rudimentary inserted on a small prominence and formed of a single segment unprovided with hairs at the tip. Clypeus broad at the front where the angles usually form acute projections. It is rounded at the sides and constricted towards the middle, forming lateral indentations. The posterior part is oval in shape with the tip acuminate and bearing spots which form a wedge-shaped pattern. According to the darkness of the colouring this may or may not be visible. The mouthparts are more or less prominent. The labrum is narrower than in the PHRYGANEIDAE, rounder, hemispherical or transverse elliptical, usually rounded at the anterior angles, arched on the sides, strongly indented in the middle of the anterior edge, widest in the middle with three pairs of hairs disposed in the shape of an arc, one pair on each side and two on the dorsal surface. Sometimes four pairs (*Halesus*), occasionally more (*Apatania*); moreover on the anterior edge two pairs of yellow curved spines, the central pair usually being very short; lateral brushes present. Mandibles dark, often black symmetrical provided with more or less rounded teeth with an internal brush and two dorsal hairs. Maxillae and labium very short, maxillary lobe usually reaching the very end of the fourth segment of the palp, provided on the inner side on top with numerous setae, hairs and several spines, cardo narrow, narrow on the outside and rounded on the inside; stipes same as in PHRYGANEIDAE but shorter. Maxillary palps of five segments, the first being large and strongly ciliated underneath, the following ones gradually decreasing in size, the last bearing several small, terminal tubercles. Labium short, labial lobe obtuse, conical. Labial palps are shorter than in the PHRYGANEIDAE of only one segment and terminated by several tactile rods.

Thoracic segments successively broader posteriorly, the metathorax almost twice as broad as the prothorax, the pronotum and mesonotum are covered by quadrangular sclerites, divided into two by a median longitudinal suture extending to the edges of the sclerite; on the anterior edge of the pronotum are short spines and slender pale hairs; on the top are longer hairs intermingled with shorter hairs, black and yellow, posterior margin black, at anterior one-third or two-fifths a transverse groove the bottom of which is often darker, moreover almost always a series of spots disposed in the shape of an 'X' on the posterior part of the median line, another larger group in middle of each of the segments of the pronotum, finally a smaller group a little inside of the centre of the side.

The sclerite of the rectangular mesonotum square anterior margin concave often widely edged with black on the posterior margin, particularly at the posterior angles; on top three groups of hairs in one of which the hair is abundant at the anterior angles, one where the hair is more sparse at the anterior margin near the centre, one group behind each side of the median line, generally no hairs at the anterior margin, at the anterior angles behind and inside almost as far as the middle of each of the parts of the segment, a group of spots forms a wedge shaped mark; in the front of this group two supplementary spots near the median line, several spots on the posterior zone, near the posterior angles a wedge-shaped dark oblique streak.

Development Cycle in the Genus Limnephilus

As a result of his previous work on the mouthparts of adult caddis flies which, contrary to the general opinion, he found to be functional, and on his large-scale light-trapping experiments, Crichton came to the opinion that the adults, in the case of the LIMNEPHILIDAE, became widely dispersed. They were also capable of living several weeks. Novak & Sehnal were able to go further and to point out the remarkable feature of the development cycle in the genus *Limnephilus* and in some related genera.

In the case of *Limnephilus rhombicus* and *Limnephilus stigma*, studied in Bohemia, egg-masses were laid in September within 10 cm. of the water's edge on the banks of small ponds. The egg-clutches were hardly ever visible from above but were deposited under overhanging banks or amongst dense grass. The young larvae are capable of independent life in the jelly egg-mass after it has been moistened but do not leave the jelly until it has been completely covered by water. The eggs of *L. rhombicus* develop in from fourteen to twenty-one days at 15° C. when partially submerged in water. The larvae of *L. stigma* generally hatched in January and February and only rarely in autumn. On the other hand, first instar larvae of *L. rhombicus* were usually found in the second half of September. Under laboratory conditions the duration of the various instars were as follows:

Limnephilus rhombicus

1st instar	50 days at 13° C.
2nd instar	—
3rd instar	20 days at 13° C.
4th instar	42 days at 11° C.
5th instar	90–100 days at 10° C.

Pupae of this species were usually found in April and May and this stage lasted ten days at 15° C. Adults appeared from the beginning of April and can be found from that date until September and even as late as November. These authors were able to maintain alive many species of *Limnephilus* for several months, indeed throughout the summer months, whereas the egg-masses were deposited within a period of a few days at the end of autumn.

The development of ovaries was examined by dissection during the period of the adult stage and it was found that, in this genus, when the females emerge in spring, they possess very imperfectly developed ovaries. Maturation of the ovaries only takes place towards the end of August and into September. It is interesting to note that the only exception to this occurred at high altitude.

It was found that the aestivation of *Limnephilus* species was a typical ovarian diapause which could be terminated by a short date photoperiod. In nature this occurs towards the end of summer. This genus, then, shows a remarkable adaptation to summer drought with the consequent drying out of the larval habitat by the phenomenon of imaginal diapause.

Detailed Descriptions of Larvae

Ironoquia dubia (Stephens)

This species has only been recorded five times in Britain, all in southern England, but it is thought to be very retiring in habit, flying at sunset then disappearing. It inhabits woodland streams, ditches and ponds in which the bottom is covered with dead leaves.

The eruciform larva is from 16 to 18 mm. in length and from 3 to 4 mm. in width. The larval case is made of pieces of dead leaf and other vegetable debris arranged rather like tiles on the roof of a house (imbricate) and the case is strongly curved and narrowing posteriorly. The abdominal gills are distinctive, each divided into about ten to twelve branches.

Apatania auricula (Forsslund)

Mosely states that the larvae of this species are to be found in cold springs but Kačalova records larvae from Latvia amongst tree-stumps, shells and stones on the open exposed shores of a lake, at a depth of from a half to two metres. She found larvae from May to July, whilst during August and September the 'quivers' (as she delightfully calls them) were closed. The adults emerged from 9 until 27 October. The larva of *A. auricula*, compared with that of *A. muliebris* MeLach, has no dark spots on the dorsal surface of the head and on the labrum there is only one brownish black stripe along the rear edge (Fig. 496). The claws are less curved and their bristles almost reach the point of the claw. The dorsal gills are present on the first segment of the abdomen.

Apatania muliebris McLachlan

The larva of this species inhabits cold springs. The larval case is from 7 to 11 mm. in length and 3 mm. wide at the front. It is constructed of sand-grains of various sizes and is conical and curved on the top surface and slightly curved at the hinder end on the under-surface (Fig. 498). The front of the head of the larva (Fig. 499) is covered with minute black spicules and on the ventral surface the margins of the genae are clear and appear to be folded (Fig. 500).

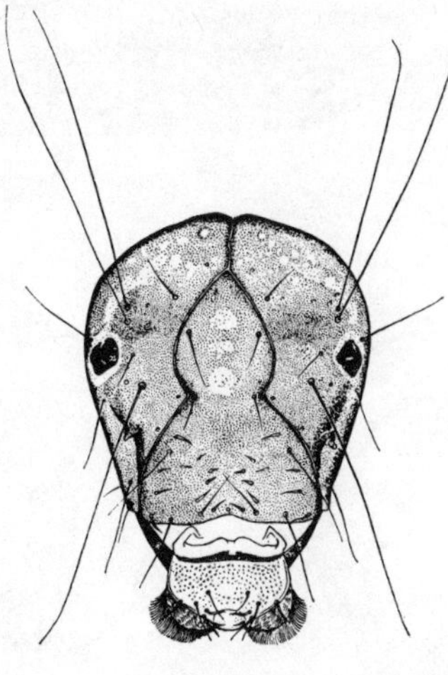

Fig. 496. *Apatania auricula.* Head of larva. After Kačalova.

Although the clypeus is of the general limnephilid type it is proportionately larger anteriorly. The articulating membrane of the labrum is large and strong (see also Fig. 496, *Apatania auricula*). The large mandibles bear no teeth but are furnished with a brush on the inner face and two long hairs on the outer. Of the latter the upper is large and black and the lower shorter and pale. The labial palps consist each of a single segment. On the thorax the two pairs of median sclerites of the metanotum are absent. The pronotum is dark brown with a number of pale spots in the posterior region on each side. The median legs are the longest. The tubercles on the first abdominal segment are well developed. The gills are mostly simple but the lateral series is absent. In the dorsal series these occur as follows: II 2.1, III 2.1, IV 0.1, V 0.1, VI 0 (1), and in the ventral series II (1) 1, III 1.1, IV 0.1, V 0.1, VI 0.1, VII 0 (1). The anal claw is without an auxiliary claw.

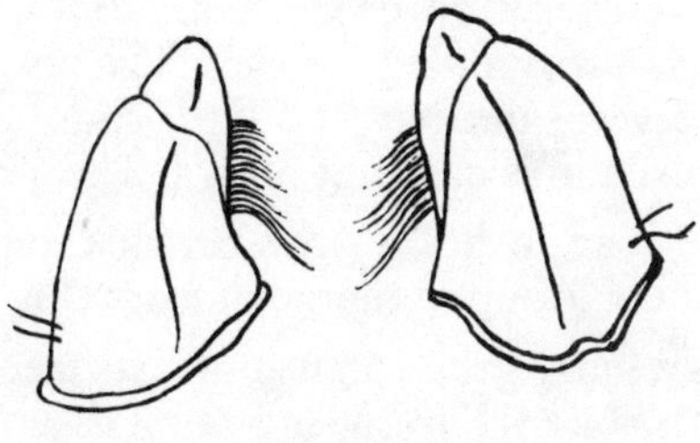

Fig. 497. *Apatania auricula.* Mandibles of Larva. After Kačalova.

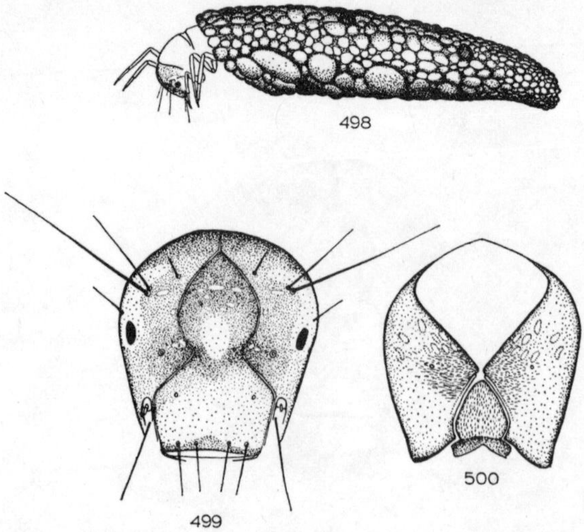

Figs. 498–500. *Apatania muliebris*. 498. Larva in case; 499. Head from the front; 500. Head from below.

Apatania wallengreni McLachlan

Immature stages unknown.

Drusus annulatus Stephens

Hanna's description of this species, made from Berkshire and Gloucestershire specimens, is as follows.

Case. Up to 14 mm. long and 3.5 mm. wide, of sand-grains and small stones (Fig. 501). The cases are curved and taper towards the posterior end. Anterior opening is oblique and the posterior small and straight.

Larva. Eruciform, up to 12 mm. long and 2 mm. wide.

Head. Hypognathous (Fig. 502), short and broad. The frontoclypeus, dark brown, has four long black setae and two colourless, all situated at its oral part. In addition, there are two setae at the constrictions of the frontoclypeus. The anterior surfaces of the genae are dark brown except for two yellowish-brown areas surrounding the eyes. The posterior surfaces of the genae are chestnut brown and have no dark spots. The gular sclerite is broader at its oral part and is chestnut brown. The genal suture is open.

Labrum. The ventral margin has a protuberance, on each side of which there are five setae and a group of marginal hairs (Fig. 503).

Mandibles. Each has a hairy brush on its inner surface and two setae on its outer surface near the base (Figs. 508, 509). The upper end is long and blade-like.

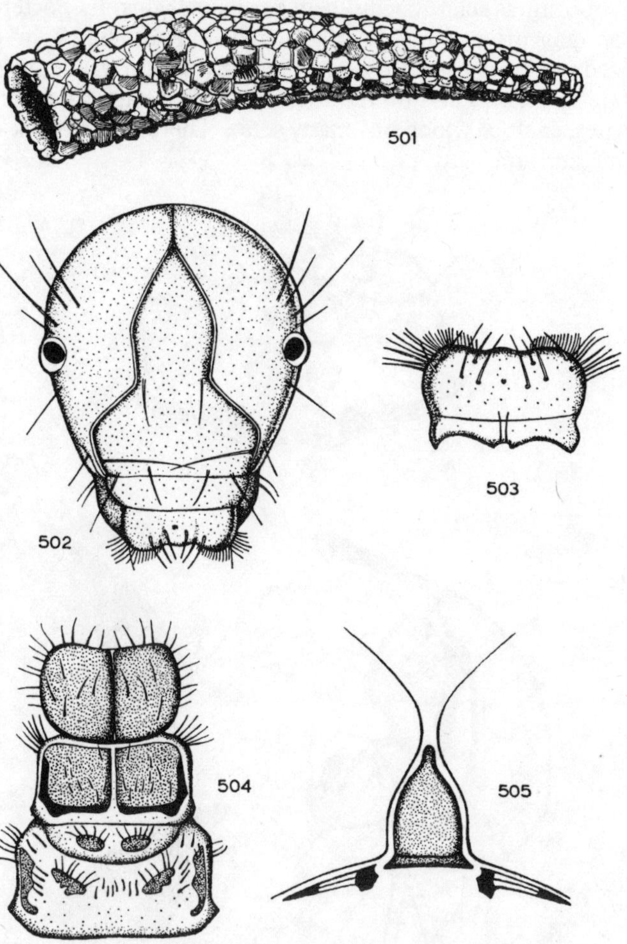

Figs. 501–505. Larva of *Drusus annulatus* Stephens. 501. Larval case; 502. Head from the front; 503. Labrum; 504. Thoracic nota from above; 505. Gular sclerite. Hanna.

Maxillae. The cardo is rod-shaped and bears a single seta. The stipes has two setae along its distal margin. The maxillary palp has five segments, of which the basal segment has many hairs. The lacinia has a few hairs and sensillae. One of the sensillae is long and has fine secondary hairs.

Thorax. The pronotum is dark brown, entirely sclerotized, and has a median longitudinal suture and a black, highly sclerotized posterior margin (Fig. 504). There are many setae on the anterior and lateral margins of the pronotum as well as on both sides of the median suture. The lateral margins of the pronotum are noticeably convex. The prosternal horn is present.

The mesonotum is sclerotized except for its margins. Its posterolateral and posterior margins are highly sclerotized and are black in colour. The sclerotized area is chestnut brown. The mesonotum has many setae on the anterolateral corners and on the chestnut-brown part. The metanotum has six sclerites, each of which has many setae. There are many scattered setae on the metanotum.

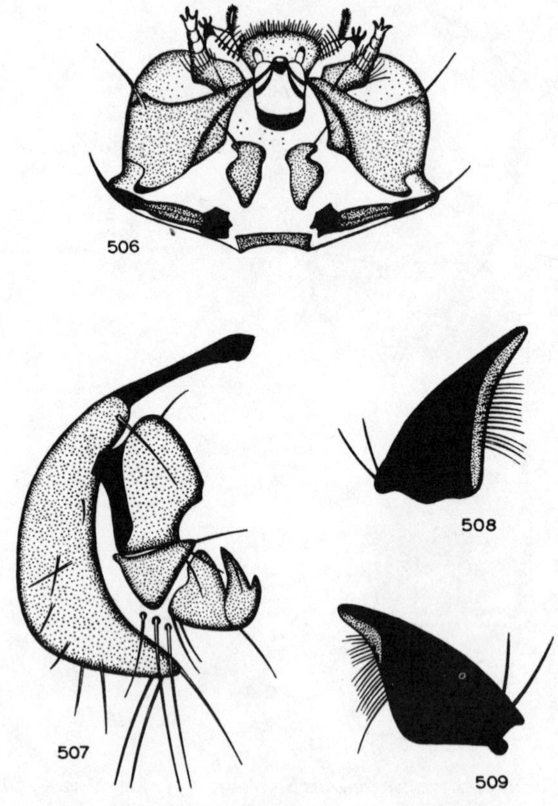

Figs. 506–509. Larva of *Drusus annulatus* Stephens. 506. Labium and maxillae; 507. Anal appendage, anal claw and supporting sclerite; 508. Left mandible; 509. Right mandible. Hanna.

Legs. The prothoracic leg is shorter than the mesothoracic and metathoracic legs, which are approximately equal in length. The second segment of the trochanters of the mesothoracic and metathoracic legs have a few hairs. All the tibiae have two spurs and a spine in between. The tarsal claw is strongly curved.

Abdomen. The dorsal protuberance on the first abdominal segment is devoid of setae, whereas each of the lateral protuberances has a small number of setae. There is a transverse row of setae on the tergum as well as on the sternum of the first abdominal segment. The gill filaments are single and are present on abdominal segments two to seven. The lateral line running from the beginning of the third to the beginning of the eighth abdominal segments, is formed of fine hairs. The anal sclerite is elliptical, yellowish-brown, has four long setae, two medium setae and a few short setae. The anal appendage has two segments and the anal claw has an auxiliary claw at its base. The anal appendage is supported by a sclerite which has a prolongation extending on to the sternum of the tenth abdominal segment. There are two long setae and a short seta between the anal appendage and the supporting sclerite.

The larvae of *Drusus*, *Mesophylax*, *Halesus*, *Chaetopteryx*, *Stenophylax* and *Micropterna* have single abdominal gill filaments and therefore may be confused. The head and pronotum are dark brown and have no spots in *Drusus* and *Mesophylax*. The mesonotum of the former genus, however, is chestnut brown. The mesonotum of *Mesophylax* is dark brown except for two lighter spots at the posterior corners (Lestage, 1921; Ulmer, 1909). The head, pronotum and mesonotum have dark spots in *Halesus*, *Chaetopteryx*, *Stenophylax* and *Micropterna*. In *Halesus* the spots are distinct, fairly large, numerous and often fuse together, whereas they are much smaller in the other three genera. In *Chaetopteryx* the dorsal gills are present from the second to the sixth abdominal segments, while the lateral gills are present on segments one to four. In *Stenophylax* there are dorsal gills on the seventh and eighth abdominal segments, whereas the lateral gills are present until the fifth segment. Hickin (1953) pointed out that the larvae of *Micropterna* are more sclerotized than those of the allied genera.

Ecclisopteryx guttulata (Pictet)

The larvae of this species are very abundant in running water in mountainous regions. The adults fly in clouds around outstanding objects on river-banks in suitable conditions. There appears to be no description made from British larvae. Lestage states that there are no spines on the median tibiae and one on the anterior tibiae.

Limnephilus affinis Curtis

This is a widely distributed species and the adult is to be found from April to October. Its larvae, however, have not been described from British material. I reared them from larvae obtained from a weedy ornamental pond in the Forest of Dean, Gloucestershire, but found that the culture was contaminated with other limnephilid species. The larval case is practically straight, cylindrical and generally made of large sand-grains, rather like that of *L. bipunctatus*, sometimes also of small leaf, twig or wood fragments disposed longitudinally or obliquely. In length it is 17 to 18 mm. and width 4 to 4.5 mm.

Limnephilus auricula Curtis

No description of the larva of this rather small abundant and widely distributed species has been made from British material. It is from 10 to 12 mm. long and 1.6 mm. wide. The case is straight, slightly narrowing posteriorly and made of small fragments of leaf, rather unequal in size, disposed longitudinally with the addition of, here and there, a few particles of sand. The anterior end is somewhat oblique whilst the posterior is transverse. It varies from 11 to 15 mm. in length and from 2.5 to 3.2 mm. in width. The case is said to resemble that of *Limnephilus incisus*.

Limnephilus bipunctatus Curtis

No description of the larva of this local species has been made from British material. In size it is from 16 to 18 mm. in length and 3 mm. in width. The case is generally large, conical and arched. The case of young larvae consists of vegetable fragments but that of older larvae consists of large sand-grains and occasionally small shells. The posterior end is rounded. It attains a length of from 18 to 20 mm. and a width of 4 mm.

Limnephilus borealis (Zetterstedt)

No description of the larva of this large species, recorded from north and western Scotland, has been made from British material. It attains a length of 28 mm. and a width of 4.5 mm. The case is straight, cylindrical or conical, and composed of rather large pieces of vegetable material placed longitudinally. The front opening is transverse whilst the hind is oblique and entirely open. It may reach a length of 58 mm. and a width of 6 mm.

Limnephilus centralis Curtis

I spent 6 April 1953 collecting with Mr. James Edwards at Chartley Moss, Staffordshire. We were endeavouring to find the larvae of *Oligotricha clathrata* Kolenati and *Rhadicoleptus alpestris* Kolenati in this classic locality; we were unsuccessful in this.

Deserting the sphagnum we netted the drainage ditches where we found larvae of *Oligotricha ruficrus* Scopoli to be quite common. We then came across a small but deep ditch in which was clear running water. The ditch was almost hidden by silver birch scrub and dead bracken fronds, but larvae, afterwards found to be of *Stenophylax lateralis* Stephens and *Limnephilus centralis* Curtis, in about equal numbers, were quite common. From larvae then collected a male adult of this latter species emerged on 26 May, which Mr. D. E. Kimmins identified for me. My drawing of the adult (Fig. 574) was made partly from this specimen and partly from specimens in the collection of Dr. C. H. Andrewes, F.R.S., to whom I am indebted for the loan of his caddis fly collection. Ulmer gives 'ditches' as the larval locality, and Mosely says of the insect that it frequents marshy

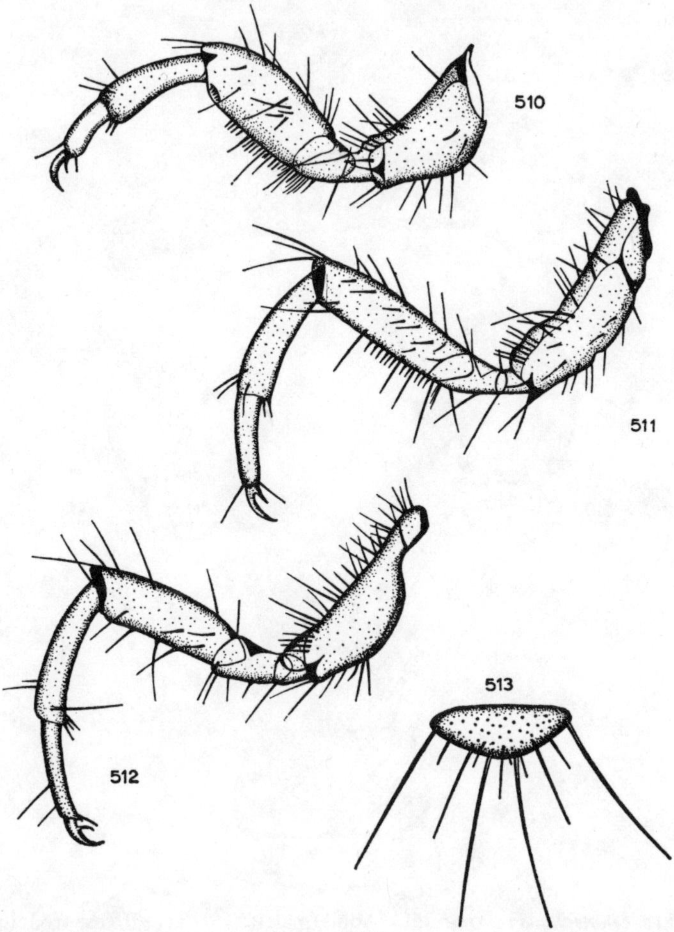

Figs. 510–513. Larva of *Drusus annulatus* Stephens. 510. Prothoracic leg; 511. Mesothoracic leg; 512. Metathoracic leg; 513. Anal sclerite from above. Hanna.

localities, so that it is of interest to note that I found the larvae in a clear trickle of water in a deep ditch, although on the edge of a sphagnum bog.

The place name, Chartley Moss, no longer appears on the one-inch ordnance survey map; it can be referred to as Anglesea Coppice, Stowe, Staffordshire, and is now a National Nature Reserve.

Case (Fig. 515). Of rounded sand-grains, oblique opening, curved, some cases narrowing more appreciably than others and some cases slightly twisting laterally. Two cases examined. When seen from below, the hinder end twists over to the right. 13–15 mm. long, 3 mm. wide at fore end. A

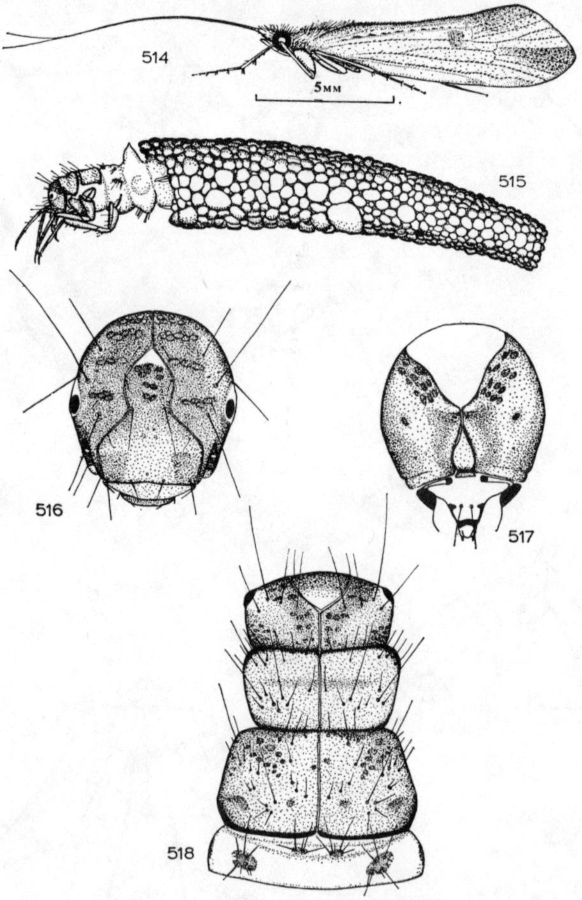

Figs. 514–518. *Limnephilus centralis*. 514. Adult (male); 515. Larva in case from the side; 516. Head of larva from the front; 517. Head of larva from behind; 518. Head and thoracic nota of larva from above.

deep notch (as is present in the case of *Limnephilus vittatus* Fab.) on the dorsal surface of the anterior end of the case is absent but a slight one was present in a few instances.

The light colour of the sand-grains stood out in striking contrast against the dead bracken and grass, on the ditch bottom when the caddis were collected.

Larva. Generally similar to *Limnephilus vittatus* Fabricius, but it can be differentiated with ease. Length up to 12.5 mm., width 2 to 2.5 mm.

Head (Figs. 516 and 517). The most conspicuous character is the absence of pigment from the triangular-shaped area at the aboral end of the clypeus. The converging longitudinal bands on the genae of *L. vittatus* (and indeed of many Limnephilids) appear in this species to be resolved into a number of transverse bands made up of darker pigmentation in the region of the muscle-spots. The pale area around the eyes is prominent. The gular sclerite is pear-shaped with a black transverse bar at the oral end. The gular sclerite of *L. vittatus* has the lateral margins parallel for some distance and the genae are contiguous on the ventral surface for a very much shorter distance than is the case in *L. vittatus.* The transverse bar at the oral end of the gular sclerite is much darker in *L. centralis* than in *L. vittatus.*

Pronotum. Transverse groove (which is very dark in *L. vittatus*) scarcely perceptible—this is the 'blackish cross' of Ulmer.

Limnephilus coenosus Curtis

No description of the larva of this local species has been made from British material. At Malham Farn in Yorkshire the larvae are to be found in very small trickles on wet moorland. The larva is 16 mm. in length and 2.5 mm. in width. The case is constructed irregularly of vegetable debris with a few sand-grains, arranged longitudinally or obliquely.

Limnephilus decipiens (Kolenati)

Generally this insect is rather local but is abundant in Hampshire and Wicken Fen and has been reported from Surrey. No description in English of the larva has yet been made. It varies in length from 16 to 23 mm. and in width from 3 to 4 mm. The case is straight, narrowing slightly and generally composed of vegetable material arranged longitudinally and often overlapping. Sometimes the case is triangular with the hinder end more transverse and not so deep as the front. There is, however, wide variation in the appearance of the case. In size the case may be as long as from 30 to 40 mm. and about 5 mm. wide. The triangular case may be 10 mm. wide and 25 mm. in length.

Limnephilus elegans Curtis

No description of British larvae of this rather local but widely distributed species appears to have been made. It is 18 mm. in length and 4 mm. wide. The straight case is cylindrical and composed of overlapping vegetable material most often of fragments of *Carex.* The posterior opening is round and very small. The length of the case may be as much as 34 mm.

Limnephilus extricatus McLachlan

Larvae of this species were collected from amongst the submerged roots of Great Willow herb (*Epilobium hisrutum* L.) at the edge of an ornamental stream at the Botanical Gardens, Edgbaston, Birmingham (1937). More larvae were collected from the roots of the same plant growing in a stream

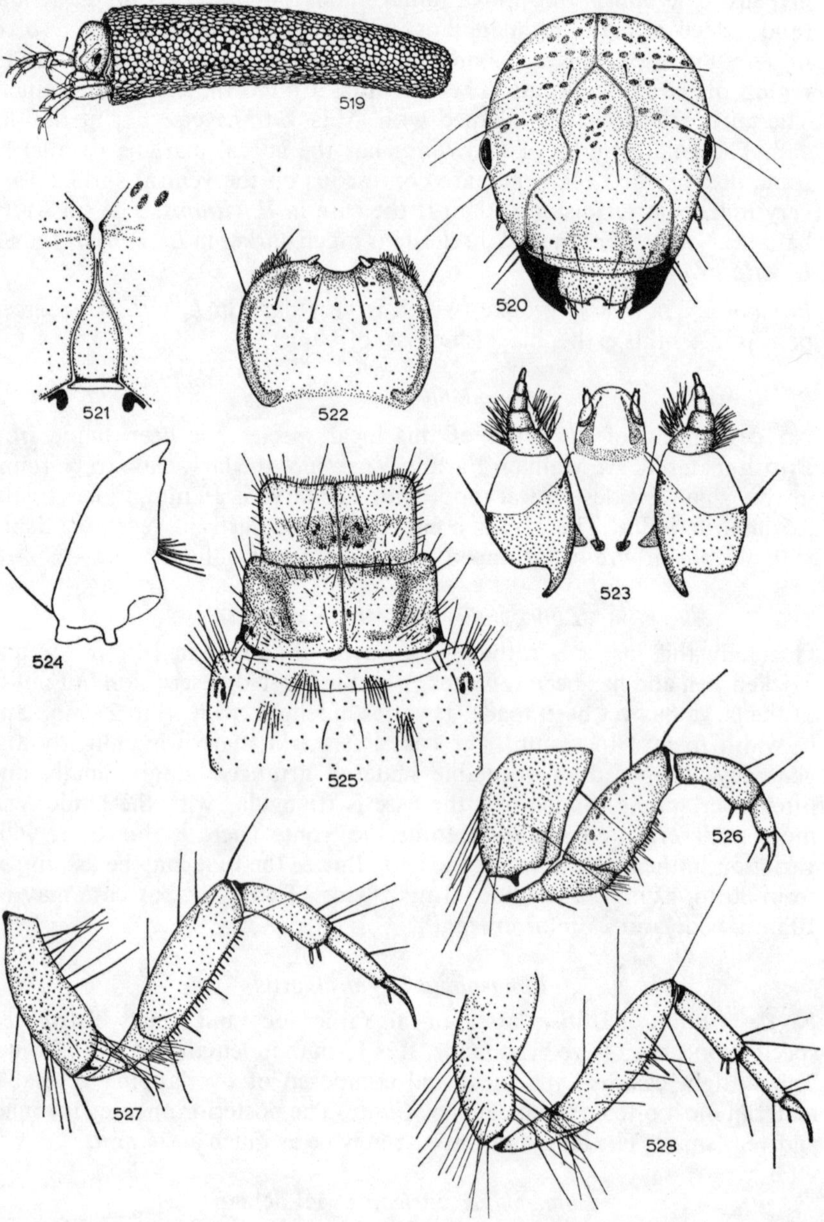

Figs. 519–528. ***Limnephilus extricatus*****.** 519. Larva and Case; 520. Head of Larva; 521. Gular sclerite; 522. Labrum; 523. Labium and Maxillae (from beneath); 524. Mandible (left); 525. Thoracic nota; 526. Prothoracic leg; 527. Mesothoracic leg; 528. Metathoracic leg.

at Chipping Campden, Glos. (1938). Adults were reared out in each case for accurate determination.

Case (Fig. 519). Made of sand-grains and small particles of hard debris. Particles of house-brick when available seem to have an attraction for this species as they are used freely. Case often has speckled appearance. Size 16 mm. in length, 3.7 mm. wide. Circular cross-section narrowing towards posterior, slightly arched. Posterior end convex.

Larva. Eruciform, head orthocentrous, abdomen cylindrical up to 14 mm. long, 2 mm. wide.

Head (Fig. 520). Slightly longer than broad. Light chestnut in colour. Transverse pattern of dark spots at aboral region of genae and clypeus. Antennae two-segmented, distal segment bristle-like.

Gular sclerite (Fig. 521). Aboral projection almost divides the genae.

Labrum (Fig. 522). Deep incision in anterior margin, lateral margins heavily sclerotised. Maxillary palp (Fig. 523) four-segmented but inserted into sclerite which, when seen from beneath, gives impression that palp has five segments.

Labium. Internal chestnut-coloured sclerites easily visible. Labial palp of two segments, but distal segment only projecting from surface of labium. The proximal segment of the palp is easily visible through the integument of the labium.

Mandibles (Fig. 524). Scoop-like, black except for the five chestnut-coloured teeth. (The teeth are not seen in the figure as they are visible only when the mandible is seen from the side.) Small bristle on each mandible on outer edge near base and a bunch of pale-coloured bristles on the inner edge of each mandible.

Thorax (Fig. 525). Pro- and mesonota sclerotized. Light greyish fawn in colour, only very lightly marked but posterior margin of mesonotum heavily marked in black with lateral spur-shaped extensions. The dorsal sclerotized patches of the metanotum are scarcely distinguishable, their positions only being defined by the groups of bristles. The lateral pair of sclerotized patches are well defined, being light chestnut in colour. Prosternal horn present but very small.

Legs (Figs. 526, 527 and 528). Prolegs shorter than meso- and meta legs. Two ventral blunt spines at distal end of tibia of all legs. Spines along ventral edge of tibia absent in prothoracic legs. In all legs articulating edges of femur and tibia black.

Abdomen. Cream to creamy-grey, cylindrical. Lateral line present on segments two to seven. The sclerotized tubercles dorsal to the lateral line vary in number from six on segment two to two on segments six and seven. Anal claw with single auxiliary claw. Sclerite on dorsum of ninth abdominal segment not defined. Gills in groups of four and three on anterior segments decreasing in size and number towards posterior end of abdomen.

529

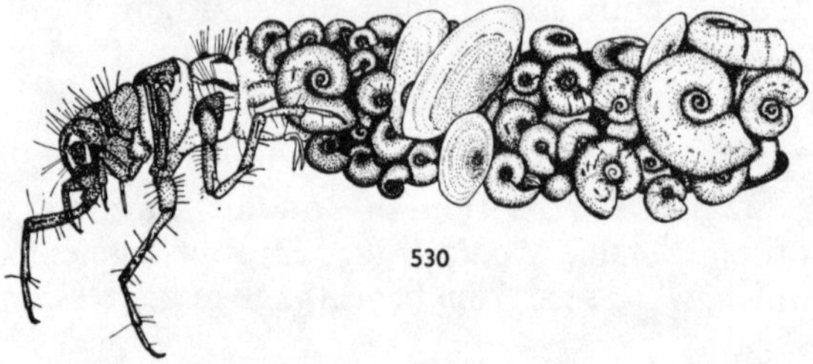

530

Figs. 529–530. *Limnephilus flavicornis*. 529. Case; 530. Larva and case.

Limnephilus flavicornis (Fabricius)

Larvae of this widely distributed species were found to be exceedingly numerous in a number of small cattle ponds in the Coleshill district of Warwickshire. In addition, numerous specimens were collected from a pool in Broadmoor Wood, Rubery, Worcestershire. Some of the larvae were reared to the adult stage and the species determination confirmed. One of the cattle ponds referred to above had deteriorated into a rubbish dump and the only living vegetation present was algal, but even so the larvae were present in very large numbers. The other pools were characterized by having an extensive marginal and floating vegetation.

Case. Cylindrical, made up of a variety of materials cemented together. The younger larvae had constructed their cases of the leaves of *Lemna* placed edgewise, older larvae had cases of the cut stalks of Potamogeton and reed, small sticks, dead leaves and other vegetable debris and snail shells. The material forming the case is usually arranged in criss-cross fashion, obliquely to the axis of the case (Fig. 529). Many of the larval cases from the pool in Broadmoor Wood obtained from submerged branches were constructed entirely of the shells of molluscs cemented together (Fig. 530). Several *Limnephilus* species use mollusc shells as case-building material. Mosely gives photographs of cases of *L. rhombius* and *L. lunatus* as well as *L. flavicornis* in which mollusc shells have been used. Anterior end oblique. Length 26 mm., breadth 7 mm.

Larva. Eruciform, cylindrical, head orthocentrous. Pro- and mesonota sclerotized, metanotum with sclerotized patches. Posterior end of abdomen obtuse. Length 20 mm., breadth 5 mm.

Head (Fig. 531). Elliptical, yellowish-brown, heavily marked with black. Clypeus at oral end almost completely black, the dark area extending into the aboral end of the clypeus as a club-shaped mark. The latter does not extend to the margin of the clypeus. Genae, adjacent to clypeus, dark, making an inverted V, oral half of mark almost black. Many small dark marks on the genae lateral to the V mark. Antennae small. Gular sclerite (Fig. 537) flask-shaped, completely separating the genae.

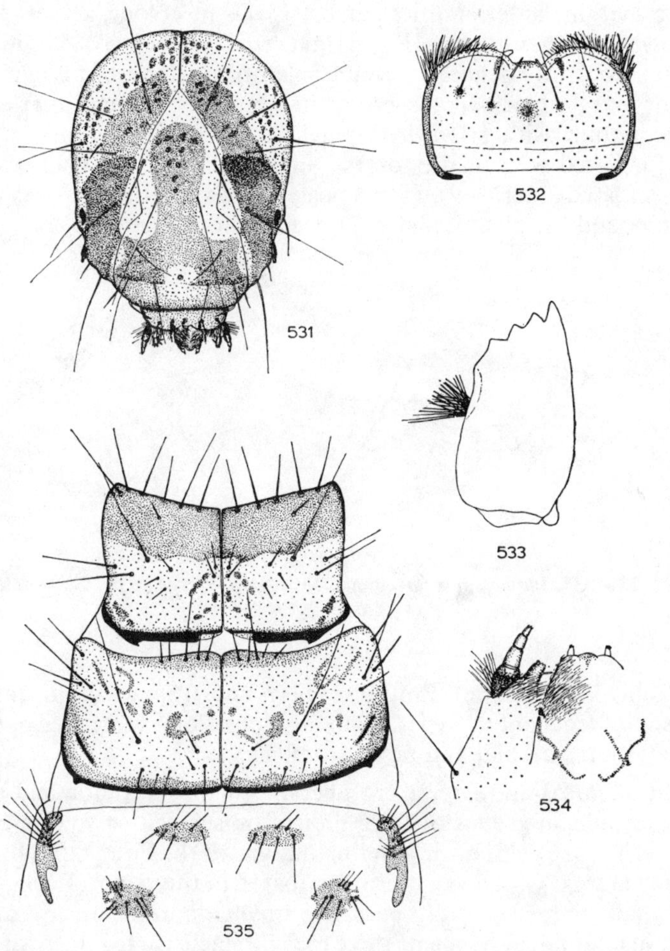

Figs. 531–535. *Limnephilus flavicornis*. 531. Head; 532. Labrum; 533. Mandible; 534. Maxilla and Labium; 535. Thoracic nota.

Mouth-parts. Labrum (Fig. 532) deeply concave on anterior margin with small cushion-like protuberance at the centre. Small pear-shaped dark mark on each side of this protuberance. Heavily sclerotized lateral margins, and a small dark mark in the centre of the transverse row of bristles.

Mandibles (Fig. 533). Obtuse, 4-toothed, brush of fine hairs on inner edge.

Maxillae (Fig. 534). Base of palp and mala sclerotized, chestnut-brown in colour.

Labium. Hairy. Two sets of bow-shaped sclerites are visible. Labial palps situated on dome-shaped protuberances.

Thorax (Fig. 535). Prothorax sclerotized, strong bristles situated along anterior margin, anterior third almost black in colour, remainder dark chestnut-brown except posterior margin, which is heavily sclerotized and black in colour. Mesothoracic notum sclerotized, anterior margin greyish, posterior margin black. Transverse variable pattern of dark marks. Metathoracic notum with sclerotized patches. Large patches in the pleural region furnished with bristles on the anterior half. Two small sclerotized patches situated anteriorly and two posteriorly, the latter half-way between the former and the pleural plates. Prosternal horn present.

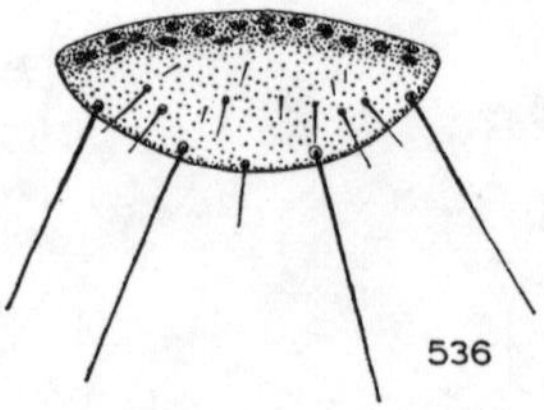
536

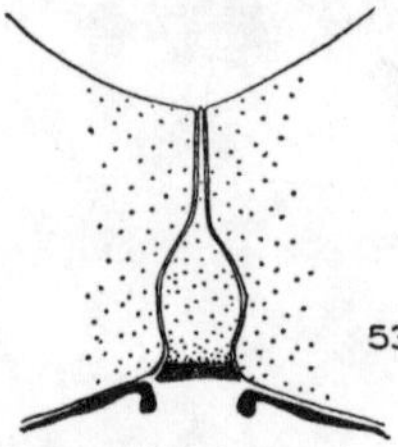
537

Figs. 536–537. ***Limnephilus flavicornis*****.** 536. Anal sclerite; 537. Gular sclerite.

Legs (Fig. 15, a, b and c). Prothoracic legs short, spines of unequal size at distal end of trochanter. Series of small spines along ventral edge of meso- and metathoracic tibial and tarsal segments.

Abdomen. Dorsal and lateral protuberances on first abdominal segment. Numerous gills in groups of three present on segments two to eight. Anal sclerite (Fig. 536) elliptical, anterior margin dark. Four large bristles on posterior margin with a small bristle situated in the centre. Each anal claw with two auxiliary claws at its base, one smaller than the other. Lateral line present in abdominal segments three to eight inclusive together with groups of from four to seven small sclerotized pustules at the anterior end of the segment.

Limnephilus fuscicornis (Rambur)

No description of the larva of this rather large, local species has been made from British material. It reaches a length of from 18 to 20 mm. and a width of 4 mm. The slightly conical case is also slightly curved. It is composed of sand-grains, never of vegetable material, and is about 20 mm. in length and 4.5 mm. wide.

Limnephilus fuscinervis (Zetterstedt)

No description of the larvae of this small species has been made from British material; indeed, it is rare in Britain but has been recorded from Scotland.

Limnephilus griseus (L.)

The larvae of this widely distributed species have not been described from British material. In length it varies from 15 to 18 mm. and from 2.5 to 3 mm. in width. The conical case is slightly arched and, in the case of young larvae, it is composed of small particles of vegetable origin. A little later it may be made up entirely or partly of sand-grains. Usually when the larvae are mature the case consists of pieces of leaf, or twig disposed longitudinally, obliquely or transversely. The case is from 14 mm. to 19 mm. in length and about 3 mm. wide.

Limnephilus hirsutus (Pictet)

The immature stages of this rather local species are unknown.

Limnephilus ignavus McLachlan

No description of this larva is available in English. The larval case is straight, only slightly conical, and composed of pieces of leaf unequally disposed longitudinally on one side or the other. The case is 16 mm. long and 3.5 to 4 mm. wide, whilst the larva may attain about 16 mm. in length and 3 mm. in width.

Limnephilus incisus Curtis

If they are disturbed the adults of this very small limnephilid advertise their presence by running up the rushes and other herbage in which they lie concealed. They are usually found in considerable numbers. The writer found them at Askham Bog, Yorks., in enormous numbers, where the bog had dried up. A description of the larva based on British material has not yet been made. According to previous authors, in size it is from 15 to 17 mm. long and from 2 to 3 mm. wide. Therefore, there must be some discrepancy as this makes the larva comparable in size with that of *Limnephilus lunatus*, which has an adult wing length of 14–15 mm., whilst *L. incisus* has a wing size of only 7–8 mm. This point needs clearing up, as does also the two illustrations of the case in Lestage which differ considerably from each other. The description by Lestage that it is only a little narrower posteriorly does not fit in with fig. 245 in his work where the case narrows considerably.

Limnephilus lunatus Curtis

Fully-fed larvae of this species were found to be abundant in a small stream at Temple Balsall, Warwickshire, on 30 June 1942. This fast-running stream is barely a yard wide for much of its length and seldom a foot in depth. There is a profuse marginal growth of such aquatic plants as *Nasturtium officinale*, *Veronica beccabunga*, etc., etc., and the larvae were confined to the submerged stems of these plants. It is apparently in a stream similar to this that *Limnephilus lunatus* larvae have caused extensive damage to watercress beds, necessitating insecticidal action by the owners. The larvae, however, are not confined to swiftly running water as they were found to be quite common in a pool in Broadmoor Wood, Rubery, Worcestershire, and are reported by Mosely as being 'found in most waters'. Larvae from both localities were reared to the adult stage and the species determination confirmed.

Case (Fig. 538). Of whole leaves, cut leaves and vegetable debris cemented together, the material usually laid with its long axis longitudinal and overlapping. Where the whole leaves of hawthorn had been used the cases could have been confused with those of *Glyphotaelius pellucidus* Retz. Length 20–23 mm., width 4 mm. Sometimes the case is constructed of broken snail-shells, as in the Hampshire chalk streams and, in addition, sand-grains are occasionally used.

Larva. Eruciform, cylindrical, head orthocentrous. Pronotum and mesonotum sclerotized, metanotum with sclerotized patches. Length 17 mm., width 3 mm.

Head (Fig. 539). Elliptical, broader aborally. Surface covered with minute spines and with numerous long hairs projecting orally. Yellowish-brown with dark chestnut-brown to black markings. Central club-shaped dark mark in the clypeus extending transversely at the oral end. Genae darker in colour towards the clypeus, but leaving three light-coloured patches. One median light patch situated in the aboral vertex of the clypeus and a pair of light bands running obliquely from the narrow part of the clypeus to the outer extremity of the oral end. An extensive pattern of dark marks situated both in the genae and the clypeus. At the aboral end of the latter, the so-called 'clypeus-mark' consists of ten to eleven dark marks arranged in a kite shape. At the narrow neck of the clypeus is a large dark mark, whilst at the oral end there are four dark marks, the pair nearer to the oral margin closer together than those farther from this margin. Antennae small but visible from the front of the head. Chaetotaxy as in figure. Genae contiguous behind head for a short distance above the gular sclerite (Fig. 540). Aborally, the margins of the genae are chestnut coloured near the median line, turning to black as they extend laterally.

Mouthparts. Labrum (Fig. 541). Centre of anterior margin deeply notched. Anterior margin hairy. A dark mark between middle bristles in the transverse row of bristles. A pair of dark marks near the anterior margin, one on each side of the notch. Mandible (Fig. 542) black, four-toothed. Two

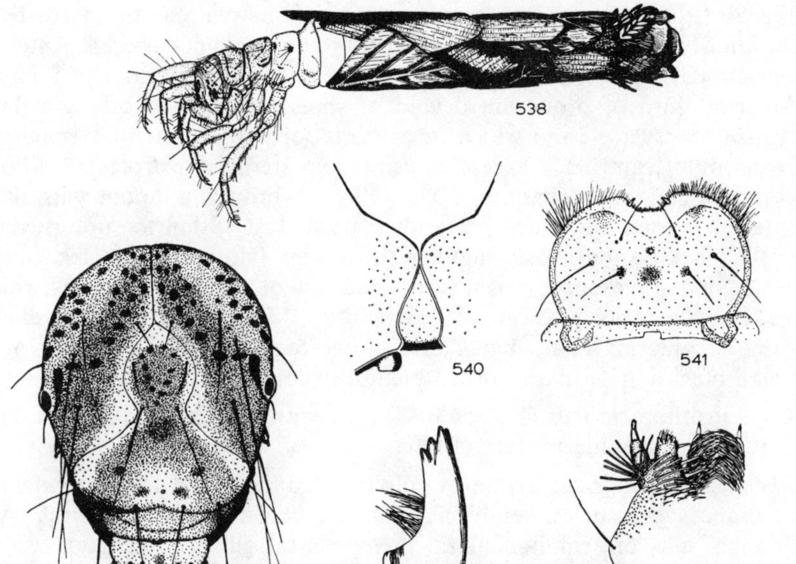

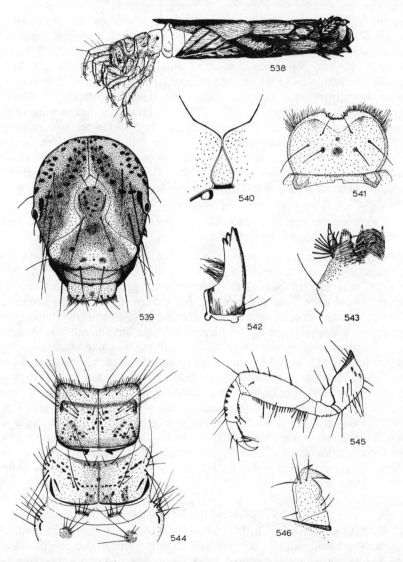

Figs. 538–546. *Limnephilus lunatus*. 538. Case; 539. Head; 540. Gular sclerite; 541. Labrum; 542. Mandible; 543. Maxilla and labium; 544. Thorax; 545. Leg (prothoracic); 546. Anal claw.

bristles at base. Maxillary palp (Fig. 543) four-segmented. Numerous blunt, yellow spines at base of palp and mala. Mala obtuse with cluster of sense organs at tip, some hair-like. Four sharp spines on mala inwardly directed. Labium hairy. Labial palp with base protuberant from labium and with cluster of slender sense organs at tip. On mentum near base of mala a group of special hairs with broad bases, projecting inwards.

Thorax (Fig. 544). Pronotum sclerotized, just as wide as the head, light chestnut-brown in colour with greyish-brown darker markings. Anterior margin dark greyish-brown edged with long bristles and short hairs. Anterior third of pronotum devoid of spots but marked off by a dark narrow transverse band which turns posteriorly at its lateral extremities. Posterior margin black except in centre. Prosternal horn present. Mesonotum wider than pronotum, light yellowish-brown in colour with dark anterior margin and black posterior margin. Just within the upturned tip of the black posterior marginal line there is an oblique black mark. In the centre of the mesonotum is a transverse row of dark brown spots which also slope anteriorly towards its extremities. Metanotum with sclerotized patches, anterior pair smaller and closer together than posterior pair. Small black lunate mark in each pleural region.

Legs. Prothoracic legs (Fig. 545) shortest with well-marked row of dark brown spots on anterior face of tibia.

Abdomen. Cylindrical, white in colour. Median dorsal and lateral protuberances present. Lateral line and sclerotized pustules present. Abdominal gills present but lateral presegmental gills absent from second segment, whilst the notal presegmental gill groups in the second to fourth segments usually consist of only two filaments. The presegmental ventral gill groups always consist of two filaments. (*Note*. In general, it is thought that the absence or presence of gills is probably a bad discriminating feature among species where the range of habitat is such as to show wide divergences in oxygenation of the water. But as Ulmer uses this for his key to the limnephilid species it is included here.) Posterior margin of anal sclerite with four long and five short bristles on anterior margin. One short bristle occurs between the central pair of large bristles and a pair of short bristles occur on each side between the outer bristles; in addition, five short bristles occur as a single transverse row at about the centre of the sclerite. Anal claw (Fig. 546) with auxiliary claw and with a few short bristles at its base.

Limnephilus luridus Curtis

No description exists of British material. The larva is up to 15 mm. in length. The case is almost straight and slightly conical and composed of leaf and twig fragments disposed either longitudinally or transversely. In size it may be up to 14 to 16 mm. in length and from 3 to 4.5 mm. in width.

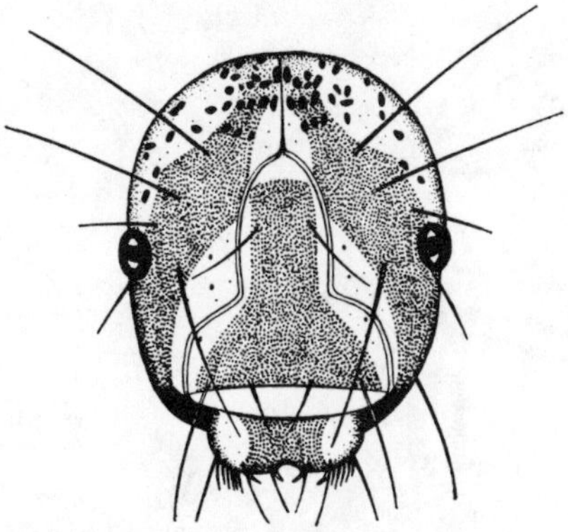
547

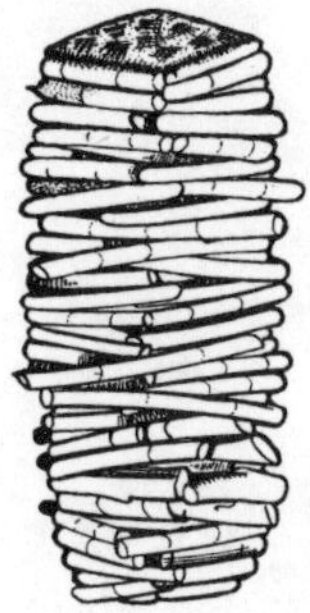
548

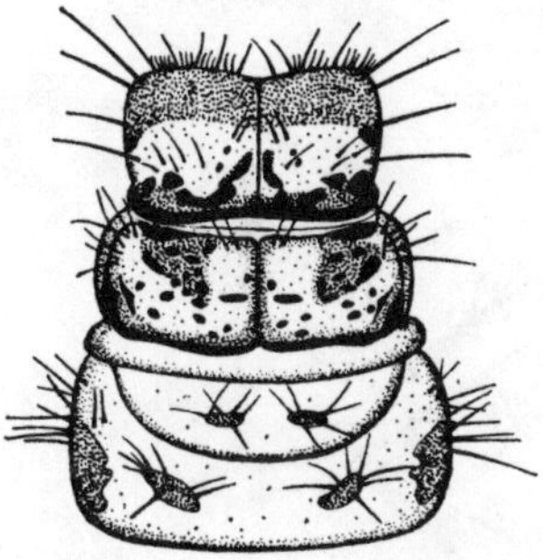
549

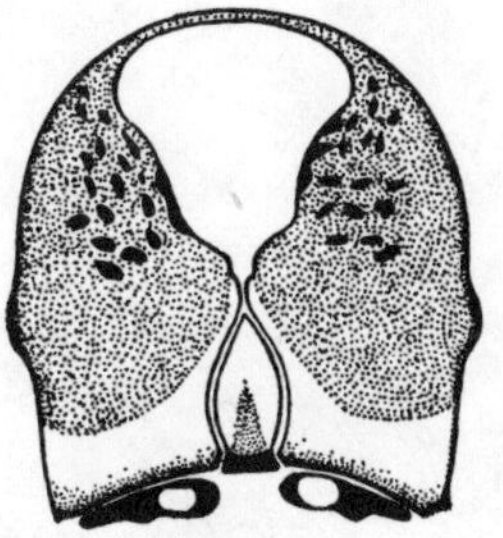
550

Figs. 547–550. *Limnephilus marmoratus* Curtis. 547. Head from the front; 548. Larval case; 549. Thoracic nota from above; 550. Head from behind. Hanna.

Limnephilus marmoratus Curtis

Hanna describes the larvae of this species, obtained from near Manchester and Reading, as follows:

Cases are up to 27 mm. in length and 8 mm. in width, and are generally built of stalks of water-plants arranged transversely or obliquely (Fig. 548). Elodea leaves, sand-grains and mollusc shells may be included in the case. Cases built entirely of seeds of water-plants are also common. On some of the cases specimens of a freshwater sponge (probably *Spongilla*) were found. The posterior opening of the case is small and lies in the middle of a silken membrane. Larva eruciform. The larvae examined are up to 28 mm. long and 4 mm. wide. Ulmer and Lestage recorded larvae up to 21 mm. long.

Head hypognathous (Fig. 547). The clypeal band is dark brown and is dilated transversely at its oral end. The dark brown bands on the genae

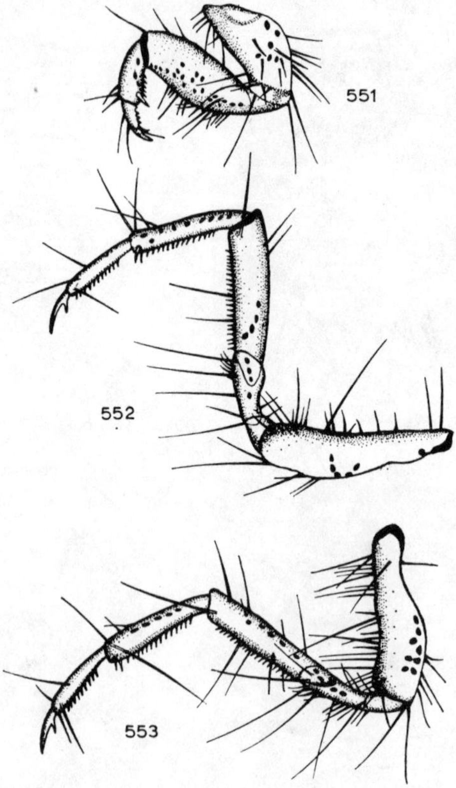

Figs. 551–553. *Limnephilus marmoratus* Curtis. 551. Prothoracic leg; 552. Mesothoracic leg; 553. Metathoracic leg. Hanna.

converge to form a V and they may touch the aboral margins of the fronto-clypeus. The genae bear dark brown spots on their anterior and posterior surfaces and along their sides. The posterior surfaces of the genae are chestnut brown except for the area surrounding the gular sclerite and the area above the labium, which are golden yellow. The gular sclerite is golden yellow and in most specimens bears a dark brown pattern and does not separate the genae completely. The genal suture may be closed or slightly open.

Labrum (Fig. 555). Ventral margin concave. On both sides of the concavity there are three long setae, two small setae and a group of marginal hairs.

Mandibles. With four blunt teeth (Figs. 554, 556), a brush on the inner surfaces and two setae on their outer surfaces near the base.

Maxilla. Both cardo and stipes bear two setae. The maxillary palp has five segments, of which the terminal bears two sensillea and the basal a group of hairs and a seta.

Thorax. Pronotum entirely sclerotized with median longitudinal suture. The posterior and latero-posterior margins of the pronotum are heavily sclerotized and are entirely black. The anterior third of the pronotum is dark brown. Posteriorly, the pronotum bears dark brown spots. The mesonotum is sclerotized except for its margins and has a median longitudinal suture. There is a dark brown pattern on both sides of the suture, which is subject to considerable variation. The posterior and latero-posterior margins of the mesonotum are heavily sclerotized. The metanotum is formed of two distinct parts and has six sclerites, each of which bears a number of setae. The prosternal horn is well developed and the prosternal sclerite is present. The mesosternum has a number of black spots divided into two groups.

Legs. The prothoracic leg is short and robust. The mesothoracic leg is slightly longer than the metathoracic leg. All the legs bear a dark brown pattern. The inner surfaces of the femur, tibia and tarsus carry small spines

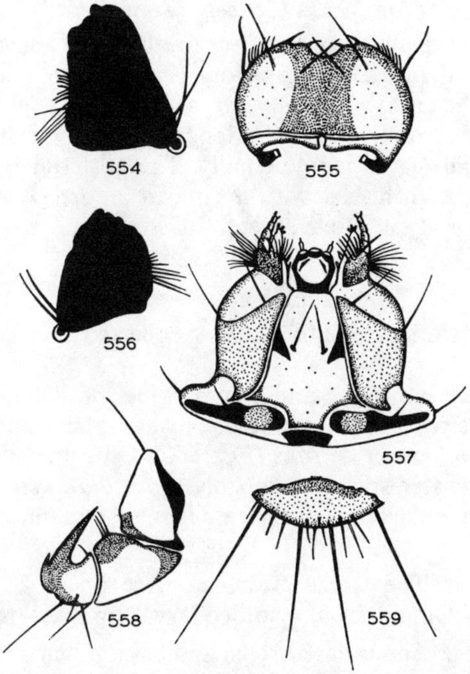

Figs. 554–559. *Limnephilus marmoratus* Curtis. 554. Left mandible; 555. Labrum; 556. Right mandible; 557. Labium and maxillae; 558. Anal appendage and anal claw; 559. Anal sclerite from above. Hanna.

on their inner margins and there are a few hairs on the second segment of the trochanter. The tibiae of all the legs carry two spurs.

Abdomen. On the first abdominal segment there are three protuberances, of which the dorsal is devoid of setae, and there are some setae around its base. The two lateral protuberances possess a small number of small setae. The gill filaments are present on abdominal segments two to seven. The lateral line running from segments three to eight is formed of fine hairs. The anal sclerite, which is elliptical, has a dark brown pattern in most of the specimens. The anal appendages have two segments each and the anal claw has an auxiliary claw at its base.

The clypeal band is considerably dilated transversely at its oral end in the larvae of *L. marmoratus* Curt., *L. politus* McLach., *L. flavicornis* (F.) and *L. stigma* Curt., and for this reason may be easily confused. The pre-segmental gills on the eighth abdominal segment, however, are absent in the larvae of *L. marmoratus*, whereas they are present in the larvae of the other three species. In *L. politus* there is a deep notch of lighter colour at oral end of the clypeal band. Ulmer and Lestage pointed out that the head of *L. flavicornis* is much shorter and broader than the head of *L. stigma*.

Limnephilus nigriceps (Zetterstedt)

No description of the larva of this very local species has been made from British material. It varies in length from 12 to 18 mm. and in width from 2.5 to 3 mm. The case, which is very variable, is straight, cylindrical or slightly conical but rarely it may be triangular in section. It is composed of vegetable material placed longitudinally. The posterior opening is closed by membrane but furnished with a circular or irregular opening. The membrane, however, is often absent. In length the case may be from 21 to 26 mm.

Limnephilus politus McLachlan

Hanna describes the larvae of this species, collected from near Manchester, as follows:

Cases up to 26 mm. long and 8 mm. wide, of stalks of water-plants arranged obliquely, bits of leaves, seeds of water-plants and mollusc shells, with a certain amount of detritus (Fig. 560). On some of the cases specimens of a freshwater sponge (probably *Spongilla*) were found (Hanna, 1953). The posterior opening of the case is small and lies in the centre of a silken membrane.

Larva. Eruciform. The larvae examined were up to 28 mm. long and 3.5–4 mm. wide. Ulmer (1909) recorded larvae up to 24 mm. long.

Head. The clypeal band is dark brown and has a notch of lighter colour at its oral end and bears a group of dark spots at its aboral end (Fig. 561). The dark brown bands on the genae converge to form a V. The genae have dark brown spots on their anterior and posterior surfaces as well as on

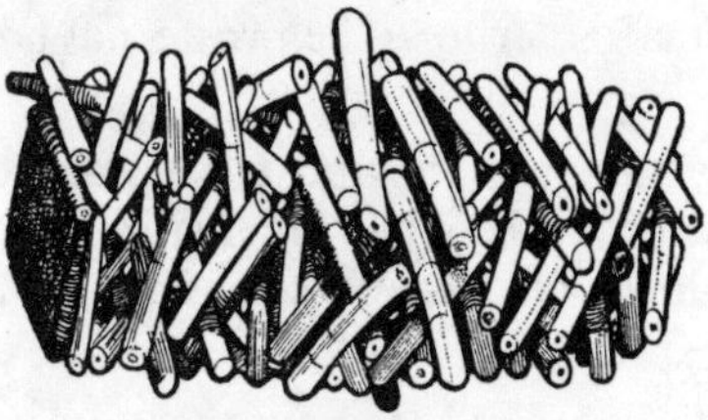

560

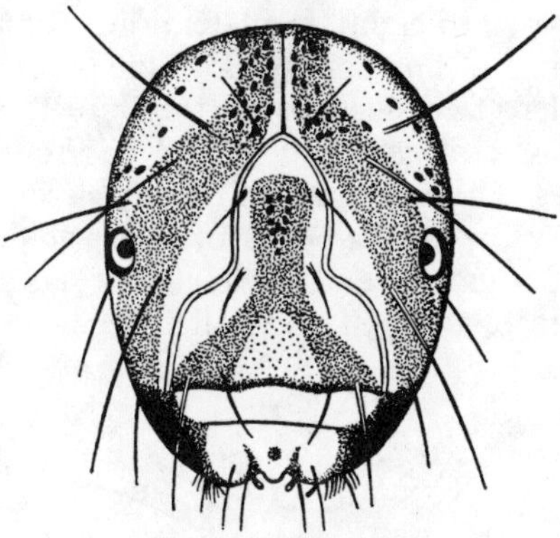

561

Figs. 560–561. *Limnephilus politus* McLach. 560. Larval case; 561. Head from the front. Hanna.

their sides. The gular sclerite is golden yellow in colour and does not separate the genae completely. The genal suture is open.

Labrum. Ventral margin bears a small protuberance. There are two dark bands and a central dark spot on the anterior surface of the labrum (Fig. 568). On each side of the labrum there are three long setae, two small setae and a group of fine hairs. The tormae are long, thin and bent inwards.

Maxilla. Both the cardo and stipes bear two setae. The maxillary palp has five segments, of which the basal one carries a group of hairs, while the apical segment bears four sensillae. The lacinia has a few sensillae and a group of hairs.

Thorax. The pronotum is entirely sclerotized and has a medium longitudinal suture (Fig. 562). Its posterior and posterolateral margins are heavily sclerotized and are entirely black. The anterior third of the pronotum is dark brown. Posteriorly the pronotum has a number of dark spots. The mesonotum is sclerotized except for its anterior and anterolateral margins and bears dark brown spots. The posterior and posterolateral corners are heavily sclerotized and are dark in colour. The metanotum has six small sclerites. The prosternal horn and the prosternal sclerite are present.

Legs. Prothoracic leg is short and robust. The mesothoracic leg is slightly longer than the metathoracic leg. The inner surfaces of the femur, tibia and tarsus carry small spines. The second segment of the trochanter carries a few hairs.

Abdomen. On the first abdominal segment there are three protuberances, of which the lateral ones bear a small number of setae. There is a group of setae round the base of the dorsal protuberance. The gill filaments are on abdominal segments two to eight. The lateral line running from segments three to eight is formed of fine hairs. Above the lateral line there is a line of three to seven sclerotized pustules on the third to seventh abdominal segments. The anal sclerite is elliptical and has four long setae and a number of short ones. The anterior half of the sclerite is dark brown and bears dark brown spots, while the posterior half is golden yellow. The anal appendages have two segments and the anal claw has an auxiliary claw at its base. The sternum of the first abdominal segment carries a group of

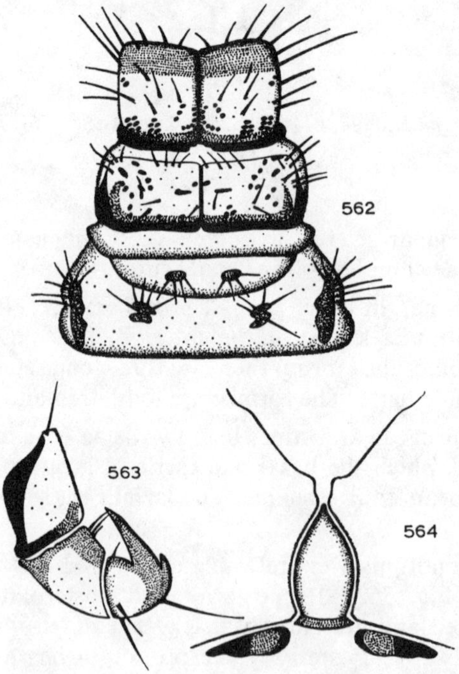

Figs. 562–564. *Limnephilus politus* McLach. 562. Thoracic nota from above; 563. Anal appendage and anal claw; 564. Gular sclerite. Hanna.

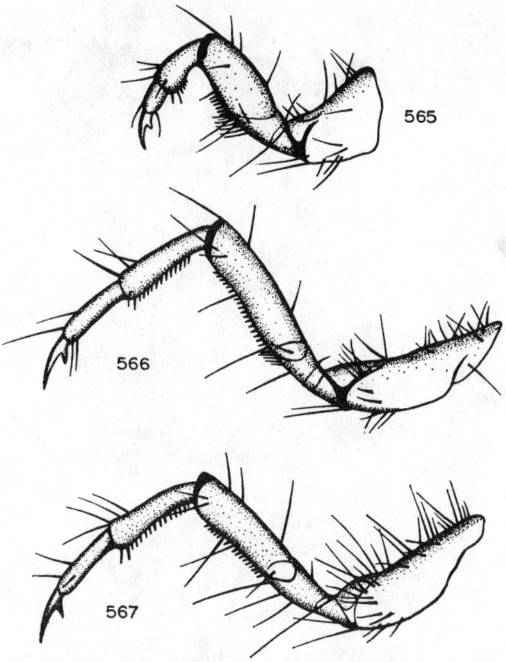

Figs. 565–567. *Limnephilus politus* McLach. 565. Prothoracic leg; 566. Mesothoracic leg; 567. Metathoracic leg. Hanna.

setae. There are elliptical sclerites on the sterna of the second to the seventh abdominal segments.

The larvae of *L. politus* McLach. may be easily confused with the larvae of *L. marmoratus* Curt., *L. flavicornis* (F.), *L. stigma* Curt. and *L. rhombicus* (L.). The notch at the oral end of the clypeal band of *L. politus* is enough to distinguish the larvae of this species from the larvae of *L. marmoratus*, *L. flavicornis* and *L. stigma*. In *L. rhombicus* there are three light coloured areas on the head. One of these areas is median and is situated at the aboral end of the clypeus. The other two areas are partly contained in the clypeus and partly in the genae.

Limnephilus rhombicus (L.)

This description is taken from larvae collected from a pond in Broadmoor Wood, Rubery, Worcestershire, and compared with larvae collected from numerous ponds and canals in Worcestershire, Warwickshire and the Lake District. For accurate determination of the species many of the larvae were reared through to the adult stage. A feature of all the habitats was still-water and an abundance of submerged and marginal vegetation, but Mosely states that *rhombicus* is found sometimes in slow-running rivers.

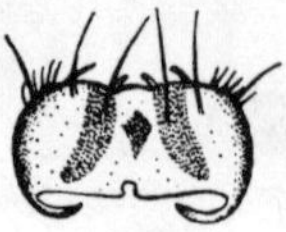

568

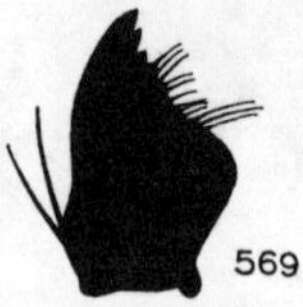

569

570

571

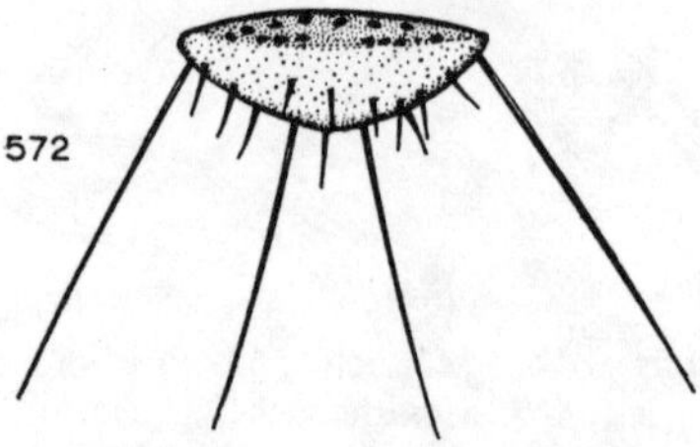

572

Figs. 568–572. *Limnephilus politus* McLach. 568. Labrum; 569. Left mandible; 570. Right mandible; 571. Labium and maxillae; 572. Anal sclerite from above. Hanna.

Case (Fig. 573). Consists of cut stems and roots of aquatic plants, vegetable debris, mollusc shells, etc., arranged tangentially to the chamber and cemented together with secretion. Hinder orifice narrowed. Case almost always has thick appearance. Length up to 27 mm., width up to 6.5 mm.

Larva. Eruciform, cylindrical, head orthocentrous. Pronotum and mesonotum sclerotized, metanotum with sclerotized patches. Length 19–25 mm., width 3.5–4.5 mm.

Head (Fig. 574). Elliptical, broader aborally. Surface covered with small spines directed orally. Yellowish brown with dark chestnut brown to black markings. Central club-shaped dark mark in the clypeus extending transversely at the oral end. A dark band, in the genae, runs adjacent to the clypeus on each side but leaving three light coloured patches which are quite discernible to the naked eye. One median light patch situated in the aboral vertex of the clypeus and a pair situated where the narrow aboral part of the clypeus widens into the broad transverse oral part. These latter light patches are partly contained in the clypeus and partly in the genae.

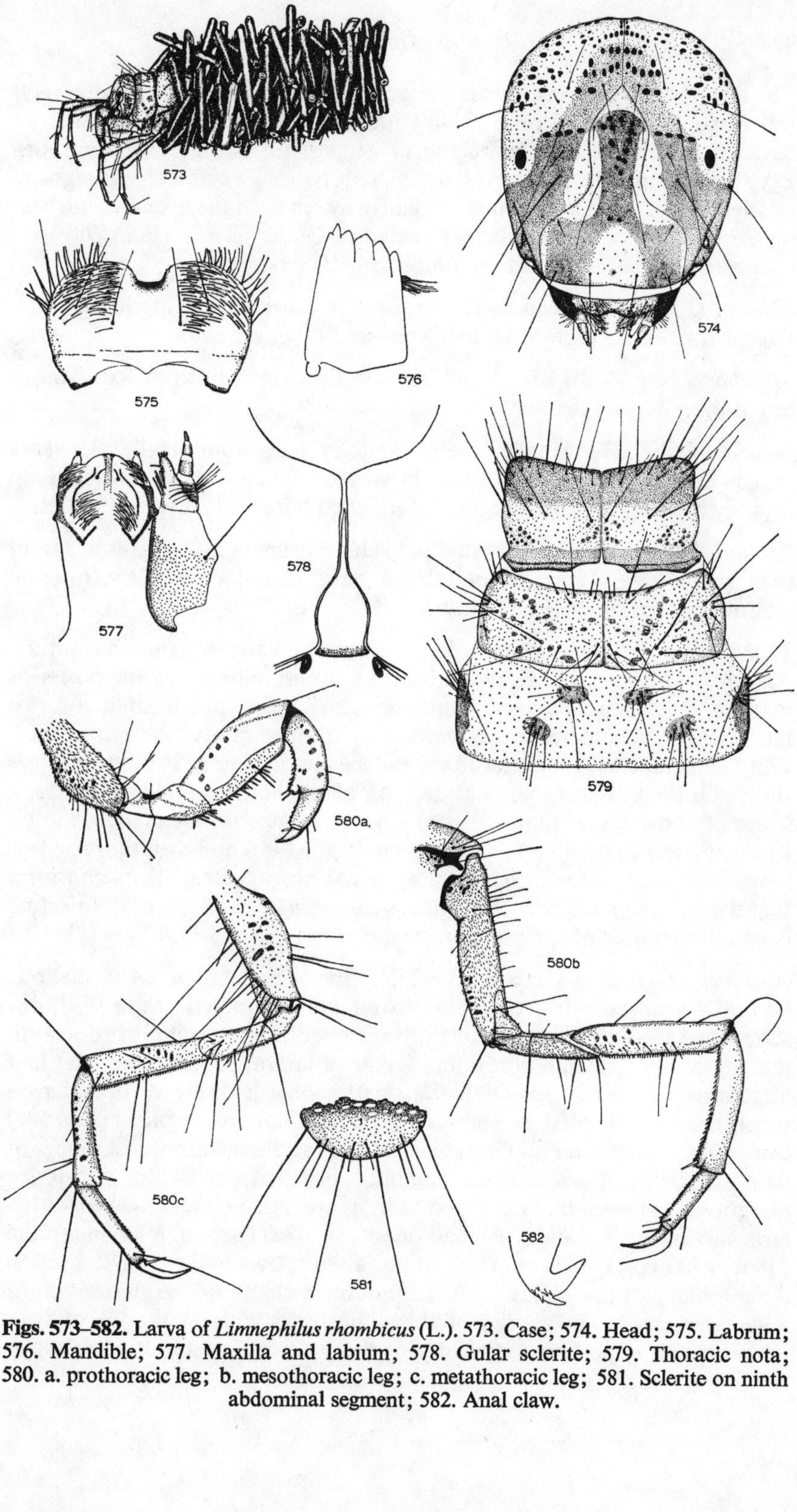

Figs. 573–582. Larva of *Limnephilus rhombicus* (L.). 573. Case; 574. Head; 575. Labrum; 576. Mandible; 577. Maxilla and labium; 578. Gular sclerite; 579. Thoracic nota; 580. a. prothoracic leg; b. mesothoracic leg; c. metathoracic leg; 581. Sclerite on ninth abdominal segment; 582. Anal claw.

The dark bands running across the genae are darker orally. A transverse row of four dark spots on each side marks a distinction between the two parts of the bands. Aboral region of genae with pattern of dark spots. Eyes distinct in light areas. Antennae small two segmented, distal segment bristle-like, situated approximately half-way between the eyes and the base of the mandibles. Chaetotaxy as in figure. Gular sclerite (Fig. 578) with long narrow aboral projection almost dividing the genae.

Labrum (Fig. 575). Deep concavity heavily marked on anterior margin. Except for central region labrum is covered by blunt hairs.

Mandibles (Fig. 576). Broad and blunt with a small bush of hairs on internal margin near the teeth.

Maxilla (Fig. 577). Smaller than maxillary palp, four bristle-like sense organs at the tip with a few blunt bristles at the base on the inside. Palp four-segmented, proximal segment furnished with several stout bristles.

Labium. Base of single-segmented palps hairy, three sense organs at top of each palp. Lateral margin of labium dark in colour and the internal sclerites of the labium are readily visible.

Thorax (Fig. 579). Pronotum sclerotized with median longitudinal suture. Anterior third dark in colour, as is also a broad band along the posterior margin. Three separate areas containing dark spots, one median and two lateral. Mesonotum sclerotized with longitudinal suture. Pattern of dark spots. Metanotum with sclerotized patches as in figure. Prothoracic legs short and thick. Black marks at division of trochanter at dorsal surface, a series of short spines along ventral edge of femur, tibia and tarsus in all legs being less in number in prothoracic legs. Meso- and metathoracic legs long, practically equal in length. Prosternal horn present. Ross considers that the horn in all larvae of the genus *Limnephilus* is short, not projecting beyond the apices of the prothoracic coxa.

Abdomen. Cylindrical creamish white, intersegmental grooves shallow. Lateral line present on the third to eighth abdominal segments. Four sclerotized tubercles on each side of segments three to eight just dorsal to the lateral line. Median dorsal and a pair of lateral protuberances on first abdominal segment. Several black bristles on each side of dorsal protuberance. Sclerite on dorsal surface of ninth abdominal segment (Fig. 581) transverse, anterior margin irregular, darker in colour but with a number of light oval marks along it. Four long black bristles inserted along posterior margin with shorter hairs grouped as in figure. Anal claw furnished with a large auxiliary claw and five small ones (Fig. 582) (Betten, referring to the LIMNEPHILIDAE of New York State, gives 'claws with a single hook'). Abdominal gills present, in groups of not more than three. Anterior gills of anterior segments in groups of three, posterior gills in groups of two. Anterior gills of posterior segments in groups of two, posterior gills simple.

Limnephilus sparsus Curtis

No description of the larvae of this very abundant and widely distributed species has been made from British material. It grows to about 17 mm. in length and 3 mm. in width and its case is somewhat larger than this, conical and curved. It is made of small particles of vegetable matter applied transversely.

Limnephilus stigma Curtis

No description has been made from British material.

The larval case is barrel shaped, constructed of vegetable debris placed tangentially to the long axis. The anterior opening is transverse and in size the case varies from 20 to 25 mm. in length and from 8 to 20 mm. in width. The mesonotum is pale orange or with a reddish tint and with the anterior margin black in the region of the angles. Gills 18+18+17+13+11+9+3=89 are more numerous than in *L. flavicornis*. On the anal sclerite there are only two or three bristles between the lateral and median bristles.

Limnephilus subcentralis (Brauer)

The immature stages of this insect, recorded from the vicinity of Loch Awe in Scotland, are unknown.

Limnephilus vittatus (Fabricius)

A number of caddis larvae collected from the canal at King's Heath, Birmingham, were confirmed as belonging to this species on examination of the adult stage to which they were reared. The water is still, with some marginal vegetation.

The case is made of sand-grains cemented together by means of secretion. It is curved and is much larger at the head end than at the anal. Anterior opening somewhat oblique. Length 12–20 mm., width 2.2 mm.

Larva (Fig. 583). Small for a species of this genus. Eruciform with head orthocentrous. Abdominal gills present. Pro- and mesonota sclerotized, metanotum partially sclerotized. Length 12 mm., width 2 mm.

Head (Fig. 584). Round. Yellowish-brown. Finely sculptured all over. Small convexities in region of eyes. Large dark mark at aboral end of clypeus in which are situated ten darker spots in the shape of a kite. Several rows of dark spots running along genae; the row nearest the clypeus on each side is joined together by a dark band. Three pairs of bristles arising from genae are very long (almost as long as length of head). Three pairs of short bristles are situate at aboral end and two pairs of longer bristles arise from the oral end of the clypeus.

Mouthparts (Fig. 586). Labrum deeply concave on anterior margin. A small spinulose protuberance in centre of concavity. Three pairs of spines set transversely across the dorsal surface of the labrum, while the ventral surface is almost completely covered with long hairs. Lateral edges of labrum heavily sclerotized and continuous with peg-like projection bent inwards and forwards. Very heavily sclerotized black bow-shaped region stretches transversely near posterior margin. Several pairs of spines along anterior margin; very often some are broken. Mandibles (Fig. 587) asymmetrical. Fringe of fine hairs along basal half of inside edge, a bristle at base of outer edge. Well-defined ridge about mid-way along outer edge.

Maxillae. With four-segmented palp.

Labium. With one-segmented palp.

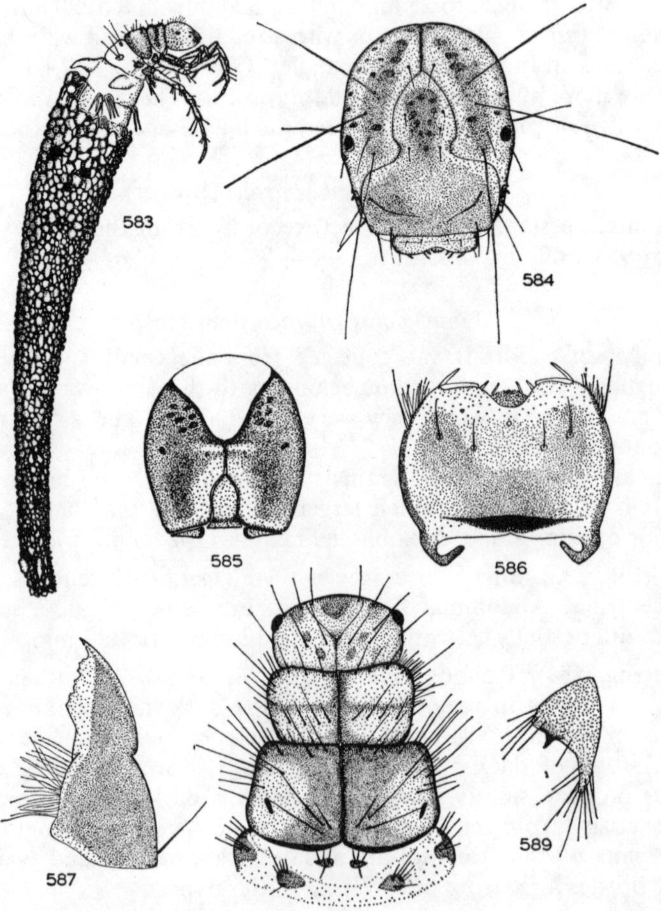

Figs. 583–589. *Limnephilus vittatus*. 583. Lateral view of larva and case; 584. Head from above; 585. Head from below; 586. Labrum; 587. Mandible; 588. Head and thoracic segments; 589. Coxal sclerite, metanotum.

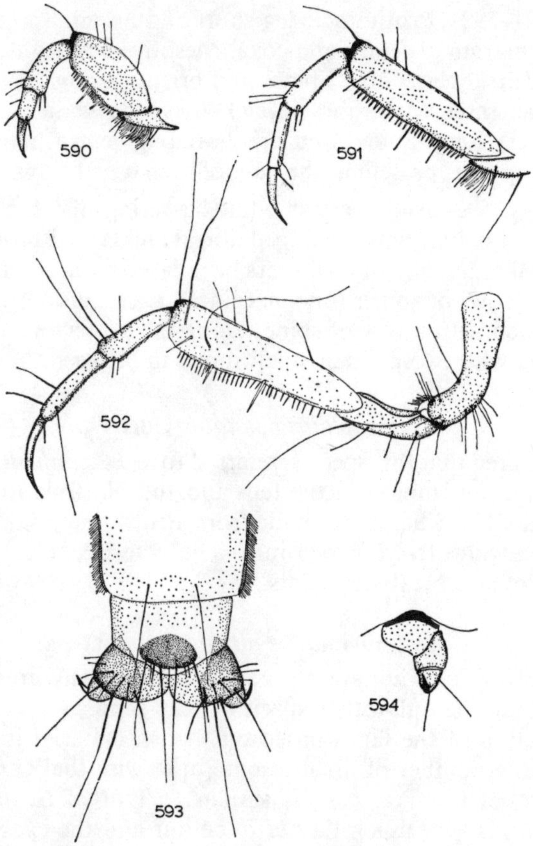

Figs. 590–594. *Limnephilus vittatus*. 590. Prothoracic leg; 591. Mesothoracic leg; 592. Metathoracic leg; 593. Anal segments; 594. Lateral view of anal claw.

Thorax (Fig. 588). Pro- and mesonota sclerotized, metanotum partially sclerotized (see Fig. 159 for coxal sclerite); prosternal spine present. Prothorax just as wide as the head, light chestnut-brown, darker along anterior and posterior margin with a dark band transversely in the middle of the segment and darker along the median line longitudinally. Longitudinal median suture present. Mesothorax wider than prothorax, dark chestnut-brown, shading lighter towards the sides. Posterior and lateral margin black, anterior margin very dark chestnut-brown, small black oblique mark on each side laterally. Mesothorax with two pairs of dorsal and two pairs of lateral sclerites all provided with spines. The inner anterior pair of dorsal sclerites is smaller. Of the two pairs of lateral sclerites, the coxal sclerites have a dark spine on the outer margin. Chaetotaxy of thorax as figure.

Legs (Figs. 591–592). Prothoracic leg short and broad. Spines and bristles along ventral margin of femur and coxa. Mesothoracic leg twice as long as prothoracic. Tarsal claw long, Spines and bristles along ventral margin of femur and trochanter. Metathoracic leg half as long again as mesothoracic. Tarsal claw very long. Spines and bristles along ventral margin of femur only. A small spine is present at the base of each tarsal claw.

Abdomen (Figs. 593–594). Greyish-white, tapering distally, gills present. Lateral fringe of hairs. Well-developed dorsal and lateral protuberances on first abdominal segment. Anal claspers have large basal sclerites each with two transverse rows of spines (one marginal). Last abdominal segment has sclerotized patch with a row of spines along the posterior margin, four of which are long with two pairs of small spines in between the outer pairs of long spines.

Limnephilus binotatus Curtis

Mosely considered that this species, referred to as *L. xanthodes* McLachlan, is local but not uncommon in the fens and the Norfolk Broads. No description of the larva has been made from British material. Length from 18 to 20 mm., width from 3 to 4 mm. The case is conical and straight, made mostly of small leaf fragments.

Grammotaulius nitidus (Müller)

This is a local species apparently confined to marshy areas. It is somewhat larger than the only other species in the genus, *G. atomarius*.

No description of the larva based on British material is available but Lepneva gives a number of illustrations (otherwise the Russian text may give difficulty). The larval case is similar to that of *G. atomarius*. The head, however, is very much darker in colour and the eyes are set much nearer the clypeus than in this latter species.

Grammotaulius atomarius (Fabricius)

Many of the low-lying fields on the north bank of the Severn estuary, which are at most only a few feet above sea-level, are ringed with ditches in which *Phragmites* grows abundantly. In such ditches, at Cleeve, near Westbury-on-Severn, Gloucestershire, which were about 6 to 9 inches deep and from

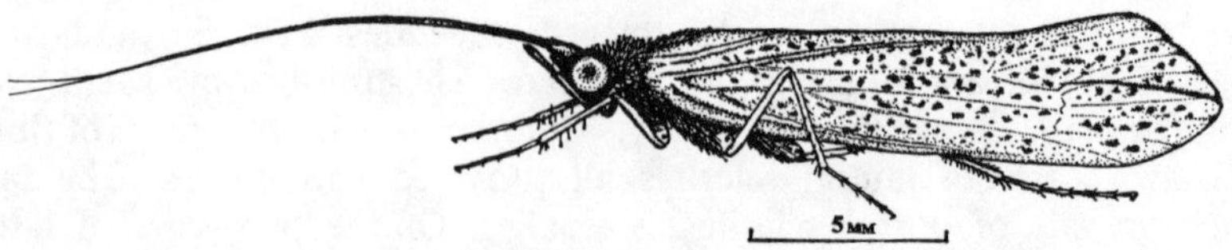

Fig. 595. *Grammotaulius atomarius* (Fabricius). Side view of adult.

1 to 2 feet across, at the end of January 1953, I found the larvae of *Grammotaulius atomarius* (Fab.) extremely common, and they were fully grown. From larvae collected at this time adults commenced to emerge from my aquarium on 6 May and continued to do so until 17 May, all emerging during the night or very early morning. I am indebted to Mr. D. E. Kimmins for the identification. It is of interest to note the colour of this insect when freshly emerged. The straw-colour is suffused with a delicate peach pink. This latter, however, is lost within an hour or two of emergence. When at rest the antennae are held pressed together. Attention also should be called to Fig. 595. The dark hair-like mark is on the hind-wings, but when the insect is at rest it shows through the rather delicate translucent wings, making it appear as though the marks were on the fore-wings. Larvae of this species will live in very unfavourable conditions; indeed, they can tolerate conditions of stagnation and putrescence which would prove fatal to all the other species of caddis of which, so far, I have had experience.

Case (Fig. 596). Constructed of large fragments of leaves of *Phragmites* applied longitudinally and fixed at the anterior end only, so that the posterior part of the leaf fragment is free. This gives a characteristic appearance to the case and often makes it very large, 50 mm. long and 7 mm. wide being not uncommon. The average was, however, just under 30 mm. in length and 5 mm. in width.

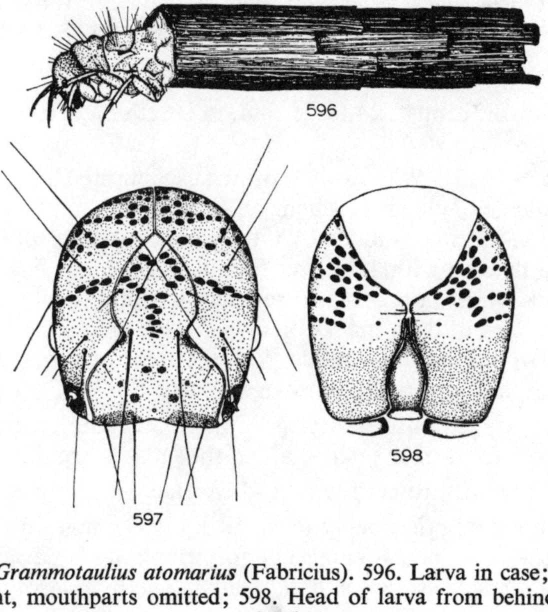

Figs. 596–598. *Grammotaulius atomarius* (Fabricius). 596. Larva in case; 597. Head of larva from front, mouthparts omitted; 598. Head of larva from behind, mouthparts omitted.

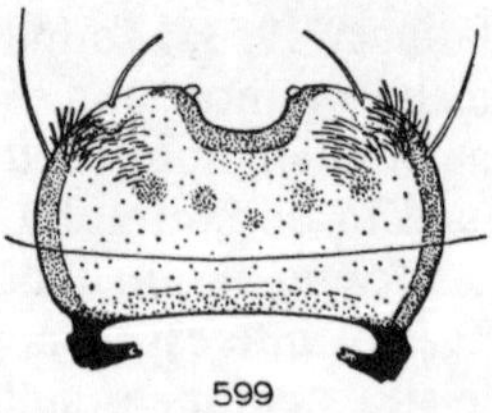

599

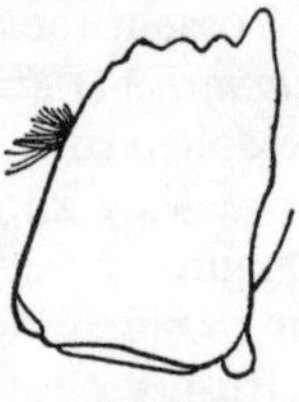

600

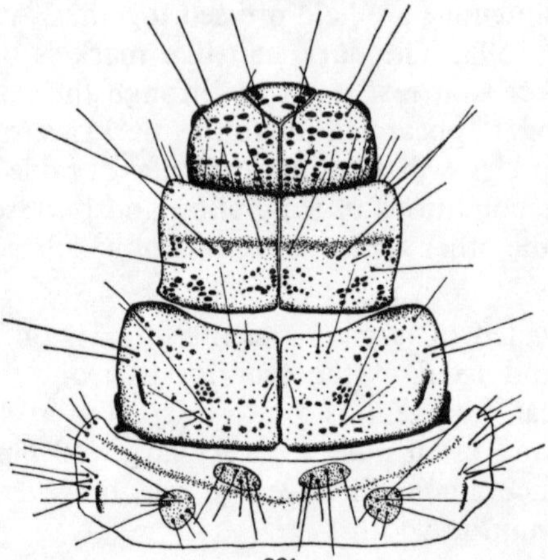

601

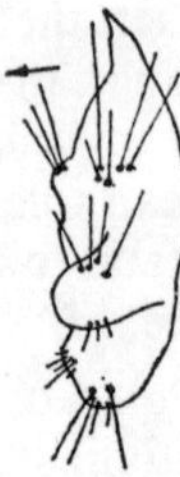

602

Figs. 599–602. *Grammotaulius atomarius* (Fabricius). 599. Labrum; 600. Right mandible; 601. Head and thoracic nota from above; 602. 1st abdominal segment from the side.

Larva. Eruciform, robust, 24 to 28 mm. in length and from 4 to 5 mm. in width.

Head. Shape as in Fig. 597. Greyish-brown except three light golden-brown areas, one at apex of clypeus (which can be seen with naked eye) and a pair on the genae in the angles made by the clypeus. A white spot at the rear of each eye can be seen with 10× lens. Small area around basal segments of antenna black whilst part of basal segment is white (easily made out with 10× lens). Pattern of dark marks on clypeus and genae shown in Figs. 597 and 598, but is somewhat variable. Ventral areas of genae greyish-brown orally, shading to light golden-brown aborally.

Gular sclerite. Edged with darker grey than genae but with light area in centre. Narrow folds of the genae along the gular margin.

Mouthparts. Do not project from head capsule.

Labrum. Interior margin deeply incised (Fig. 599), margin heavily sclerotized, rich chestnut in colour except for triangular area adjacent to the incision on each side, which is light yellow. The hairs around these latter areas are blunt.

Mandibles (Fig. 600). Black, four teeth arranged transversely, brush of hairs on each mandible on inner edge.

Maxillae and *labium* short and squat. Labial palp with proximal segment bulbous with sclerotized ring around its base. Thoracic segments typically Limnephilid, marking shown in Fig. 601, but fold between anterior and posterior pair of sclerites on mesonotum is yellowish or chestnut coloured. Prosternal horn yellow.

Legs. Also typically Limnephilid. Lestage gives the length of leg ratio as 10:14:15. The dark spots are well pronounced. A good description is given by Lestage.

Abdomen. Creamy-white to greyish-white in colour. Intersegmental grooves dark grey, fairly well defined. Protuberances on first abdominal segment well developed, dorsal protuberance conical, terminating in a sharp point directed anteriorly; at the sides of the base are thirteen black hairs of which three are at the front and five at each side. On the sternite is situated a well-defined protuberance which carries black hairs, as do the lateral protuberances (Fig. 602). Lateral line present from the third to the seventh abdominal segments inclusive; two, three or four sclerotized bifid organs present from the second to the seventh segments. Filiform abdominal gills are present on second to eighth segments and are mostly united in groups of two or three. The total number of gills is somewhat variable, but the number 84 given by Ulmer appears to be about average, and they are disposed as follows: 18+18+15+12+9+9+3.

Anal sclerite. Four large black bristles on posterior margin with one smaller bristle in centre, and two yet smaller bristles between the outer pairs; a few variably situated dark marks on the anterior margin. Anal claws with one auxiliary hook and the basal lobes bearing a number of dark spots on their ventral surface. Where the sclerotized plates meet the posterior margin of the ninth segment a pair of orange-coloured arrowhead-shaped marks are situated (easily visible with the naked eye).

Glyphotaelius pellucidus (Retzius)

Larvae of this widely distributed Caddis Fly were collected from a small pool in Brathay Quarry Wood, near Wray Castle, Ambleside, on 27 October, 1945. This pool, from which the larvae of *Oligotricha ruficrus* (Scopoli) and *Phryganea striata* L. were obtained in 1943, is overhung around its whole margin by oak, alder, sycamore and willow, and the cases of the *pellucidus* larvae were made up from circles cut from the fallen leaves of these trees.

The larvae from Brathay Quarry Wood were very numerous, fifty-three being collected from about two square feet of the pond margin, where the bottom was covered completely by fallen leaves from the surrounding trees.

Case. This is quite characteristic of the larvae (Fig. 603). It consists of pieces of the leaves of trees often cut in almost perfect circles, or quite commonly whole leaves of some trees are used (hawthorn). The pieces of cut leaf are arranged in two horizontal series between which a tube is constructed of small pieces of vegetable debris, cemented together with the secretion from the modified spinning gland situated in the ligula. The anterior piece of cut leaf in the dorsal series overlaps the entrance to the tube; the anterior piece in the ventral series is situated just posterior to the entrance. From the dorsal aspect of the cased larva nothing can be seen of the larva as it moves jerkily across the submerged leaves under its mantle.

Larva is of eruciform type. Head and thorax light brown, marked with dark brown. Abdomen creamish, with gut contents showing through integument. Length 23 mm., width 4 mm.

Fig. 603. *Glyphotaelius pellucidus.* Larval case from beneath.

Head (Fig. 604). Orthocentrous. Oval, light brown, strongly marked with dark brown. Several transverse rows of dark parts on dorsal surface. Aboral apex and areas adjacent to lateral excisions of clypeus light. Oral region of clypeus very darkly marked. Eyes prominent; antennae reduced. Chaetotaxy as in Fig. 604. Anteclypeus pale.

Labrum (Fig. 605). Transverse, light yellowish-brown, more darkly marked in chestnut brown around margin; four long, black bristles arranged transversely across the middle.

Mandible (Fig. 606). Chestnut brown with black transverse band. Brush of fine bristles unequal in length on inner margin (and strong bristle at base of outer margin) of both mandibles.

Maxilla (Fig. 607). Small. Palp, four segmented, bent inwards, with bunch of sense organs at tip, and a prominent circular sense organ at base of penultimate segment. At base of palp is situated a bunch of bristles as long as the palp. A group of sense organs at tip of maxilla, with four blade-like bristles directed inwardly.

Labium. Labial palps of only one segment apparent. Two rod-like organs at tip of palps.

Gular sclerite (Fig. 608). Small and pear-shaped, not completely dividing the genae, but extent of aboral apex somewhat variable.

Thorax (Fig. 609). Prothoracic notum sclerotized, light chestnut in colour, with anterior third darker in colour. Median longitudinal suture present, posterior margin black. Mesothoracic notum sclerotized, posterior margin black. A pattern of dark spots as in the figure. Metathoracic notum with small sclerotized patches.

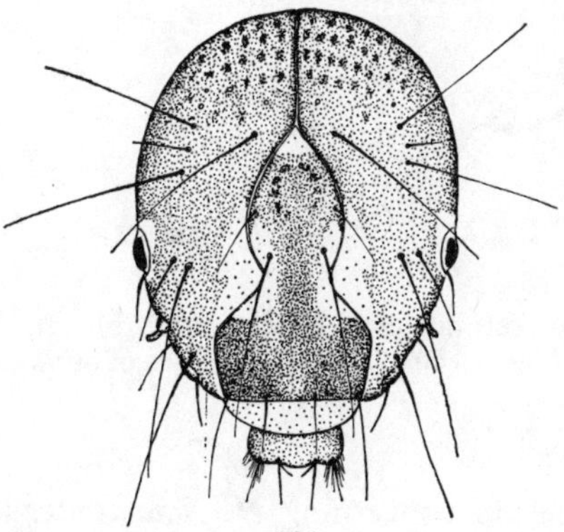

604

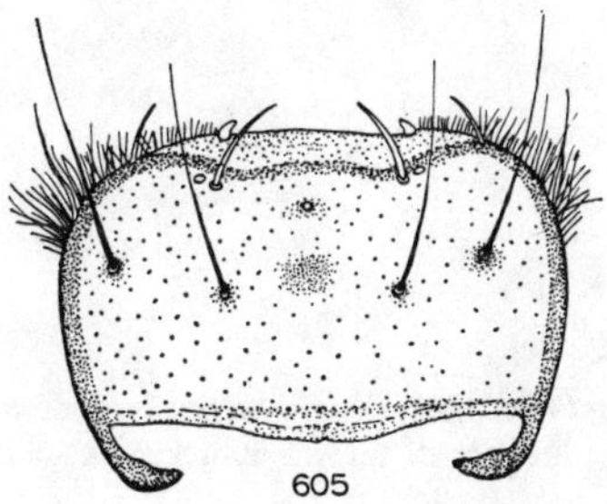

605

Figs. 604–605. *Glyphotaelius pellucidus*. 604. Head of larva; 605. Labrum of larva.

Legs. Prothoracic legs (Fig. 610, a) short and deep, trochanter divided, meso- and metathoracic legs (Fig. 610, b) long and thin.

Abdomen. Protuberances on first abdominal segment present. Lateral line present on segments three to eight inclusive. Basal lobes of anal claspers heavily marked in dark brown—readily visible to the naked eye. Anal claw with single auxiliary claw which is more readily discerned when viewed from the side. Filiform abdominal gills present in bunches of three and two, united at their bases, on segments two to eight inclusive. On the eighth segment, the gills consist of a united pair on the dorsal surface of each side and adjacent to the anterior margin of the segment.

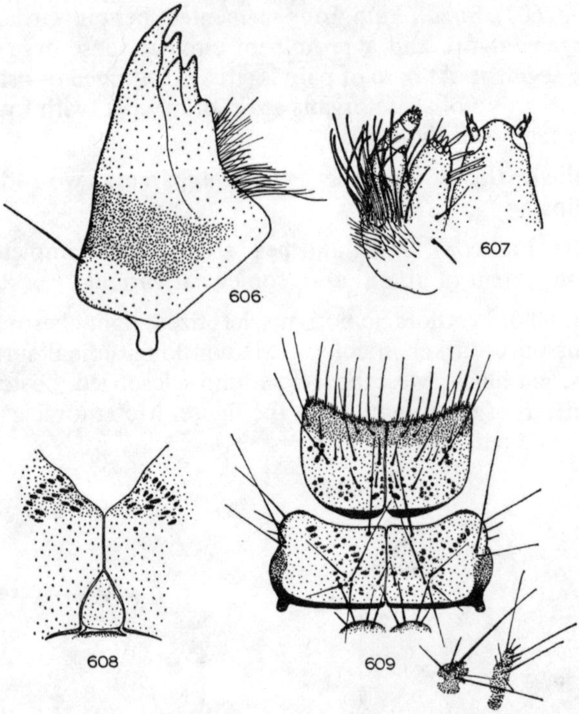

Figs. 606–609. *Glyphotaelius pellucidus*. 606. Left mandible of larva; 607. Maxilla and labium of larva; 608. Gular sclerite of larva; 609. Thoracic sclerites of larva.

Lestage has collected the information concerning *N. punctatolineatus* and *G. pellucidus*, and states that gills are present on segments two to seven inclusive in the latter species, whilst in the former additional gills are present on the eighth segment. This certainly does not correspond with what I have found in *pellucidus*.

Nemotaulius punctatolineatus (Retzius)

Three larvae of this species were collected by H. P. Moon from the north-west end of Lake Windermere in 1934, but the late M. E. Mosely refused to accept the identification because 'the adult had never been collected in Britain'. E. C. Pelham-Clinton, however, collected a single adult at light at Aviemore, Inverness-shire in 1965. Its British status rests on these specimens and throughout its range it is rare. It is a large species, the length of the forewing measuring about 24 mm. Illustrated descriptions of the larva are given by Wesenberg-Lund and Lepneva. The young larva makes a case of pieces cut from the leaves of *Potamogeton* in the

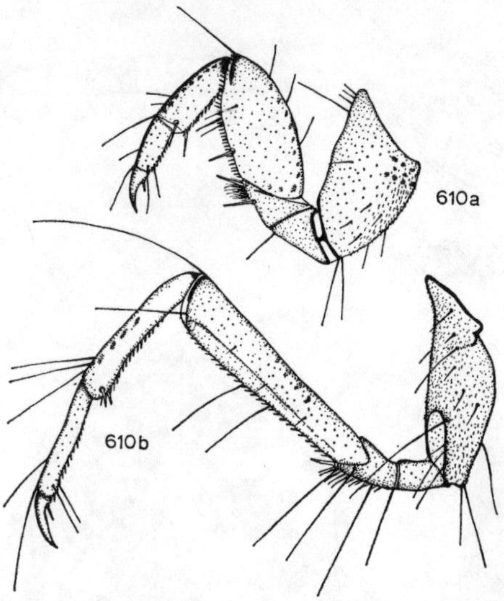

Fig. 610. *Glyphotaelius pellucidus:* a. prothoracic leg of larva; b. mesothoracic leg of larva.

manner of *Glyphotaelius pellucidus*. It may be distinguished from this latter species, however, by the wide oral part of the clypeus not being uniformly dark, and also by possessing pronounced 'furcal bands'.

Anabolia brevipennis (Curtis)

Only nine specimens of this species are known to have been taken in Britain and all during the last century from Yorks., Cheshire and Norfolk. No description from British larvae has been made. The larvae inhabit pools with leaf-covered bottoms and attain a length of 10 to 12 mm. and a width of 2 mm. The head is of a uniform dark brown colour except for a whitish zone around the eyes and a lighter transverse area occupying a posterior-lateral position in which is situated a series of spots. The abdominal gills are in groups of two or three or more and the abdominal claws are only feebly sclerotised. The triangular-section larval case may be

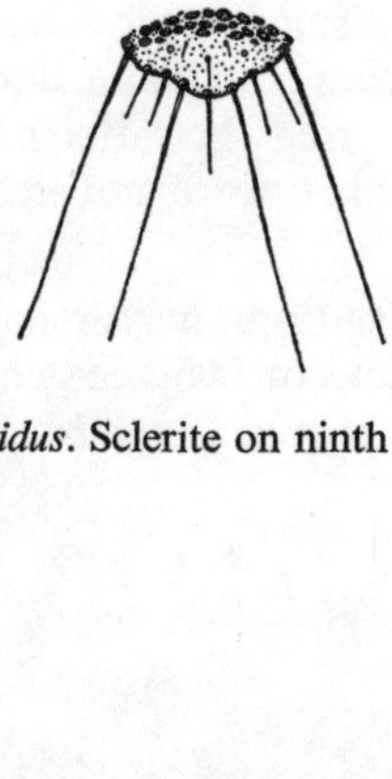

Fig. 611. *Glyphotaelius pellucidus*. Sclerite on ninth abdominal segment of larva.

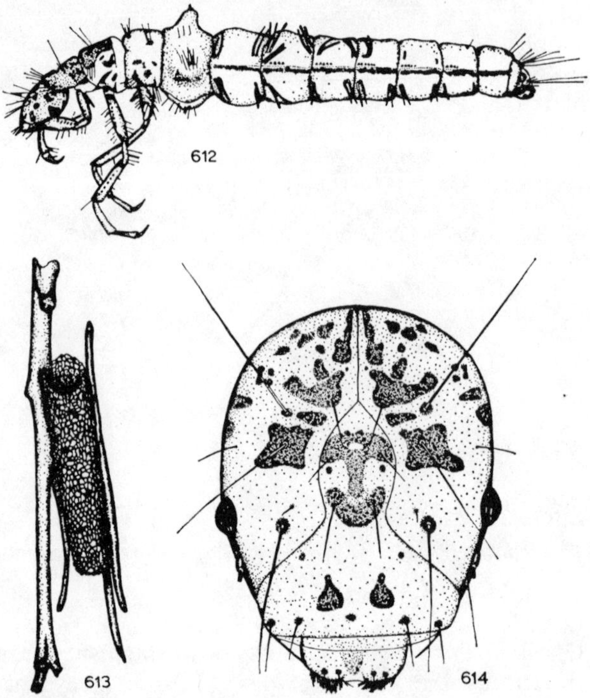

Figs. 612–614. *Anabolia nervosa.* 612. Larva (lateral view); 613. Case; 614. Head.

up to 20 mm. in length and 6 mm. in width. It is deeper at the front than behind and consists of a silken tube secreted by the larva covered with pieces of leaf of uniform size. Because of this latter character the case is of uniform breadth which serves to distinguish it from those of *Limnephilus decipens* and *L. nigriceps* which, in both cases taper posteriorly.

Anabolia nervosa (Curtis)

Two streams, one at Temple Balsall, Warwickshire, and the Dowles Brook, near Bewdley, Worcestershire, were found to be very thickly populated with larvae of *Anabolia nervosa*. Several were reared to the adult stage for the determination to be confirmed. Both streams are rapidly flowing, but larvae were also collected from a stagnant flood pool on the River Rea at Cleobury Mortimer, Shropshire. The latter larvae appeared quite healthy although the head and thorax were covered with adherent mud, and festooned with green algae.

Case (Fig. 613). Conical, of sand-grains cemented together. Anterior opening oblique. One or more sticks or twigs cemented longitudinally to conical sand-grain portion. Length 26 mm., breadth 6 mm. (sand-grain portion);

the sticks may increase the over-all length to as much as 60 mm. Ulmer gives dimensions slightly in excess of these. It is probably that my specimens tend to be small. In the young larvae the case consists of small pieces of cut stalk, stick and vegetable debris arranged longitudinally with a large stick cemented to the side.

Larva (Fig. 612). Eruciform, cylindrical, head orthocentrous, pro- and mesonota sclerotized, metanotum with scattered sclerotized plates. Abdomen white, obtuse. Length 18 mm., breadth 4 mm.

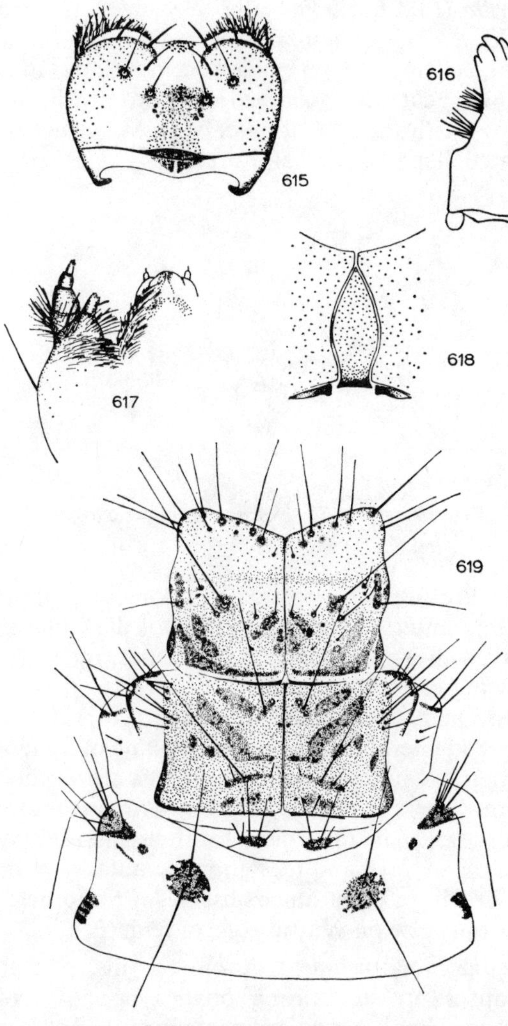

Figs. 615–619. *Anabolia nervosa*. 615. Labrum; 616. Mandible; 617. Maxilla and labium; 618. Gular sclerite; 619. Thorax.

Head (Fig. 614). Elliptical, yellowish-brown with dark reddish-brown almost black markings. The extent of the marking is variable but the marking given in the figure is about average. At the aboral end of the clypeus is a T-shaped mark with a yellow mark at the confluence of the strokes. At the base of the T is situated a bow-shaped mark. Between the two marks is a small dark discrete spot on each side. Two pear-shaped dark marks at oral end of clypeus. On the genae a number of irregularly shaped marks make up an inverted V. Chaetotaxy as in Fig. 614. Antennae very small. Gular sclerite (Fig. 618) not completely dividing genae.

Mouthparts. Labrum (Fig. 615). Partially sclerotized along outer margin and centre. Besides a series of spines (three pairs along outer margin and three other pairs), a fringe of fine hairs occurs along anterior margin. Mandibles (Fig. 616) acute, four-toothed with two brushes of hairs along inner edge and two bristles at base of outer edge. Maxillae hairy (Fig. 617), palp four-segmented. Tip of mala palpiform.

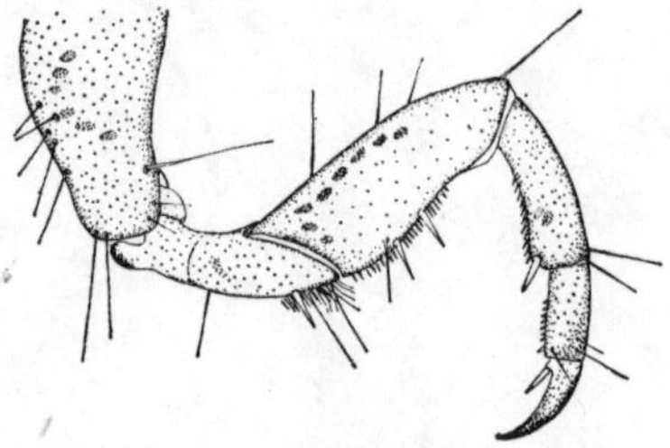

Fig. 620. *Anabolia nervosa.* Prothoracic leg.

Thorax (Fig. 619). Prothorax sclerotized with longitudinal median suture, strong bristles along anterior margin. Pattern of dark marks in posterior half of prothorax with a medium row of bristles anterior to the latter. Prosternal horn present. Mesothoracic nota sclerotized with median longitudinal suture and about equal in size to prothorax. Unsclerotized pleural regions extend laterally. A pattern consisting of oblique dark marks situated in mesothoracic notum. The latter marks are variable in intensity but the figure shows average marking. Mesothorax with a few sclerotized patches, three pairs of which are furnished with bristles. It may be necessary to soften the sclerotized pro- and mesothoracic nota by boiling in caustic soda solution before the pattern can be observed. Prothoracic leg (Fig. 620) with spines of unequal size on ventral edge of femur.

Abdomen. First abdominal segment devoid of gills, median dorsal and lateral protuberances present, strong bristles present around base of median protuberance. Small sclerotized patch furnished with strong bristles on each side in pleural region. Lateral line present consisting of fine hairs, and in segments three to seven a series of small pustules five to seven in number, with sclerotized extremities dorsal to the lateral line and in the anterior half of the segments. Abdominal filiform gills present. Anal claws small, surmounted by a small auxiliary claw (Fig. 622). A bow-shaped sclerite at base of claw. Extremity of ninth segment sclerotized on dorsal surface with a series of bristles of unequal length along posterior margin (Fig. 621).

Rhadicoleptus alpestris (Kolenati)

The larva of this very local species is unknown.

Potamophylax cingulatus (Stephens)

In July 1951 I collected a number of caddis larvae from a small fast-flowing stream at Nutley in Ashdown Forest, Sussex, at about 300 feet above sea-level. From some of these I reared one of each sex of the species *Potamo-*

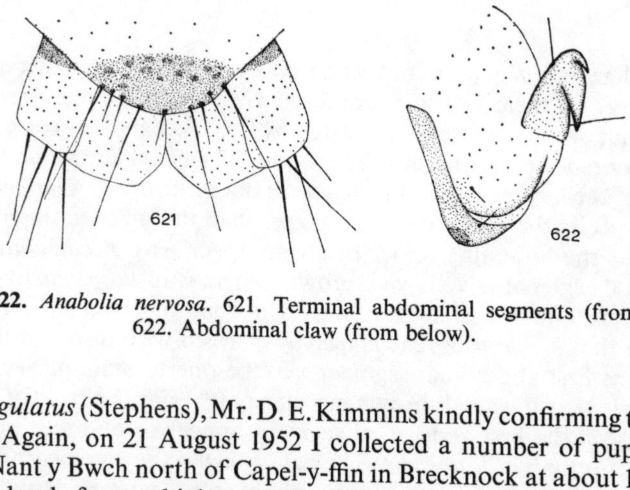

Figs. 621–622. *Anabolia nervosa.* 621. Terminal abdominal segments (from above); 622. Abdominal claw (from below).

phylax cingulatus (Stephens), Mr. D. E. Kimmins kindly confirming the identification. Again, on 21 August 1952 I collected a number of pupal cases from the Nant y Bwch north of Capel-y-ffin in Brecknock at about 1400 feet above sea-level, from which one of this species emerged on 29 August. Mr. Kimmins, who has reared this species from pupal cases obtained from a small stream draining into Lake Windermere (at about 130 feet above sea-level), believes that this species may be confined to the smaller streams, whilst *P. latipennis* (Curtis) may prefer the larger streams. This view is confirmed by my experience. For the purpose of this description I have re-examined my original material of *P. latipennis* and fresh larvae collected from Dowles Brook, Worcestershire, on 8 April 1953.

Larva. So very like *Potamophylax latipennis* (Curtis) that great difficulty is experienced in separating it from this latter species. The patterns of dark marks on head and thoracic segments are so variable that no reliance can be placed on these. In *latipennis*, however, the colour of the head is much

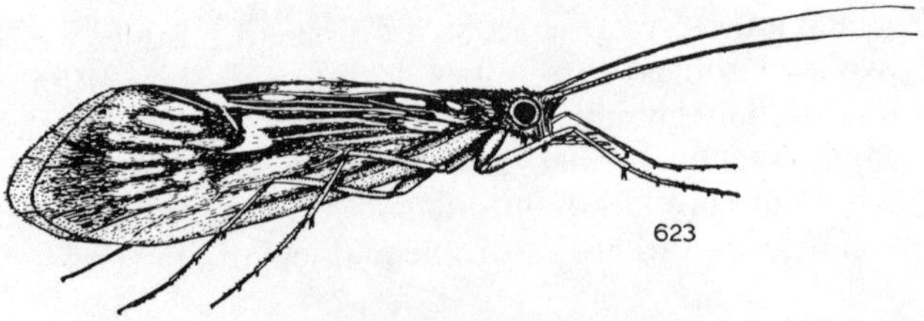

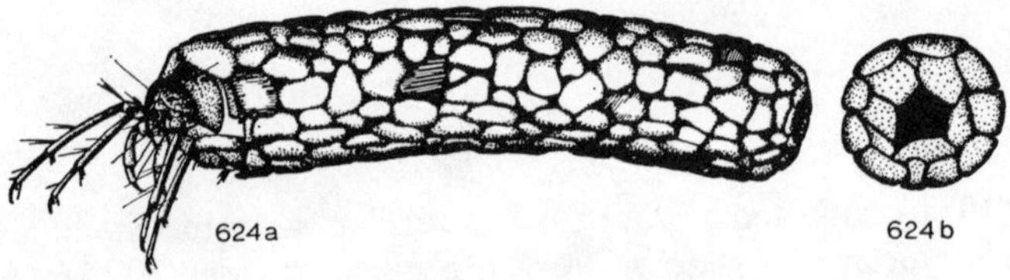

Figs. 623–624. *Potamophylax cingulatus* (Steph.). 623. Adult (length: head to wing tips 1 in.); 624. a. larva in case; b. larval case from behind.

darker than in *cingulatus*, but when cleared in potash solution and seen from the front there is a light-coloured area on each gena near the genal suture, but separated from the latter by a dark band on each side. This light-coloured area is absent in *cingulatus*. The genal suture on the ventral surface of the head is closed in *cingulatus* but quite open in *latipennis*, while in both species the gular sclerite is darker than the surrounding genal areas except for the bounding folds. In all my specimens of *cingulatus* the first abdominal segment is yellowish-brown, whereas in *latipennis* it is grey. A possible explanation of this, however, is that the stream from which I obtained the *cingulatus* larvae is heavily charged with iron and the coloration of the first abdominal segment may be due to staining. Perhaps it is due to this that the small brown circular patches at the base of the spines and hairs on the first abdominal segment appear to be more pronounced in *cingulatus* than in *latipennis*. The pale areas in the clypeus and the genae which have been given by previous authors as a means of identification are shown in Fig. 625, but I am by no means satisfied that reliance on this character will give successful determination in every case. Length 20 mm., width 4 mm.

Case (Fig. 624). Length 22 mm., width 6 mm. Slightly curved, flattened on ventral surface, almost uniform width when larva fully grown, of small stone chips. The cases of the Brecknock specimens were of small round pebbles and larger sand grains. Hinder end of case rounded and partially closed.

This species 'congregates'—a number are found under one particular stone in the stream whereas apparently identical situations elsewhere are not favoured.

Potamophylax rotundipennis (Brauer)

A description of the larva of this local species has not yet been made from British material. The larval case is curved, slightly conical, convex behind and the posterior end partially closed with sand grains but allowing a small circular opening. The anterior end is oblique. In the young larvae the case is not convex behind but is strongly acuminate. The case is constructed of sand grains, often smaller behind and larger in front and small stones may be attached at various positions on the case prior to pupation.

Potamophylax latipennis (Curtis)

Described from larvae collected from Dowles, Worcestershire, some of which were reared to maturity and also compared with larvae collected from the Lake District (Grizedale Tarn).

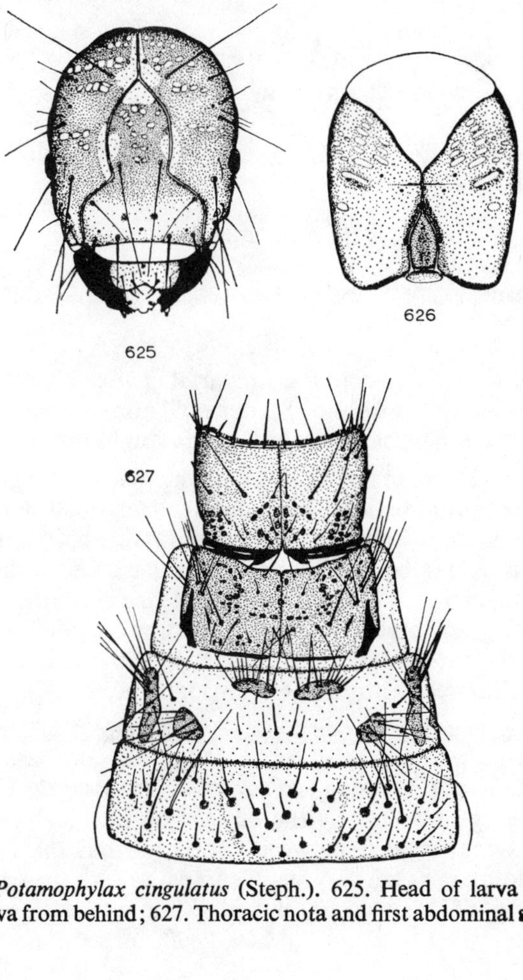

Figs. 625–627. *Potamophylax cingulatus* (Steph.). 625. Head of larva from the front; 626. Head of larva from behind; 627. Thoracic nota and first abdominal segment of larva.

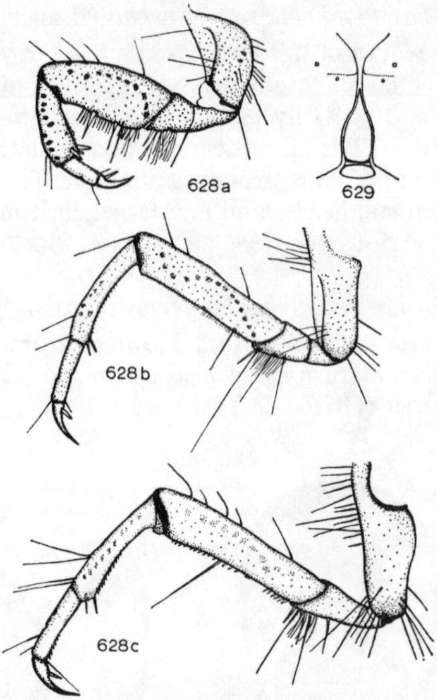

Figs. 628–629. *Potamophylax cingulatus* (Steph.). 628. Legs of larva; a, pro-, b, meso- and c, metathoracic; 629. *Potamophylax latipennis* (Curtis). Gular sclerite.

In Dowles Brook (a fast-running stream) it is most abundant, sheltering in small groups under the slabs of Bunter sandstone which form the bed of the stream. There is no phanerogamic vegetation in this water.

Case. About 20–22 mm. in length and is slightly curved longitudinally with slight dorso-ventral flattening. It is composed of small flattened pebbles and fragments of stone and large sand-grains (Fig. 630). Smaller pebbles and sand-grains tend to be placed on the ventral surface of the case and the larger pieces of stone are placed at the sides near the ventral surface. There is a small opening at the posterior end, and, previous to pupation, the case is attached by the anterior end to the underside of a large stone. The *larva* is uniformly cylindrical with the head hypognathous.

Head (Fig. 631). The clypeus has a kite-shaped pattern of dark spots at the aboral end with a pair of large bristles at the base of the pattern. At the oral margin of the clypeus there is a row of six bristles; the outermost pair of bristles are just embraced by the outcurving of the clypeus. A pair of smaller bristles are situated between the latter and the former pair of bristles. The mouthparts are very small compared with the size of the larva. The labrum (Fig. 632) is strongly indented, the indentation being heavily sclerotized. A small blunt spine is situated at each end of the sclerotized part. A row of six bristles is located transversely in the middle of the exposed part of the labrum. The mandibles, which are slightly asymmetrical (Fig. 633), are broad and heavily sclerotized. There are four large teeth and two much smaller teeth. A tuft of hair is situated near the small teeth and directed inwards. The maxillae and labium (Fig. 634) are extremely small. The maxillary palp is four-segmented and on the maxilla are several tufts

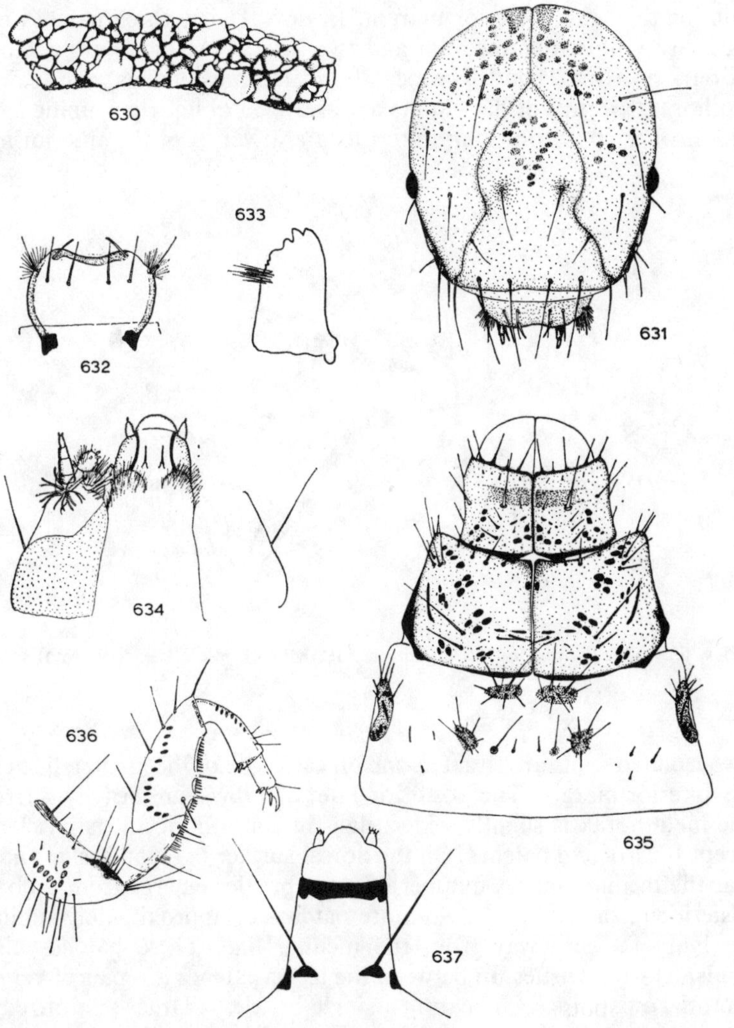

Figs. 630–637. *Potamophylax latipennis*. 630. Case; 631. Head; 632. Labrum; 633. Right mandible; 634. Labium and maxilla; 635. Head and thorax; 636. Prothoracic leg; 637. Ventral view of labium.

of hairs and modified setae which are probably sense organs. On the oral side of the labium is a pair of spines, set on small protuberances, whilst on the ventral surface of the labium there is a pair of large spines as long as the labium itself, and projecting from a pair of sclerotized parts near the base (Fig. 637).

Thorax. The terga of pro- and mesothorax are completely sclerotized whilst the metathoracic tergum is only partially sclerotized (Fig. 635). A dark transverse band runs across the prothoracic tergum slightly nearer the anterior than the posterior margin. In dorsal view three patches of dark spots are visible, one medium and two lateral. All these are posterior to the dark transverse band. The mesothorax is almost half as wide again as the prothorax. Groups of dark spots are arranged obliquely running from the middle of the posterior margin to the lateral vertex of the anterior margin.

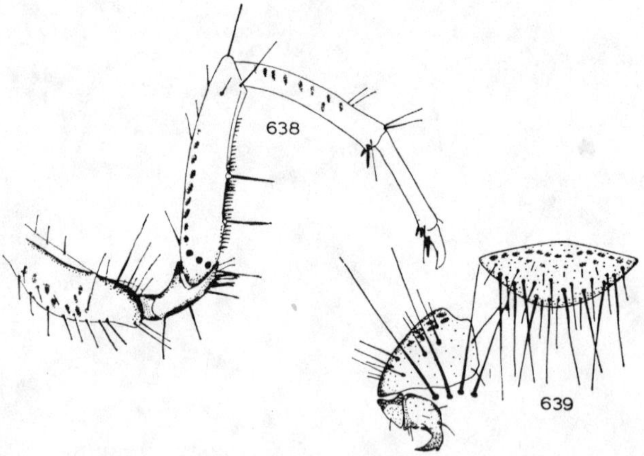

Figs. 638–639. *Potamophylax latipennis*. 638. Mesothoracic leg; 639. Anal sclerites.

Two isolated spots are situated one on each side of the median line close to the anterior margin. The posterior lateral vertices are heavily sclerotized. The metathorax is slightly wider than the mesothorax. It is unsclerotized except for isolated patches. On the dorsal surface two patches lie anteriorly near the median line. A number of stout bristles emerge from each patch. Posterior to the latter are two more patches of approximately similar size but lying farther away from the median line. These patches also are furnished with bristles. In between the latter patches is a pair of very small sclerotized patches each bearing a single bristle. A larger sclerotized patch occupies the pleural regions of the segment. These patches bear a bunch of bristles anteriorly and in addition are marked by a short series of dark spots. A dark spot in the central part of the patch is elongated.

Legs. The prothoracic legs (Fig. 636) are approximately two-thirds the length of the mesothoracic legs (Fig. 638), but are considerably broader. Meso- and metathoracic legs are approximately equal in length but meso-

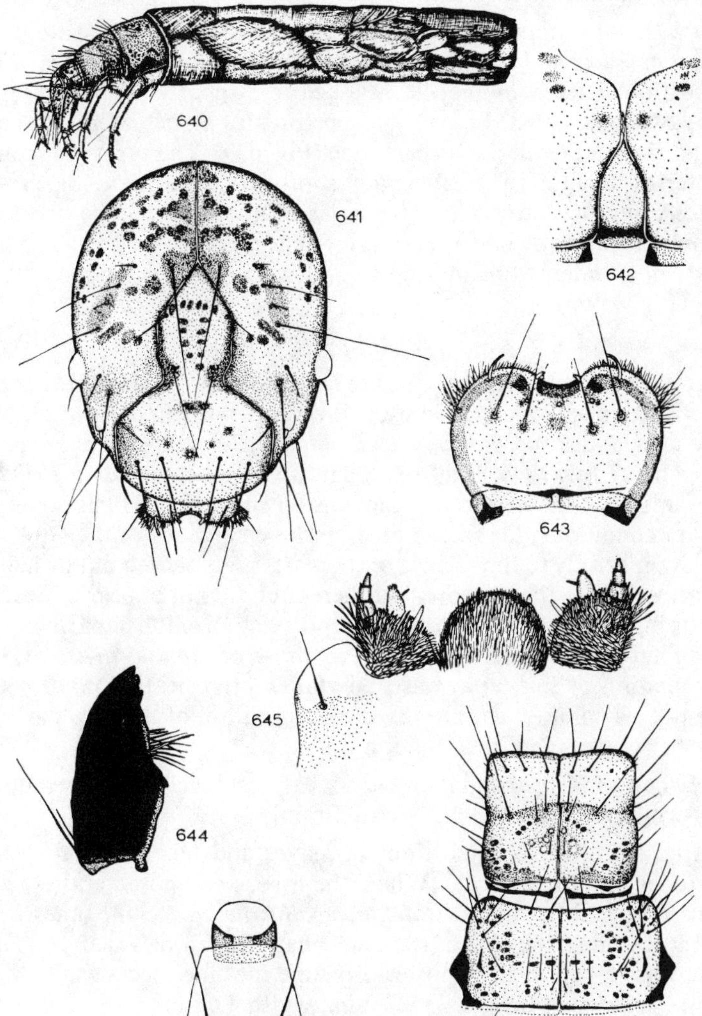

Figs. 640–647. *Halesus digitatus* (Schranck). 640. Larva in case; 641. Head of larva; 642. Gular sclerite; 643. Labrum; 644. Left mandible from above; 645. Maxillae and labium from above; 646. Labium from beneath; 647. Thoracic nota.

thoracic legs are slightly broader. The most heavily sclerotized segment is the trochanter in each case. In all legs there are two distal spurs on the two segments of the tarsus and two marginal spurs on the tibia. A spine is present at the base of each claw.

Abdomen. Hooks on first abdominal segment are absent. Gills are present, but their position and numbers are variable, and thus have little value in identification. The abdominal claws are closely opposed to the last segment and are two-segmented. Four very stout bristles are situated between the claw and the lateral sclerite on each side (Fig. 639). The superanal plate has the anterior margin of greater radius of curvature than the posterior margin. Along the latter there is a series of about nine large bristles and some smaller ones. Along the anterior margin there is a row of faintly marked spots indented in the centre and with a fainter and smaller row behind (Fig. 639).

Halesus digitatus (Schranck)

The habitat at Nutfield, Surrey, where larvae of this species were obtained, is described in my description of the larvae of *Chaetopteryx villosa* (Fabricius) (1948. *Proc. R. ent. Soc. Lond.* (A), **23**, 59). Twelve larvae were collected on 27 July 1946, and from them two pupae emerged on 25 October of the same year. As very often happens when rearing Trichoptera under aquarium conditions, the pupae died before emergence of the adults. The pupae swam briskly to the surface of the water and crawled out by means of stalks provided for the purpose, but thereafter did not appear to be able to break the pupal integument. However, by dissection of the pupal integument the imaginal genitalia were made out as those of *Halesus digitatus.*

Examination of this larva was undertaken after heating in 10 per cent caustic potash solution as the very dark coloration of the fore part of the body otherwise renders examination difficult.

Larva. Eruciform. Head and thoracic nota almost black in colour, abdomen creamy-white. Length 20 mm., width 4 mm.

Case (Fig. 640). Made of portions of leaves and other vegetable debris, small stones, sand-grains, etc. Where the case is composed entirely of leaf portions it resembles the case of *Limnephilus lunatus*. Sometimes a small stick is stuck on to the side of the case when it resembles that of *Anabolia nervosa.* Case tapers slightly posteriorly, and in some specimens very slight curvature is noticed. Length 23–27 mm., width 4.6 mm.

Head (Fig. 641). Orthognathous, dark golden brown with large number of small dark brown marks. Lateral margins of aboral part of clypeus parallel, giving it square appearance, surmounted by the triangular aboral vertex. Antennae small; marks and chaetotaxy as shown in Fig. 641.

Gular sclerite (Fig. 642). Not completely dividing the genae. Where the genae meet, however, remnants of the attenuated vertex of the gular sclerite are seen.

Labrum (Fig. 643). Deeply notched and sclerotized anterior margin. Lateral margins sclerotized. The median marginal spines (on each side of the excision) short and blunt. The outer anterior spines curved and blunt. The lateral spines moderate in length and sharp.

Mandibles (Fig. 644). Black, four teeth with some slight asymmetry. Group of light-coloured hairs on inner margin of each mandible. One long and one short black hair at the base on the outer side of each mandible.

Maxillae (Fig. 645). Palp four-segmented proximal segment squat. Maxilla hairy, hairs at base of palp long, light in colour and blunt. Large blade-shaped spine present.

Labium. Globular and hairy. Labial palp two-segmented, distal segment small, proximal segment large but partly inserted into the labium which, on account of the globular shape of the labium, appears to arise on the ventral surface.

Thorax (Fig. 647). Prosternal horn present, dark golden brown in colour. Pronotum sclerotized and extending over pleural region, strongly convex mid-dorsally immediately posterior to a transverse furrow. Dark golden brown shading almost to black along margins. Some variation in the pattern of marks and not always symmetrical. Median longitudinal suture present. Mesonotum completely sclerotized with median longitudinal suture. Dark golden brown, black in the area of the lateral projections and at the lateral extremities of the anterior margin. In addition to a pattern of dark brown marks, a lunate black mark is present near each lateral projection. Metanotum only partially sclerotized. Anterior pair of dark brown kidney-shaped sclerites are divided from posterior pair by a transverse band of small marks. Posterior sclerites have pattern of light marks as shown in the figure and a pair of small dark marks are situated near the inner extremity of each of the posterior sclerites.

Legs (Fig. 648 a, b and c). Prothoracic legs short, meso- and metathoracic legs about equal in length. On all legs two spines at distal end of tibia and tarsus with small, short, blunt spine directed downwards between each pair in the tarsus, but in the tibia the spine is larger. In meso- and metathoracic legs posterior edge of tibia and tarsus with row of short spines; in prothoracic legs a row of short spines on tibia only.

Abdomen. First abdominal segment greyish with a number of short stout bristles arising from small sclerotized patches on ventral surface. Protuberances small. Remainder of abdomen creamy-white, cylindrical. Lateral line present on the second to the seventh segments inclusive. The associated groups of sclerotized tubercles are present also in the anterior part of each of segments two to six inclusive. The number of sclerotized points appears to be eight on each side of the segments. Sternal ovoid plates are present on segments two to seven (Fig. 650) but one larva examined showed no plate on the second segment. The sternal plates become progressively larger posteriorly and have a thin sclerotized brown margin. Filiform abdominal gills present. Sclerotized tergite present on ninth abdominal segment (Fig. 651). Anal claw with additional single auxiliary claw.

Nielsen worked on the immature stages of this species. Exact head

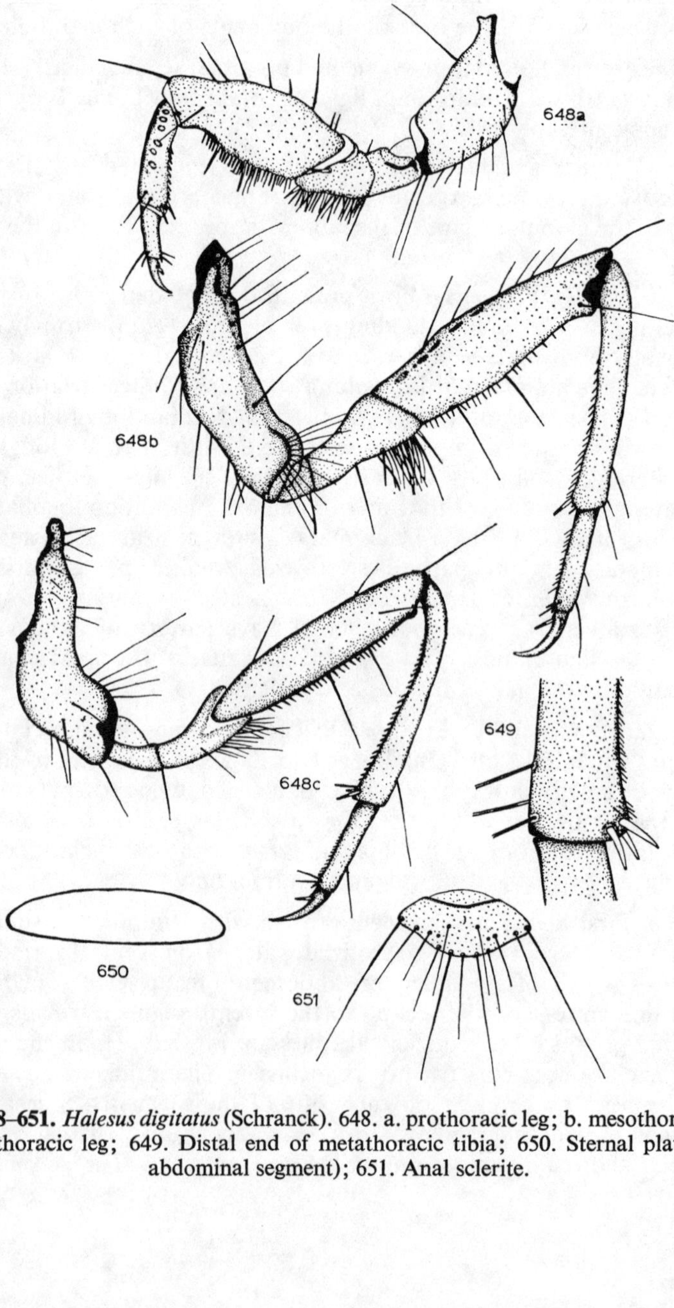

Figs. 648–651. *Halesus digitatus* (Schranck). 648. a. prothoracic leg; b. mesothoracic leg; c. metathoracic leg; 649. Distal end of metathoracic tibia; 650. Sternal plate (third abdominal segment); 651. Anal sclerite.

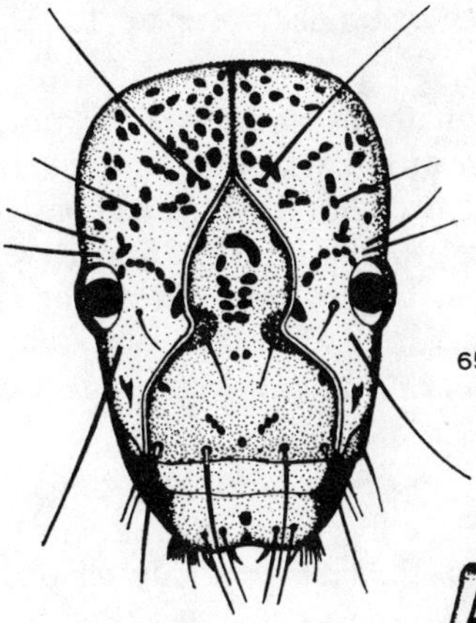

652

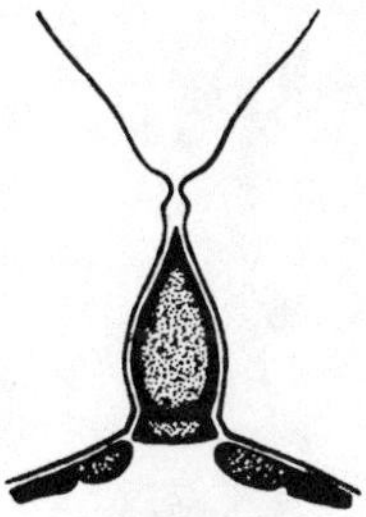

654

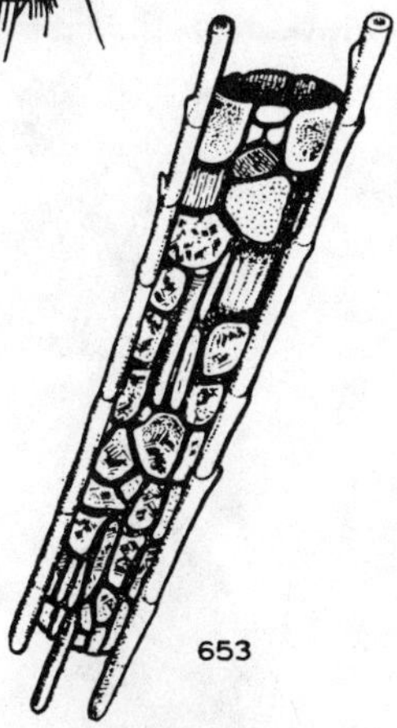

653

Figs. 652–654. Larva of *Halesus radiatus* (Curtis). 652. Head from the front; 653. Larval case; 654. Gular sclerite. Hanna.

measurements and length and breadth of the larva elicited valuable information concerning the life-cycle and the incidence of the five stadia at different collecting times.

Halesus radiatus (Curtis)

Hanna gives the following description from larvae collected near Reading where they were in company with the larvae of *Limnephilus marmoratus* Curt., *Limnephilus flavicornis* (F.) and *Anabolia nervosa* (Curt.).

Case (Fig. 653). Up to 32 mm. long and 6 mm. wide of pieces of bark, stems and leaves of water-plants. Anterior opening of the case oblique, posterior opening small and straight. Cases have one to three long sticks along their whole length. Sticks may reach 38 mm. in length. In some, sticks were absent, but large pieces of bark had been substituted.

Larva. Eruciform. Larvae examined were up to 22 mm. in length and 5 mm. in width.

Head (Fig. 652). Hypognathous and ovoid. Fronto-clypeus yellowish-brown or chestnut-brown with black patterns. Four long setae at oral end of the fronto-clypeus and two short ones at its constrictions. The anterior surfaces of the genae are yellowish-brown and have large distinct black spots. Eyes surrounded by lighter areas. Posterior surfaces of the genae dark brown except for two areas above the labium and two areas on both sides of the occipital foramen. Latter areas bear black spots. Gular sclerite does not separate the genae completely.

Labrum. The ventral margin of the labrum (Fig. 655) has a concavity, on each side of which there are five long setae, a short seta and a group of marginal hairs. Margins of the labrum highly sclerotized and darker than

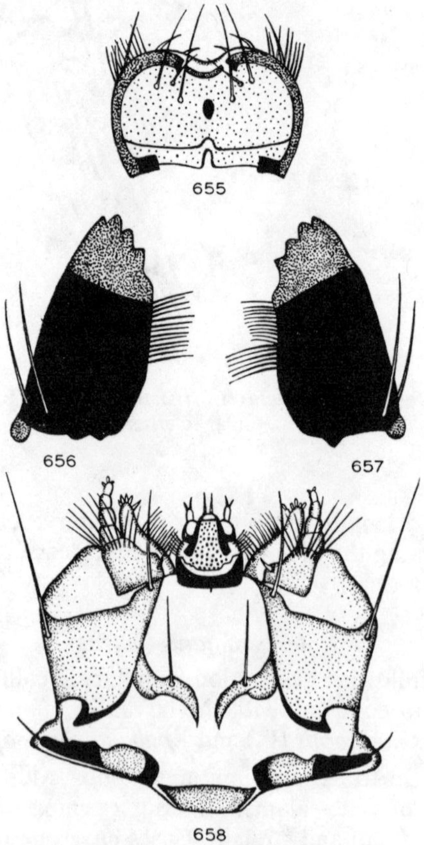

Figs. 655–658. Larva of *Halesus radiatus* (Curtis). 655. Labrum; 656. Left mandible; 657. Right mandible; 658. Labium and maxillae. Hanna.

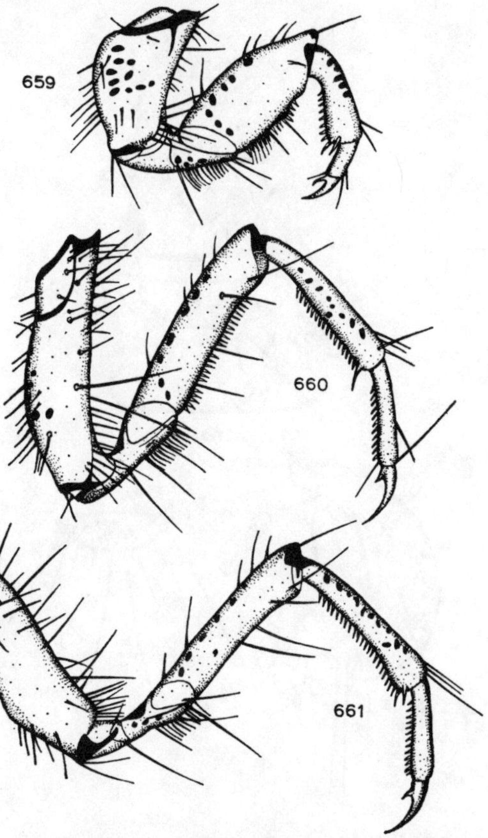

Figs. 659–661. *Halesus radiatus*. 659. Prothoracic leg; 660. Mesothoracic leg; 661. Metathoracic leg. Hanna.

the rest of the labrum. Labrum with a dark brown central spot. The tormae are thick and short and are bent inward.

Mandibles. Each has four teeth, a hairy brush on inner surface and two setae on outer surface near base (Figs. 656, 657).

Maxilla. Cardo small with two setae. Distal margin of the stipes has two setae. Maxillary palp has five segments, of which the basal segment has many hairs. Lacinia furnished with a few sensilla and many hairs (Fig. 658).

Thorax. Pronotum yellowish-brown, entirely sclerotized, and a median longitudinal suture and many black spots. Posterior and posterolateral margins of pronotum highly sclerotized and dark brown. Pronotum has groove along anterior third and two dark brown areas on anterior half. Pronotum has fine hairs and few setae along anterior margin. Some setae on anterolateral corners and on both sides of median suture. Mesonotum

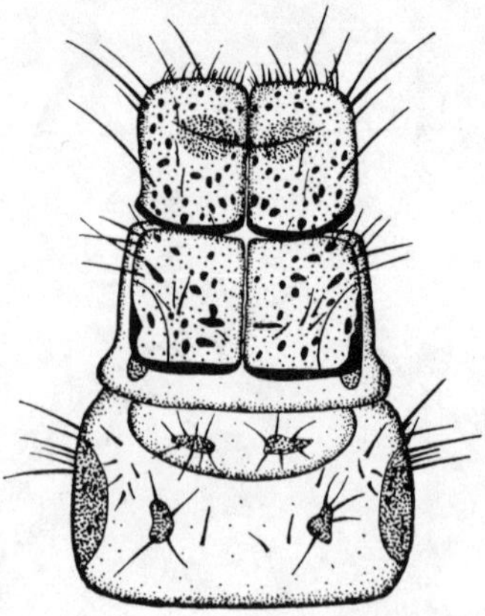

662

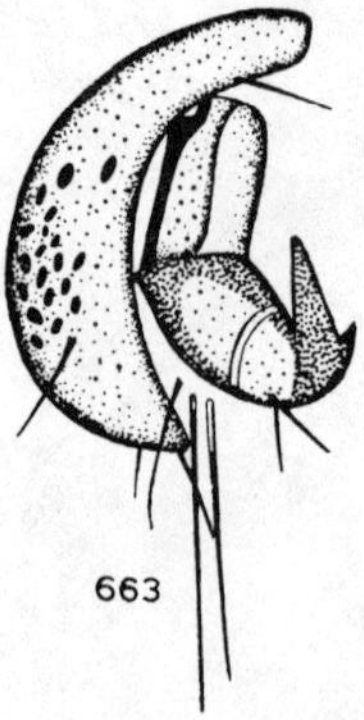

663

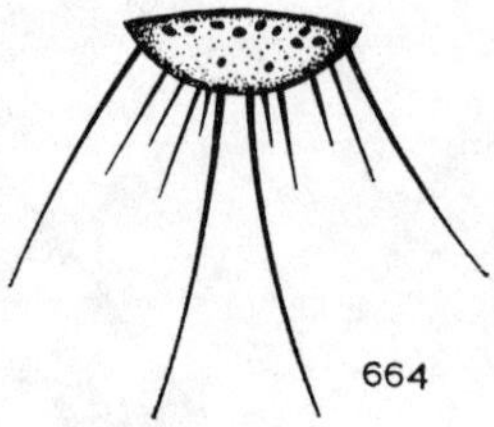

664

Figs. 662–664. Larva of *Halesus radiatus* (Curtis). 662. Thoracic nota from above; 663. Anal appendage, supporting sclerite and anal claw (side view); 664. Anal sclerite from above. Hanna.

sclerotized except for lateral and posterior margins. Sclerotized area chestnut brown except for posterior corners which are golden yellow and have a black spot. Posterior and posterolateral margins of sclerotized area even more sclerotized and black in colour. Many black spots on both sides of the median suture. Metanotum has six small sclerites, each of which has some setae. Outermost sclerites of metanotum have dark brown spots. Prosternal horn present and well developed.

Legs. Prothoracic leg short and robust (Fig. 659), mesothoracic leg (Fig. 660) slightly longer than metathoracic leg (Fig. 661). Femora, tibiae and tarsi of all legs bear small spines. Two additional long spines and a few hairs on the femora of prothoracic legs. Second segments of the trochanters of all legs have group of fine hairs. Dark brown spots on coxae, trochanters, femora and tibiae of all legs. Two spurs on the tibiae of each leg.

Abdomen. On first abdominal segment three protuberances, dorsal one devoid of setae, two lateral ones have a few setae. Tergum and sternum of first abdominal segment bear a group of setae, at the bases of which there are dark brown spots. Gill filaments single and present on abdominal segments one to seven. Lateral line running from beginning of third to the

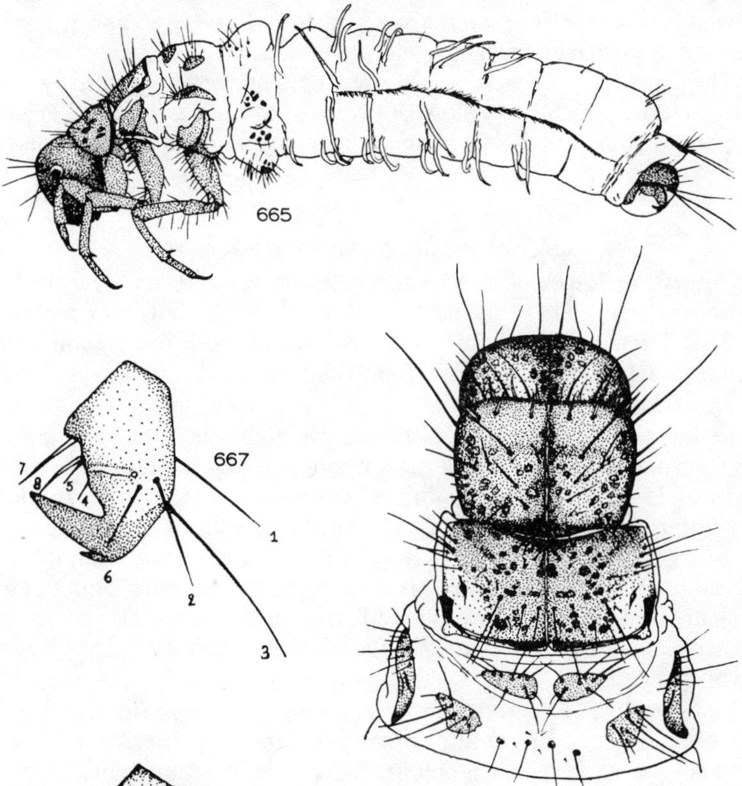

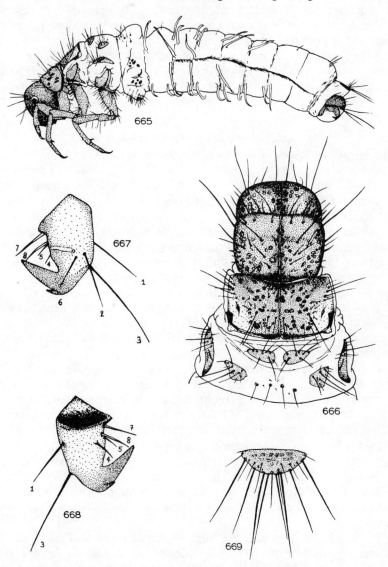

Figs. 665–669. *Melampophylax mucoreus* (Hagen). 665. Lateral view of larva; 666. Head and thoracic nota from above; 667. Outer face of left anal claw; 668. Inner face of left anal claw; 669. Sclerite of ninth abdominal segment. Frochot.

beginning of eighth abdominal segment of fine hairs. Above lateral line dark brown sclerotized pustules in groups of four to nine on abdominal segments three to seven. In a few specimens pustules absent on sixth segment. Elliptical pale yellow sclerites on sterna of second to seventh abdominal segments. In one specimen sclerite on sternum of second abdominal segment absent. Anal sclerite (Fig. 663), elliptical, yellowish-brown spots along anterior margin and four long setae and a few short setae along posterior margin. Anal appendage of two segments and anal claw has auxiliary claw at base. Anal appendage supported by golden-yellow sclerite which has yellowish-brown spots and a few setae. Two long setae and a short seta between anal appendage and supporting sclerite.

Melampophylax mucoreus (Hagen)

The immature stages of this local species have not been described from British material. The only detailed description is that of Frochot who obtained larvae from a number of localities in the Bourguignonne region in southern France. They were found only in cold, fast-flowing rivers in chalky districts.

The larval case (Figs. 670, 671) is distinctly arched, narrows only very slightly from front to rear and the anterior opening is oblique. It reaches a length of 15–17 mm. and a width of 4–5 mm. It is composed of globular sand-grains which are finer and somewhat better arranged on the concave ventral surface, which results in this being rather smoother than the others. These latter have an irregular appearance due to the coarser but flat grains which are added later. There is a tendency for the larval case of the earlier stages to contain rather coarser material, sometimes with small fragments of wood added.

The larva (Fig. 665) measures approximately 15 mm. in length and the sclerotized areas are generally somewhat dark in colour, usually a dark brownish red against which colour the muscle insertion spots hardly ever show. The posterior angles of the mesonotum, on the other hand, are light yellow and contrast well against the general dark colour.

On the head, all the primary hairs are present (Fig. 676). The fourteenth hair, inserted over the eye as is frequently the case in the LIMNEPHILIDAE, is the largest. This is the case also in *Allogamus auricollis* (Pictet). The sixteenth hair is always dark, fairly thick and straight. The fifth hair of the fronto-clypeus is relatively long, reaching the base of the first and second hairs.

The black, fairly narrow and slightly curved mandibles are remarkable in lacking teeth (Figs. 677–682), whereas all species so far studied of *Stenophylax* and *Halesus* possess dentate mandibles. A transverse furrow, as also found in *Stenophylax*, is present. The internal brush of hairs is short and composed of only a few hairs.

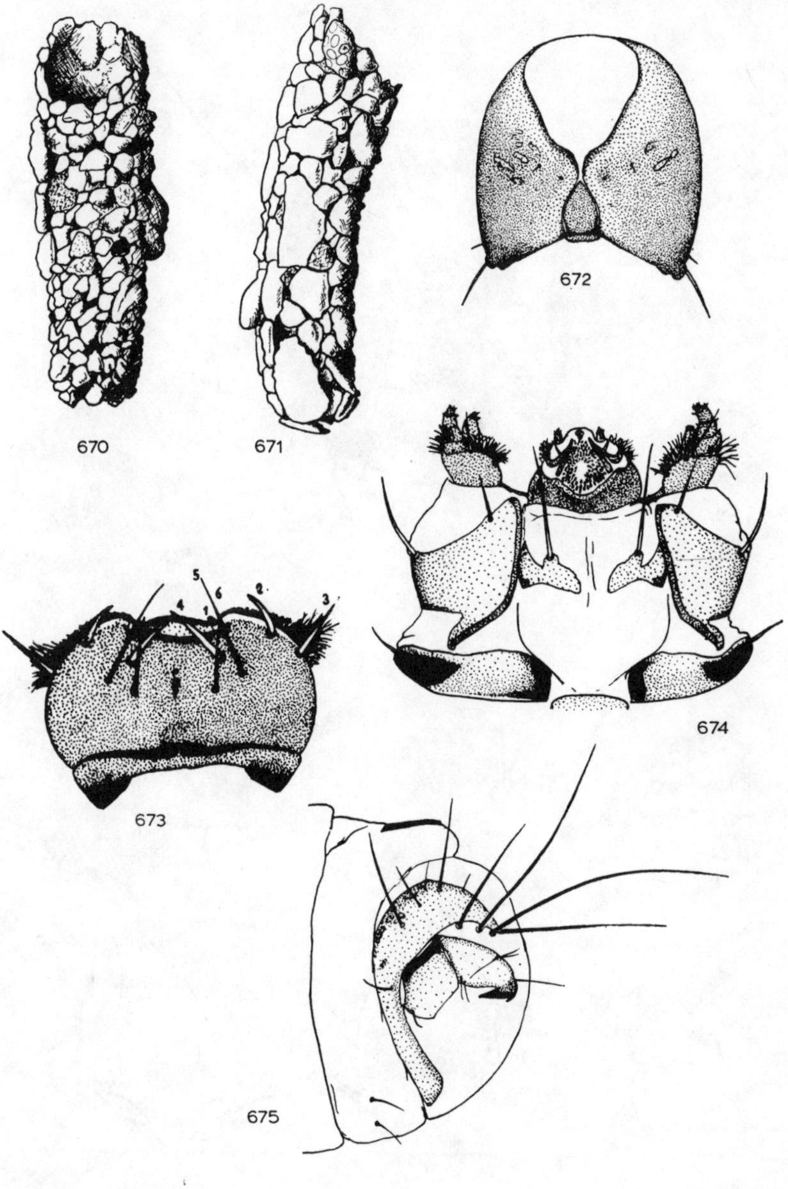

Figs. 670–675. *Melampophylax mucoreus* (Hagen). 670. Case of last stage larva from below: 671. Case of last stage larva from the side; 672. Head from below; 673. Labrum dorsal view; 674. Maxillae and labium ventral view; 675. Posterior segments of abdomen lateral view. Frochot.

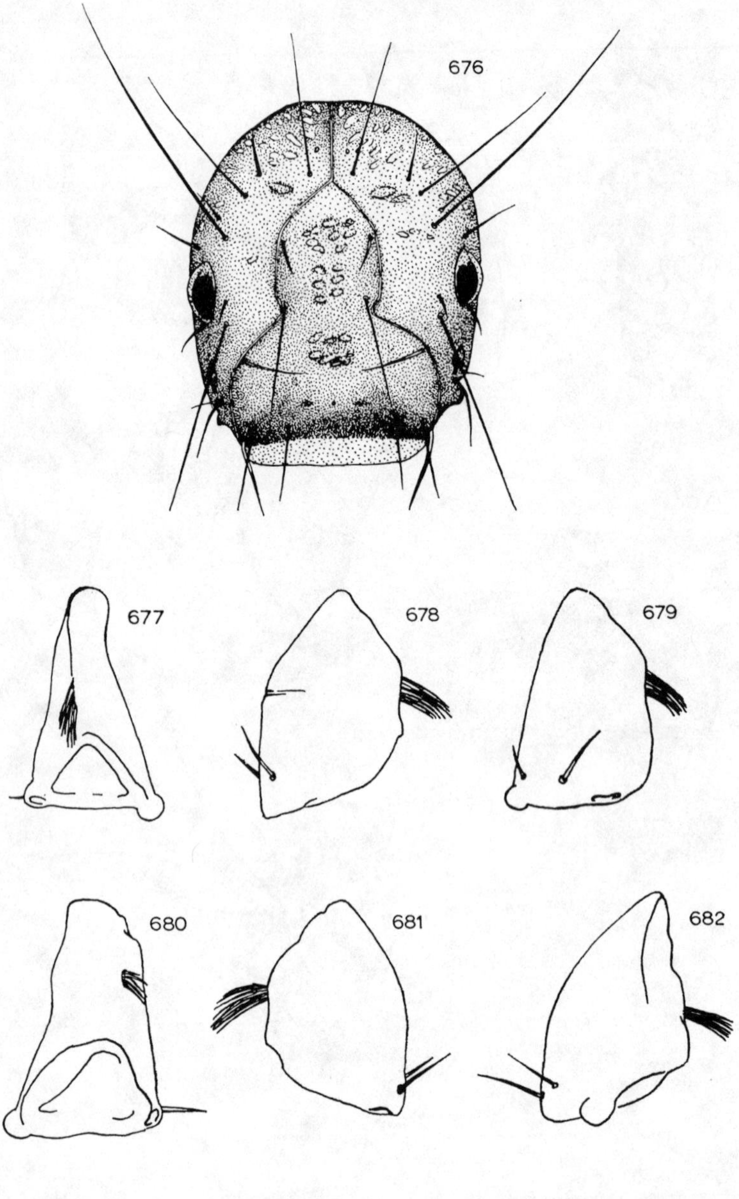

Figs. 676–682. *Melampophylax mucoreus* (Hagen). 676. Head from the front; 677. Left mandible internal face; 678. Left mandible anterior face; 679. Left mandible external face; 680. Right mandible internal face; 681. Right mandible anterior face; 682. Right mandible posterior face. Frochot.

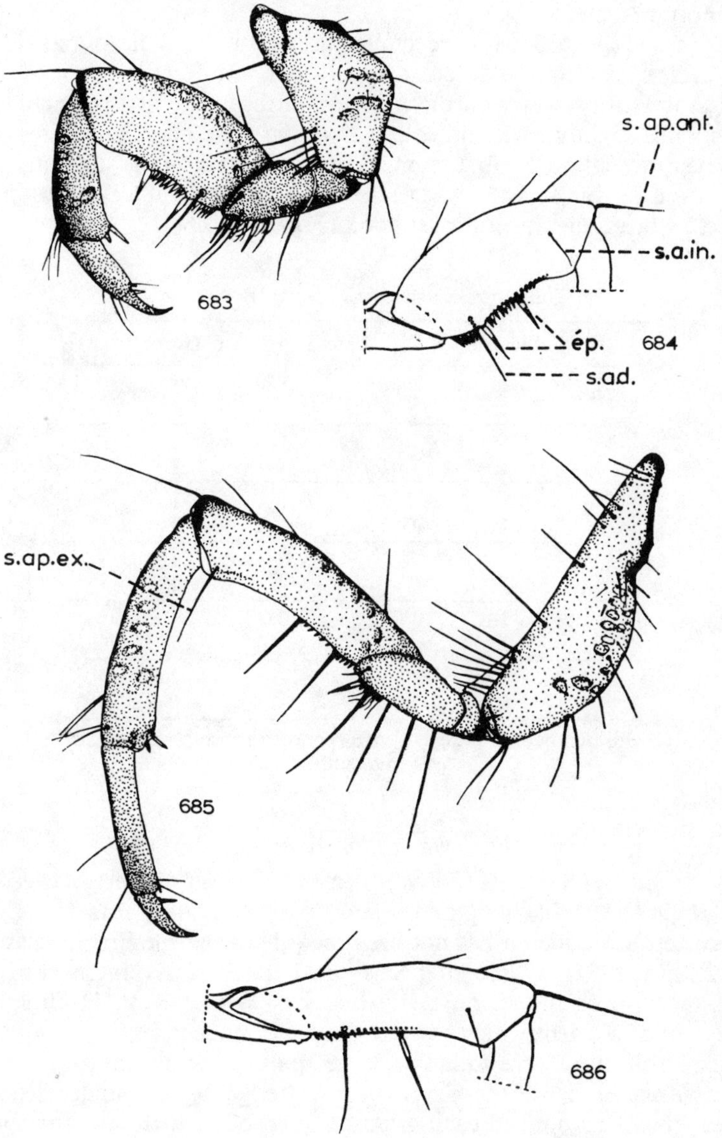

Figs. 683–686. *Melampophylax mucoreus* (Hagen). 683. Left prothoracic leg external face; 684. Femur of prothoracic leg internal face; 685. Left mesothoracic leg external face; 686. Femur of left mesothoracic leg internal face. Frochot.

The pronotum and mesonotum are of the same colour (Fig. 666) as the head except for a large light yellow mark at the posterior corners of the mesonotum.

The legs (Figs. 683–688) are similar in colour to the head and thoracic nota and, excepting the anterior apical hair, the femurs bear only three (more rarely four) hairs on the anterior surface. The first segment of the abdomen is slightly darker in colour than the remainder. The filiform gills are never present on the first segment but are found on the sixth and sometimes even on the seventh where there is some individual variation. The anal claw bears one large dorsal hook (Figs. 667, 668).

Gill arrangement. *Melampophylax mucoreus*

	Ventral	Lateral		Dorsal
2	1		1 (0)	1
	1	1		1
3	1		1	1
	1	1		1
4	1	1	1	1
	1	(0) 1		1
5	1		0 (1)	1
	1	(1) 0		1
6	1		0 (1)	1
	1			1
7	1 (0)			1 (0)
	0 (1)			0 (1)

Gill numbers in brackets correspond to those observed less frequently.

Enoicyla pusilla (Burmeister)

Enoicyla pusilla (Burmeister) was collected a number of times between 1868 and 1879 by McLachlan and Fletcher in localities near the City of Worcester. Since then it has not been recorded from the British Isles until, on 22 April 1957, my daughter Sari and I collected several larvae from a small meadow on the Worcestershire edge of Wyre Forest. The first larvae were found in a tent which my daughter had pitched in the meadow, but subsequently they were found in large numbers at the bases of tufts of grass within 10 feet of the hedge, or forest edge, usually under dead oak-leaves which surrounded each grass tuft and often within mouse tunnels. When the particular habitat had been recognized the larvae were found very easily—about 200 being observed within an hour.

Enoicyla pusilla is unique among the British Trichoptera in possessing terrestrial larvae and pupae, and in the adult female being apterous. Moreover, the genus *Enoicyla* is unique among the world Trichoptera in

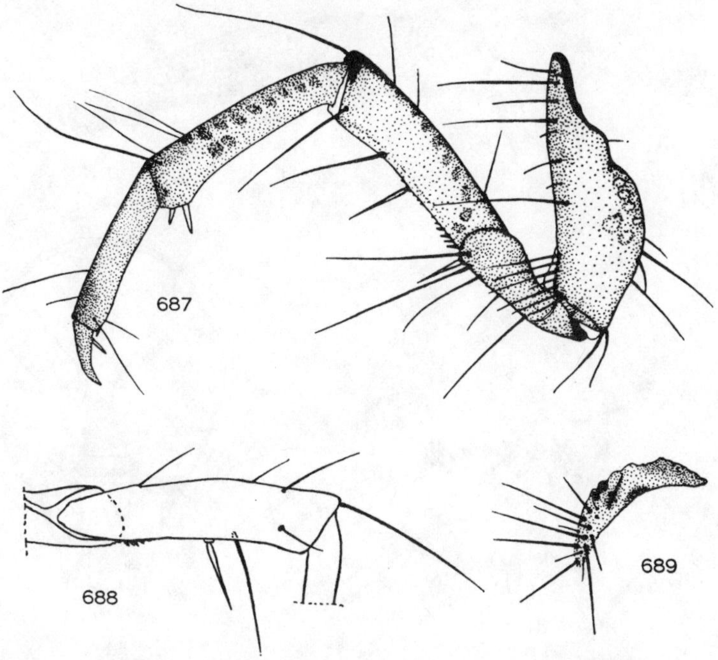

Figs. 687–689. *Melampophylax mucoreus* (Hagen). 687. Left metathoracic leg external face; 688. Femur of left metathoracic leg internal face; 689. Left metathoracic spur. Frochot.

possessing these characteristics. *E. pusilla* appears to have an erratic and discontinuous distribution in Europe, and there are two other species, *E. reichenbachi* Kolenati and *E. costae* McLachlan, not so far found in the British Isles. Döhler has given an account of the genus and reviewed the literature.

Larva (Fig. 690). To the naked eye an apparently typical Limnephilid larva; up to 7.0 mm. in length and 1.1 mm. in width; head and pronotum chestnut-brown, legs light yellowish-brown, rest of body greyish-white.

Case. Up to 10 mm. in length and 2 mm. in width; composed of sand-grains with small pieces of vegetable debris (fragments of dead leaves), conical and arched.

Head (Figs. 691 and 692). Aboral region covered with spicules, from each of which an extremely fine hair arises; cervical foramen encircled by a deep furrow, more apparent when head is viewed from below; genae widely separated at the ventral suture.

Labrum (Fig. 693). With light golden-brown areas between the dark brown margins and darker central area.

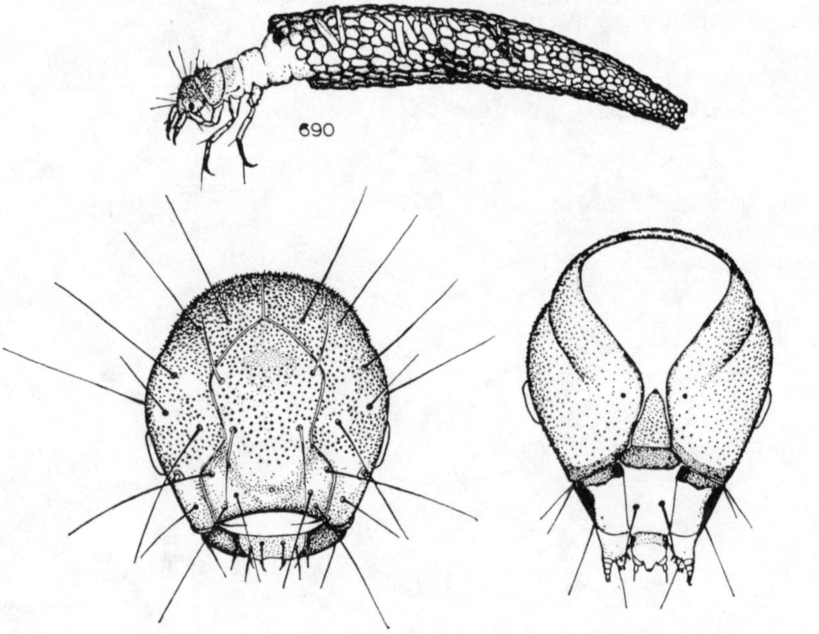

691 692

Figs. 690–692. *Enoicyla pusilla* (Burmeister). 690. Larva in case; 691. Head from the front; 692. Head from behind.

Mandibles (Fig. 694). With internal brush of hair and external pair of bristles present on each.

Pronotum (Fig. 695). Spicules, similar to those on the head, present in a transverse band; pattern of dark marks as shown in figure; prosternal horn absent.

Legs (Figs. 696 and 697). The continuous series of spines on ventral margin of tibiae, present in most LIMNEPHILIDAE, absent.

Abdomen. Dorsal protuberance on first abdominal segment, lateral protuberances not prominent; gills and lateral tubercles absent; two stout spines present, each arising from a disc-like base on central surface of each abdominal segment; anal sclerite not well defined, but numerous long stout black spines present, occurring also on the anal lobes; anal claw (Fig. 698) small, auxiliary claw absent; black mark on basal lobe.

The food and the general biology of the terrestrial limnephilid *Enoicyla pusilla* calls for special mention. This has been studied by Rathjen, Kelner-Pillault and Vander Drift & Witkamp. The adults emerge during September, October and November, and whilst the male lives for about two weeks, the apterous females live for only about five days. Fifty eggs, in a gelatinous mass, are laid by the female shortly after fertilization and the eggs hatch in about three weeks. Immediately on hatching the larva commences the construction of its case which at the end of the first day is about 1.5 mm. in length. Vander Drift & Witkamp state that the average length of the case has increased to 3.3 mm. in January to 4.5 mm. in March, and in June the larva is fully grown with an average case length of 6.7 mm. Most growth of the larva occurs during the months of March, April and May, and the larva feeds on the softer tissues of dead oak-leaves, mosses and algae, moving around in the litter layer and able to pull the case along by utilizing the tunnels made by mice and voles around the bases of old grass tufts.

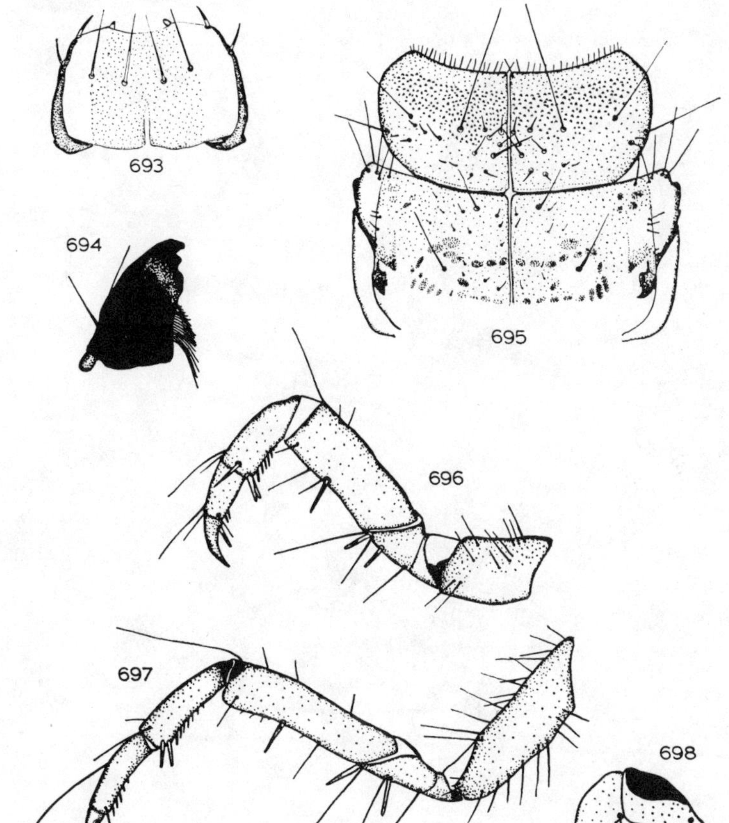

Figs. 693–698. *Enoicyla pusilla* **(Burmeister). 693. Labrum; 694. Left mandible; 695. Pro- and mesonotum; 696. Prothoracic leg; 697. Mesothoracic leg; 698. Anal claw.**

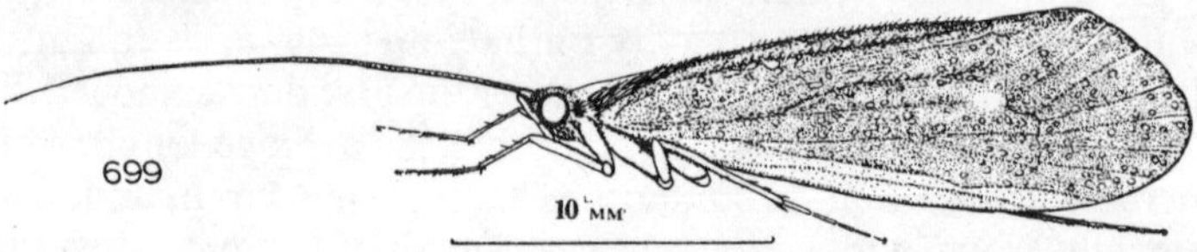

699

700

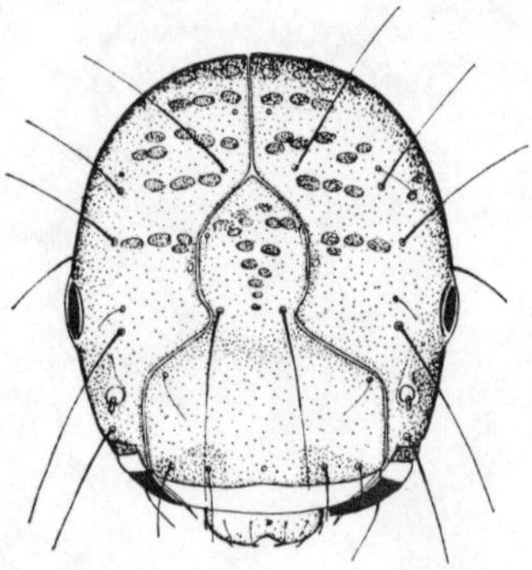

701

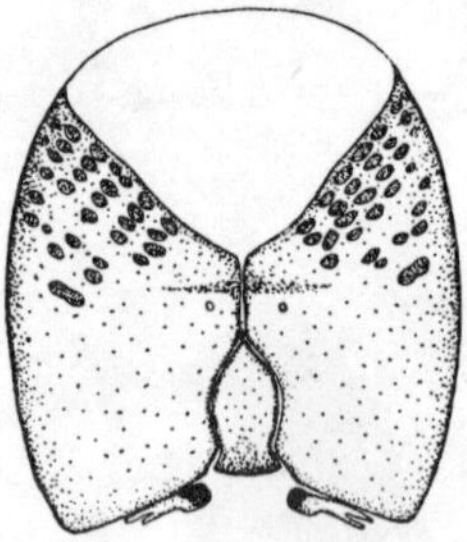

702

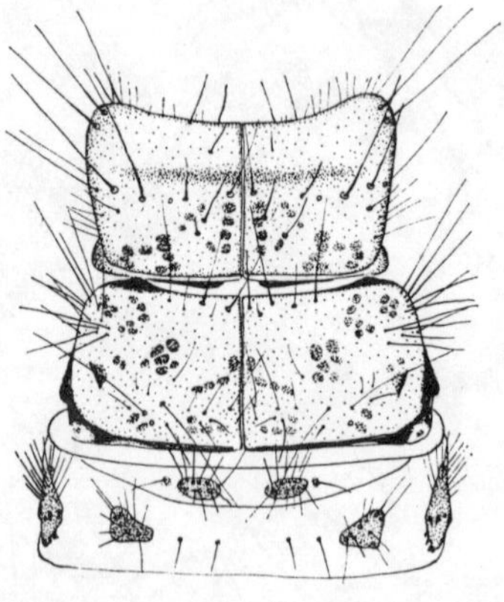

703

Figs. 699–703. *Stenophylax lateralis* (Stephens). 699. Adult; 700. Larva in case from the side; 701. Head of larva from the front; 702. Head of larva from behind; 703. Thoracic nota of larva.

At the end of June a silken wall is spun across the posterior aperture of the case and the larva then burrows into the litter layer and soil. The anterior aperture is then closed with silk and pupation takes place during August and September. The pupal stage lasts about a month when the pupa cuts a hole in the silken plug in the anterior end and crawls to the surface.

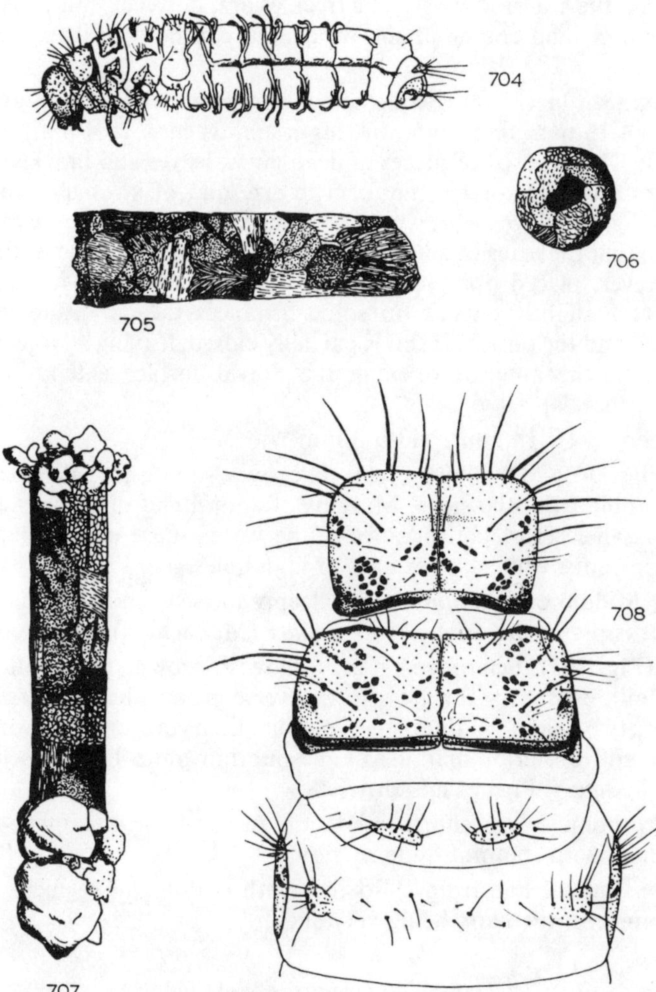

Figs. 704–708. *Stenophylax permistus* McLachlan. 704. Lateral view of larva; 705. Larval case; 706. Hind view of larval case; 707. Larval case at pupation; 708. Thoracic nota from above. Frochot.

Stenophylax lateralis (Stephens)

Earlier in this chapter, describing the larvae of *Limnephilus centralis* Curtis, I gave details of my finding the larvae of *Stenophylax lateralis* (Stephens) at Chartley Moss (Anglesea Coppice, Stowe, Staffordshire) with Mr. James Edwards, on 6 April 1953. From larvae then collected and maintained in an aquarium, a female *Stenophylax lateralis* emerged on 25 May, Mr. Kimmins identifying it.

Mosely, in his description of the adult (Fig. 699), refers to the 'brown freckles' on the anterior wings. The freckles are, however, much paler than the general ground colour of the wings and can be described as 'cream-yellow'.

Case. Eighteen larval cases collected are very similar in appearance, and range from 16 mm. to 19 mm. in length and 3½ mm. to 4 mm. in width. (Fig. 700). They consist of pieces of dead grass, leaves and bracken fronds; the pieces are more or less uniform in size and of strongly contrasting colours—light yellowish brown to very dark brown, all the pieces fitting closely so that the exterior surface of the case is smooth. Some of the pieces are, however, placed obliquely. The case is circular in cross-section and is only very slightly curved (in some instances the curvature is hardly apparent), and the posterior end is partially closed, leaving a large irregular opening. At the anterior opening the dorsal surface extends over the ventral surface.

Larva. Length 16–18 mm., width 3 mm.

Head (Figs. 701–702). Orange chestnut, eyes situate in pale areas more readily visible from the sides. Margins of aboral end of clypeus approximately parallel before convergence to the vertex. Base of antennae white, and conspicuous white area at base of mandibles.

Labrum. Middle of anterior margin deeply incised, and the small lateral dark marks present in *Stenophylax sequax* (McLachlan) are absent.

Thorax (Fig. 703). Sclerotized parts yellowish-brown, pronotum of uniform colour, except for the greyish transverse groove, but in *S. sequax* the posterior two-thirds is darker in colour than the anterior one-third. Anterior and posterior margins of mesonotum more heavily sclerotized than in *S. sequax*. Prosternal horn yellow.

Legs. Prothoracic legs short and stout, meso- long and meta- slightly longer than meso- but not quite so stout.

Abdomen. Lateral line from third to eighth abdominal segment but not very prominent, the hairs being very short.

Stenophylax permistus McLachlan

The larva of this widely distributed species has not been described from British material. Until recently it was unknown but Frochot has described it from larvae obtained from the Dijon area. They were found in small, fast-flowing forest streams carpeted with dead leaves on which the larvae subsisted as well as constructing their cases of them. The streams were shallow, received very little light and supported no vegetation.

The larval case (Figs. 705–707) is straight, of circular cross-section, and slowly tapering. The rather irregular anterior opening is perpendicular and the posterior opening is rounded and pierced by a small irregular orifice. The average length of the case was in the region of 17 to 18 mm. and the width was 4 to 5 mm. at the front. In addition to leaves, as already mentioned, small pieces of bark and other vegetable material are used. This

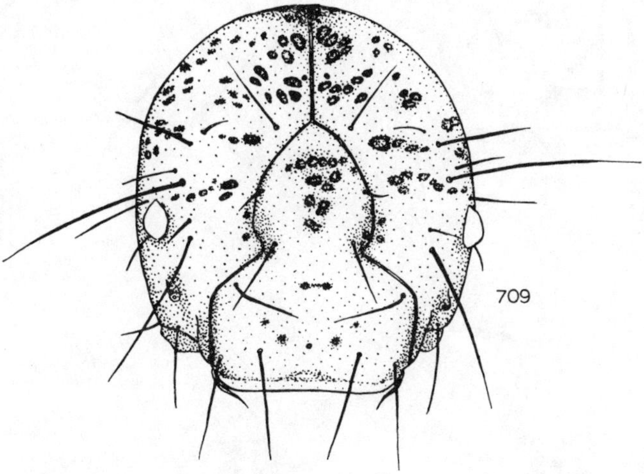
709

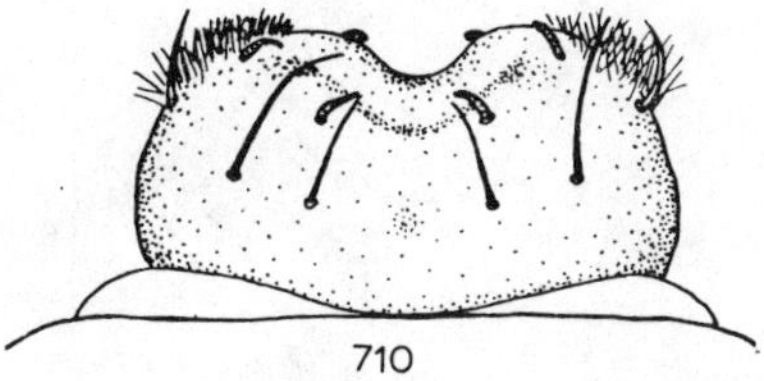
710

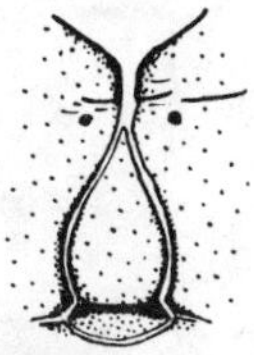
711

712

713

Figs. 709–713. *Stenophylax permistus* McLachlan. 709. Head from the front; 710. Labrum from above; 711. Gular sclerite (Submentum) and adjoining areas of the genae; 712. Left mandible anterior face; 713. Left mandible internal face. Frochot.

has the effect of giving the case a very dark appearance. In constructing the case, the larva sets the pieces of material end to end so precisely that the surface of the case is very smooth. The larvae break the leaves up into small pieces before utilizing them. Just prior to pupation the larva burrows vertically into the stream bottom and fixes small stones or plugs of other material to the two ends. It then spins a web-like membrane of silk at each end at the inside of each plug. The membranes have a number (often about twenty) of small orifices.

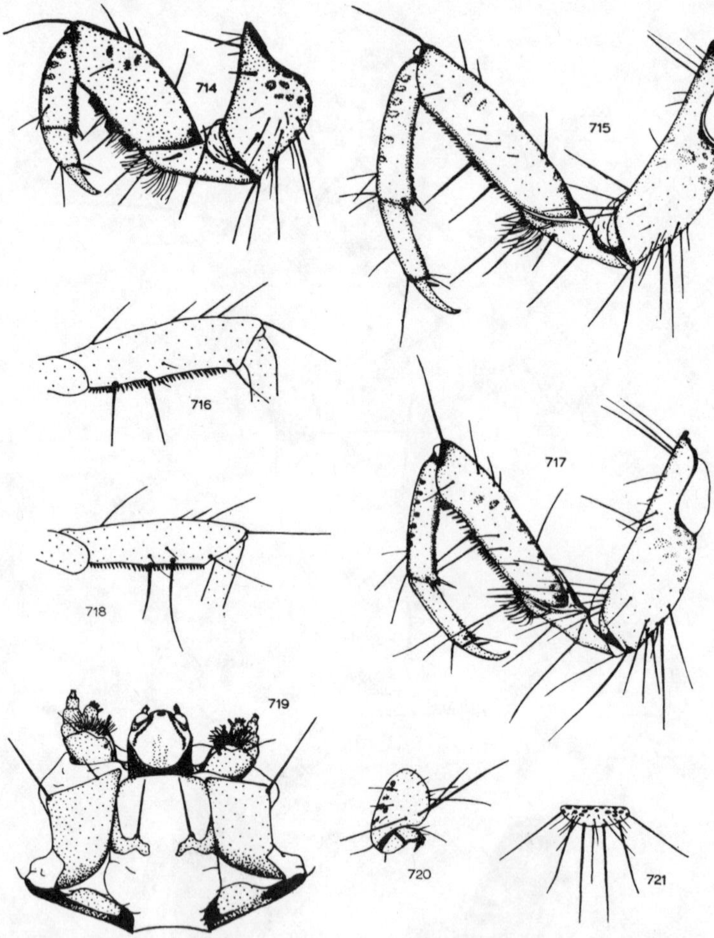

Figs. 714–721. *Stenophylax permistus* McLachlan. 714. Right prothoracic leg internal face; 715. Right mesothoracic leg internal face; 716. Femur of the right mesothoracic leg external face; 717. Right metathoracic leg internal face; 718. Femur of the right metathoracic leg external face; 719. Maxillae and labium from below; 720. Left anal appendage external face; 721. Sclerite on ninth abdominal segment. Frochot.

The length of the larva (Fig. 704) averages 20 mm. The head is relatively large compared with *S. lateralis* and *S. sequax*. The sclerotized parts are fairly pale reddish brown in colour and although the usual dark brown marks can be seen common to the LIMNEPHILIDAE there appear to be no characteristic patterns. The hairs of the head (Fig. 709) follow the usual pattern but many are rather short. However, the hair arising just above and slightly behind the eye is well developed and is the largest of the cephalic hairs. The gular sclerite (Fig. 711) (the submentum of Fotius and Frochot) is of the same pale colour as the adjoining areas of the genae, whereas in *Stenophylax vibex* (Curt.) it is distinctly darker.

Each mandible (Figs. 712, 713) has four teeth, a small brush of yellow hairs on the internal surface and two well-separated hairs on the outer surface. Of the latter, the anterior is very much the larger.

The thorax (Fig. 708) follows fairly closely the general limnephilid pattern but it should be noticed that the two hairs on the sides of the median suture of the pronotum just behind the transverse furrow are fairly short, whereas they are long in *Stenophylax vibex*. The legs (Figs. 714–718), again, follow the general pattern.

In the abdomen the first segment is darker in colour and the very distinct lateral line extends from the beginning of the third segment to the end of the eighth. The gills are numerous but very variable. The second, third and fourth segments all bear six pairs of gills and the fifth to eighth segments bear less, but it is in these segments that variation exists. Usually, the dorsal series extend farther to the rear than the laterals. The eighth segment is more often lacking in gills, but not always so.

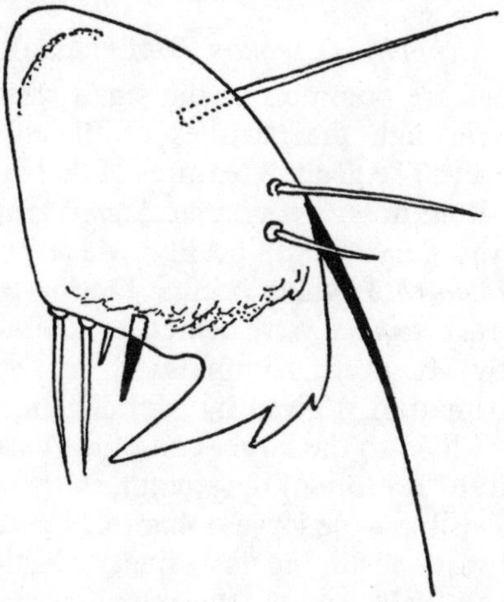

Fig. 722. *Stenophylax permistus* McLachlan. Left anal claw external face. Frochot.

The tergal plate of the ninth segment (Fig. 721) bears seven large black hairs on the posterior margin the two largest being those on each side of the median hair.

The anal appendages are well developed and three-segmented (Figs. 720–722). The basal segment is a large sclerotized plate with brown markings towards the front and about fifteen black, various sized hairs situated mostly towards the posterior margin. Two of the hairs are directed forwards and thus overlie the ninth segment. The second segment is short and bears a blackish band and a small black hair at its base. The third segment bears the anal claw which is furnished with an auxiliary hook and the latter sometimes but not always bears a further hook at its base. The chaetotaxy of the third segment of the anal appendage is shown in detail in Fig. 722.

Gill arrangement. *Stenophylax permistus*

	Ventral	Lateral		Dorsal
2	1		1	1
	1	1		1
3	1		1	1
	1	1		1
4	1		1	1
	1	1		1
5	1		0 (1)	1
	1	(0) 1		1
6	1			1
	1	0 (1)		1
7	1			1
	1			0 (1)
8				0 (1)

Stenophylax sequax (McLachlan)

Larvae of this species are common in the small streams which run in a southerly direction through the parishes of Bletchingley. Nutfield and Outwood in east Surrey. The general features of these streams have already been described but it is noteworthy that *Stenophylax sequax* occurs in abundance within the very small locality where the rare Psychomyid Caddis *Tinodes pallidula* McLachlan occurs. During the summer of 1950 a number of *Stenophylax sequax* were reared to the adult stage and their identity confirmed by Mr. D. E. Kimmins.

A note on case formation is given by McLachlan, who says that when the larger stones are added to the larger end of the case, the larva sinks the case into the slime at the bottom of the stream, so that only the stones used to close the case are visible—the larva doing this by turning in its case and digging a hole, afterwards resuming its ordinary position. I was not able to observe this in the Nutfield stream, the pupal cases being mostly found under flat stones on a rocky bottom.

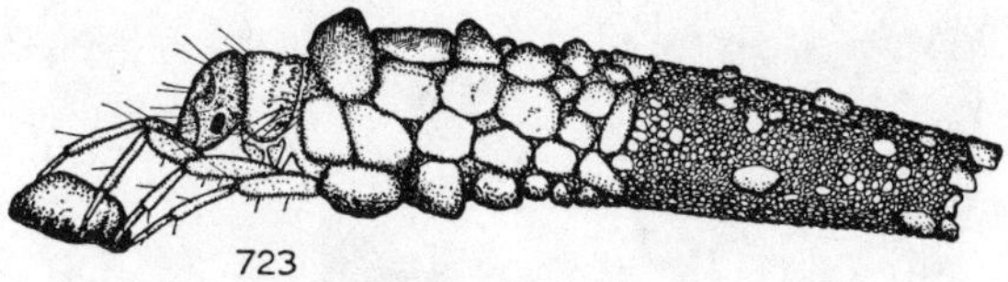
723

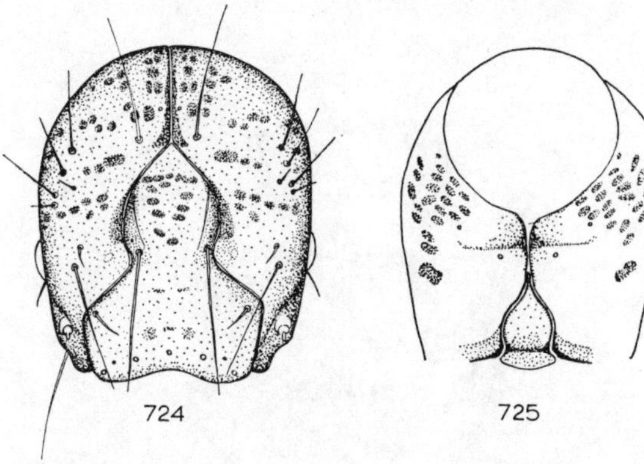
724 725

Figs. 723–725. *Stenophylax sequax* (McLachlan). 723. View of larva in case; 724. Head of larva from the front, labrum and mandibles not shown; 725. Head of larva from behind.

Larva. Very like larva of *Stenophylax* and *Halesus*. Length up to 19 mm. and width 3–4 mm. Appears to be always more heavily sclerotized than those species of related genera which I have seen. (This refers, of course, to the darkness of colour of the sclerotized areas and not to their extent or distribution.)

Case (Fig. 723). Of sand-grains at hinder-end and small stones at fore-end. Tapering, slightly curved and only very slightly flattened.

Head (Figs. 724, 735). Anterior margin of clypeus deeply emarginate. Surface of genae and clypeus papillate—this feature, commonly found in the LIMNEPHILIDAE, is pronounced in this species. The genae on the ventral surface thrown into a transverse fold lying between the pairs of bristle pits. *Gular sclerite* does not completely divide the genae. A number of 'muscle' spots prominent in this region. A pair of spots on each side contiguous and situated on oral side of transverse fold.

Labrum (Fig. 726). Varying from light to dark chestnut with inturned horns at base, black and with a small black streak at each side.

Mandibles (Fig. 725). Each mandible has brush of hairs on inner face and one large and one small bristle at base of outer face.

Maxillae. Chestnut coloured.

Labium. Heavily sclerotized—mostly light to dark chestnut colour (Fig.

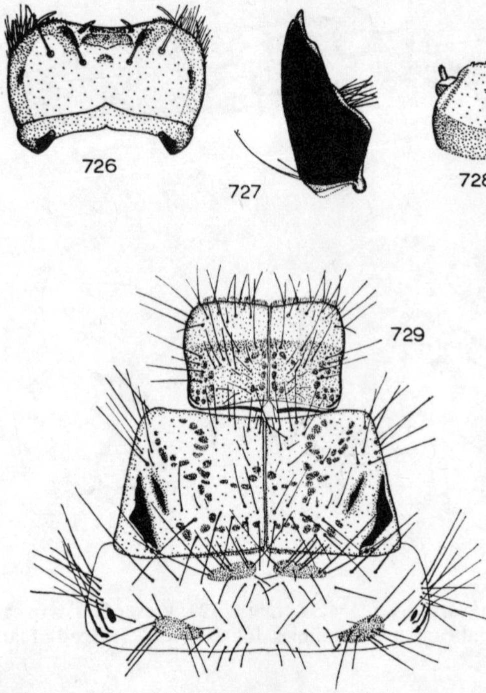

Figs. 726–729. *Stenophylax sequax* (McLachlan). 726. Labrum of larva; 727. Right mandible of larva from beneath; 728. Labium of larva from beneath; 729. Thoracic nota of larva.

728). Ring-like sclerite, dark chestnut in colour, envelopes the labium below the insertion of the labial palps. Palps of two segments, the proximal being bulbous, the distal cylindrical, both coloured light chestnut.

Prosternal horn. Strongly developed, chestnut coloured.

Anal claw. With a single auxiliary claw, the larger bristles set in reddish-brown raised pits.

Stenophylax vibex Curtis

This large light-coloured caddis fly was familiar to me for three years before I discovered its larvae. Each year from the beginning to the middle of June these caddis alighted on the lit windows of my home at Bletchingley, Surrey (550 feet a.s.l.), but the nearest stream which their larvae could inhabit is just over one mile away. This is the stream at Nutfield, inhabited by *Halesus digitatus*, *Plectrocnema conspersa*, *Tinodes pallidula* and other species of caddis, but not apparently by *Stenophylax vibex*. However, on 15 February 1948 I did collect a number of larvae, some of which subsequently emerged on 6 June and were determined as *S. vibex* (confirmed

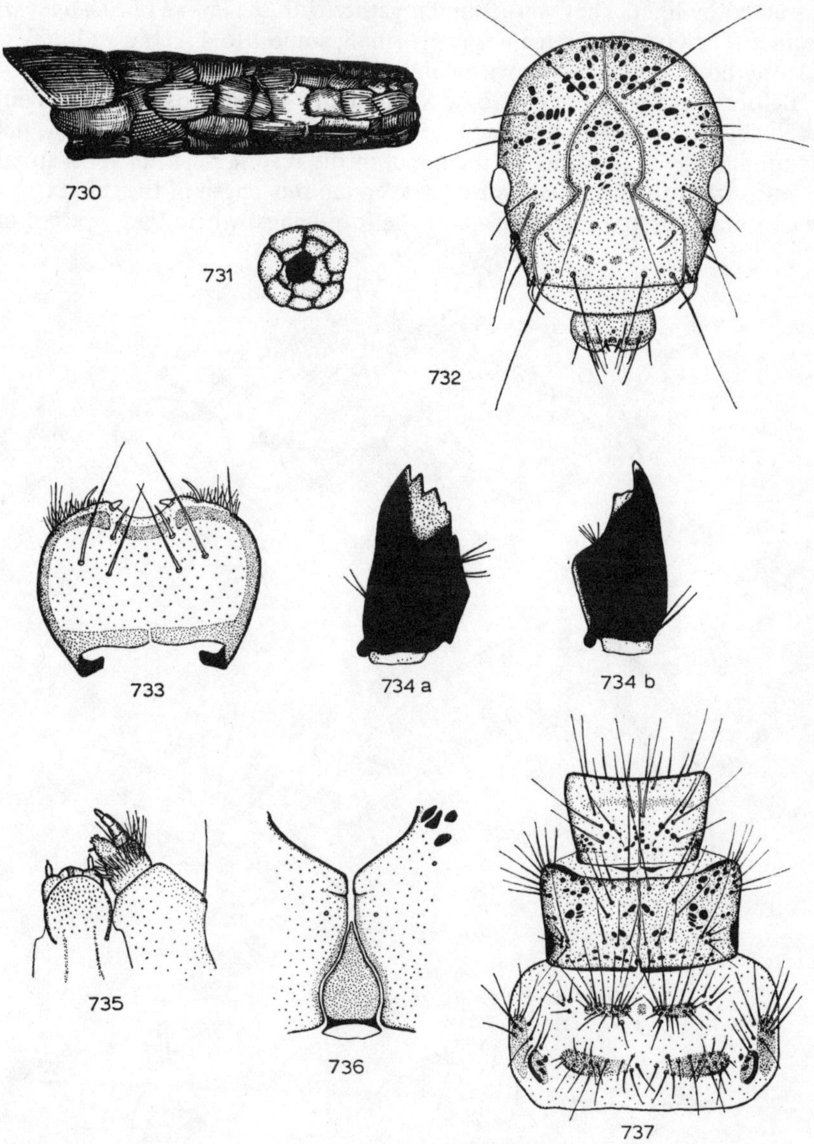

Figs. 730–737. *Stenophylax vibex*. 730. Larval case; 731. Posterior end of case; 732. Head; 733. Labrum from above; 734. a and b. mandible; 735. Labium and right maxilla; 736. Gular sclerite; 737. Thoracic nota.

by the late M. E. Mosely). These came from a stream near Outwood, Surrey, 220 feet a.s.l., which was just two miles from my home, to which the fly was attracted by light. They were found together with the larvae of *Stenophylax sequax*. The Outwood stream is fairly small, some 2 to 4 feet in width, with a stony bottom alternating with patches of clay.

Before describing the larvae of *S. vibex* one characteristic of the living adult should be mentioned, and that is the streak of light golden buff which dominates the appearance of the insect in the resting position. This streak extends from the head backwards, traversing the length of the thorax, and is continued by the hind margin of the fore-wings, where they overlap on

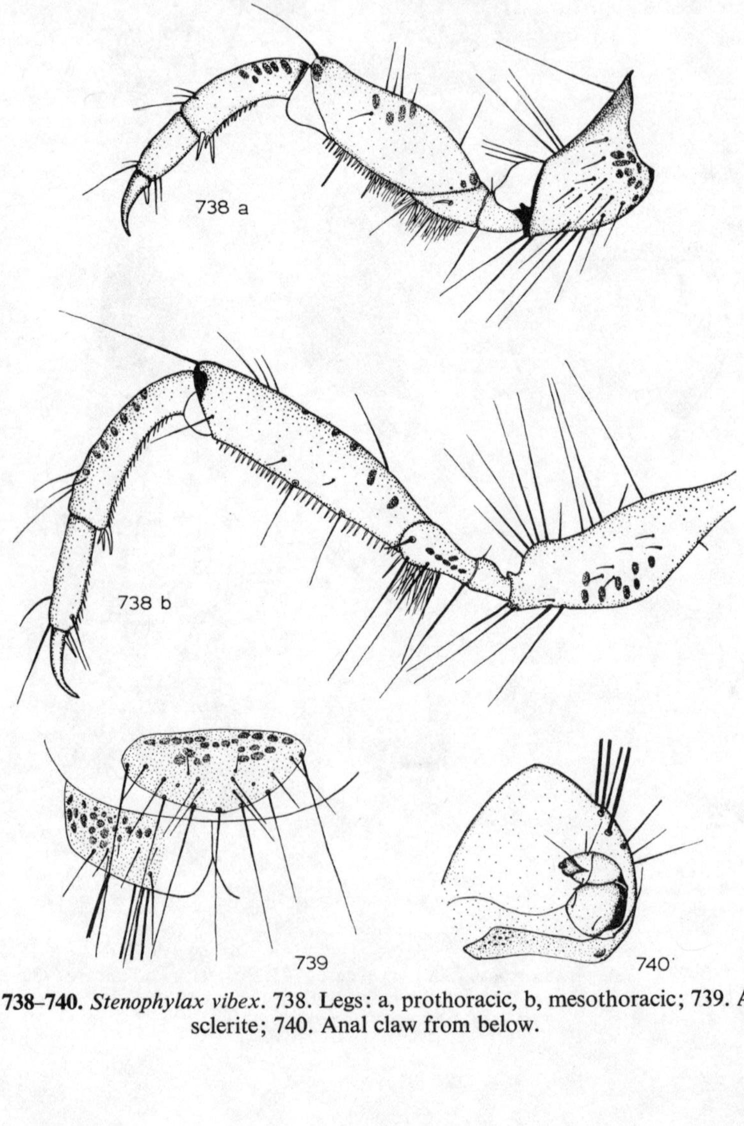

Figs. 738–740. *Stenophylax vibex*. 738. Legs: a, prothoracic, b, mesothoracic; 739. Anal sclerite; 740. Anal claw from below.

the dorsal surface of the abdomen. The streak appears to be a good example of disruptive coloration.

Case (Fig. 730). Made of plate-like portions of vegetable debris, such as parts of dead leaves with no projecting edges, brown or black in colour. Length up to 12.5 mm., conical, tapering in width from 4.8 mm. at anterior end to 3.5 mm. at posterior end. Posterior end rounded and partly closed, leaving small opening only (Fig. 731). No curvature of the case is apparent, nor is the ventral surface flattened.

Larva. Eruciform. Head orthognathous. Length 12.3 mm., width 4.0 mm.

Head (Fig. 732). Light brown shading to dark brown, with pattern of black spots (not usually symmetrical) forming transverse bands. Eyes well defined, antennae very small.

Labrum (Fig. 733). A deep excision on anterior margin; lateral margins and a small patch below the excision dark in colour.

Mandibles (Fig. 734 a and b). Four-toothed; each has a small brush of hairs on inner margin.

Maxilla (Fig. 735). Hairy at base, with outer distal part darker in colour and bearing a bristle which lies inside the four-segmented palp. Two thick light-coloured spines arise from the base of the mala.

Labium globular, with proximal segment of two-segmented labial palp globular.

Gular sclerite (Fig. 736). Not completely dividing the genae, together with the adjacent margins of the genae darker in colour than the ventral area of the genae. A small fold occurs at the ventral genal suture.

Thorax. Pro- and mesothoracic tergites sclerotized with pattern of dark spots as shown in Fig. 737. Pronotum constricted transversely near hind margin. Prosternal horn present. Metathoracic tergum partially sclerotized with bristles more numerous and larger than in *Potamophylax latipennis.* Prothoracic legs (Fig. 738) short; meso- (Fig. 738) and metathoracic legs long and similar; a few dark marks on the distal segment of the trochanter.

Abdomen. Greyish white, but first abdominal segment slightly darker and browner in colour, protuberances present on this segment but not large. A number of bristles present on the ventral surfaces of the first abdominal segment, each arising from a small dark circular patch. Filiform gills present. Anal sclerite (Fig. 305) with pattern (somewhat variable) along anterior margin; the basal lobe of the anal claw also with a pattern of dark marks. Anal claw (Fig. 306) with one auxiliary claw.

Mesophylax aspersus (Rambur)

The immature stages of this large, rare, cave-dwelling species are unknown. It is a Mediterranean species of which only three British specimens are known, one taken in 1883 and two in 1961. It is probable that it does not breed in Britain and that the specimens taken are involuntary immigrants.

Mesophylax impunctatus McLachlan

No description of British material has been made. The eruciform larva is from 15 to 16 mm. in length and 2–5 mm. in width. Head dark brown (a little reddish) but lighter on posterior half of clypeus where a cuneiform pattern of spots can be distinguished. Mandibles with five unequal teeth. Pronotum orange brown with X-shaped mark distinct but sometimes darker in colour. Transverse furrow distinct, blackish at the centre. Anterior third of pronotum same coloration as remainder. The median suture often margined with dark brown, which, with the transverse furrow,

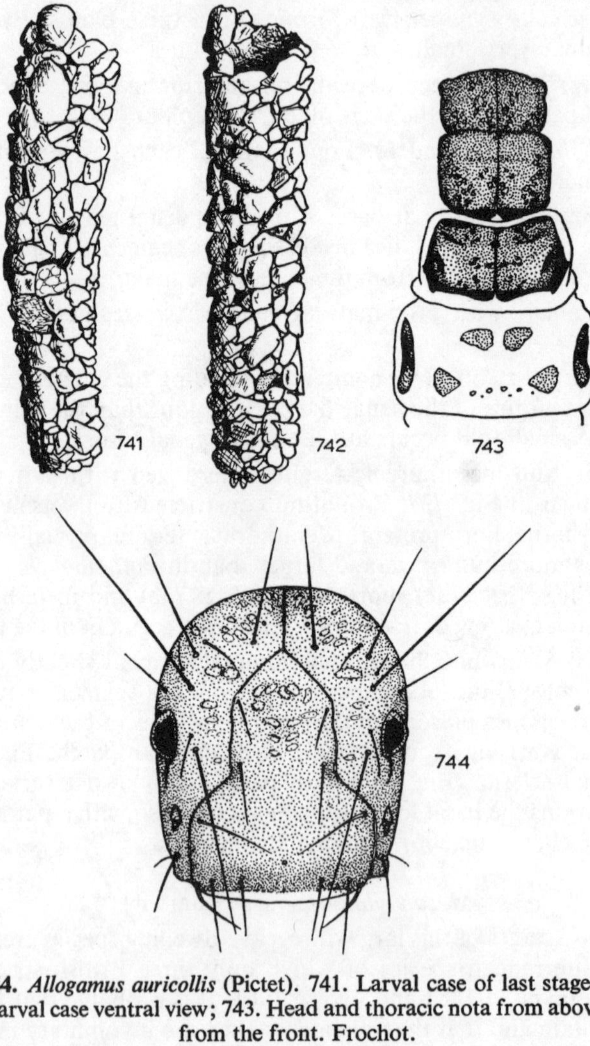

Figs. 741–744. *Allogamus auricollis* (Pictet). 741. Larval case of last stage larva lateral view; 742. Larval case ventral view; 743. Head and thoracic nota from above; 744. Head from the front. Frochot.

gives the appearance of a cross. Mesonotum brownish-orange, the median area darker than the sides. Anal claw with dorsal auxiliary claw. Case slightly curved and conical of mineral material but often with added vegetable material.

This rare species in Britain is thought to inhabit rather slow-moving water and drainage ditches in or in the vicinity of caves but has also been collected along the shores of Lake Windermere.

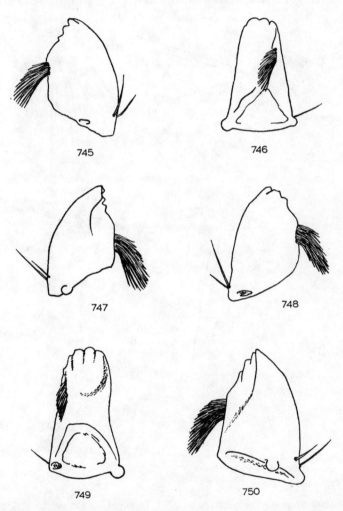

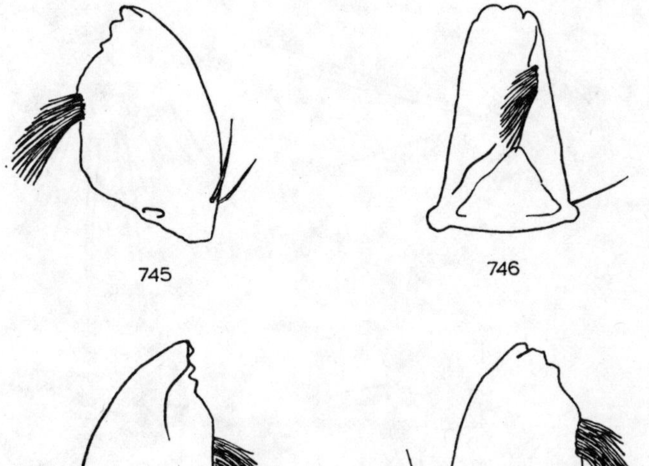

Figs. 745–750. *Allogamus auricollis* (Pictet). 745. Right mandible anterior face; 746. Right mandible internal face; 747. Right mandible posterior face; 748. Left mandible anterior face; 749. Left mandible internal face; 750. Left mandible posterior face. Frochot.

Allogamus auricollis (Pictet)

The larva of this species is not known from British material and was not described until 1963, when Frochot gave a description from larvae collected in the same localities as he found *Melampophylax mucoreus* (Hagen), i.e. cold, fast-flowing rivers in chalky areas.

The larval case (Figs. 741, 742) is rather long and slender, being 15–17 mm. in length and 3–4 mm. in width. It is only slightly arched and composed of flattish small stones or large sand-grains (averaging about half to three-quarters of a millimetre in diameter). The grains are arranged in an

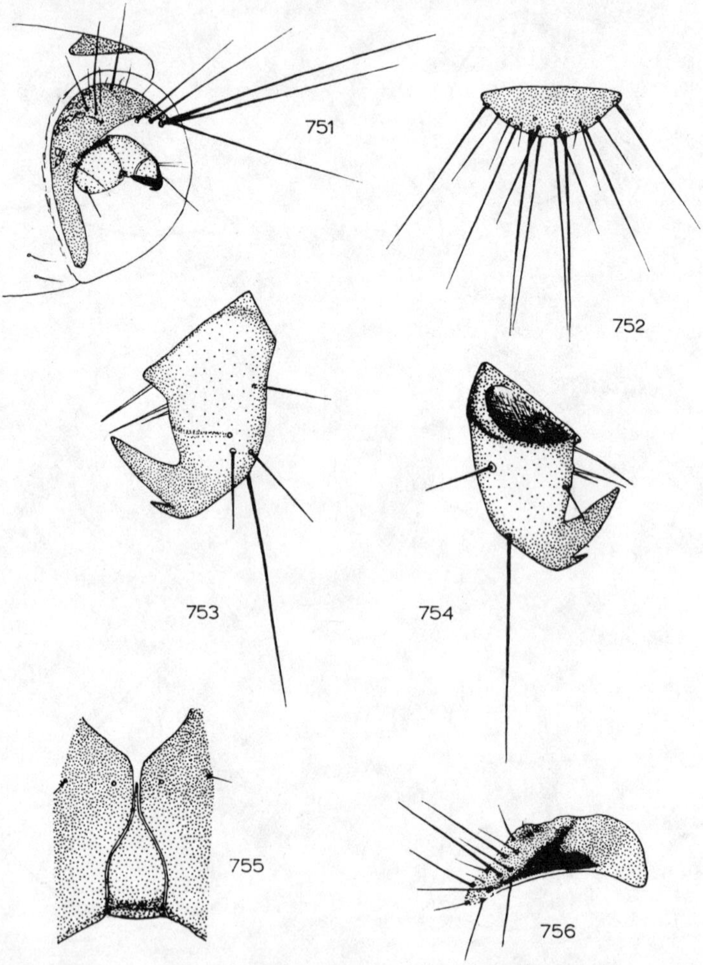

Figs. 751–756. *Allogamus auricollis* (Pictet). 751. Terminal segments of the abdomen from the side; 752. Sclerite on ninth abdominal segment; 753. Left anal claw external face; 754. Left anal claw internal face; 755. Gular sclerite (submentum) and adjacent areas of the genae; 756. Left metathoracic shield. Frochot.

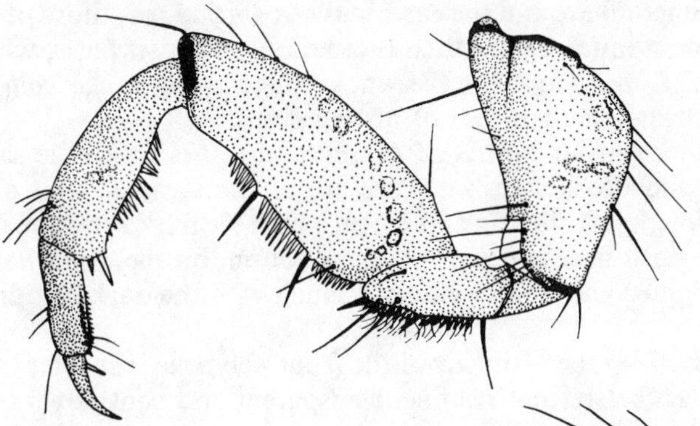

757

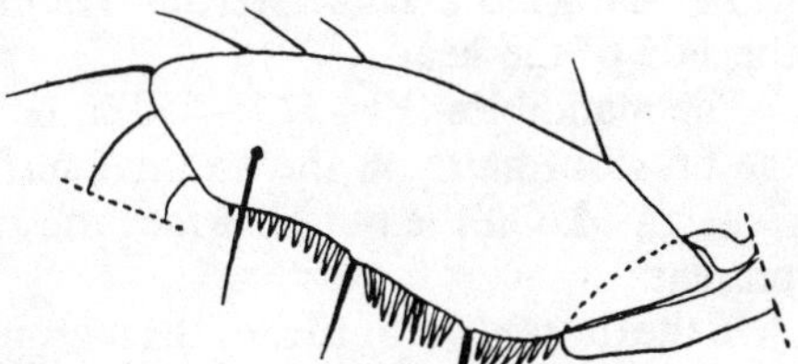

758

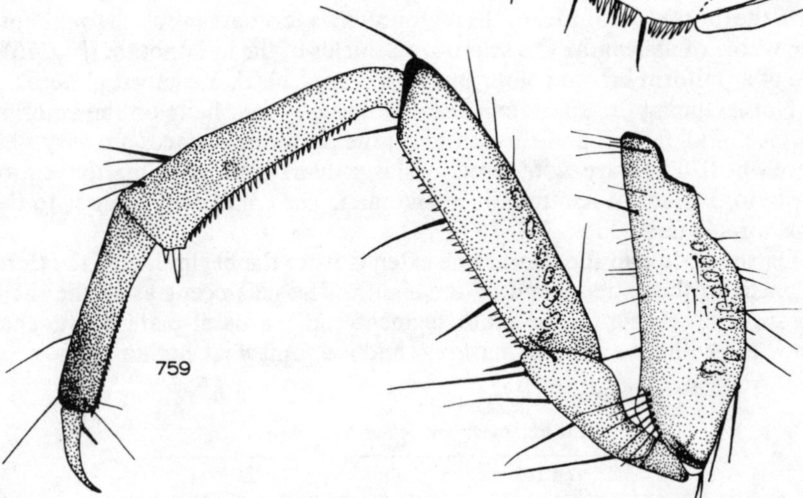

759

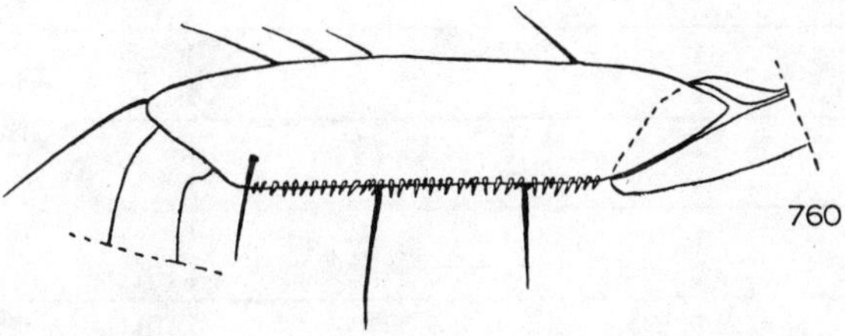

760

Figs. 757–760. *Allogamus auricollis* (Pictet). 757. Left prothoracic leg external face; 758. Femur of right prothoracic leg internal face; 759. Left mesothoracic leg external face; 760. Femur of left mesothoracic leg internal face. Frochot.

homogeneous manner all around the case, without sorting, i.e. those of the dorsal surface are not much bigger than those of the other surfaces, which is not the case in *M. mucoreus*. On the whole, the case of *A. auricollis* is finer and a little straighter than that of *M. mucoreus*.

The larva is rather similar to that of *M. mucoreus*. It is about the same size, 16–17 mm., and the sclerotized parts are of the same general dark brown colour although just slightly darker, on which the dark brown marks hardly show. The posterior angles of the mesonotum, on the other hand, are only slightly lighter and do not contrast much with the dark colour of the tergites.

On the head (Fig. 744) the fifth hair of the fronto-clypeus is shorter than in *M. mucoreus* and the sixteenth is fine, transparent and bent down over the side of the head.

The mandibles (Figs. 745–750) are furnished with four obtuse teeth and the brush of hairs on the internal surface is long and broad, being rather fan-shaped. On the outer surface the anterior is the larger of the two hairs present.

In the thorax, the border of the mesonotum is very dark in colour for almost the whole of its length. The sclerotized shields of the mesonotum (Fig. 756) are of a uniform brown colour but with a large black Y-shaped pattern.

Not including the apical hair there are usually four hairs on the anterior surface of all femurs and the combs on the posterior surfaces are very well developed. They are composed of large hairs and take up the entire posterior surface of femurs, tibiae and tarsi. The comb even extends to the prothoracic tarsi (Figs. 757–760).

On the abdomen the lateral line extends from the beginning of the third segment to the extreme end of the eighth. The gills occur as in the table below. The sclerite on the ninth segment and the basal plate of the anal claws are without any clear markings and are somewhat similar to those of *M. mucoreus*.

Gill arrangement. *Allogamus auricollis*

	Ventral	Lateral		Dorsal
2	1		1	1
	1	1		1
3	1		1	1
	1	1		1
4	1		1	1
	1	1		1
5	1			1
	1			1 (0)
6	1			1 (0)
	1			
7	1			0 (1)
	1			

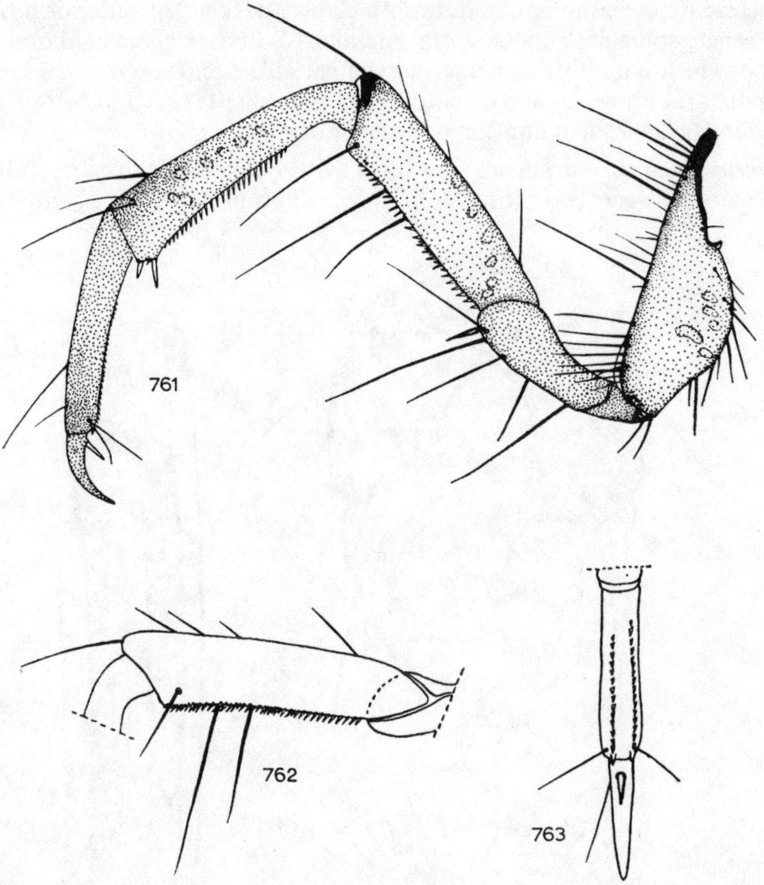

Figs. 761–763. *Allogamus auricollis* (Pictet). 761. Left metathoracic leg external face; 762. Femur of right metathoracic leg internal face; 763. Tarsus of left prothoracic leg posterior face. Frochot.

Hydatophylax infumatus (McLachlan)

Hanna gives the following account obtained from larvae collected near Manchester.

Case slightly curved ventrally up to 20 mm. long and 7 mm. wide. Dead vegetable material with three or, less commonly, four stalks attached along whole length of case. Stalks up to 40 mm. in length, while Ulmer (1909) has recorded them up to 60 mm. in length. All cases examined were cylindrical. Posterior opening of case smaller than anterior.

Larva eruciform. Larvae examined measured up to 23 mm. long and 4.5 to 5 mm. wide. Ulmer recorded up to 25 mm.

Head hypognathous, ovoid. Fronto-clypeus darker than anterior surfaces of genae, some dark spots at oral end and 10–13 dark spots at aboral end. Genae bear a number of black spots of variable size. Posteriorly, gena has lighter area above labium on each side. Gular sclerite chestnut brown, does not separate genae completely. Genal suture open.

Labrum. Ventral margin has slight concavity, on each side five setae. Group of hairs between two most lateral setae. Labrum has brown band which

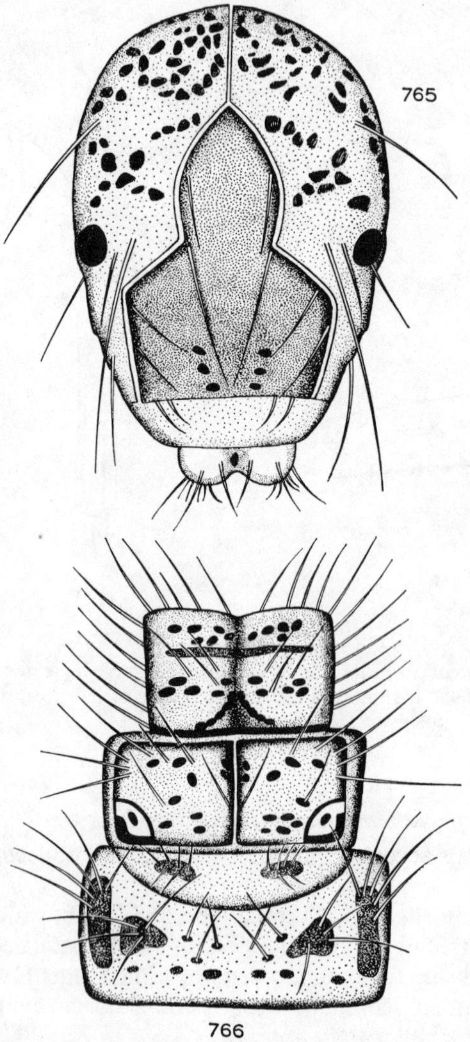

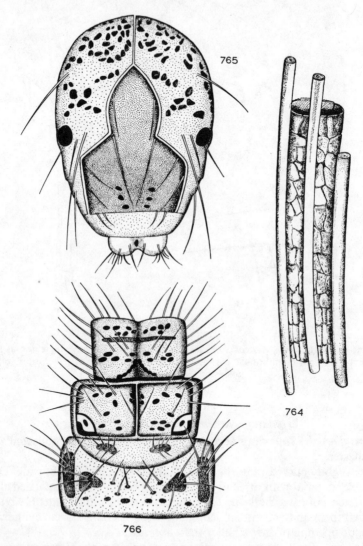

Figs. 764–766. *Hydatophylax infumatus* (McLachlan). 764. Larval case; 765. Head from the front; 766. Thoracic nota from above. Hanna.

carries a central black spot and three dark patches at ventral margin. Lateral margins of labrum heavily sclerotized.

Mandible. Bears four teeth, brush on inner surface and single seta on outer surface near base.

Maxilla. Cardo small and bears single seta. Stipes bears long lateral and short median seta. Maxillary palp has five segments. Lacinia carries few sensillae and many hairs.

Thorax. Pronotum entirely sclerotized and has median longitudinal suture,

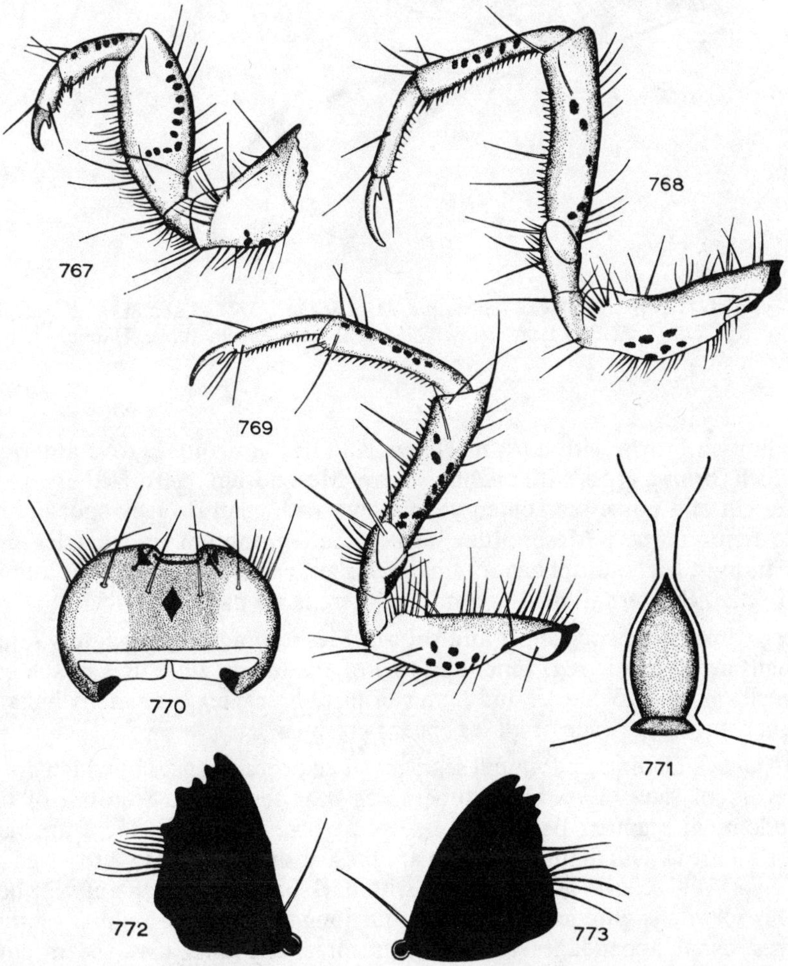

Figs. 767–773. *Hydatophylax infumatus* (McLachlan). 767. Prothoracic leg; 768. Mesothoracic leg; 769. Metathoracic leg; 770. Labrum; 771. Gular sclerite; 772. Left mandible; 773. Right mandible. Hanna.

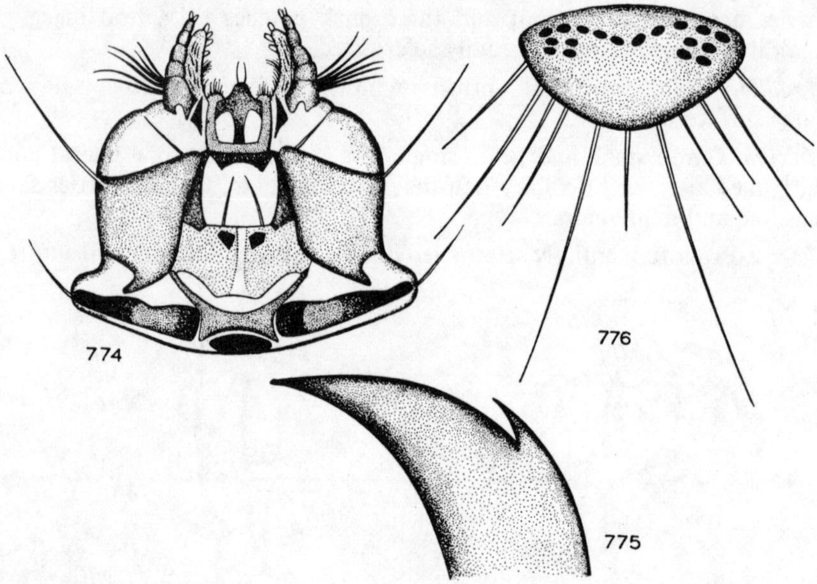

Figs. 774–776. *Hydatophylax infumatus* (McLachlan). 774. Labium and maxillae; 775. Anal claw from below; 776. Anal sclerite from above. Hanna.

yellowish brown with dark markings, and has horizontal grove anteriorly which forms a cross with median suture. Mesonotum sclerotized except for margin and possesses golden yellow area with central black spot at each posterior corner. Mesonotum darker than pronotum and carries dark markings. Metanotum has six small sclerites, each of which carries number of setae. Prosternal horn present but there is no prosternal sclerite.

Legs. Prothoracic leg short and robust. Mesothoracic leg is slightly longer than metathoracic leg. Inner surfaces of the femur, tibia and tarsus bear small spines. Trochanter and femur of prothoracic leg bear many hairs on inner surfaces. Tibiae of all legs bear two spurs.

Abdomen. On first abdominal segment three protuberances of which dorsal devoid of setae. Lateral protuberances bear few setae. Sternum of first abdominal segment bears many setae at bases of which there are dark brown areas. Gill filaments which are present on segments two to seven are single. Anal sclerite golden yellow with dark brown spots on anterior half. This sclerite is elliptical and bears four long setae and a number of short ones. Anal appendages are two-segmented and anal claw bears small auxiliary claw at base. Lateral line running from segments three to seven formed of fine hairs. Above lateral line is line of seven to nine sclerotized pustules on third to seventh abdominal segments.

Black spots on anterior surfaces of genae of larva of *Hydatophylax infumatus* are much larger than those on the genae of the other species of *Hydatophylax* and *Stenophylax*, and therefore may be easily confused with those of *Halesus*. Ulmer (1909) pointed out that the chitinous shields of the pronotum are distinctly separated in *Halesus*, while they are very close in *H. infumatus*. Ulmer and Lestage stated that the presegmental gills were absent on the second abdominal segment in *Hydatophylax infumatus*, while they were present in *Halesus*, but on the other hand they were present on most of the specimens of *H. infumatus* here examined.

Chaetopteryx villosa (Fabricius)

Larvae of this widely spread caddis fly were found to be common in 1945 and 1946 in a small stream at Nutfield, Surrey. This stream flows with moderate speed, and is characterized by having thick marginal and overhanging vegetation. The larvae were more abundant where the stream was less turbulent, and where the stream bottom alternated with large flat stones and patches of mud. The depth of water was about 8 inches, and larvae of *Halesus digitatus* (Schrank) were found commonly in the same habitat. Larvae of *Plectrocnemia conspersa* (Curtis) were also very common and constructed their nets in the more turbulent portions of the stream where it rippled over the stones.

Two of the larvae of *Chaetopteryx villosa* were reared to the pupal stage, and by dissecting away the pupal integument the underlying imaginal genitalia were made out.

This species was reared also in 1937 from larvae collected from a small stream—scarcely more than a trickle—in Broadmoor Wood, Rubery, Worcestershire, and also from larvae collected from a small stream in Painswick, Gloucestershire, in 1937.

Case (Fig. 777). Of large sand-grains and small chips of stone, slightly arched, up to 20 mm. long and up to 4 mm. wide. It should be noted, however, that adults of this species show a wide variation in size; such a disparity should be seen in the full-fed larvae. No satisfactory explanation has so far been put forward to account for this phenomenon. Other materials for case-making are said, on occasion, to be used—vegetable debris, cut roots, leaves and snail-shells, longitudinally or obliquely placed.

Larva. Eruciform, up to 15 mm. long and up to 3 mm. wide.

Head (Fig. 778). Orthognathous, dark golden brown with darker brown markings, eyes situated laterally, antennae small. Concavity between oral and aboral parts of the clypeus fairly deep. Clypeus mark triangular consisting of about twelve dark brown marks. In the region of the concavities a group of four dark marks is situated in two pairs and in the oral region a further four dark marks extend transversely. In the genae a wide dark band extends on either side of the clypeus, joining dorsally by the mid-genal

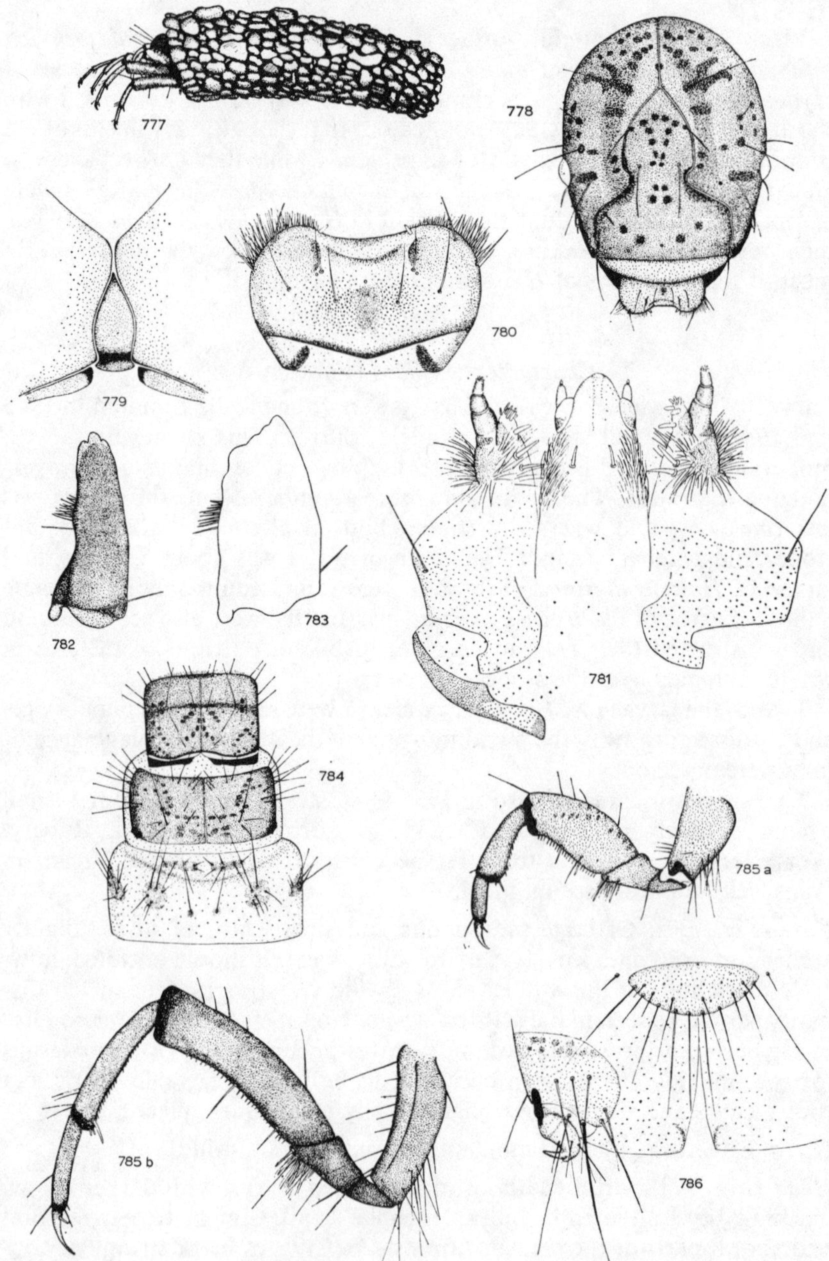

Figs. 777–786. Larva of *Chaetopteryx villosa* (Fabricius). 777. Lateral view in case; 778. Head; 779. Gular sclerite; 780. Labrum; 781. Maxillae and labium; 782. Right mandible from above; 783. Right mandible from the inside; 784. Thoracic nota; 785. a. prothoracic leg; b. mesothoracic leg; 786. Anal segments of abdomen.

suture. A number of darker marks is contained within the bands. Chaetotaxy as in the figure.

Labrum (Fig. 780). Golden brown with greyish brown shallow excised anterior margin. Posterior margin black, posterior incurved angles black. Apart from the darker margins, labrum has four areas darker in colour, a large central area made up of a number of small dark spots, a transverse concave ridge situated just posterior to the anterior margin and a pair of bulbous prominences directed inwardly. Chaetotaxy well defined as shown in Fig. 780.

Mandibles (Figs. 782 and 783). Long, black, each with three teeth, narrow, when seen from above, and with five teeth, one very small, and broad when seen from the inner face. A group of small bristles, with an isolated longer bristle situated on a convexity on the inner edge of each mandible. A long outwardly directed bristle at the base of each mandible. Lestage *in* Rousseau states that there are two short bristles on the 'dorsal side'. I never found two bristles. (There does appear to be a tendency for the bristles to break off, however, in some species.)

Maxillae (Fig. 781). Maxillary palp of four segments, the sclerotization of the proximal segment not completely encircling it. The maxilla has a large golden brown sclerotized plate on the outer side with two large inwardly directed spines on the inner side. The mala also has a golden brown sclerotized plate on the outer side with a single large blunt bristle on the inner side. It is furnished at the tip with several sensory organs, some of which are large.

Labium (Fig. 781). Labial palp two-segmented with large terminal rod-like sense organ. Proximal segment large and bulbous with golden brown sclerotized patch at the base. The spinneret is barely perceptible at the tip of the ligula.

Gular sclerite (Fig. 779). Not completely dividing the genae. Small concavities in the lateral margins at the anterior end fit in the anterior ventral apices of genae. (Note: In the figure the anterior part is shown lowermost in order to correspond with the figure of the head, which is best observed under the low power of the microscope in this position.)

Thorax (Fig. 784). Pro- and mesonota sclerotized, metanotum with sclerotized patches. Pleural margins of pronotum parallel, posterior margin very heavily sclerotized with double black bands, the outer (considerably larger) divided, the inner divided only by the median longitudinal suture. A transverse narrow dark band is situated one-third of the length of the pronotum from the anterior margin. Mesonotum wider than pronotum with median longitudinal suture. Posterior lateral apices of mesonotum black. Metathoracic segment wider than mesothorax. Anterior pair of sclerites transverse, each bearing six bristles. An integumentary fold in the metanotal membrane divides these latter sclerites from the remainder, which consist of a pair of oblique pleural sclerites, a pair of circular sclerites on the notum, each containing about six bristles of unequal length. Between these latter sclerites is situated a pair of bristles with a small golden brown area at the base.

Legs (Figs. 785 a and b). Prothoracic legs short. Anterior extremities of femur, tibia and tarsus black. Meso- and metathoracic legs long, approximately equal in length. Anterior extremities of femur, tibia and tarsus dark in colour. Trochanter in all legs divided into two segments, the proximal segment of which is devoid of hairs and bristles.

Abdomen. Cylindrical, creamy white, protuberances on first abdominal segment small. Lateral line present from third to eighth segment and made up of black hairs. A series of small sclerotized protuberances or 'points' associated with the lateral line is definitely present although very indistinct. There appear to be four (on each side) in the anterior part of the third segment (just dorsal to the lateral line), three in the fourth segment and two in the fifth segment. Single filiform gills present on second to seventh segments inclusive. On the seventh segment ventral gills are present only. The sclerite on the dorsal surface of the ninth segment (Fig. 786) is greyish brown in colour with many unequal hairs. Anal claw (Fig. 786) with auxiliary hook and large basal segment sclerotized and furnished with several long bristles.

Key for Identification of Larvae of LIMNEPHILIDAE

The following key is based on that of Lestage, which Mr. R. Edwards kindly rearranged for me. This key cannot be said to be satisfactory, and certain species are omitted from it. Many of the earlier descriptions on which it is based were not made in sufficient detail so that there is much interspecific overlapping. With the present inadequate state of knowledge the reader is advised to make comparisons of the text figures when making an identification.

1. Pronotum clearly dark on the anterior third part (in front of the transverse furrow); median area clear; head having furcal bands contrasting with the pale colour of the head; on the clypeus a dark longitudinal band well marked, more or less broader towards the front — 2

— Pronotum not as dark on the third anterior as on the median area — 8

2. Clypeal band slightly extended in front, allowing a large part of the basic colour between this band and the furcal bands to be seen; distal spur on the anterior femora much longer than the proximal spur; an additional hair between these two spurs; distal calcariform hair of the median and posterior femora much bigger than the proximal — 3

— Clypeal band much wider towards the front where it almost entirely covers the clypeus and all that is left of the basic colour is a narrow line between the clypeal and the furcal bands — 4

3. Head short and broad, larger than in the following species; furcal bands not following exactly the furcal lines; transverse furrow of the pronotum not deeper than the third anterior which is generally darker (brown–black) than the

following species; sclerite of the ninth segment furnished with numerous hairs (generally twenty-four); 19 to 21 mm. long and 3½ to 4½ mm. wide. Larval case variable, generally straight, or (old larvae) very slightly narrowing towards the rear, composed either of fine or coarse vegetable matter disposed transversely or obliquely, or of sand-grains (in which case the larval case is cylindrical) or of shells; anterior opening oblique; length 18–27 mm, width 5–6½ mm.
LIMNEPHILUS RHOMBICUS

— Head smaller and more elongated; furcal bands following exactly the furcal lines; transverse furrow, at least in the centre, always darker than the anterior third, which is brown; sclerite of the ninth tergite furnished with fifteen hairs at the most; 16–23 mm. long and 3–4 mm. wide. Larval case generally straight, slightly conical, of vegetable matter, disposed longitudinally or imbricate, sometimes triangular, then posterior part narrower and less deep than anterior; length 30–40 mm., width approximately 5 mm,; triangular case; width up to 10 mm. and 25 mm. in length LIMNEPHILUS DECIPIENS

4. Second abdominal segment furnished with presegmental lateral gills, grouped in twos or more often in threes 5

— Second segment without presegmental gills; head yellowish, furcal bands brown or brown–black occasionally confluent with pale brown pleural bands. (In certain cases, the clypeal band can be as obviously extended in front as in *L. flavicornis* but the absence of the lateral presegmental gills in the second segment will always indicate that the larva is that of *L. nigriceps*, as this characteristic is always constant); pronotum yellowish; mesonotum darker, brown, growing much darker at the anterior border and particularly at the anterior corners, and furnished with clear black spots; sclerite of the ninth tergite with 9–17 hairs; 12–18 mm. long and 2½–3 mm. wide. Larval case very variable, straight, cylindrical or slightly conical, rarely triangular, composed of vegetable matter generally disposed longitudinally; posterior well sealed with a membrane (often absent), furnished with a rounded or irregular opening, length 21–26 mm. LIMNEPHILUS NIGRICEPS

5. Hypostome dark; pronotum brownish-black on the third anterior, not darker in the centre of the transverse furrow than in the median area 6

— Hypostome pale; pronotum dark on the anterior third; transverse furrow darker than the median area of the pronotum. (The pale larvae of *L. nigriceps* belong to this group) 7

6. Clypeal band having a deep indentation in the anterior border where the basic colour can be clearly seen; eighth segment almost always furnished with dorsal and ventral presegmental gills; 18–24 mm. long and 4–4½ mm. wide. Larval case straight or slightly arched, conical, of vegetable matter disposed transversely or obliquely, fine or coarse, and in this case the tube is rough; anterior opening oblique; posterior extremity straight and closed with a vegetable plug and a membrane which has a rounded opening; length approximately 25 mm., width 6–7 mm. LIMNEPHILUS POLITUS

— Clypeal band not or only slightly indented on the anterior border; eighth segment generally without gills; 17–21 mm. long and 4 mm. wide. Larval case straight, rarely slightly arched, cylindrical, of vegetable matter disposed more often transversely, rarely longitudinally or obliquely, and of rectangular form, fairly broad, making the surface of the tube quite even; anterior and posterior extremities as in previous species, length 19–30 mm.; width 5–10 mm.
LIMNEPHILUS MARMORATUS

7. Head short and broad, brown or dark brown on whole of posterior part of the ventral surface; mesonotum brown and almost black near the anterior corners; gill formula: 18+17+15+13+10+6+3=82; five hairs between the lateral and median hairs of the ninth abdominal sclerite; 20–24 mm. long and $3\frac{1}{2}$–$4\frac{1}{2}$ mm. wide. Larval case as *L. rhombicus*, but generally broader and often with finer matter at hinder end, disposed transversely, longitudinally or obliquely; in the case of mature larvae, the larval case irregular, almost straight, oblique at front, straight at back, wider in front, and sealed at the posterior extremity by the usual plug and membrane; length 20–25 mm. LIMNEPHILUS FLAVICORNIS

— Head long and narrow, without a single dark mark underneath; mesonotum pale yellow or slightly rose tinted, with the anterior border black near the anterior corners, gills more numerous: 18+18+17+13+11+9+3=89; two or three hairs only between the lateral and median hairs of the sclerite of the ninth tergite; size and form as previous species. Case straight, its widest part in centre or behind centre, sometimes composed of vegetable matter disposed perpendicularly to the axis superimposed, sometimes similar to that of the *L. flavicornis*, but constructed exclusively of vegetable; anterior opening almost straight; posterior extremity as in previous species; length 20–25 mm., width 8–20 mm. LIMNEPHILUS STIGMA
(In certain individuals of *L. stigma* the clypeal band is not as obviously extended in front, but it is, however, more so than in *L. rhombicus*)

8. Clypeus with a definite pattern, often similar to that of *L. rhombicus*, always narrow in front and not covering all the clypeus; clear furcal bands; sometimes, however, replaced by spots and dots; head generally very pale (the pale-headed species, decorated on the clypeus with a pattern slightly darker, fusing in the front with the furcal bands and of cuneiform shape, do not belong to this group, but include the larvae which are slightly coloured and whose furcal bands and design of clypeus are only indicated by spots and marks (Ulmer)); pleural bands absent, replaced by a series of dark dots, transverse furrow of the pronotum darker in the middle 9

— Head entirely dark, or with clypeal band joined, at least on anterior part, with furcal bands, in such a way that whole clypeus is covered. (In certain slightly coloured individuals, the cuneiform mark on the clypeus is sometimes apparent; even in this case, the dark colouring is not visible between the clypeal and furcal bands) 11

9. No lateral presegmental gills on second abdominal segment; dorsal presegmental gills of 2–4 segments grouped generally in twos; this also applies to ventral presegmental gills; no dark spots on third anterior of the pronotum; posterior margin of mesonotum entirely black; pleural spots and occipital spots small; 17 mm. long and 3 mm. wide. Case straight, slightly conical, rarely of sand-grains, almost always of vegetable matter, variable in size and disposed obliquely, or particularly longitudinally, sometimes imbricate, length 20–23 mm., width 4 mm. LIMNEPHILUS LUNATUS

— Second abdominal segment with lateral presegmental gills; presegmental gills of the dorsal series of the 2–4 segments grouped in threes, as those of the ventral series 10

10. Pronotum without spots on third anterior, mesonotum pale yellow, brown on posterior border (rarely with fine black border), with black lateral line near the black posterior corners; pleural and occipital spots small; 28 mm. long and

$4\frac{1}{2}$ mm. wide. Case straight, cylindrical or conical, sometimes swollen in the centre, of vegetable matter in large quantity disposed longitudinally; anterior opening oblique, posterior extremity straight, entirely open; length up to 58 mm., width up to 6 mm. LIMNEPHILUS BOREALIS

— Pronotum furnished with large clear spots on anterior third; mesonotum largely black at posterior border, pleural and occipital spots very large; 18–20 mm. long and 4 mm. wide. Larval case much smaller, conical, slightly arched, never composed of vegetable matter but of sand-grains; length approximately 20 mm., width $4\frac{1}{2}$ mm. LIMNEPHILUS FUSCICORNIS

11. Clypeal band fused with furcal bands only in front, so that the outline is clearly defined and basic pale colour is still visible in space near the posterior corners of clypeus and on each side of angles of the furcal bands; transverse furrow of the pronotum brown at least in centre 12

— Clypeal band completely fused with furcal bands and entirely covering the clypeus so that the top side of the head is generally completely dark and no background colour to be seen (except occasionally in the posterior corners of clypeus, as in *L. bipunctatus*, for example)

12. Spotted design on head, clypeal and furcal bands not very dark; anterior area of clypeus furnished with dark spots; pronotum at brown not dark brown 13

— Darker forms; anterior area of clypeus always without spots; pronotum generally dark brown; the Λ mark on the clypeus entirely dark; spots invisible; pronotum yellow, transverse furrow brownish, third anterior not darker, mesonotum darker than pronotum, larger, 18–20 mm. long and 3–4 mm. wide. Case conical, straight, smooth, mostly of small fragments of leaves, obliquely cut off at the anterior edge; length 20 mm., width about 4 mm.
LIMNEPHILUS BINOTATUS

13. Heavily sclerotized parts, brown, anterior border of the mesonotum, particularly in the anterior corners, dark brown. Case almost straight, cylindrical, generally of coarse sand-grains (as in *L. bipunctatus*), sometimes of fine vegetable matter (fragments of leaves, bark or wood) disposed longitudinally or obliquely; length 17–18 mm., width $4–4\frac{1}{2}$ mm. LIMNEPHILUS AFFINIS
(Silfvenius's 1906 description was from an exuvium)

— Heavily sclerotized parts, greyish-brown, head almost concolorous on top, blackish, reddish-brown underneath, only anterior border of mesonotum brown, anterior corners not dark; up to 15 mm. long. Case almost straight, slightly conical, even, composed of vegetable matter (fragments of leaves or bark), fairly broad (4 mm. long), disposed longitudinally or transverse, length 14–16 mm., width $3–4\frac{1}{2}$ mm. (As Ulmer states, the larva of *L. luridus* would be better placed in the L. centralis group) LIMNEPHILUS LURIDUS

14. Head relatively pale (yellowish or pale reddish) so that the pattern of spots (the Λ mark on the clypeus, furcal and pleural dots) visible 15

— Head much darker, the areas of clypeus and furcal bands entirely concolorous, sombre (brown or black); clypeus usually without spots and having at most a pale mark at the posterior corners 18

15. Posterior femora furnished with a spur and a black calcariform hair 16

— Posterior femora furnished with two calcariform hairs 17

16. Median femora with two spurs; mesonotum near the posterior corners, with black lateral band extending to dorsal extremity in front of centre of segment: 16–17 mm. long and 3–4 mm. wide. Case smooth, conical, arched, of fine particles of vegetable matter disposed transversely; length 16 mm., width 3–4 mm. LIMNEPHILUS SPARSUS

— Median femora with single spur and calcariform hair; mesonotum with simple black lateral border; 16–18 mm. long, 3 mm. wide. Case generally bigger, conical, arched; the mature larvae have cases of coarse sand-grains (more rarely of minute shells); the young larvae have cases of vegetable fragments; length 18–20 mm., width 4 mm. LIMNEPHILUS BIPUNCTATUS

17. Head pale brown, very short, slightly longer than wide; pro- and mesonotum yellowish-brown, the first with black posterior corners, anterior spurs of the metanotum indistinct, the two other pairs not very distinct, anterior tibiae without spines; 11–14 mm. long, 2 mm. wide approximately. Case almost cylindrical, slightly narrow at the back, of sand-grains; posterior extremity convex and rounded; length up to 16 mm., width 3.7 mm. LIMNEPHILUS EXTRICATUS

— Head excessively short and broad, reddish brown (darker than in *L. sparsus*); pronotum reddish brown with transverse furrow brown on first third; mesonotum margined in black, except on indented anterior median; anterior spurs of the metanotum small and oval, the following small and triangular, the laterals (crescent shaped) large and having a black median mark; anterior tibiae entirely furnished with yellow spines; 18 mm. long and 4 mm. wide. Larval case cylindrical, straight, of vegetable matter (fragments of *Carex* in particular); imbricate; posterior opening rounded, very small, length up to 34 mm. LIMNEPHILUS ELEGANS

18. Median and posterior femora with only one yellow spur and a black calcariform hair 19

— Median and posterior femora with two calcariform black hairs 20

19. Pronotum with dark cross formed by medio-longitudinal line and the transverse furrow which are dark; spurs of the anterior femora slender, distal spur very long and almost as large as the additional hair between the proximal spur and the proximal extremity; additional hair of the posterior femora five times as long as the spur; tarsal claws subequal with tarsi, anal claws with two small dorsal hooks; 10–12 mm. long, 2 mm. wide. Case smooth, conical, narrow, very narrow at the rear, of fine sand-grains, with vegetable particles, anterior opening oblique, often deeply furrowed on ventral side and sometimes also less deeply on the dorsal surface; length 15–25 mm. long, width 2.2 mm. anterior, 1 mm. posterior LIMNEPHILUS VITTATUS

— Pronotum without black cross; spurs of the anterior femora large, very short, subequal; additional hair nearer the proximal spur and almost as big as the spurs; additional hair of the posterior femora at the most three times as long as the spur; anterior claws often longer than tarsi, middle and hind claws are two-thirds the length of their tarsi; anal claws with single dorsal hook; 10½–12½ mm. long and 2–2½ mm. wide. Case smooth, conical, arched, of sand-grains; anterior opening oblique, posterior extremity straight, with usual membrane; length 11–14 mm., width 2½–3 mm. LIMNEPHILUS CENTRALIS

20. Posterior legs with excessively long hairs (distal terminal hairs of the trochanters and femora, on internal and external surfaces, and sometimes external hairs of tibiae and tarsi); distal calcariform hair at least twice as long as proximal; length 15–18 mm., width 2½–3 mm. Case conical, slightly arched, either (young larvae) of fine vegetable particles, later entirely or partly sand-grains, or (mature larvae) sometimes entirely vegetable-matter (pieces of leaves or bark) disposed longitudinally, obliquely or transversely; length 14–19 mm., width approximately 3 mm. LIMNEPHILUS GRISEUS

— Posterior legs without these long hairs, the calcariform distal hair only a little longer than the proximal 21

21. Additional hair of the anterior femora situated between the two spurs in the centre; head darker than the pro- and mesonotum which is yellowish-brown; anal claws with two dorsal unequal hooks (Silfvenius); 10–12 mm. long and 1.6 mm. wide. Larval case smooth, straight, a little narrow at the rear, composed of fragments of small leaves, slender, unequal, disposed longitudinally (this larval case is very similar to that of the LIMNEPHILUS INCISUS, and, added sometimes, here and there several sand-grains, posterior extremity straight, without vegetable plug, open or sealed by the usual membrane; anterior opening slightly oblique; length 11–15 mm., width 2½–3.2 mm. LIMNEPHILUS AURICULA

— Additional hair of the anterior femora situated nearer the base in relation to the proximal spur; head, pronotum and mesonotum concolorous, dark brown; anal claws with only one large dorsal hook (Ulmer); approximately 16 mm. in length, 3 mm. in width. Case smooth, straight, slightly conical, of fragments of brownish unequal leaves disposed longitudinally (this larval case has the exact appearance of that of *L. binotatus*); length 16 mm., width 3½–4 mm. LIMNEPHILUS IGNAVUS

Note: The term 'calcariform', which Lestage uses extensively, denotes a movable spur or spine-like process.

BETTEN, C. 1934. The Caddis Flies or Trichoptera of New York State. *Bull. N.Y. St. Mus.* **292**, 308.

BRINDLE, A. 1961. The Larval Taxonomy of the British Trichoptera. I. *Entomologist's Rec. J. Var.* **73**, 114–25.

BRINDLE, A. 1964. *Entomologist's Rec. J. Var.* **76**, 289–92.

CRICHTON, M. L. 1961. Observations on the Longevity and Dispersal of Adult Limnephilidae (Trichoptera). *XI Int. Congr. Ent. Wien*, 1960, **1**, 366–71.

DÖHLER, W. 1914. Beitrage zur Systematik und Biologie der Trichopteren. *Sber. naturf. Ges. Lpz.* **30**, **35**, **39**, **41**, 39, 51.

DÖHLER, W. 1915. Beiträge zur Systematik und Biologie der Trichopteren. *Sber. naturf. Ges. Lpz.* **41**, 91.

DÖHLER, W. 1920. *Zool. Anz.* **51**, 1–13.

ESBEN-PETERSEN, P. 1916. *Vaarfluer in Danm. Fauna* **19**, 50, fig. 43 *b*.

FAKUBISIAKOWA, J. 1931. Les Trichopteres du Lac Kiekrz (Pologne occid). *Bull. Soc. Amis Sci. Lett. Poznań* (*B*) **5**, 44–8.

FOTIUS-JABOULET, M-C. 1961. *Travaux du Laboratoire de Zoologie et de la Station Aquicole Grimaldi de la Faculté des Sciences de Dijon.* No. 40–1961.

FROCHOT, B. 1962. La Larve de *Stenophylax permistus* McL. *Travaux du Laboratoire de Zoologie et de la Station Aquicole Grimaldi de la Faculté des Sciences de Dijon*. No. 42–1962.

FROCHOT, B. 1963. Trois nouvelles larves du genre *Halesus*. *Travaux du Laboratoire de Zoologie et de la Station Aquicole Grimaldi de la Faculté des Sciences de Dijon*. No. 49–1963.

GRANDI, G. 1951. *Introduzione allo Studio della Entomologia* **2**, 53. Bologna.

HANNA, H. M. 1953. Sponges on Caddis Cases. *Entomologist's mon. Mag*. **89**, 145.

HANNA, H. M. 1956. The Larva of *Limnephilus marmoratus* Curt. *Entomologist's Gaz.* (in the press).

HANNA, H. M. 1957. *Entomologist's Gaz.* **8**, 218–22.

HANNA, H. M. 1961. The Larva of *Drusus annulatus* Stephens (TRICHOPTERA; LIMNEPHILIDAE). *Entomologist's Gaz.* **12**, 36–41.

HICKIN, N. E. 1942. Larvae of the British Trichoptera. 1. *Stenophylax stellatus* Curtis. *Proc. R. ent. Soc. Lond.* (A) **17**, 9–11.

HICKIN, N. E. 1943. Larvae of the British Trichoptera. 8. *Limnephilus flavicornis L. Proc. R. ent. Soc. Lond.* (A) **18** (1–3), 6–10.

HICKIN, N. E. 1943. *Proc. R. ent. Soc. Lond.* (A) **18**, 11–14, 66–8, 72–4.

HICKIN, N. E. 1946. *Proc. R. ent. Soc. Lond.* (A) **21**, 61–5.

HICKIN, N. E. 1946. Larvae of the British Trichoptera. *Trans. R. ent. Soc. Lond.* **97**, 187–212.

HICKIN, N. E. 1948. Larvae of the British Trichoptera. 25. *Limnephilus rhombicus* L. *Proc. R. ent. Soc. Lond.* (A) **23** (4–6), 54–6.

HICKIN, N. E. 1948. *Proc. R. ent. Soc. Lond.* (A) **23**, 12–13, 59–61.

HICKIN, N. E. 1949. Larvae of the British Trichoptera: *Halesus digitatus* (Schranck). *Proc. R. ent. Soc. Lond.* (A) **24**, 56–9.

HICKIN, N. E. 1950. Larvae of the British Trichoptera. 31. *Stenophylax vibex* Curtis. *Proc. R. ent. Soc. Lond.* (A) **25**, 107–10.

HICKIN, N. E. 1952. *Caddis. A Short Account of the Biology of British Caddis Flies, with special reference to the immature stages*. Methuen, London.

HICKIN, N. E. 1953. Larvae of the British Trichoptera: *Micropterna sequax* McLachlan. *Proc. R. ent. Soc. Lond.* (A), **28**, 163–5.

HICKIN, N. E. 1954. *Proc. R. ent. Soc. Lond.* (A) **29**, 55–8, 89–92, 96–8, 145–6.

HICKIN, N. E. 1958. *Proc. R. ent. Soc. Lond.* (A), **33**, 176–178.

KAČALOVA, O. L. 1961. *Fauna of the Soviet of Latvia*. III. Riga.

KELNER-PILLAULT, S. 1960. Biologie, Ecologie d'*Enoicyla Pusilla* Burm. (Trichopteres, Limnephilides) *Ann. Biol.* **36** (1–2), 51–9.

KIMMINS, D. E. 1944. Supplementary Notes on the Trichoptera of the English Lake District. *Entomologist* **77**, 81.

KIMMINS, D. E. 1956. British Trichoptera (Caddis Flies) A Modified Family-Key, and a Key to the Genera of the Family LIMNEPHILIDAE, with a Check-List of the Species of LIMNEPHILIDAE. *Entomologist's Gaz.* **7**, 29–38.

KIMMINS, D. E. 1963. *Entomologist's Gaz.* **14**, 24–8.

LEPNEVA, S. 1966. Larvae of the Trichoptera, Integripalpia. *Fauna USSR*. N.S. **95**, 145, 167. (In Russian.)

LESTAGE, J. A., *in* ROUSSEAU, E. 1921. *Les Larves et Nymphes Aquatiques des Insectes d'Europe*.

McLACHLAN, R. 1874–1884. *A Monographic Revision and Synopsis of the Trichoptera of the European Fauna*.

MOON, H. P. 1936. The Shallow Littoral Region of a Bay at the North-West End of Windermere. *Proc. zool. Soc. Lond.* Part 2, 491–515.

MORTON, K. J. 1884. *Entomologist's mon. Mag*. **21**, 125.

MOSELY, M. E. 1921. *The Dry-Fly Fisherman's Entomology*, 42.

MOSELY, M. E. 1939. *The British Caddis Flies* (*Trichoptera*). London.

NIELSEN, A. 1942. *Über die Entwicklung und Biologie der Trichopteren. Sonderdruck aus den Archiv für Hydrobiologie*, 255, 489.

NOVAK, K. & SEHNAL, F. 1963. The Development Cycle of Some Species of the Genus Limnephilus (Trichoptera). *Acta Societatis entomologicae Čechoslovenicae* **60** (1–2), 68–80.

PELHAM-CLINTON, E. C. 1966. *Nemotaulius punctatolineatus* (Retzius), A Caddis Fly new to the British Isles (TRICHOPTERA, LIMNEPHILIDAE). *Entomologist's Gaz.* **17**, 5–8.

PETERSEN, E. 1914. *Ent. Meddr.* **10**, 141.

RATHJEN, W. 1939. *Z. Morph. Okol. Tiere* **35**, 14.

ROSS, H. H. 1944. *The Caddis Flies or Trichoptera of Illinois*, 178.

ROUSSEAU, E. 1921. *Les Larves et Nymphes Aquatiques des Insectes d'Europe*, 680.

SILFVENIUS, A. J. 1903–4. *Acta Soc. Fauna Flora fenn.* **25** (4), 9, 27.

SILFVENIUS, A. J. 1904. Über die Metamorphose einiger Phryganeiden und Limnephiliden. *Acta. Soc. Fauna Flora fenn.* **27** (2).

SILFVENIUS, A. J. 1906. *Acta Soc. Fauna Flora fenn.* **28** (4), 44, 45, 49.

SILTALA, A. J. 1905–6. *Acta Soc. Fauna Flora fenn.* **27**, 31.

SILTALA, A. J. 1908. *Acta Soc. Fauna Flora fenn.* **31** (3), 14–17.

SILTALA, A. J. 1921. *Les Larves et Nymphes Aquatiques des Insectes d'Europe*, 766.

STRUCK, R. 1899. *Neue und Alte Trichopteren-Larvengehäuse. Ill. Z. Ent.* **4**, 264.

STRUCK, R. 1903. Beiträge zur Kenntnis der Trichopterenlarven. *Mitt. geogr. Ges. naturh. Mus. Lübeck* **17**, 41–124, table 4, fig. 11.

THIENEMANN, 1905. *Zool. fahrb. Abt. für System.* **22**, fig. 21.

ULMER, G. 1903. *Metamorph. Trichopt.*, 50 et sqq., 375, figs. 29, G, 39.

ULMER, G. 1904. *Allg. Z. Ent.* **9**, 55, 57.

ULMER, G. 1909. *Die Süsswasserfauna Deutschlands* **5–6**.

WALTON, C. L. 1939. *Ann. Rept. Agric. Hort. Res. Sta. Long Ashton* 1939, 77–8.

WESENBERG-LUND, C. 1910. Über die Biologie von *Glyphotaelius punctatolineatus* Retz. Nebst Bemerkungen über das freilebende Puppenstadium der Wasserinsekten. *Int. Revue ges. Hydrobiol. Hydrogr.* **3** (1/2), 93–114.

16

LEPTOCERIDAE

Only species in the sub-family LEPTOCERINAE are represented in Britain. A number of generic names have changed since Mosely's handbook. A check-list of British species is given below, the names as appearing in Mosely's handbook being shown on the right-hand side. The species *Triaenodes simulans* was not included in the British list by Mosely, nor was *Triaenodes reuteri* which Kimmins at first thought was *simulans. Interjectus* (McLachlan) is now considered to be only a variety of *albifrons.*

ATHRIPSODES

1. *albifrons* (Linné)	=*Leptocerus albifrons* Linné
albifrons var. *interjectus* (McLachlan)	=*Leptocerus interjectus* McLachlan
2. *alboguttatus* (Hagen)	=*Leptocerus alboguttatus* Hagen
3. *annulicornis* (Stephens)	=*Leptocerus annulicornis* Stephens
4. *aterrimus* (Stephens)	=*Leptocerus aterrimus* Stephens
5. *bilineatus* (Linné)	=*Leptocerus bilineatus* Linné
6. *cinereus* (Curtis)	=*Leptocerus cinereus* Curtis
7. *commutatus* (Rostock)	=*Leptocerus commutatus* McLachlan
8. *dissimilis* (Stephens)	=*Leptocerus dissimilis* Stephens
9. *fulvus* (Rambur)	=*Leptocerus fulvus* Rambur
10. *nigronervosus* (Retzius)	=*Leptocerus nigronervosus* Retzius
11. *senilis* (Burmeister)	=*Leptocerus senilis* Burmeister

MYSTACIDES

12. *azurea* (Linné)
    azurea var. *albicornis* Mosely
13. *longicornis* (Linné)
14. *nigra* (Linné)

TRIAENODES

15. *bicolor* (Curtis)
16. *conspersus* (Rambur)
17. *simulans* Tjeder
18. *reuteri* McLachlan

EROTESIS

19. *baltica* McLachlan

ADICELLA
20. *filicornis* (Pictet)
21. *reducta* (McLachlan)

OECETIS
22. *furva* (Rambur)
23. *lacustris* (Pictet)
24. *notata* (Rambur)
25. *ochracea* (Curtis)
26. *testacea* (Curtis)

LEPTOCERUS

27. *interruptus* (Fabricius)	=*Setodes interrupta* Fabricius
28. *lusitanicus* (McLachlan)	=*Setodes lusitanica* McLachlan
29. *tineiformis* Curtis	=*Setodes tineiformis* Curtis

SETODES

30. *argentipunctellus* McLachlan	=*Setodes argentipunctella* McLachlan
31. *punctatus* (Fabricius)	=*Setodes punctata* Fabricius

Adult LEPTOCERIDAE may be identified by the following characters. The antennae are extremely long and slender. In the male they may be as much as three times the length of the wings but they are generally shorter in the female. The basal segment is bulbous and rarely longer than the head. The maxillary palpi are very long and hairy, all the segments being comparatively long. The fourth and fifth segments are usually thinner than the others and the fifth is flexible and sometimes the outer part of the fourth. The labial palpi are small. The densely pubescent anterior wings are long and narrow with the discoidal cell always closed. The posterior wing is always much shorter than the anterior and a frenulum occurs along the costal margin consisting of a row of short hooks which fit into a narrow fold in the hind margin of the anterior wing. The discoidal cell is open in the posterior wing.

Distribution

Athripsodes albifrons. Very widely distributed and generally abundant (M.E.M.); Berks. (M.I.C. *et al.*); Northumberland (G.N.P.). var. *interjectus*. Galway (M.E.M.).

Athripsodes alboguttatus. Large rivers and lakes. R. Thames. Rather local (M.E.M.); Berks. (M.I.C. *et al.*); Northumberland (G.N.P.).

Athripsodes annulicornis. Widely distributed often with *cinera* (M.E.M.); Northumberland (G.N.P.).

Athripsodes aterrimus. Very widely distributed and found on almost every kind of water, running or still (M.E.M.); Berks. (M.I.C. *et al.*); Northumberland (G.N.P.).

Athripsodes bilineatus. Very local, rarely appearing in large numbers (M.E.M.).

Athripsodes cinereus. Very widely distributed, large rivers and lakes. Abundant wherever it occurs. Countless numbers on Thames and tributaries (M.E.M.); Symond's Yat, Glos. (A.F.P.); Berks. (M.I.C. *et al.*); Yorks. (P.F.H.); Northumberland (G.N.P.).

Athripsodes commutatus. Rather local (M.E.M.); Northumberland (G.N.P.).

Athripsodes dissimilis. Rather local species, generally taken singly (M.E.M.); Yorks. (P.F.H.); Northumberland (G.N.P.).

Athripsodes fulvus. Still waters, marshes and fens. Widely distributed. Galway (M.E.M.); Berks. (M.I.C. *et al.*); Yorks. (P.F.H.).

Athripsodes nigronervosus. Large rivers and lakes. Widely distributed (M.E.M.); Berks. (M.I.C. *et al.*); Yorks (P.F.H.); Northumberland (G.N.P.).

Athripsodes senilis. Very local, ponds and lakes (M.E.M.).

Mystacides azurea. Widely distributed and very abundant, running and still water (M.E.M.); Symond's Yat, Glos. (A.F.P.); Berks. (M.I.C. *et al.*); Yorks. (P.F.H.); Northumberland (G.N.P.).

Mystacides longicornis. Widely distributed and very abundant, lakes and ponds (M.E.M.); Berks. (M.I.C. *et al.*); Yorks. (P.F.H.); Northumberland (G.N.P.).

Mystacides nigra. Rather more local than other species in the genus (M.E.M.); Yorks. (P.F.H.); Northumberland (G.N.P.).

Triaenodes bicolor. Widely distributed and abundant where it occurs, lakes and ponds (M.E.M.); Berks. (M.I.C. *et al.*).

Triaenodes conspersus. Rather local. Running water (M.E.M.); Berks. (M.I.C. *et al.*).

Triaenodes simulans. Aberfoyle, Perthshire (K.J.M.).

Triaenodes reuteri. Saltmarsh Spurn, Yorks. (E.C.P-C.).

Erotesis baltica. Probably local. Not uncommon in Hants along banks and parts of R. Test (M.E.M.).

Adicella filicornis. Local, generally found with *reducta* (M.E.M.).

Adicella reducta. Very widely distributed. Abundant at springs and boggy patches (M.E.M.).

Oecetis furva. Rather more local than *ochracea*, ponds and lakes (M.E.M.).

Oecetis lacustris. Fairly widely distributed, ponds and lakes (M.E.M.). Berks. (M.I.C. *et al.*).

Oecetis notata. Local, slow and rather deep streams (M.E.M.).

Oecetis ochracea. Widely distributed, ponds and lakes (M.E.M.); Berks. (M.I.C. *et al.*); Northumberland (G.N.P.).

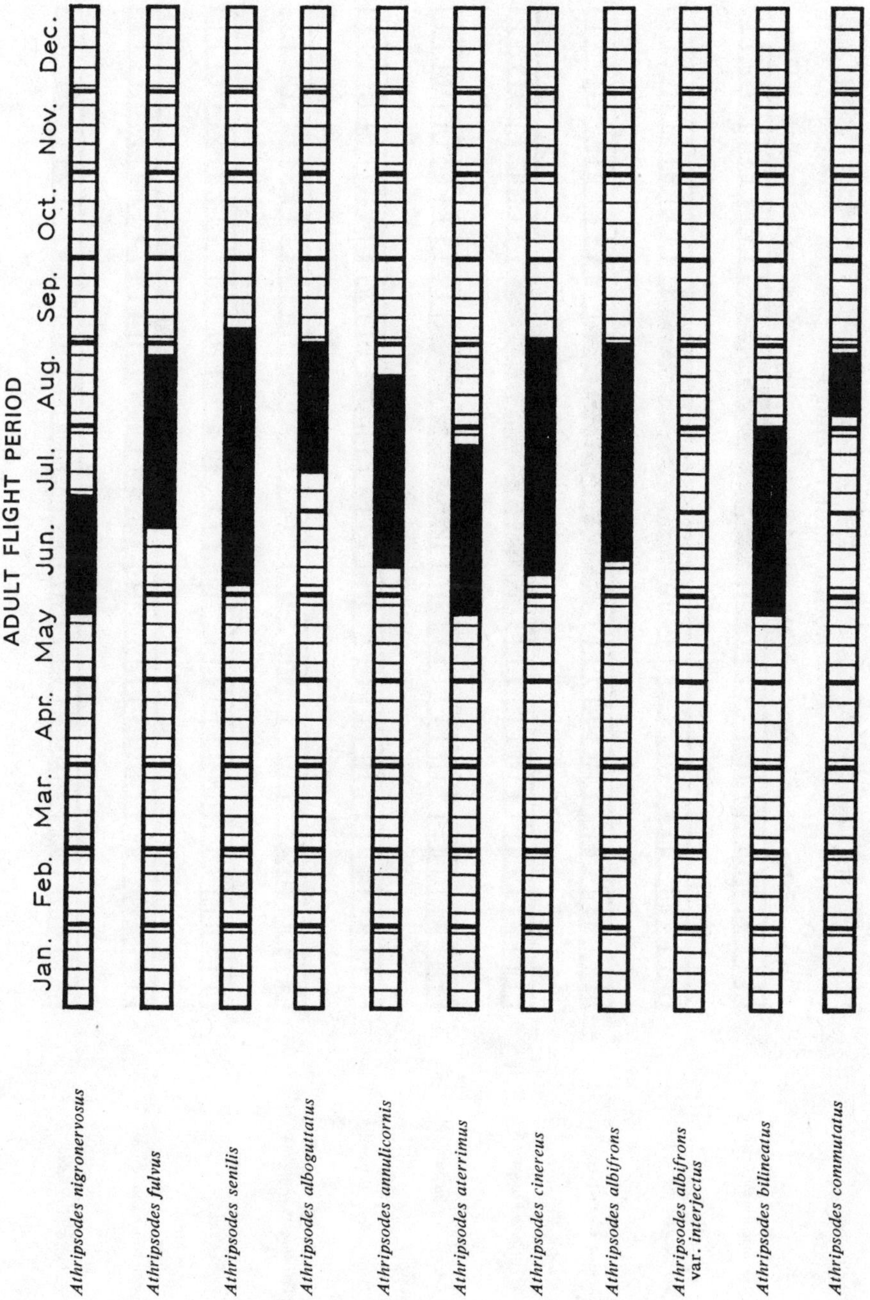
ADULT FLIGHT PERIOD
Jan.
Feb.
Mar.
Apr.
May
Jun.
Jul.
Aug.
Sep.
Oct.
Nov.
Dec.
Athripsodes nigronervosus
Athripsodes fulvus
Athripsodes senilis
Athripsodes alboguttatus
Athripsodes annulicornis
Athripsodes aterrimus
Athripsodes cinereus
Athripsodes albifrons
Athripsodes albifrons var. *interjectus*
Athripsodes bilineatus
Athripsodes commutatus

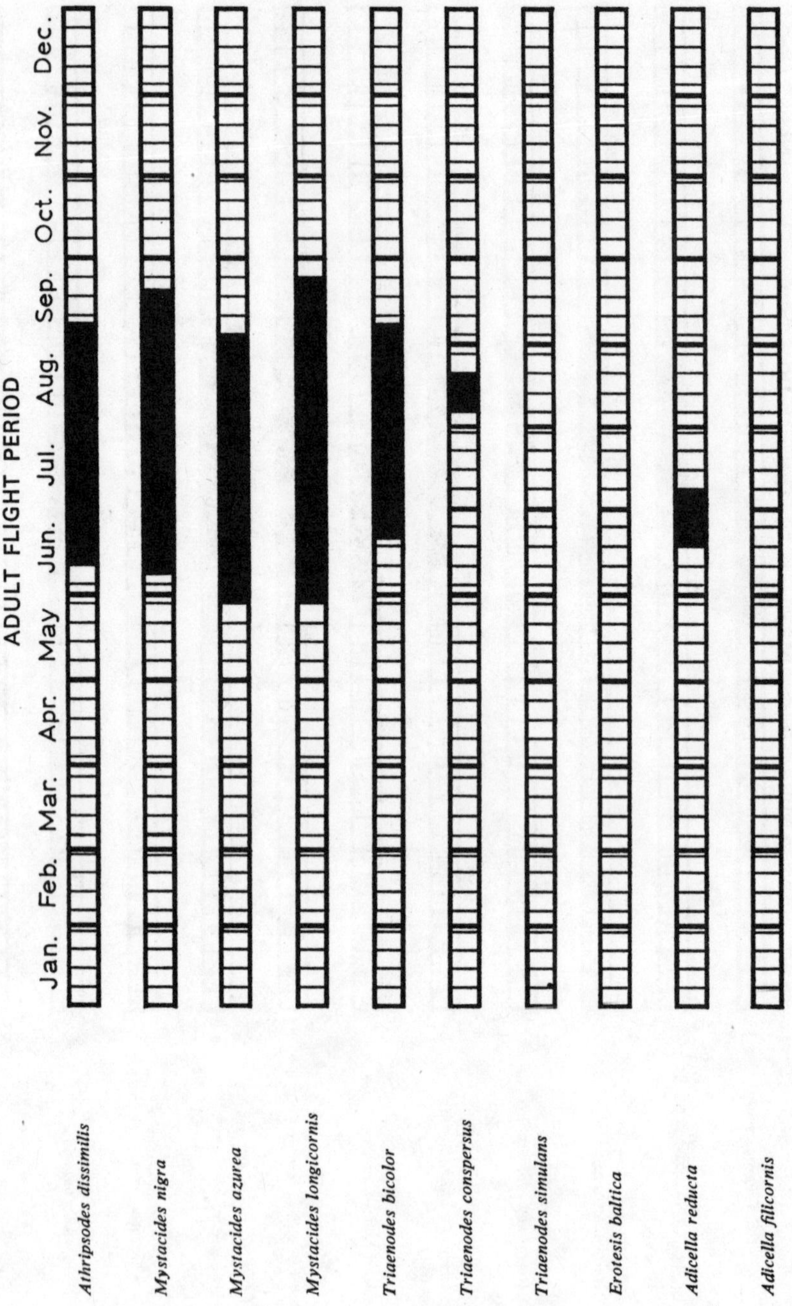
ADULT FLIGHT PERIOD
Jan.
Feb.
Mar.
Apr.
May
Jun.
Jul.
Aug.
Sep.
Oct.
Nov.
Dec.
Athripsodes dissimilis
Mystacides nigra
Mystacides azurea
Mystacides longicornis
Triaenodes bicolor
Triaenodes conspersus
Triaenodes simulans
Erotesis baltica
Adicella reducta
Adicella filicornis

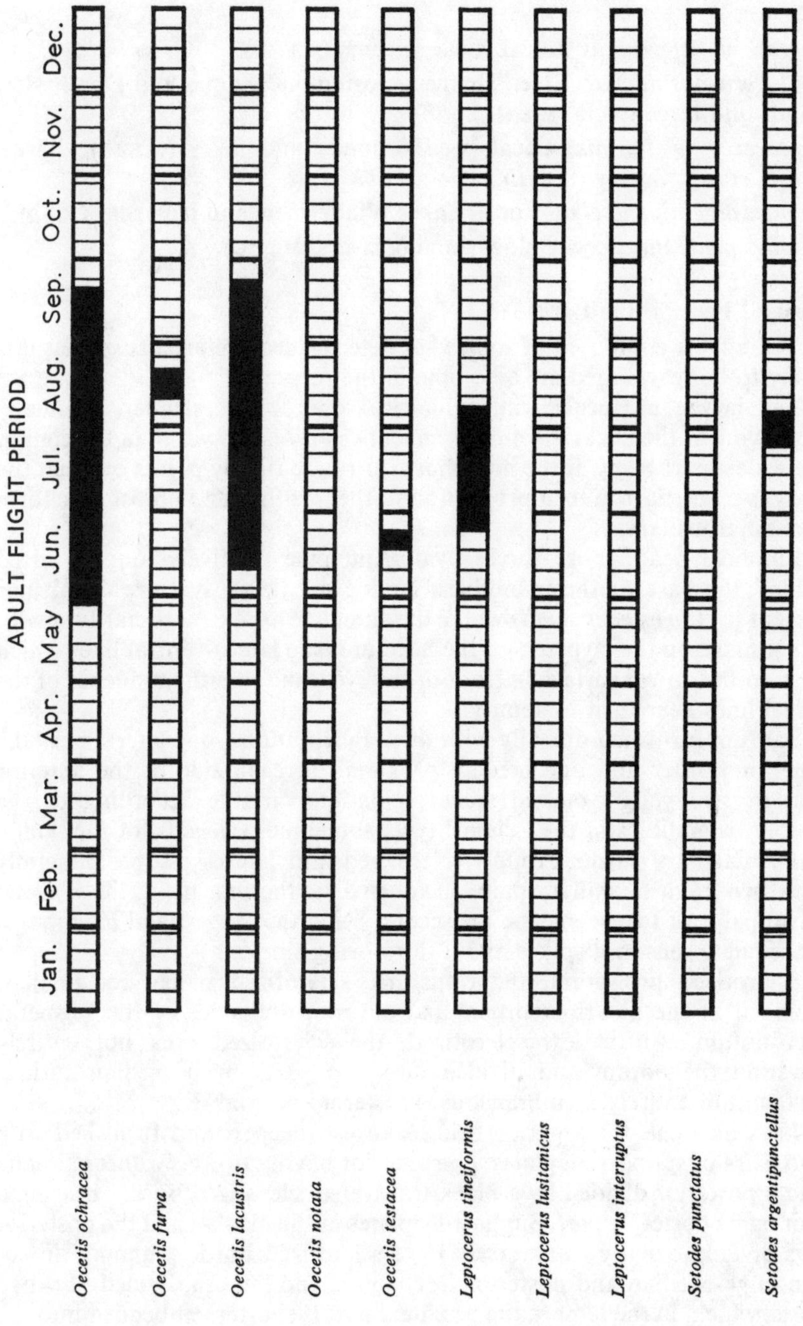

LEPTOCERIDAE. *Adult Flight Period taken from published records of individual dates.*

Oecetis testacea. Widely distributed but not abundant, streams and rivers (M.E.M.).

Leptocerus interruptus. Local, clear streams (M.E.M.).

Leptocerus lusitanicus. River Thames, Goring-on-Thames and Dorchester, ponds and large rivers (M.E.M.).

Leptocerus tineiformis. Local, weedy ponds and lakes (M.E.M.); Berks. (M.I.C. *et al.*); Surrey (N.E.H.).

Setodes argentipunctellus. Local, Lakes Windermere and Killarney (M.E.M.).

Setodes punctatus. Local, slow-running rivers (M.E.M.).

General Form of the Larvae

This family is composed of somewhat heterogeneous elements so that it is likely to be rearranged at some time in the future.

The larvae are eruciform, cylindrical or conical, slender, gradually narrowing at the back, the ninth segment clearly narrower than the eighth, the widest part being at the metathorax. Brindle rightly points out that the body is much narrower in proportion to the length than in most eruciform trichopterous larvae.

Elliptical head, more rarely oval. Antennae relatively long, inserted behind the base of the mandibles, having a pale fairly large distal hair (except in *Athripsodes senilis* where the antennae are extraordinarily small). No X mark on the clypeus; on the head are pale latero-ventral lines which pass on the dorsal surface in front of the eyes, and join the branches of the furcal lines near their extremity.

Labrum furnished dorsally with three indentations, one lateral near the anterior border and one median proximal on each side of the anterior border, two yellow curved spines; ciliation variable according to the species. Mandibles of the 'scissor' type, sometimes (*Oecetis*) of the 'knife' type; maxillary cardo strongly sclerotized and hairless; stipes generally with two hairs; maxillary palps in four parts, the first being the biggest; labial palps of two segments, the second being the longest and having two biarticulate sensorial spikes and a third being simple.

Sclerotized pronotum, the transverse sclerotized patch, rectangular, rounded at the anterior corners and covering the sides of the segment; mesonotum a little less sclerotized, the sclerotized area not entirely covering the notum and divided into two parts by a median suture; metanotum entirely membraneous. Prosternal horn absent.

Legs unequal, the anterior being shortest, deepest and furnished with two pairs of supporting plates, one anterior having two very unequal hairs and a posterior divided by a black transverse sclerotized border bearing a hair on its posterior part. Supporting plates of the median and the posterior legs divided equally transversely by a sclerotized border running dorso-ventrally. Median and posterior trochanters and femora divided into two parts, which, in the femora the proximal part is shorter, embedded into the trochanter, and separated from a much longer distal part by a membraneous area. Posterior legs nearly three times longer than the anterior, furnished, as the others, with dark hairs on the upper distal edge of the anterior and median trochanters and the anterior femora are feathery pale hairs; no conical tubercle on the distal part of the tibiae; at the most, on the anterior tibiae, an obtuse projection; never two spurs at the top of the tibiae; posterior tibiae entire or in two parts.

First abdominal segment furnished with three protuberances, the laterals densely spined. Lateral lines, when present, are furnished with very short hairs on segments three to seven.

Gills sometimes present, sometimes not. When present, they are more or less developed, simple or grouped. Anal appendages short, terminal claws short and wide.

The larval case is, generally, clearly conical, narrow, straight or curved, larger than the larva, but closely adjusted, and composed of fine grains of sand, or of vegetable matter; sometimes entirely secreted.

A Simple Key to the Larvae of Genera of LEPTOCERIDAE, *given by Brindle, based on Lestage, is given below:*

1. Abdominal gills in tufts, each gill dividing into six or more filaments — *Athripsodes*
— Abdominal gills single or absent — 2
2. Mandibles with apex long and pointed sharply — *Oecetis*
— Mandibles broader and shorter — 3
3. Posterior legs with fringes of long setae — 4
— Posterior legs without such fringes — 5
4. Gills prominent on most abdominal segments, case of vegetable material — *Triaenodes*
— Gills present on no more than two segments, often more reduced; case of secretion only — *Leptocerus*
5. Head pattern distinctive; case of vegetable material arranged partly transversely, partly in a circular manner around the case — *Erotesis*
— Head pattern otherwise; case not so constructed — 6
6. Head yellow with prominent black longitudinal bands, at least along anterior part of fronto-clyeus — *Mystacides*
— Head yellowish or brownish without black bands — 7
7. Head and pronotum uniformly reddish — *Adicella*
— Head and pronotum with darker spots and bands — *Setodes*

Detailed Descriptions of Larvae

Athripsodes albifrons (L)

Immature stages unknown.

Athripsodes alboguttatus (Hagen)

Immature stages unknown.

Athripsodes annulicornis (Stephens)

A description based on British material is not available. The case is composed of sand-grains, fine underneath, larger at the sides and on top. It is strongly curved and somewhat flattened dorso-ventrally. Abdominal gills are present on segments two to seven. The pale pronotum is without spots.

Athripsodes aterrimus (Stephens)

A large number of larvae were collected at King's Heath, Birmingham, several of which, on being reared to the adult stage, were determined as *Athripsodes aterrimus* (Steph.). The larvae were dragging their cases over the gravelly bottom at the sides of the canal. In this still water there is present a fair amount of aquatic and marginal vegetation.

Case (Fig. 787). Sand-grains cemented together with secretion, curved and tapering posteriorly. Opening at posterior end oblique. Length and width at posterior end varies within fairly wide limits as, in many cases, the older (posterior) end of the case breaks off. Length up to 18 mm.; width at anterior end up to 3 mm.

Larva. Eruciform with abdomen uniformly cylindrical. Length 11 mm., width 2 mm.

Head (Fig. 788). Orthocentrous, elliptical, about one and a half times as long as wide, with outer edges of genae parallel. Light yellowish-brown with heavy black markings. Clypeus large and divided into a wide transverse oral part and a narrow longitudinal aboral part not strongly marked. Clypeus mark consists of four black spots at aboral end, two arranged in the median line longitudinally (represented by Esben-Petersen by four smaller spots), with one mark on each side. At oral end three pairs of spots arranged transversely. At the narrowest part of the clypeus is a pair of black marks. Genae marked with a sinuous black band running on each side of the clypeus and tapering towards the oral end, with several black marks at the side and a group near the vertex of the clypeus. Gular sclerite (Fig. 789) tapers acutely aborally. Antennae two-segmented, distal segment long and provided with an apical bristle. Labrum notched at points where marginal bristles arise. A small dark greyish-brown mark occupies a median position near the anterior margin. Right mandible with three teeth, left mandible with two. Maxillary palp four-segmented, exactly same length as maxillary lobe (to tip of large sense organs). Base of maxillary lobe (Fig. 792) provided with long bristles, a few of which are fairly stout. Labium with tuft of fine hairs on each side near the base of the labial palps. Lateral margins of labium sclerotized at the base.

Thorax. Prosternal horn absent. Only pro- and mesonota sclerotized (Fig. 793), but sclerotization extends over pleural region in prothorax only. Pro- and mesonota with median longitudinal sutures and approximately equal in width. A pair of black marks on each side of the pronotum laterally and a single median mark on the posterior margin, divided by the median suture. A pair of smaller black marks on each side of the posterior half of the notum. Mesonotum slightly darker in colour and with black marks arranged in three groups. A pair of hook-shaped marks anteriorly

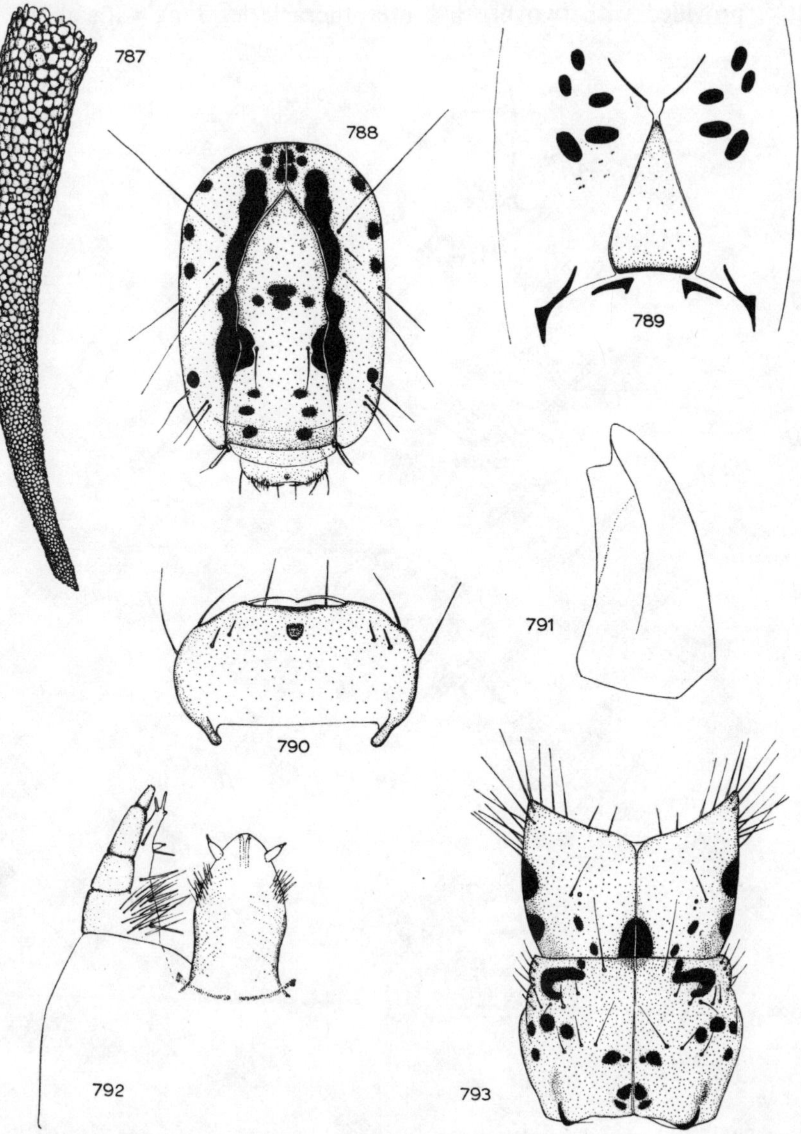

Figs. 787–793. Larva of *Athripsodes aterrimus*. 787. Case; 788. Head; 789. Gular sclerite; 790. Labrum; 791. Mandible; 792. Maxilla and labium; 793. Pro- and mesonota.

(sometimes made up of small separate spots), a series arranged in a transverse concave band and four black marks in the median line near the posterior margin.

Legs (Fig. 794–6). Prothoracic legs short and broad, held bunched up at the sides of the head. A free plate-like sclerite arises from the coxa pleurite. It is provided with two bristles. Mesothoracic leg long with about six

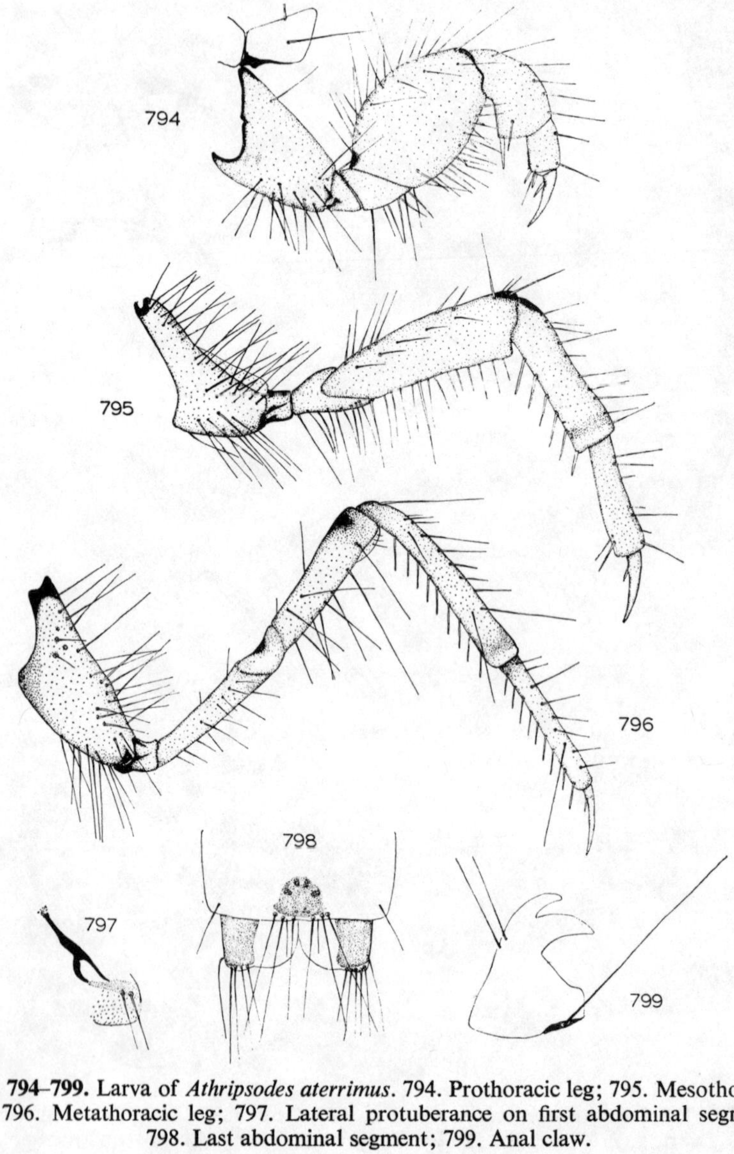

Figs. 794–799. Larva of *Athripsodes aterrimus*. 794. Prothoracic leg; 795. Mesothoracic leg; 796. Metathoracic leg; 797. Lateral protuberance on first abdominal segment; 798. Last abdominal segment; 799. Anal claw.

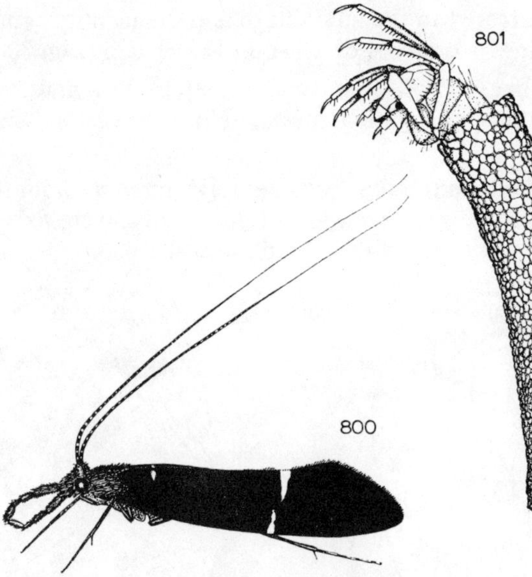

Figs. 800–801. *Athripsodes bilineatus* L. 800. Side view of adult; 801. Larva in case.

evenly spaced spines on ventral edge of the tibia. Metathoracic leg longer than mesothoracic. Dark marks at the extremities of the segments of the leg give a banded appearance. Femur partially divided at the proximal end.

Abdomen. Uniformly cylindrical. Creamy-white in colour. Lateral and median dorsal protuberances on first abdominal segment. Lateral protuberances are partially sclerotized, provided with a few bristles and further supported by a spur-shaped black sclerite (just visible to the naked eye). Gills are present only on the first three abdominal segments and on the first of these there is no dorsal series. The gills consist of small tufts of filaments united to a common base or stalk. Anal sclerite (Fig. 798) with four darker marks at anterior margin and with several long bristles near the posterior margin. Anal claws short and broad with two hooks.

Athripsodes bilineatus (L.)

For three years I observed a swarm of adults of *Athripsodes bilineatus* (L.) towards the end of July at the Nutfield Stream, Surrey, where I have collected *Tinodes pallidulus* McL. It was not, however, until June of 1952 that I found the larvae. Nine larvae that I collected were distributed over about thirty-five yards of stream, where they were crawling up the sides of stones over which the water was moving rapidly. Mosely considered this species to be very local in Britain. In view of Ulmer's note that this species inhabits fast clear streams, more usually in mountains, it is interesting to record that the Nutfield locality is only some 200 feet above sea-level. Four of the larvae were reared to the adult stage and the identity confirmed by Mr. D. E. Kimmins. A drawing of an adult is included (Fig. 800).

Larva. Somewhat like *Athripsodes aterrimus* (Steph.). Pro- and mesonotum sclerotized, the latter very slightly larger than the former. Length 8–9 mm., width 1.3–1.4 mm.

Case of light coloured sand-grains (considerably lighter in colour than those of other species inhabiting the same locality), conical and curved (Fig. 801). Length about 12 mm., width about 2 mm. at wider end.

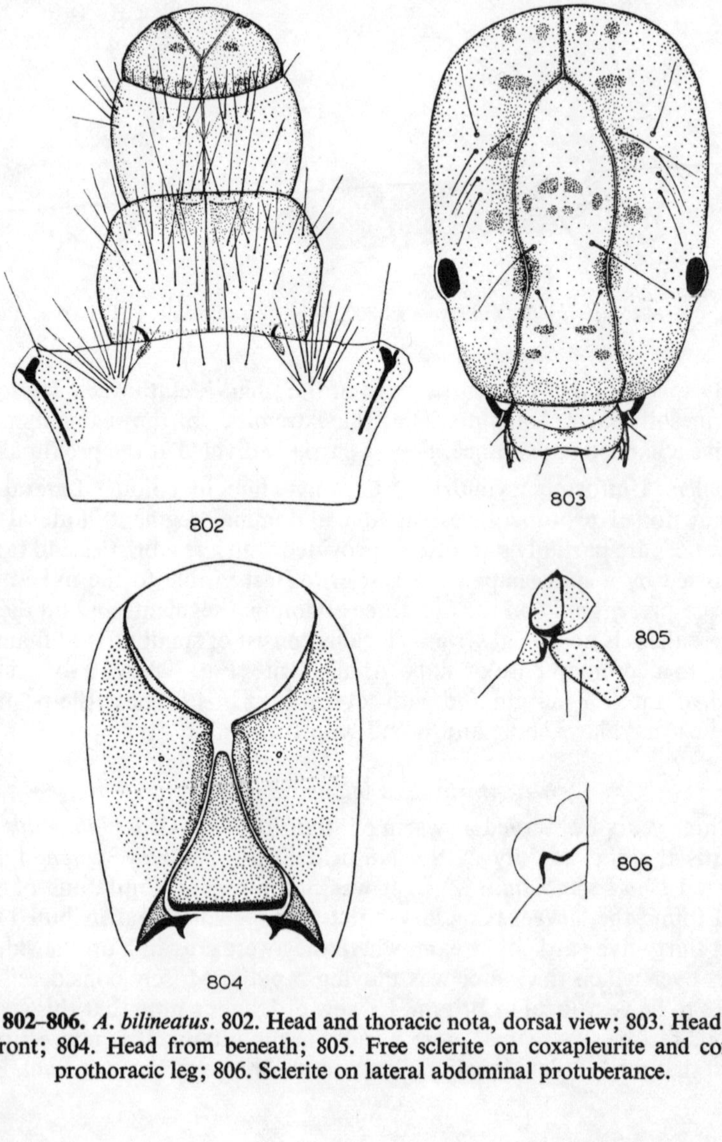

Figs. 802–806. *A. bilineatus*. 802. Head and thoracic nota, dorsal view; 803. Head from the front; 804. Head from beneath; 805. Free sclerite on coxapleurite and coxa of prothoracic leg; 806. Sclerite on lateral abdominal protuberance.

Mystacides azurea L.

Macdonald has described the larva of this species from a Scottish locality, at depths from 0.3–4.5 m. They were very abundant between 2 and 3 m. on a substratum of sand and *Isoetes*, and there was a considerable amount of vegetable debris.

Larval case. The case (Fig. 807) is composed of vegetable fragments and a few small stones, often with one or two small twigs or stem portions running longitudinally along the case, extending beyond either or both ends. The length, excluding the projecting twigs, may reach 15 mm. and the width 2 mm.

Larva. The larva is eruciform, with the pro- and mesonotum entirely sclerotized, but the metanotum is soft except for a pair of small sclerotized areas. The abdomen is white and has no gills. When fully grown the larva is about 12 mm. in length and 1 mm. in width.

Head. The head (Fig. 808) is long, with small ovoid and larger rectangular black markings on a light golden yellow background. The clypeus is twice as long as broad, and posteriorly it bears an oval black mark (a pair of spots fused) with a transverse row of three or four spots behind it. Anteriorly, there are two pairs of spots, the hinder pair being sometimes indistinct. The anterior margin is bounded by a wide black transverse band. On each side of the coronal suture there is a series of large rounded black spots, extending along the parietals to the frontal sutures. From the centre of each of the latter sutures to the anterior margin of the parietals runs a wide black band. Each parietal has an irregular broad light brown band on the posterior two-thirds surrounding some of the black spots. On the genae are numbers of black spots extending to the ventral side of the head. The chaetotaxy is as shown in the figure. The antennae are two-segmented, the proximal segment being large and bulbous, the distal, long and narrow and terminating in a seta. They arise on the anterior borders of the pareitals (Fig. 808). The labrum (Fig. 809) is sclerotized on its anterior lateral margins and has an anterior median concavity, at the base of which arises a pair of setae. Running in a single transverse row are three pairs of setae, and there are two pairs of spines on the anterior margin, which is notched at their insertions. The mandibles are asymmetrical. The right mandible (Figs. 811, 812) has three teeth on its upper and two teeth on its lower sides. The left mandible (Figs. 813, 814) bears three teeth on both upper and lower sides. Each mandible has a pair of setae arising from the outer margin. The maxilla (Fig. 810) bears a four-segmented palp, each segment being sclerotized on its outer surface, but the basal segment has an irregularly shaped sclerite whose inner proximal part is produced backwards. The distal segment has a few sense-organs at its tip. The maxillary lobe bears a cluster of sense organs distally and one or two on its inner border. Proximally there are three inner spines and a series of fine hairs. The labial palps (Fig. 810) bear a small papilla distally. In cleared, mounted

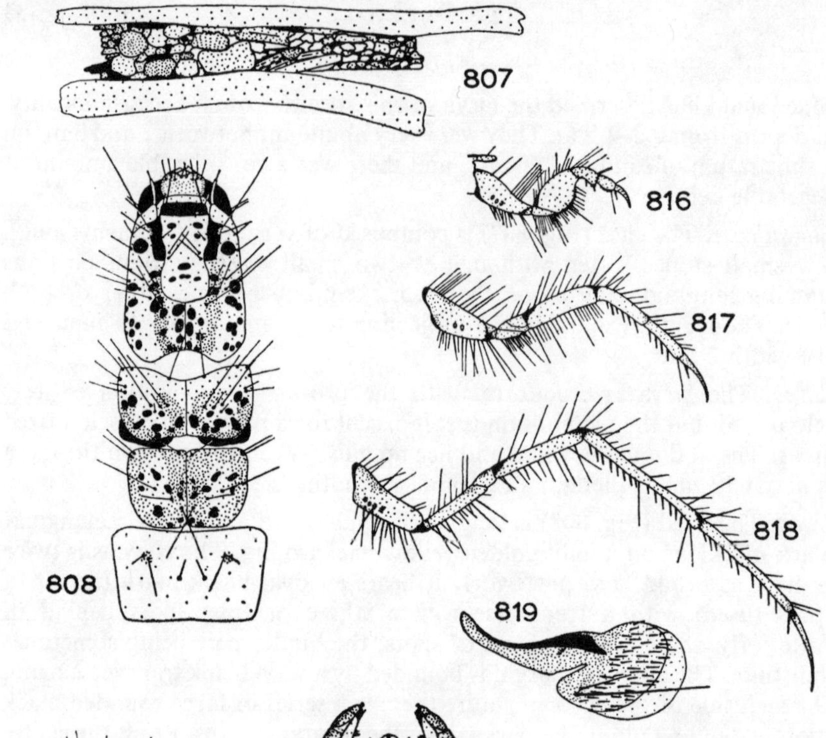

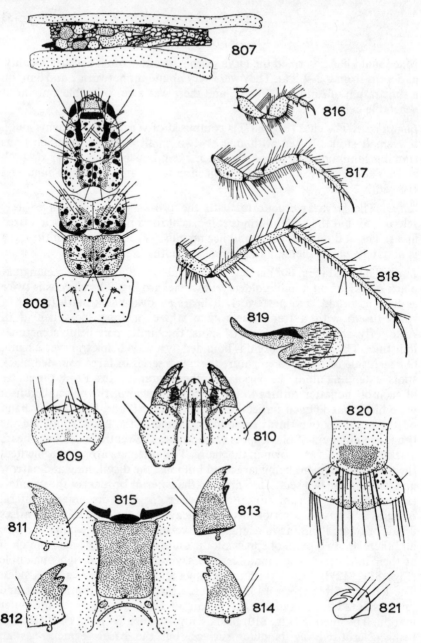

Figs. 807–821. *Mystacides azurea* (L.). 807. Case; 808. Head and thorax; 809. Labrum; 810. Labium and maxillae; 811. Right mandible (dorsal view); 812. Right mandible (ventral view); 813. Left mandible (dorsal view); 814. Left mandible (ventral view); 815. Gular sclerite; 816. Right prothoracic leg; 817. Right mesothoracic leg; 818. Right metathoracic leg; 819. Right lateral sclerite of first abdominal segment; 820. Anal sclerite and appendages (dorsal view); 821. Anal claw. Macdonald.

specimens the silk press is clearly seen below the prementum. The gular sclerite (Fig. 815) is rectangular and divides the genae, and has two small elliptical sclerites and a narrow, transverse, bow-shaped sclerite proximally.

Thorax (Fig. 808). The pronotum is sclerotized and is golden yellow in colour with dark markings. Except medially, the posterior border is black. A suture divides the pronotum longitudinally and on each side of it there is a brown horseshoe mark, the free ends directed forwards with one limb running near and diverging from the median suture, the other extending along the pleural border. Neither extends as far as the anterior margin. Superimposed on this horseshoe mark and within it, are a number of black spots. There is no prosternal horn. The mesonotum is sclerotized and is a little narrower than the pronotum. A suture divides it longitudinally into two sclerites, each of which is marked with light brown, only one small golden yellow area remaining in the centre, in and around which are about six black spots. On each side of the suture on the posterior margin are two additional spots. The metanotum is soft except for two small rounded lightly sclerotized patches, each situated midway between the centre and the lateral margins of the segment and bearing two setae. The prothoracic leg (Fig. 816) is short, with a broad femur, and the distal end of the tibia is slightly dilated. The trochanter, femur and tibia each bears two spines on the under surface. In addition, the trochanter has a row of setae; the femur, setae and a series of fine hairs; and the tarsus, a series of small spinules on the lower margin. The mesothoracic leg (Fig. 817) is longer with large numbers of setae on the coxa, trochanter and femur. In the latter segment the setae are regularly placed and are of uniform size, with the exception of four which are much longer than the rest. The under surface of both the tibia and the tarsus bears a series of four or five spines. The metathoracic leg (Fig. 818) is the longest, with irregularly placed setae on the coxa, trochanter and femur, and a large series of equidistant spines on the tibia and tarsus, each of which has its distal half attenuated. The tarsal claw of each leg is relatively long and has a basal spine.

Abdomen. The abdomen is white in colour and has no gills. The first segment has a dorsal and two lateral protuberances, each of the latter bearing a sclerite with two setae and a patch of small, forwardly directed spines (Fig. 819). There is a lateral row of about twenty small, elevated, chestnut-coloured pustules on the eighth segment, in line with the lateral line of fine hairs on segments three to seven inclusive. The anal sclerite (Fig. 820), lying dorsally on the ninth segment, is half-moon shaped and brown in colour. On the posterior margin of this segment there is a series of setae. On the proximal segment of each anal appendage there are six or seven dark spots and five or six strong setae. The anal claw (Fig. 821) is strongly curved and bears three mid-dorsal denticles, of which the middle one is smallest.

Mystacides longicornis (L.)

On 30 May 1952, at 7 p.m., I visited the Bournville Yachting Pool, Birmingham, which has been a fruitful source of the Caddis *Molanna angustata* Curt. (1946, *Proc. R. ent. Soc. Lond.* (A) **21**, 55), and there I found adults of *Mystacides longicornis* L. flying strongly in many thousands. My identification was kindly confirmed by Mr. D. E. Kimmins. The larvae were present in large numbers, crawling up the vertical algae-covered concrete sides of the pool. Vertical seams in the concrete were utilized by the larvae as a site for pupating, the cases being fixed by their hinder (narrower) ends.

This caddis is the 'grouse-wing' of the fly fisherman and I include a drawing of it (Fig. 822). It should be noted that no trace of the dark bands can be seen in rubbed specimens.

Larva. Much like *Mystacides nigra* (L.) and *M. azurea* (L.), differing, however, as follows:

Head (Fig. 824) more heavily marked in black than in *M. azurea* but not so heavily marked as in *M. nigra* in the region of the vertex of the clypeus, and the black transverse band on the oral end of the clypeus in the latter insect is replaced by four small black spots in the present species. On the ventral surface of the genae are about five black spots on each side at the aboral end (Fig. 825), and the marking of the pro- and mesonotum is more heavy than in *M. nigra*, although resembling it (Fig. 826). Lateral sclerite of first abdominal segment rather more strongly curved behind

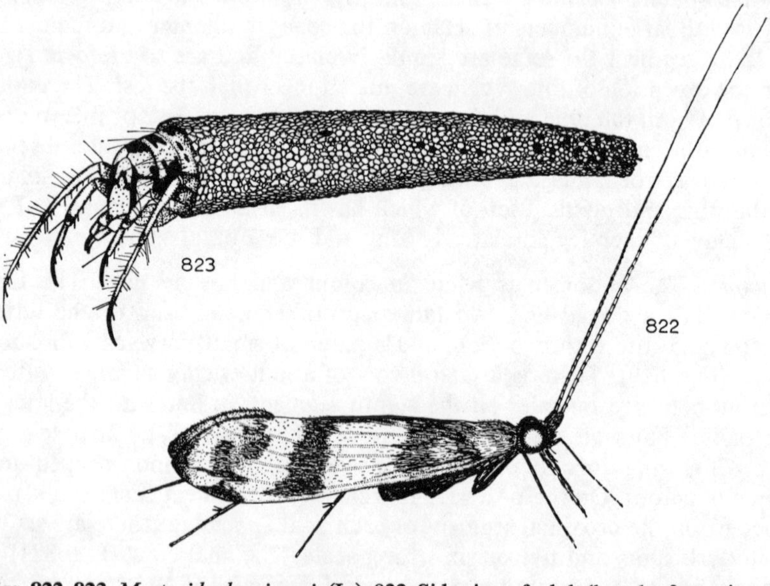

Figs. 822–823. *Mystacides longicornis* (L.). 822. Side view of adult (length of anterior wing 9 mm.); 823. Larva in case.

than in the two compared species (Fig. 827), and it should be noted that the spines project anteriorly, not as might be understood from my figure of this sclerite in *M. nigra* (L.). Gills present on second to seventh or eighth abdominal segment. Length 11–12 mm., width 1.5 mm. Length of case 15 mm., width 2 mm.

A convenient key for differentiating the larvae of the three British species of *Mystacides* is as follows:

1. Heavy black bands on genae bordering clypeus and large black mark in middle of mesonotum ... 2
— Black band on the genae bordering only the oral half of the clypeus, black marks in middle of mesonotum very small ... *azurea* (*L.*)
2. About five black spots present on ventral surface of genae on each side of gular sclerite at aboral end. Gills present on the second to the seventh or eighth abdominal segments ... *longicornis* (*L.*)
— Black spots absent on ventral surface of genae. Gills present on the second to fourth abdominal segments ... *nigra* (*L.*)

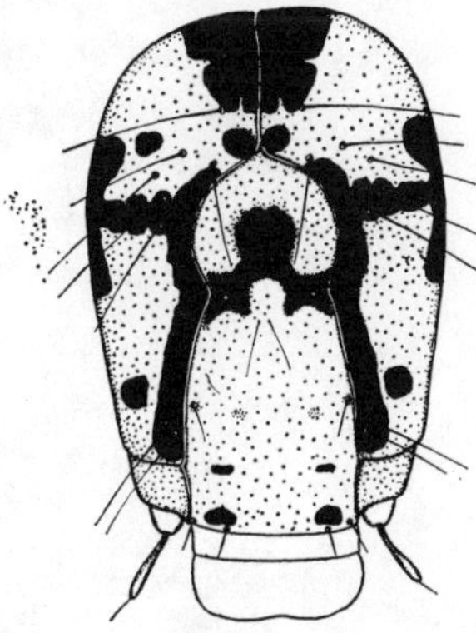

824

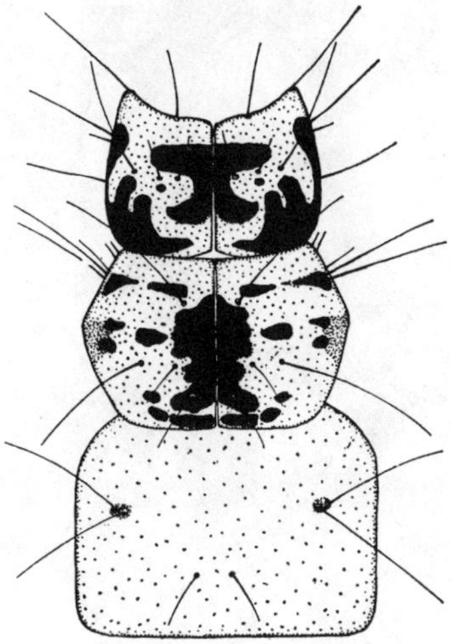

826

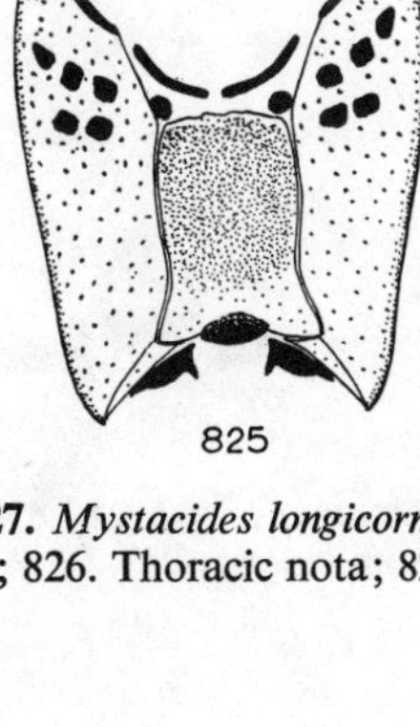

825

827

Figs. 824–827. *Mystacides longicornis* (L.). 824. Head from the front; 825. Head from behind; 826. Thoracic nota; 827. Lateral sclerite of first abdominal segment.

Mystacides nigra (L.)

Larvae of this species are found in rivers, streams, lakes and ponds. The specimens from which this description is taken were collected from a small fast-running stream at Temple Balsall, Warwickshire. The larvae were present in considerable numbers on the lower parts of stems of marginal

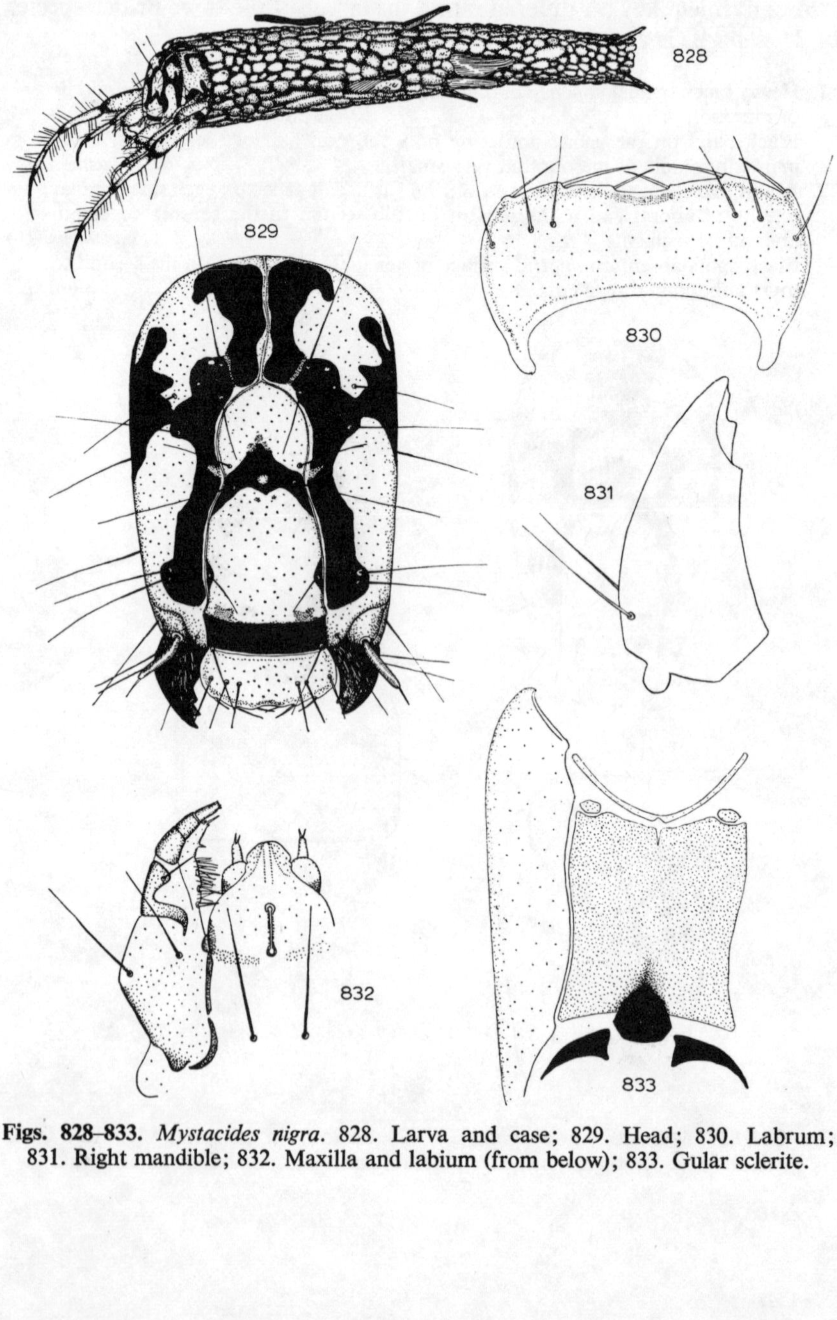

Figs. 828–833. *Mystacides nigra*. 828. Larva and case; 829. Head; 830. Labrum; 831. Right mandible; 832. Maxilla and labium (from below); 833. Gular sclerite.

plants. They were also collected from Wray, Lake Windermere, where they were found crawling over algae-covered stones at the lake edge; and the case was constructed of small flakes of stone.

Case (Fig. 828). Cylindrical, tapered and very slightly curved. It is constructed of sand-grains, cemented together with secretion, to which are attached small portions of hard vegetable debris usually dark in colour.

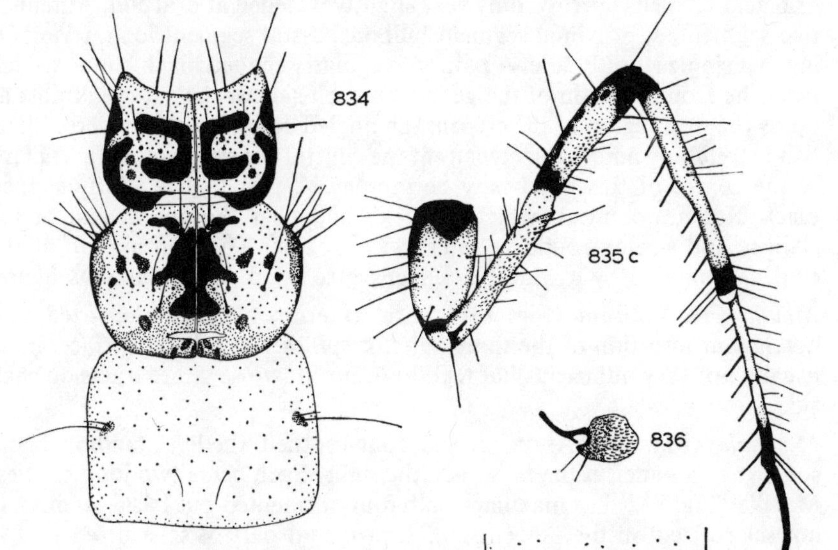

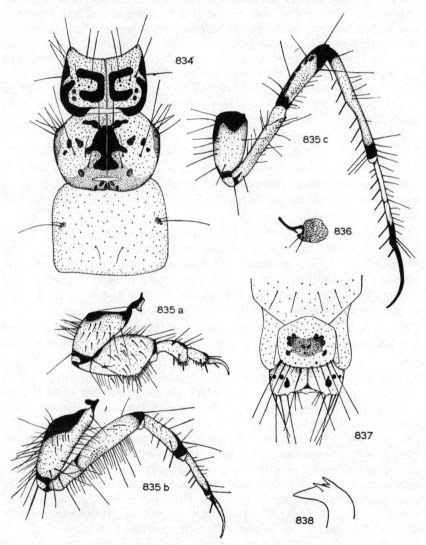

Figs. 834–838. *Mystacides nigra.* 834. Thorax; 835. Legs: a. prothoracic; b. mesothoracic; c. metathoracic; 836. Lateral sclerite on first abdominal segment; 837. Anal segments; 838. Anal claw.

It is also reported that the case is occasionally constructed entirely of vegetable debris. Length up to 15 mm., width 2.0–2.5 mm.

Larva. Eruciform, head procentrous. Pro- and mesonota sclerotized, metanotum with a pair of very small sclerotized patches. Abdomen tapering slightly from second segment. Length 12 mm., width 1.5 mm.

Head (Fig. 829). Long, light golden yellow strongly marked in dark brown or black. Clypeus narrow, only very slightly widened at oral end. Antennae two-segmented, proximal segment bulbous, distal segment long, terminating in a single stout bristle. A pair of irregularly shaped black bands stretch from the frontal region of the genae, on each side of the suture, almost as far as the oral region of the clypeus, an angled connecting transverse black band stretching across the clypeus at the central constriction. A light mark in the centre of this band may be present. A pair of irregularly shaped black blotches connect with the black band near the aboral end of the clypeus and stretch round to the cervical region. The oral margin of the clypeus bounded by a wide black transverse band. Chaetotaxy as figure.

Mouthparts. Labrum (Fig. 830) with anterior margin sclerotized and notched at insertion of the three pairs of spines. On ventral surface, hairs usually not very numerous but tend to occur in two separate tracts on each side.

Mandibles (Fig. 831). Asymmetrical, four-toothed, the left mandible being somewhat broader at the base than the right. Each bears two long bristles. Maxilla (Fig. 832) has maxillary palp four-segmented but basal segment is not sclerotized on the inner face and proximal part of sclerotized part is produced inwardly. Maxillary lobe almost as long as palp. Labial palps have bulbous base dark in colour at the proximal end.

Gular sclerite (Fig. 833). Quadrangular with two small distinct elliptical sclerites and a single narrow transverse bow-shaped sclerite at the proximal end.

Thorax (Fig. 834). Pronotum sclerotized, golden yellow heavily marked in black. Posterior margin with wide black border except for small central portion. This border turns anteriorly along the pleural region and may contain several light-coloured areas. A large X-shaped black mark in the centre of the pronotum divided longitudinally by the suture. Prosternal horn absent. Mesonotum sclerotized, little wider than pronotum, golden yellow, heavily marked with black. Posterior margin and pleural region greyish-brown. Central black mark divided by longitudinal suture and has three wing-like projections on each side. The position of other smaller dark marks is shown in Fig. 834. Metanotum not sclerotized except for a pair of very small patches, from each of which arise two bristles.

Legs (Fig. 835, a, b and c). Prothoracic short, femur broad, tibia with lower margin dilated at distal extremity, each tipped with black on its upper surface. Mesothoracic leg longer, femur, tibia and tarsal segment tipped with black. Femur with region of small spines occurring singly, and in groups of two, three and four. Metathoracic leg very long; segments tipped with black. Tarsal claw long. Trochanter and femur with regions of small grouped spines. Distal half of tibia attenuated.

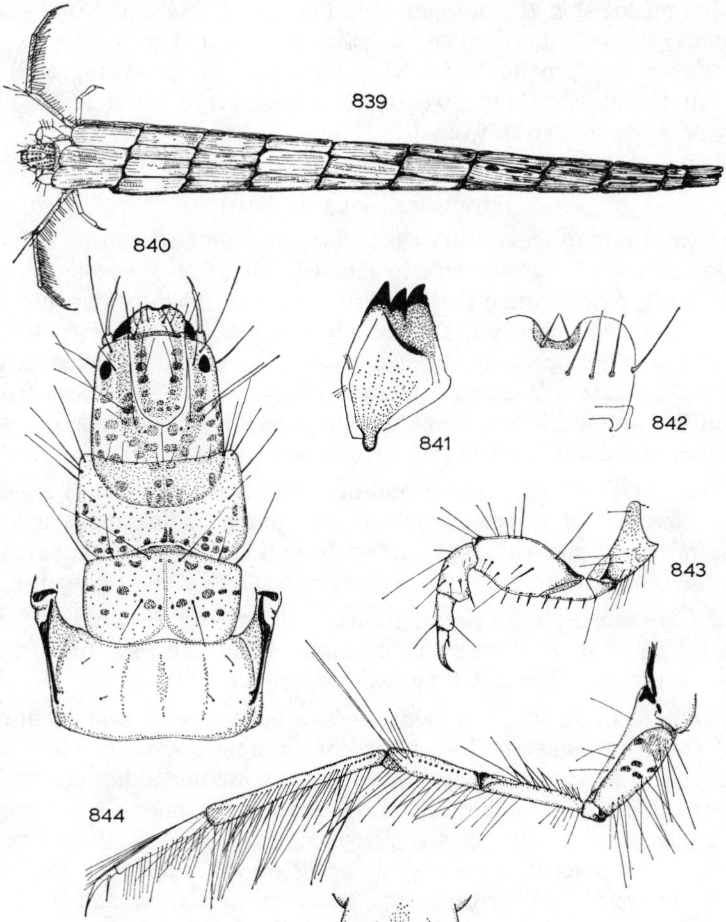

Figs. 839–846. *Triaenodes bicolor*. 839. Larva in case; 840. Head and thorax; 841. Mandible; 842. Labrum; 843. Prothoracic leg; 844. Metathoracic leg; 845. Anal segments; 846. Antenna.

Abdomen. White in colour. First abdominal segment narrow. Lateral protuberances on first abdominal segment with sclerite and patch of posteriorly projecting spines (Fig. 836). Dorsal protuberance present. Lateral line not prominent. Lateral row of about eighteen small chestnut-coloured pustules on eighth segment in line with lateral line. Anal sclerite irregularly shaped, darker in colour at lateral extremities. Transverse row of dark marks on proximal lobes of anal claws. Anal claws (Fig. 838) very small, furnished with three auxiliary claws, one being minute. Gills occur singly in segments two to five.

Triaenodes bicolor (Curtis)

The larva of this species is very easily distinguished both morphologically and on account of its characteristic habit of swimming. It is able to propel itself with its case through the water by means of rapid movements of its metathoracic legs, which are especially adapted for the purpose. The following description of the larva is based on material collected at Clay Pond, Wray Castle, Windermere, and compared with specimens from a pool in Broadmoor Wood, Rubery, Worcestershire. Both these pools are still water and contain much phanerogamic vegetation.

Case (Fig. 839). The case is long and tapering. It is composed of regularly shaped pieces of leaf or small whole leaves. All the fragments are arranged in a spiral whorl, the whole case being strongly reminiscent of a miniature *Phryganea grandis* case. Size up to 35.0 mm. in length, 2.0–2.5 mm. in width.

Larva. The head is almost prognathous. A prominent feature of the larva is the length of the metathoracic 'swimming' legs and their hair fringes. Size of larva 12.0–13.0 mm. long, 1.5 mm. broad.

Head (Fig. 840). Very bright golden yellow marked with prominent dark spots. Two dark bands run longitudinally along the clypeus in which some of the dark spots are confluent. The dark bands are united posteriorly by a transverse band of four dark spots. A larger but similar shaped pair of dark bands runs parallel to the first-mentioned bands just outside the clypeus. Other patches of dark spots occur on the genae, and four large spots lie between the four transverse spots on the clypeus and the larger pair of dark bands where they run together. The antennae are two-segmented. The proximal segment is bulbous, whilst the distal segment is long, slightly incurved and surmounted with a strong bristle (Fig. 846).

Mouthparts. The mandibles (Fig. 841) are asymmetrical. The maxillary palps are three-segmented, whilst the ligula is broad. The single-segmented labial palps are very small. All the mouthparts are very small in comparison with the size of the larva. The labrum is notched, the concavity being sclerotized (Fig. 842).

Thorax (Fig. 840). Pro- and mesothorax strongly sclerotized, deep golden yellow in colour and marked with prominent dark spots. Metathorax unsclerotized but with long dark coxal sclerites with which the swimming legs articulate. The prothorax has patches of dark spots situated along the posterior margin. Several stout bristles are placed on the anterior angles of the segment. In the mesothorax a short row of dark spots lies along the anterior margin, together with a row of about six long stout bristles. In addition a transverse row of spots lies across the middle of the segment and divides into two a little way from the median line.

Legs. Prothoracic legs short and broad. Femur has three marginal spines and the trochanter four. Tarsal segments short, proximal segment very broad (Fig. 843). Metathoracic legs long with two rows of hairs running along whole length (Fig. 844).

Abdomen. Dorsal and lateral hooks present on first segment of the abdomen. Single gills are usually present on the dorso-lateral and ventro-

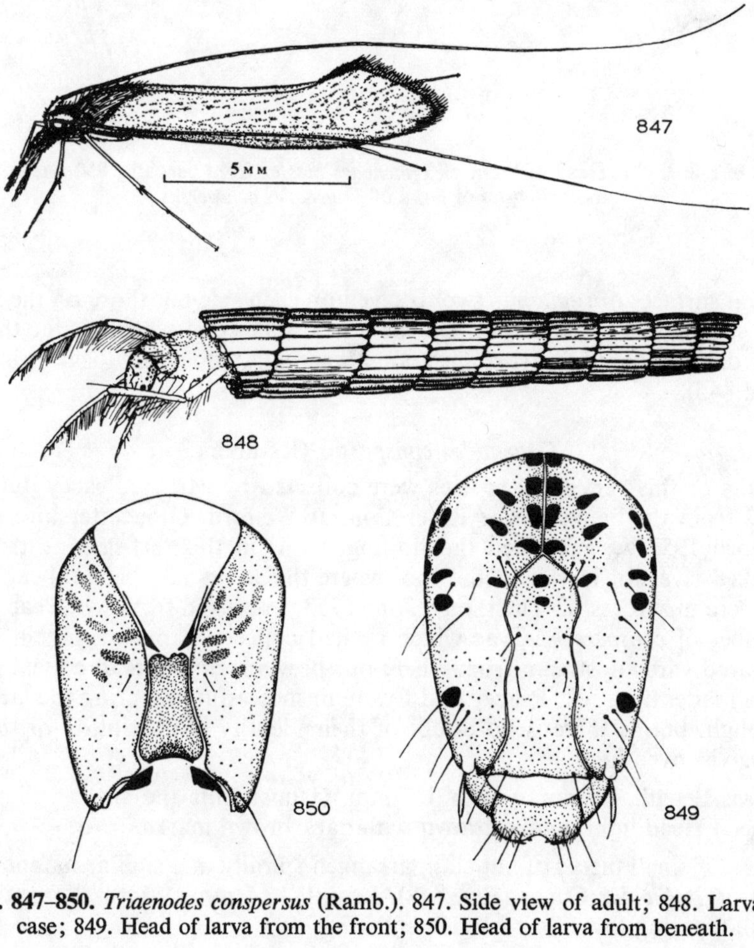

Figs. 847–850. *Triaenodes conspersus* (Ramb.). 847. Side view of adult; 848. Larva in case; 849. Head of larva from the front; 850. Head of larva from beneath.

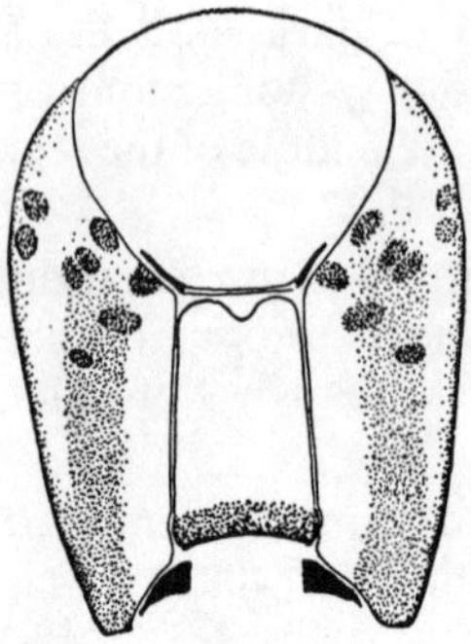

851

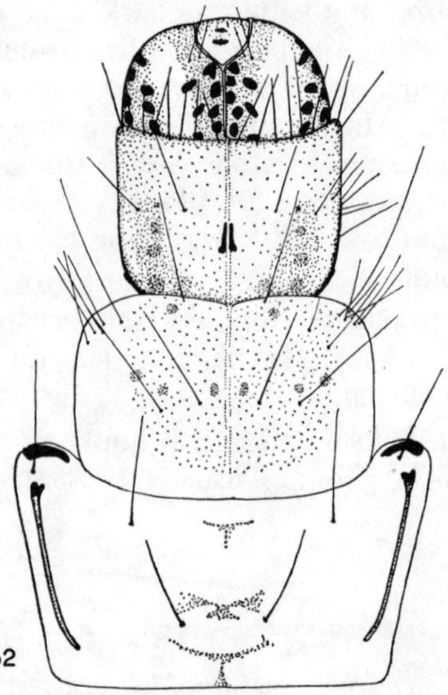

852

Figs. 851–852. 851. Head of larva of *Triaenodes bicolor* from beneath; 852. Head and thoracic nota of larva of *Triaenodes conspersus*.

lateral surfaces of segments two to seven on each side but those on the last two segments may be absent. On the dorsal surface of segment nine there is a dark-coloured heart-shaped anal sclerite. It does not bear bristles (Fig. 845).

Triaenodes conspersus (Rambur)

Adults of this very local species were collected by Mr. A. Peacey during 1952 from the banks of the River Coln at Welford, Gloucestershire. On 9 April 1953 we examined the trailing weeds in this fast-flowing trout-stocked river for larvae, at the spot where the adults had been taken, but we were unsuccessful. Then, on 9 June 1953, I received from Mr. Peacey a number of *conspersus* larvae which he had collected from a point a few hundred yards upstream from where our previous attempts to collect had been made. Both Mr. Peacey and I were unsuccessful in rearing the larvae through, but as there is no doubt of their identity a description of these larvae is given here.

Larva. Length 12 mm. Width 1.5 mm. (Ulmer puts the size as slightly larger.) Head light golden brown with dark brown marks.

Case. Of small pieces of cut stalk arranged spirally in a similar manner to that of *T. bicolor* Curtis (Fig. 848). Length 17.5 mm. Width at front 2.4 mm., tapering towards posterior end.

Larva very similar to that of *Triaenodes bicolor*. It has also the habit of being able to swim through the water with its case, by means of rapid beats of its long, heavily fringed metathoracic legs. The larvae may, however, be easily differentiated as follows: the marginal bands of dark marks on either side of the clypeus in *bicolor* are absent in *conspersus*. Substantial differences exist between the two species when the head is viewed from beneath. In *conspersus* the gular sclerite is dark grey in colour with concave lateral margins, and the width at the aboral end is much less than that at the oral end. Whereas in *bicolor* only the oral end of the gular sclerite is dark grey, the lateral margins are straight and the width at the aboral end is little less than that at the oral end. The longitudinal dark bands stretching almost the whole length of the underside of the head in *bicolor* are absent in *conspersus*.

Key for Separation of Two Known Species of Triaenodes

1. Head yellow with prominent dark spots and dark longitudinal bands extending to anterior border of head (Fig. 840). Common in lakes and ponds *bicolor*

— Head yellowish brown with dark spots confined to occipital area, no dark band extending down to anterior border of head (Fig. 849). Local in running water which is rich in vegetation *Conspersus*

Triaenodes simulans Tjeder

The larva of this species is unknown.

Triaenodes reuteri McLachlan

The confusion which has existed concerning this and the preceding species has been resolved by Mr. E. C. Pelham-Clinton. *T. reuteri* is a saltmarsh species collected at Spurn, Yorks., in 1963. Brindle has described the larva and given the following key for separating the three known larvae of *Triaenodes*.

Key to Known Larvae of Triaenodes

1. Head broader, and with longitudinal dark bands on the dorsal surface of the genae and on the fronto-clypeus; ventral apotome short and broad; cases longer and more tapering; in fresh water habitats, in ponds and lakes rich in aquatic vegetation *bicolor* (Curtis)

— Head narrower and more elongate, without any longitudinal dark bands; cases shorter, more parallel-sided and truncate posteriorly; in lotic water or in saline habitats 2

2. Ventral apotome short and broad; head more heavily marked with dark spots, five being on the fronto-clypeus; characteristic of saline habitats *reuteri* McLachlan

— Ventral apotome long and narrowed posteriorly; head less heavily marked with spots, only two dark spots on fronto-clypeus (Hickin, 1954); in rivers *conspersus* (Rambur)

Erotesis baltica McLachlan

No description of the larva of this species based on British material is available.

Mosely states that it abounds along parts of the River Test in Hampshire and although it is usually thought to be associated with rivers the author has collected it from the shores of Lake Windermere. Lestage states that the distinctive case is constructed of small lengths of vegetable matter arranged in more or less circular bands along one side of the case and transversely along the other. The junction of the two arrangements make a zig-zag ridge along the case. It may be up to 10 mm. in length.

The larva is rather like that of *Triaenodes bicolor* but is smaller, being only about 8–5 mm. in length. It differs also in possessing a characteristic head pattern, being brown with the eyes situated in yellow patches and a light longitudinal band running the length of the clypeus dilated in the centre. This light band extends to the top of the head where it dissipates. The apical segment of the antenna is of exceptional size, often being upwards of 0.25 mm. in length. Gills are absent.

Adicella filicornis (Pictet)

The larva of this species was described by Morton and the description is given here in his original words:

Larva slender, with reddish head, prothorax and legs, the rest of the body whitish.

Head elongate, oval, sparingly behaired; clypeus short and broad; under-piece large, subquadrangular, angles rounded, slightly narrower in its posterior part.

Antennae very highly developed, long, fusiform, placed on a rounded base, obliquely truncate at apex, which bears a single spinous hair.

Labrum transverse, sides rounded, anterior margin rounded and excised; disc with six or so strong hairs.

Mandibles rather short; right deeply toothed, left less conspicuously so.

Maxillae elongate, with three-jointed palpi; labium sub-conical.

Prothorax, seen from above, little broader than the head; pronotal plate pilose, broadly transverse, nearly straight in front, sides slightly rounded, posterior margin shallowly excised.

Mesothorax broader, the notal plate indistinct, because con-colorous with the metanotum, which is membranous.

The legs, which are pilose, are comparatively strong, and there is not such great disparity between the length of the pairs as in some other genera, the distal joints, especially of the fore-legs, spined internally.

Abdomen. First segment broadest of all, gradually tapering thereafter; the first segment bears the usual protuberances, dorsal and lateral, the latter in the shape of a rounded space covered with points and posteriorly running into a long black chitinous band. Tracheal branchiae apparently ill-developed, filamentose, single. Fringe also ill-developed, composed of minute hairs, modified on eighth segment to a row of strong points. Free part of anal limbs short, provided with double hooks; about the anal region are several series of regular spines or spinous hairs.

The cases are strongly curved and tapering, 8 to 10 mm. long; they are russet coloured, sometimes partially blackish, and are so smooth that they look as if altogether membranous; they seem to be composed of very fine sand or mud fixed to a strong inner silken tube. They are usually found attached to mosses growing in and around trickling springs, which the insect haunts along with *Beraea*, *Crunoecia*, and *Diplectrona*.

Morton goes on to say that the head and prothorax are 'unicolorously reddish' and unmarked.

Adicella reducta (McLachlan)

Immature stages unknown.

Oecetis furva (Rambur)

This is a local species also found in lakes and ponds. No description is available based on British material. It is a small species, the larva measuring only from 7 to 9 mm. in length. The case is straight, constructed of pieces of vegetable matter arranged transversely. The head of the larva is pale yellow ornamented with numerous dark spots and marks. There are two transverse marks in the posterior angles of the clypeus and six spots in two longitudinal series in the anterior region. The mesonotum is brownish yellow ornamented with rather indistinctive spots and a clear triangular mark at each of the anterior angles.

Oecetis lacustris (Pictet)

This is the smallest of the British *Oecetis* species, is fairly widely distributed and inhabits lakes and ponds. No description is available based on British material.

The case is generally composed of fine sand-grains, strongly curved and extremely tapering. The anterior end also narrows slightly and is sometimes furnished with rather larger sand-grains or bits of vegetable material. The larva measures from 8 to 10 mm. in length and from 1.4 to 1.8 mm. in width. The head is pale yellow with a number of dark marks (sometimes simply yellow). A long median mark is present between the eyes whilst in the posterior region of the clypeus there is a bicoloured mark generally paler in the centre, appearing to be formed of five small confluent spots.

The pronotum is pale yellow, brownish on the posterior half and ornamented, as is the mesonotum, with a number of brownish, generally indistinct, marks.

Oecetis notata (Rambur)

Immature stages unknown.

Oecetis ochracea (Curtis)

A description of the larva of this species based on British material is not available. It is widely distributed in Britain, frequenting lakes and ponds. Whereas larvae of all other genera in the LEPTOCERIDAE subsist on vegetable matter, larvae of the genus *Oecetis* are said to feed on small arthropods.

The larval case of *O. ochracea* is made of sand-grains intermixed with rather larger pieces of mineral matter or bits of vegetable matter, and is conical but only slightly curved. The head is yellow and ornamented with a series of six spots arranged in a semi-circle on the fronto-clypeus. There is a large spot situated between the eyes which is surrounded by dark spots. The pronotum is pale yellow with a large median transverse 'smoky' band which sometimes reaches the anterior border. The mesonotum is pale yellow in front and at the sides with a small yellowish median zone ornamented with several dark brown spots.

The larva is 14 mm. in length.

Oecetis testacea (Curtis)

Immature stages unknown.

Key to known Larvae of Oecetis

1. Claws of middle and posterior legs each with a prominent basal spine; two or three setae on each side of metasternum; head and pronotum pale yellow with numerous dark spots; cases straight, of pieces of vegetable material arranged transversely. Local, in lakes and ponds — *furva*

— Claws of middle and posterior legs with rudimentary basal spines; numerous setae on each side of metasternum; cases curved, usually of sand-grains — 2

2. Prosternum with numerous setae near anterior coxae; head pale yellow with dark spots, pronotum pale yellow, darkened on posterior half with brownish spots; case strongly curved. Widely distributed in lakes and ponds — *lacustris*

— Prosternum with a single seta near anterior coxae; head yellow with a series of six spots arranged in a semicircle on fronto-clypeus; a large spot between the eyes surrounded by dark spots; pronotum pale yellow with large dark transverse median band; cases weakly curved. Frequent, in lakes and ponds — *ochracea*

Leptocerus interruptus (Fabricius)

Larvae of this species have not been described from British material.

It is local in distribution and found in running water, generally clear streams. The larval case is made entirely of secretion, is from 12 to 15 mm. in length and is only slightly curved. (It is more curved than that of *L. tineiformis*, however.) The posterior legs are furnished with swimming fringes and Ulmer thought that the median legs were rather more robust than those of *L. tineiformis*.

Leptocerus lusitanicus (McLachlan)

Immature stages unknown.

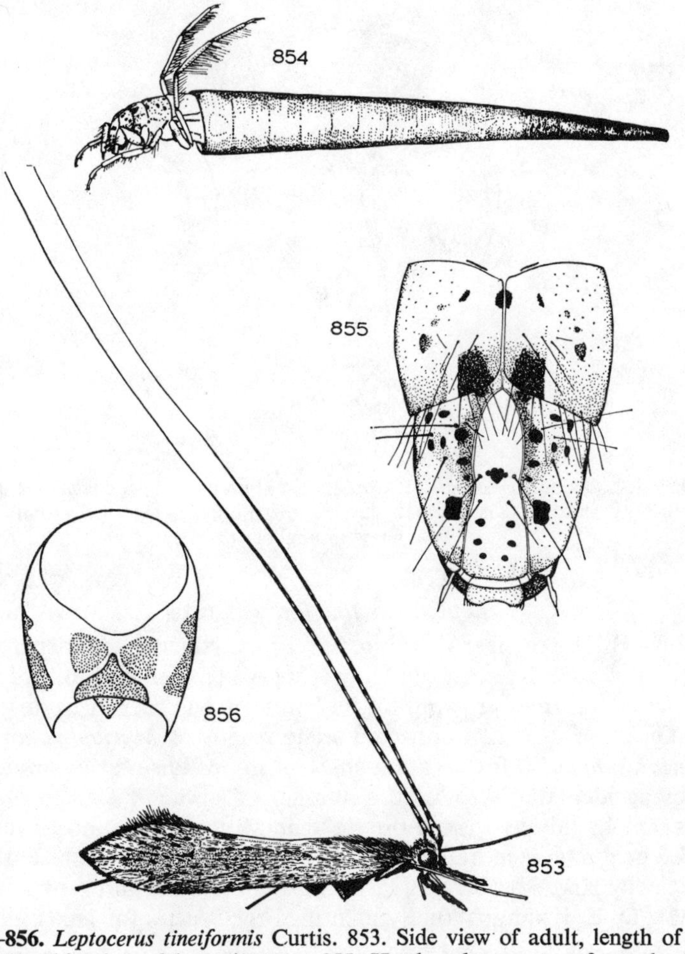

Figs. 853–856. *Leptocerus tineiformis* Curtis. 853. Side view of adult, length of body 8 mm.; 854. Side view of larva in case; 855. Head and pronotum from the front; 856. Head from below.

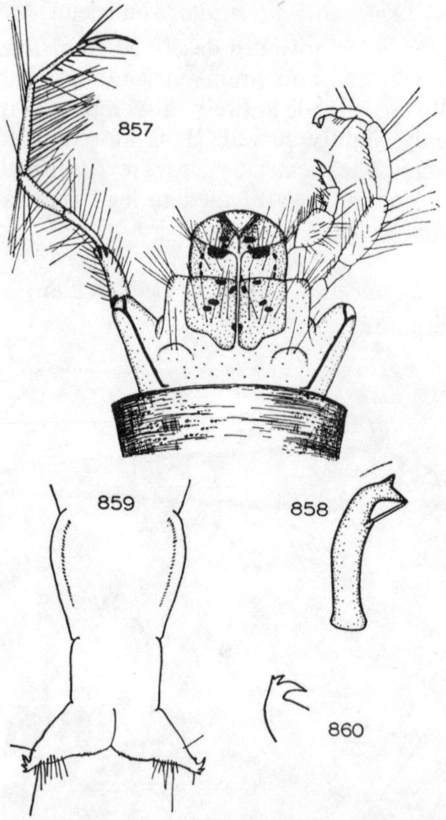

Figs. 857–860. *Leptocerus tineiformis* Curtis. 857. Thoracic nota and legs (one leg of each segment only shown); 858. Tarsal claw of mesothoracic leg; 859. Anal segments; 860. Claw of anal clasper.

Leptocerus tineiformis Curtis

Old Bury Hill Lake, near Dorking, Surrey, is artificially constructed by the damming of a small stream. Judging from the large dimensions of many of the ornamental trees growing on its banks, it has been there a very long time. On 28 May 1952 I collected some fronds of *Myriophyllum* sp. and *Ranunculus aquatilis* for aerating some of my caddis-rearing aquaria, and thereby accidentally introduced a number of larvae of *Setodes tineiformis* Curtis. On 13 July I revisited the Lake and found that almost every frond of weed held one or more larvae of this caddis. It must be present there in fantastically large numbers. A series was reared through to the adult stage and Mr. D. E. Kimmins kindly identified the species for me (Fig. 853).

Case. Yellowish-brown, of secretion only except at posterior end, where some dark-coloured material appears to be admixed. Anterior end transparent, through which the green abdomen of the larva can be seen. When empty, very like a hedgehog quill in appearance, only very slightly curved. Length 12–15 mm. Just over 1 mm. wide at anterior end (Fig. 854).

Larva. Length 8 mm., width 0.75 mm. Head orthocentrous, considerably retracted into prothorax. Pro- and mesonota sclerotized.

Head. Golden yellow. Pattern of dark brown spots as shown in Fig. 855. Eyes appear to be situated closer to the mouth than in *Setodes argentipunctella.* Antennae well developed; distal half of distal segment blade-shaped.

Gular sclerite. It is difficult to make out exact limits; it may, however, be

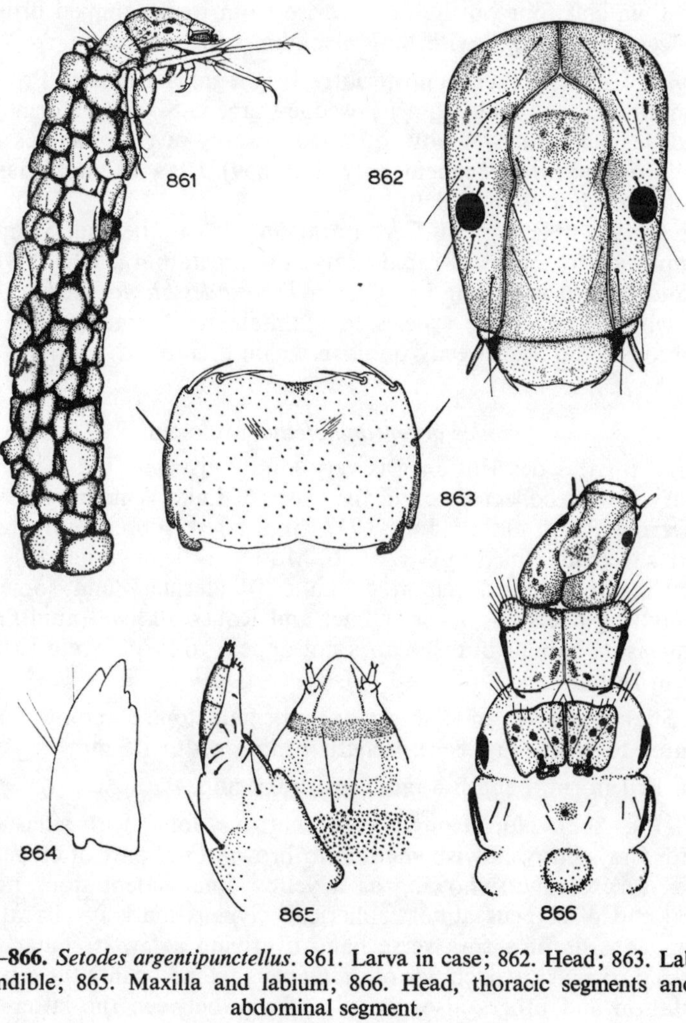

Figs. 861–866. *Setodes argentipunctellus.* 861. Larva in case; 862. Head; 863. Labrum; 864. Mandible; 865. Maxilla and labium; 866. Head, thoracic segments and first abdominal segment.

the light brown triangular area lying immediately below the paired light brown areas (Fig. 856).

Pronotum. Golden yellow with two black areas near centre of anterior margin.

Mesothorax. Half as wide again as pronotum.

Mesonotum. Slightly darker in colour and having a pattern of black spots (Fig. 857).

Legs. Prothoracic legs short; meso-legs intermediate in size with characteristic tarsal claw (Fig. 858). Tip of claw bifid, and tip of tarsal bristle just touches tip of nearest claw. A series of eight bristles on the ventral surface of the tibia and four on the tarsus arise from wedge-shaped projections. Meta-legs long, fringed with long black hairs for swimming.

Abdomen. Conical, tapering posteriorly, bright green in colour. Protuberances on first abdominal segment wedge-shaped. Single filiform gills on second segment only. A row of about twenty-seven tubercles situated laterally on the ninth segment only (Fig. 859). Claw of anal clasper with two auxiliary hooks (Fig. 860).

The larva is able to swim freely, dragging its case behind. Swimming is accomplished by means of rapid beats of the metathoracic legs, which are long and fringed with long hairs, as in *Triaenodes bicolor* Curtis. It is this habit which enables the species to maintain itself amongst the surface vegetation in the deep weedy pools in which it is found.

Setodes argentipunctellus McLachlan

Material for this description was very kindly given by Dr. T. T. Macan. The larvae were collected round the shore of Lake Windermere, south of Watbarrow Point, on 13 July 1937, and a few subsequently reared to maturity were identified by Mr. M. E. Mosely as belonging to this species. It inhabits the littoral region of Lake Windermere and some of the neighbouring lakes and tarns. Ulmer and Rousseau give rapid mountain streams as its habitat but this does not appear to be characteristic of this insect in Britain.

Case. Slightly curved, of sand-grains and small stone chippings and flakes cemented together (Fig. 861). Length 7 mm., width 1.5 mm.

Larva. Eruciform. Length 6 mm., width 0.9 mm.

Head (Fig. 862). Not strongly orthocentrous, long, with sides tapering towards oral end, yellowish shading to brown. Oral part of clypeus long; sutures very distinct, showing as a yellow line, indentations not deep. Aboral end of clypeus almost spherical, clypeus mark in aboral part of clypeus consists of a transverse band overlying a few irregularly placed spots. A dark mark at each side of the clypeus extending into the genae at the indentation and others also about half-way between the latter and the apex of the clypeus. Antennae with bulbous proximal segment and comparatively long filiform distal segment.

Mouthparts. Labrum (Fig. 863) with two pairs of inwardly directed light yellow spines on the anterior margin and with a bristle on each lateral margin. More heavily sclerotized on the lateral margins and curving inwards posteriorly and terminating in a short, rounded projection. Mandibles (Fig. 864) four to five toothed with two bristles on outside edge. Maxillary palp (Fig. 865) four-segmented, maxillary lobe long and slender, almost as long as the palp, and with a few claw-like spines at the base. Tips of palp and lobe provided with sense-organs. Labium with a transverse sclerotized bar curving posteriorly on the ventral surface. Base of labium spinulose with two long bristles.

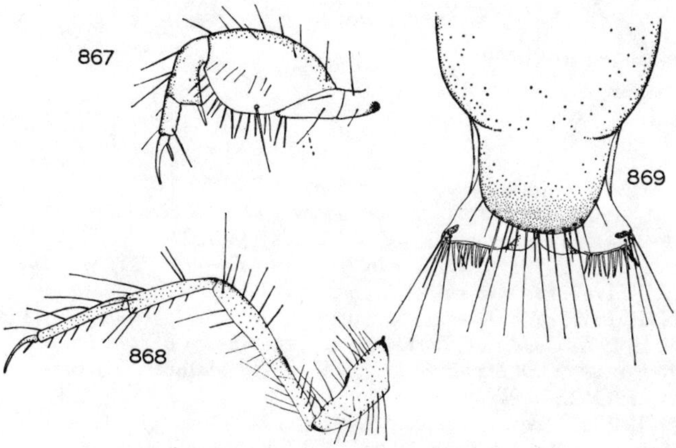

Figs. 867–869. *Setodes argentipunctellus.* 867. Prothoracic leg; 868. Metathoracic leg; 869. Anal segments.

Thorax (Fig. 866). Pronotum sclerotized with a median longitudinal suture and two lateral sutures cutting off the anterior angles. Each angle provided with hairs. Two or three dark marks along each side of the median suture, lateral margin heavily sclerotized, black in colour. Mesonotum sclerotized, consisting of two almost rectangular plates (divided by a median suture) with two transverse rows of bristles and one transverse row of dark spots concave anteriorly and a larger spot situated in the middle of each sclerite near the posterior margin. In the concavity of the posterior margin is a pair of dark marks (sometimes confluent) on each side with a few very small hairs. Coxa pleurites of meso- and metathorax black, heavily sclerotized. Metanotum unsclerotized but with a faint dark mark in the centre.

Legs. Prothoracic legs (Fig. 867) very short and broad, usually held bunched up under the head. Tibia with an angle projecting ventrally and provided with a small spine. Meso- and metathoracic legs (Fig. 868) long and thin with a few hairs but not constituting a 'swimming fringe'. Metathoracic legs slightly longer than the mesothoracic.

Abdomen. Long, greyish-white to yellow (with a reddish tinge in some specimens), tapering distally. Slight lateral and dorsal protuberances on the first segment. Last segment (Fig. 869) much narrower than the preceding with a number of long bristles along the posterior margin. Anal claws represented only by a small sclerite at the tip of the anal lobe, apparently completely embedded. Four or five large bristles near each sclerite and a number of yellow spines (about fifteen) along the posterior margin of the lobe.

Setodes punctatus (Fabricius)

Immature stages unknown.

BRINDLE, A. 1962. *Entomologist's Rec. J. Var.* **74**, 148–55.

BRINDLE, A. 1966. The Larva of *Triaenodes reuteri* McLachlan (TRICHOPTERA, LEPTOCERIDAE). *Entomologist's Rec. J. Var.* **78**, 130–133.

ESBEN-PETERSEN, P. 1916. *Vaarfluer* in *Danmarks Fauna* **19**, 31, figs. 25 and 26.

HICKIN, N. E. 1942. *Proc. R. ent. Soc. Lond.* (A) **17**, 12–13.

HICKIN, N. E. 1943. *Proc. R. ent. Soc. Lond.* (A) **18**, 69–71, 106–8, 109–11.

HICKIN, N. E. 1952. *Caddis. A Short Account of the Biology of British Caddis Flies, with Special Reference to the Immature Stages.* Plate IV. Methuen. London.

HICKIN, N. E. 1952. *Caddis*, pl. IV.

HICKIN, N. E. 1953. *Proc. R. ent. Soc. Lond.* (A) **28**, 74–6, 111–13, 114–16.

HICKIN, N. E. 1953. The Courtship of a Caddis Fly, *Mystacides nigra L. Entomologist's mon. Mag.* **89**, 71.

HICKIN, N. E. 1954. *Proc. R. ent. Soc. Lond.* (A) **29**, 153–5.

KIMMINS, D. E. 1949. Some Changes in Generic Names in the Family LEPTOCERIDAE (Order Trichoptera). *Entomologist* **82**, 201.

KIMMINS, D. E. 1964. The British Species of the Genus *Athripsodes* with Comments on the Species described by Stephens (TRICHOPTERA, LEPTOCERIDAE) *Entomologist's Gaz.* **15**, 153–80.

KIMMINS, D. E. 1964. *Triaenodes simulans* Tjeder in Britain (TRICHOPTERA, LEPTOCERIDAE). *Entomologist.* **97**, 40–4.

LESTAGE, J. A. *in* ROUSSEAU, E. 1921. *Les Larves et Nymphes Aquatiques des Insectes d'Europe*, 605, 613, 617, 623, 631–4, 644, 645, 650, 651.

MACDONALD, W. W. 1950. *Proc. R. ent. Soc. Lond.* (A) **25**, 19–22.

MORTON, K. J. 1904. *Entomologist's mon. Mag.* **40**, 82–4.

MORTON, K. J. 1906. *Triaenodes reuteri* McLach. A Species of Trichoptera new to Britain. *Entomologist's mon. Mag.* **43**, 270–1.

MOSELY, M. E. 1939. *The British Caddis Flies*, 153, 157, 160.

PELHAM-CLINTON, E. C. 1966. *Triaenodes reuteri* McLachlan, a Species to be Restored to the British List; and a Redescription of *T. simulans* Tjeder (TRICHOPTERA, LEPTOCERIDAE). *Entomologist* **99**, 47–50.

SILFVENIUS, A. J. 1905–6, *Acta Soc. Fauna Flora fenn.* **27** (6), 57, 58, 61, 74, fig. 19.

SILFVENIUS, A. J. 1905–6. *Acta Soc. Fauna Flora fenn.* **27(b)**, 48, 53.
SILFVENIUS, A. J. 1905–6. *Acta Soc. Fauna Flora fenn.* **27** (6), 81, 84.
SILFVENIUS, A. J. 1906–7. *Acta Soc. Fauna Flora fenn.* **27** (6), 77, 82.
SILTALA, A. J. 1906. *Acta Soc. Fauna Flora fenn.* **27**, 71.
STRUCK, R. 1903. *Beiträge zur Kenntnis der Trichopterenlarven*, 68.
THIENEMANN, A. 1905. *Zool. fahrb., Abt. für System.*, figs. 14–16, 76–77.
THIENEMANN, A. 1905. *Zool. Jb.* (*Syst.*) **22**, figs. 75–78.
ULMER, G. 1903. *Metamorph. Trichopt.*, 100, 101, fig. K, 1, 19, 41, 64; 102, 103, figs. 89, 103, 120; 105, 107, fig. 31, 68; 108, 109.
ULMER, G. 1903. *Über die Metamorphose der Trichopteren*, 101, 105.
ULMER, G. 1904. *Allg. Z. Ent.* **9**, 261.
ULMER, G. 1909. *Die Süsswasserfauna Deutschlands* **5–6**, 246, 247, 248, fig. 373, a and b; 249, 250, 251.
ULMER, G. 1909. *Metamorph. Trichopt.*, 202.
WESENBERG-LUND, C. 1908. *Int. Revue ges. Hydrobiol. Hydrogr.*, 589.
WESENBERG-LUND, C. 1911–12. *Int. Revue ges. Hydrobiol. Hydrogr. Biolog. Supplem. ser. III, fig.* 26.

17

MOLANNIDAE

There are only two British species in the family MOLANNIDAE. These are *Molanna angustata* Curtis and *Molanna palpata* McLachlan. Adults of this family may be identified easily by their moderate size (the length of the anterior wing of *M. angustata* is 13 mm. in the male and 15 mm. in the female, whilst in *M. palpata* it is 10 mm. in the male and 14 mm. in the female), the curious way in which the long narrow anterior wings are folded round the abdomen when the insect is at rest and the unusually long legs. The body is usually held at an angle to the surface on which the insect is sitting. The wings are smoky-grey and the neuration is conspicuous. The insect resembles a piece of dried grass stalk. At dusk *M. angustata* performs complicated gyrations over the water surface.

Distribution

Molanna angustata. Very widely distributed (M.E.M.).

Molanna palpata. Widely distributed throughout Scotland. Recorded from the Hebrides (M.E.M.). The specimens in my collection are from County Cavan, Ireland. I can find no description of the larva of this species. As the shield-shaped form of the larval case is widespread in this genus the case of *palpata* is likely to be similar.

Lestage gives a general account of the larvae of MOLANNIDAE to include *Molannodes*, which latter genus is not known to occur in Britain. The case with its wing-like lateral extensions, the characteristic pattern on the mesonotum, the shape of the gular sclerite and the folds of the ventral margins of the genae, together with the unique tarsal claws of the metathoracic legs, serve to make the identification of the larva of the known British species a simple task.

Molanna angustata Curtis

This caddis is a widely distributed species, occurring commonly where its typical larval habitat is found. The latter consists of lakes, pools and slow-moving rivers and streams, with a bottom consisting of patches of sand and light gravel. In many of such localities phanerogamic vegetation is absent.

In 1943, it was found in unusual abundance in an ornamental pool used for model yacht racing at Bournville, Birmingham. This pool is slightly less than two acres in extent, with a depth of 1 ft. 8 in. at the sides, sloping to 2 ft. 3 in. at the centre. At the time of the emergence of the *M. angustata* adults, the pupal skins were piled up against one end of the pool in many

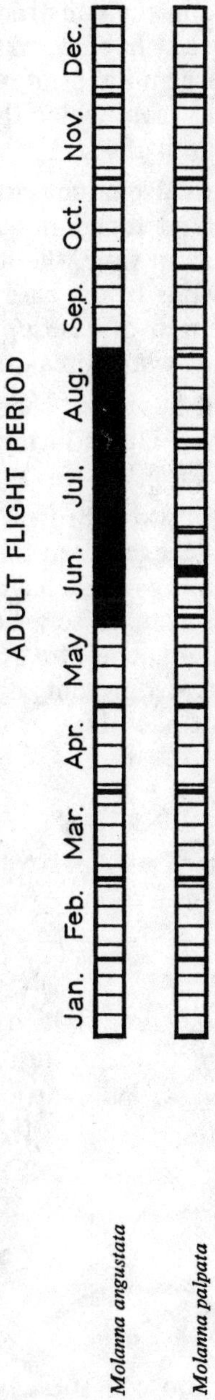

MOLANNIDAE. *Adult Flight Period taken from published records of individual dates.*

tens of thousands, and gave the appearance of a band of foam about 30 ft. long, and about 3 to 4 in. deep. Much of the pool bottom consists of the old disused larval cases, and it seems highly probable that some of the old cases are utilized by the larvae in the construction of new cases.

During the winter and spring any haphazard sweep with a pond-net along the pool bottom will bring a large number of old cases and a few inhabited cases to the side. The specimens from which the following description was made were collected on 26 March 1944.

Case (Figs. 870–872). Of unusual characteristic design, and constructed entirely of sand-grains, cemented together with secretion, shield-shaped and convex, with a central conical tube, the anterior opening of which is about 3 mm. diameter. The wings of the case extend forwards as a hood over the anterior opening. Length of case up to 26 mm., width 12 mm. Very often, however, the lateral wings break off or become damaged.

Larva is of eruciform type, but has some characteristics of the sub-eruciform type (cf. PHRYGANEIDAE). Up to 17 mm. in length, and 2.7 mm. in width (Fig. 873).

Head (Fig. 874). Long, light yellowish-brown, with a black band running from the dorsal cervical area of the genae to the aboral apex of the clypeus, then dividing with an arm running on each side of the clypeus. A narrow black bar divides the clypeus transversely near the oral end. Anterior margin of clypeus excised. Antennae of two segments, proximal large and bulbous, with pit-like sense organs present; distal small, and surmounted by a bristle; anteclypeus pale. Chaetotaxy as in Fig. 874.

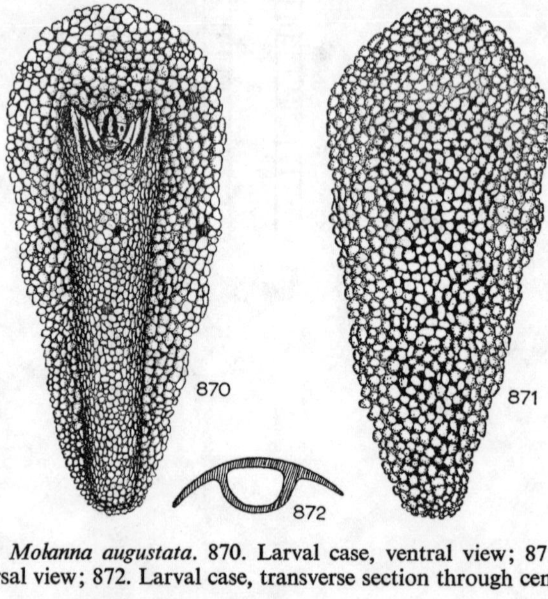

Figs. 870–872. *Molanna augustata.* 870. Larval case, ventral view; 871. Larval case, dorsal view; 872. Larval case, transverse section through centre.

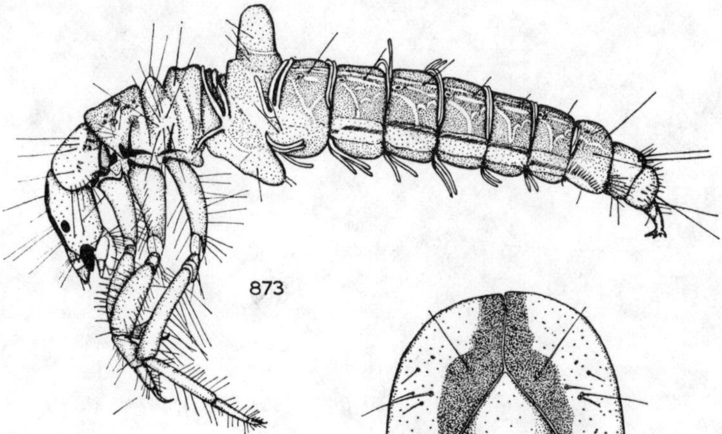

873

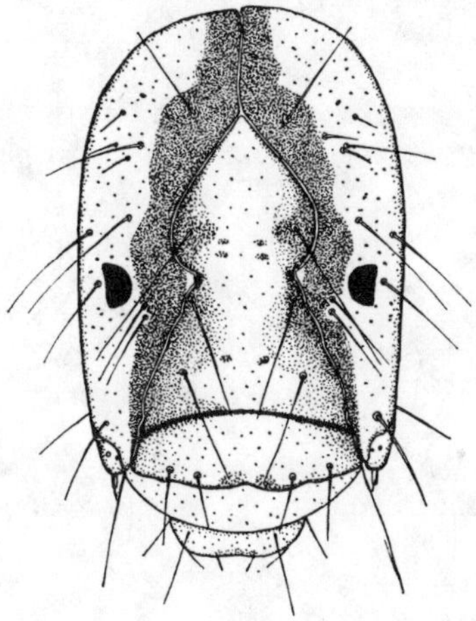

874

Figs. 873–874. *Molanna angustata*. 873. Lateral view of larva; 874. Head of larva.

Labrum (Fig. 875). Excision on lateral margin shallow, with pair of short blunt spines at centre, and longer inwardly-directed spines at outer angles. Hairy tracts restricted to ventral surface.

Mandible (Fig. 877). Both mandibles with two stout bristles at the base of the outer edge. Brush of hairs on inner edge absent. Left mandible with four blunt teeth. Right with two.

Maxillae (Fig. 876). Maxillary palp three segmented and equal in length to the mala to which it is closely adpressed. Hairy tract at base of maxilla confined to a small area. Two groups of ten long black spines on ventral surface of mentum.

Labium. Labial palps with one segment only apparent, but with sense organs giving bifid appearance at the tips.

Gular sclerite (Fig. 878). Bell-shaped, completely dividing the genae, a black transverse bar across the oral end. The regions of the genae, adjacent to the

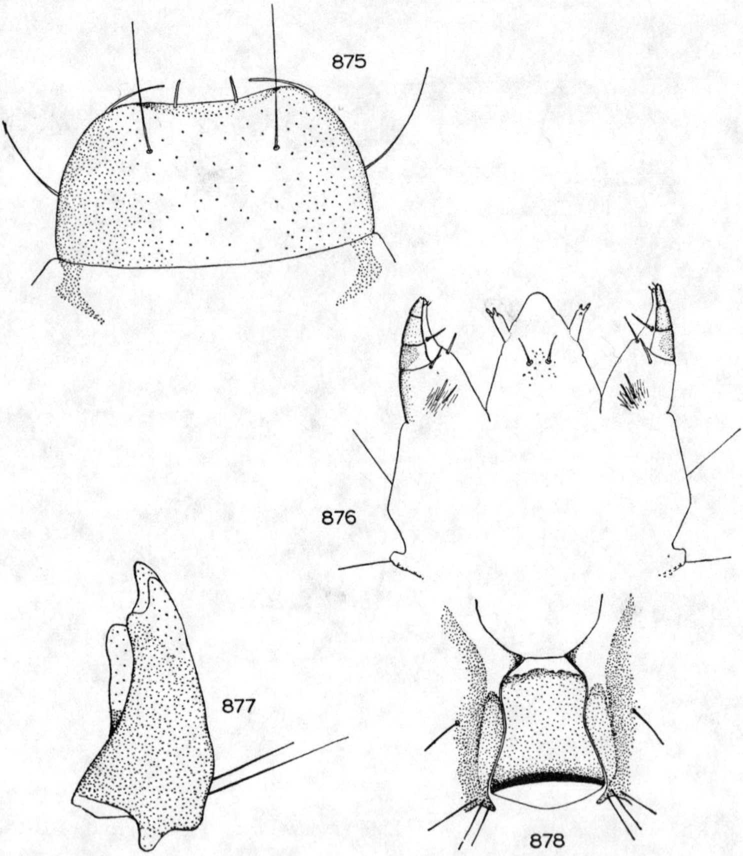

Figs. 875–878. *Molanna angustata*. 875. Labrum of larva; 876. Maxillae and labium of larva; 877. Left mandible of larva from beneath; 878. Gular sclerite of larva.

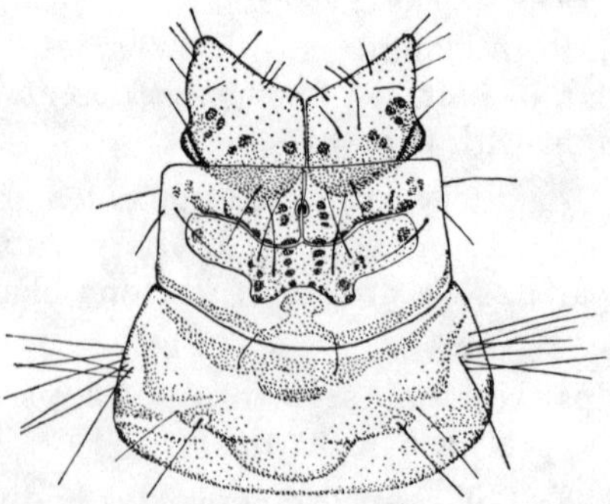

Fig. 879. *Molanna angustata:* thoracic nota of larva.

gular sclerite, show fold-like darker coloured regions, somewhat variable in size and depth of colour.

Thorax. Prothoracic notum sclerotized, divided by a longitudinal median suture. Mesothoracic notum with two large sclerotized areas divided by a transverse suture, the anterior area subdivided by a longitudinal suture. The posterior area is not subdivided, and has a pair of laterally projecting lobes on each side, the anterior being much larger than the posterior. The metathoracic notum is unsclerotized.

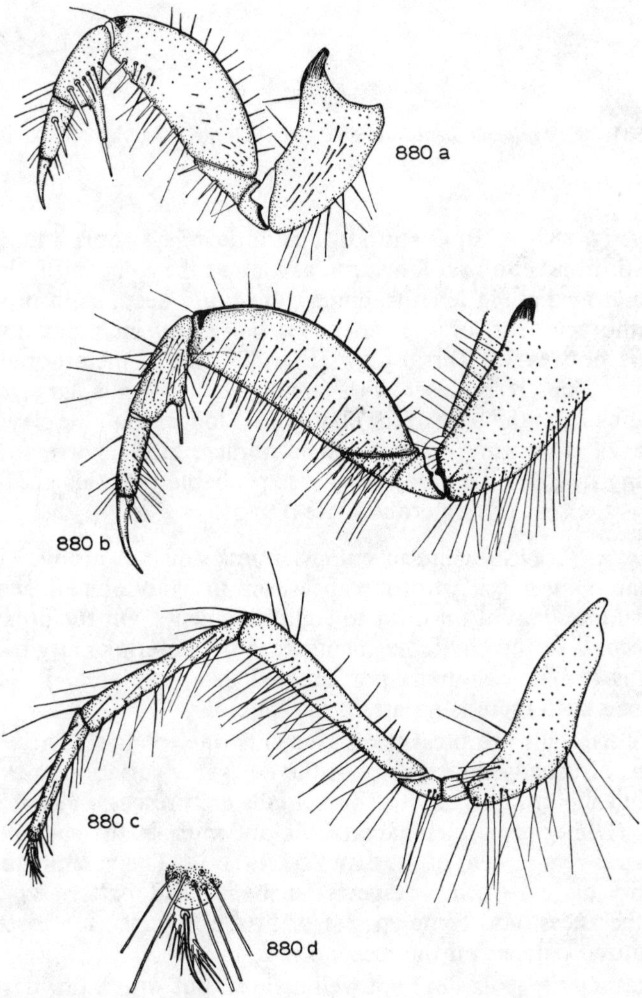

Fig. 880. *Molanna angustata.* a. prothoracic leg of larva; b. mesothoracic leg of larva; c. metathoracic leg of larva; d. tarsal claw of metathoracic leg.

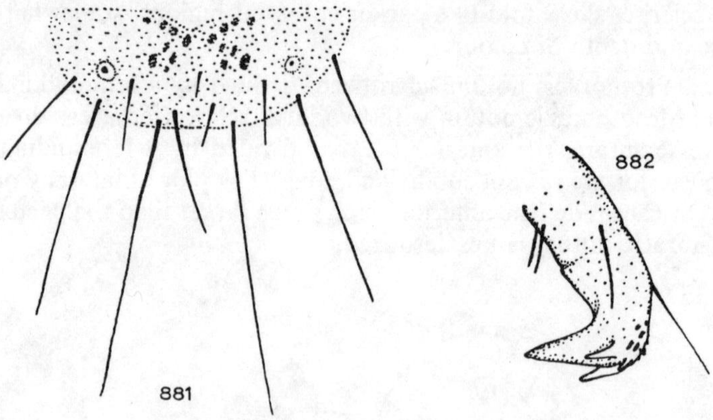

881

882

Figs. 881–882. *Molanna angustata.* 881. Anal sclerite of larva; 882. Anal claw of larva.

Legs (Fig. 880, A, B, C and D). Prothoracic legs short and deep. Ventral spine of tibia set on prolongation as long as the spine itself. Mesothoracic legs intermediate in length; femur robust and deep, as in prothoracic leg. Metathoracic legs long, slender and hairy. An incipient division of the tibia is noticeable. Tarsal claw (Fig. 880, D) of metathoracic legs quite characteristic. It is beset with small spines, with a large curved spine emerging from the base of the claw, and is longer than the claw itself. When the larva is crawling only the metathoracic tarsi emerge from the overhanging hood of the case; thus, it is probable that this specialized tarsal claw is the seat of numerous sense-organs.

Abdomen. Greyish-white in colour. Intersegmental grooves well defined. Median and lateral protuberances on first abdominal segment large. Lateral line present on third to eighth segments. On the dorsal surface of the second to the eighth segments near the anterior margin on each side there is a dark coloured spot situated in a pale area. In each of these segments also there is a pattern of white bars.

Filiform gills are present on segments one to eight. These are in small groups of usually three gills, but there is some variation in the number of gills in each group. Only the dorsal gills are present on segments one and eight. (Ulmer gives the shape of the abdomen in the subfamily MOLANNINAE as reminiscent of the PHRYGANEIDAE. This resemblance, however, is quite close in other respects, such as head pattern and abdominal protuberances, and some species of PHRYGANEIDAE are stated to have sclerotized patches on the mesonotum.)

Anal sclerite (Fig. 881) not well defined, but with a few dark spots. The eye-like marks, a pair of which are found on each abdominal notum, are situated also on this sclerite. The anal claws (Fig. 882) are furnished with three auxiliary claws, and in addition a number of much smaller spine-like claws at their base. The basal lobes of the claws (not shown in the figure) are also furnished with a number of curved brown spines.

Molanna palpata McLachlan

Immature stages unknown.

ESBEN-PETERSEN, P. 1916. *Danm. Fauna* **19**. Vaarfluer, 25.
HICKIN, N. E. 1946. *Proc. R. ent. Soc. Lond.* (A) **21,** 55–60.
LESTAGE, J. A., *in* ROUSSEAU, E. 1921. *Les Larves et Nymphes Aquatiques des Insectes d'Europe*, 656.
MOSELY, M. E. 1939. *The British Caddis Flies,* 140.
SILFVENIUS, A. J. 1905–6. *Acta Soc. Fauna Flora fenn.* **27**, 38.
ULMER, G. 1903. *Metamorph. Trichopt.* 98.
ULMER, G. 1909. *Die Süsswasserfauna Deutschlands* **5–6**, 242.
WESENBERG-LUND, C. 1908. *Int. Revue ges. Hydrobiol. Hydrogr.*, 601.

18

BERAEIDAE

The trichopterous family BERAEIDAE appears to have a fairly scattered world distribution but the number of genera and species is small. Kimmins gives two genera involving but three species in the Australasian region. Three species, all in the genus *Beraea*, are found in North America although confined to the eastern part. They occur throughout the Palaearctic region with the exception of Manchuria and Japan, and are present in the African region but absent from the whole of the Oriental region. This family is usually placed between the SERICOSTOMATIDAE and the MOLANNIDAE.

BERAEA
1. *maurus* (Curtis)
2. *pullata* (Curtis)

ERNODES
3. *articularis* (Pictet)

BERAEODES
4. *minutus* (L.)

The adults of BERAEIDAE are characterized as follows. All four species are black in colour and are small. The largest species has an anterior wing length of 6 mm. in the male, but mostly the anterior wing length is only from 4 to 5 mm. The moderately stout antennae are about as long as the wings and the basal segment, which is stout and hairy, is usually as long and sometimes longer than the head. The tibial spur formula is 2, 2, 4. The head is much broader than long and bears two large hairy warts. The front part of the head is produced between the antennae.

Ocelli are absent. The maxillary palpi are very long, stout and hairy and are held ascending by the living insect.

The basal segment is short and the remaining segments long whilst the terminal segment is inflexible. The labial palpi also are long, stout and hairy. The wings are very hairy and, in the case of the male, with the exception of *Ernodes*, there is either a basal fold in the anterior wing known as a callosity, or a pencil of hairs is present. The callosity may be associated with a fringed process or seed-like scales. Discoidal cells are absent.

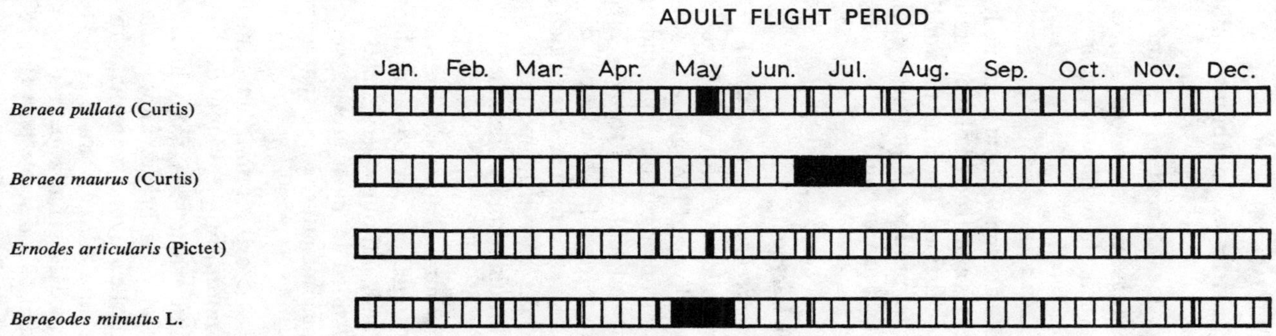

BERAEIDAE. *Adult Flight Period taken from published records of individual dates.*

Distribution

Beraea maurus. Rather local. Often found in herbage bordering rocky springs or small waterfalls (M.E.M.).

Beraea pullata. Widely distributed inhabiting shallow water and marshy ground (M.E.M.).

Ernodes articularis. Very local. Small springs and trickles (M.E.M.).

Beraeodes minutus. Amongst the masses of underwater fibrous roots of willow trees under banks near the outflow of stream into lake. Glos. (N.E.H.).

General Description of Larvae

The larvae of BERAEIDAE are small, some being no longer than 9 mm. They inhabit tubular curved cases constructed of sand-grains. The lateral margins of the gular sclerite are indistinct. The hind part of the pronotum is folded and projects forwards to produce a process which is furnished with a number of hairs. The legs are unequal; in the case of *Bereodes minuta*, excessively so.

Detailed Descriptions of Larvae

Beraea maurus (Curtis)

On 9 April 1953, when collecting with Mr. Peacey at Cherrington, near Stroud, we found a larva of what I thought to be a Beraeid on a submerged branch in a small stream running into a lake, and on 24 May Mr. Peacey collected another dozen and sent them to me. Three specimens remained alive in a small aquarium for a few weeks but did not survive. Then in February 1954 Mr. Peacey sent a further small batch from the same locality and from these two adults emerged, one on 19 June and the other on 25 June 1954. Mr. Kimmins kindly identified them as *Beraea maurus* (Curtis).

Case (Fig. 883). Length 7 mm., diameter at anterior end 1.1 mm., of small sand-grains cemented together, curved and diminishing towards the hinder end; posterior end, although very small in transverse section, is further almost entirely filled in by a greyish secretion, to which a few minute grains of sand become attached, so that a small crescentic opening only remains.

Larva. Length 6 mm., width 0.9 mm., widest at about first abdominal segment; head and pronotum bright brick-red.

Head (Fig. 884). Round when seen from in front, covered with fine sculpturing (Fig. 887); clypeus broad, almost triangular; a furrow just ventral to the eye runs to the antenna; antenna of three segments, basal one of which is not distinct, and a bristle arises at the base of the small distal segment; on ventral surface the genae appear to be united, the characteristic gular sclerite being absent as a discrete part (Fig. 885).

Labrum. Robust, with a number of blunt horny tubercles situated aborally.

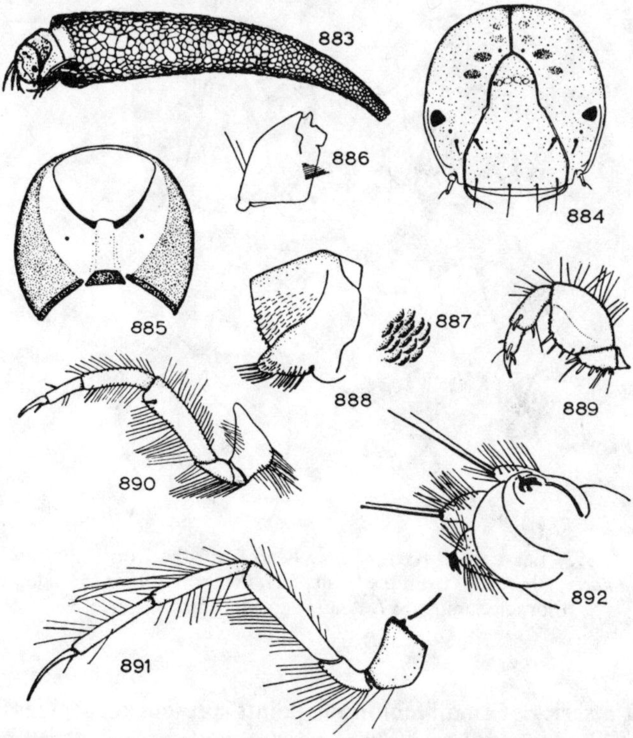

Figs. 883–892. Larva of *Beraea maurus*. 883. Larva in case; 884. Head from the front; 885. Head from behind; 886. Left mandible; 887. 'Honeycomb' sculpturing of head; 888. Pronotum from side; 889. Prothoracic leg; 890. Mesothoracic leg; 891. Metathoracic leg; 892. Anal segments partly from beneath.

Mandibles (Fig. 886). A single brush of hairs on inner surface of each mandible, and two bristles on outer surface.

Maxilla. Palp four-segmented; two processes at distal end two-segmented; two blade-like bristles project inwards and a series of bristles extend the length of the maxilla on the inner face, becoming stouter aborally.

Pronotum. Prominent lateral wing-like projections present, extremities beset with bulbous based spines (Fig. 888).

Meso- and metanota larger than pronotum, mesonotum with anterior transverse row of black hairs.

Legs. Prominent spines on ventral edge of prothoracic leg, meso- and metathoracic legs hairy (see Figs. 889, 890 and 891 for comparative size).

Abdomen. White, segments gradually decreasing in size posteriorly, intersegmental grooves shallow; abdominal protuberances present on first segment, dorsal not prominent, lateral disc-like; gills absent; lateral line

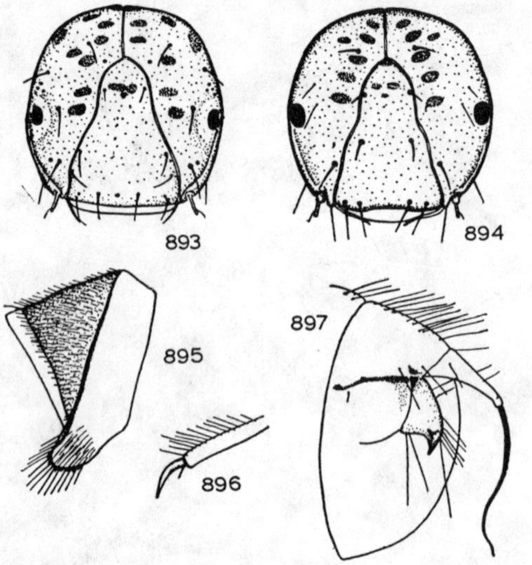

Figs. 893–897. 893. Larva of *Beraea pullata:* head from the front; 894–897. Larva of *Ernodes articularis*; 894. Head from the front; 895. Prothorax from the side; 896. Metathoracic tarsus; 897. Anal segments from the side.

absent but a series of small 'chitinous points' present on segments two to eight and on segment nine an oblique series of orange-coloured tubercles. *Anal claws* with two unequal auxiliary claws (Fig. 892); on dorsal surface anal segment is produced into a pair of lobes, each having two large and several smaller spines. (My specimens each had two large spines but Ulmer states that three may be present.)

Beraea pullata (Curtis)

A number of larvae were collected from the derelict Thames–Severn Canal at Brimscombe, near Stroud, Glos., when I visited this locality with Mr. Peacey on 16 April 1955. The larvae were found amongst the roots of the marginal vegetation. An adult emerged from the aquarium on 15 May 1955, and was identified by Mr. Kimmins. When freshly emerged the wings are violet-black.

Nielsen (1942) has given a very detailed description of the larva of this species, with many figures, so that it is now proposed to deal with it only briefly.

Larva similar to that of *B. maurus*, but colour of head and pronotum slightly more orange; 'muscle spots' on head more prominent and two brushes present on inside edge of each mandible; the most distinctive differences, however, are the presence of only one large black bristle on each anal lobe, and of two brushes of hairs on the inner face of each mandible.

Ernodes articularis (Pictet)

In spite of the widest search for British examples of larvae of *Ernodes articularis* Pictet, I have been unsuccessful. Mosely stated that this species was very local in this country, though widely distributed in alpine regions on the Continent. The late Dr. Georg Ulmer was, however, kind enough to send me some German specimens. The figures and references to this species in this paper refer to these specimens.

The larva strongly resembles that of *Beraea maurus* but differs as follows:

All bristles on front of head (Fig. 894) light coloured; pronotum (Fig. 895) with a wide light creamy-yellow anterior margin; central transverse triangular part dark reddish-brown covered with fine light-coloured hairs; posterior part light creamy-yellow and separated from central part by a straight dark well-marked ridge. Metathoracic tarsal claw shorter and stouter than in *Beraea* (less than half length of tarsal segment) (Fig. 896). Anal lobes (Fig. 897) narrow and with only a few small hairs and a single large black bristle (curved in my specimens); anal claw bears only a single auxiliary claw.

Beraeodes minutus (L.)

Larvae were collected on 17 April 1955, from Toadsmoor, Gloucestershire, where they were present in very large numbers amongst the masses of underwater fibrous roots of willow trees under the banks near the outflow stream of the lake. I was visiting Toadsmoor with Mr. A. F. Peacey in search of the larvae of *Ernodes articularis* Pict., adults of which had previously been collected in this locality by Mr. Peacey, to whom I am greatly indebted for placing at my disposal his very wide knowledge of Gloucestershire insects. Adults of *Beraeodes minutus* commenced to emerge from my aquaria on 30 April 1955, and continued to do so until 23 May. Mr. D. E. Kimmins kindly confirmed the identification.

Case (Fig. 898). Conical, curved, up to 10 mm. long and 1.5 mm. wide, of very fine sand-grains, the surface having a smooth appearance apparently due to a large proportion of silk-like secretion.

Larva. 7–9 mm. long, 1.0–1.2 mm. wide; tarsal claws of prothoracic and mesothoracic legs normally held close to the mouthparts (Fig. 899); metathoracic legs very long and slender, the tarsal claws being particularly slender.

Head (Figs. 899–900). Clypeus and adjacent areas of genae mostly black; clypeus covered with fine hairs, each arising from a small white spot; edges of clypeus parallel until converging to the vertex; distal segment of antenna slender, bearing a bristle near the tip; genae apparently fused at mid-ventral line (it cannot be said for certain that any part of what I have previously called the gular sclerite has taken part in this fusion).

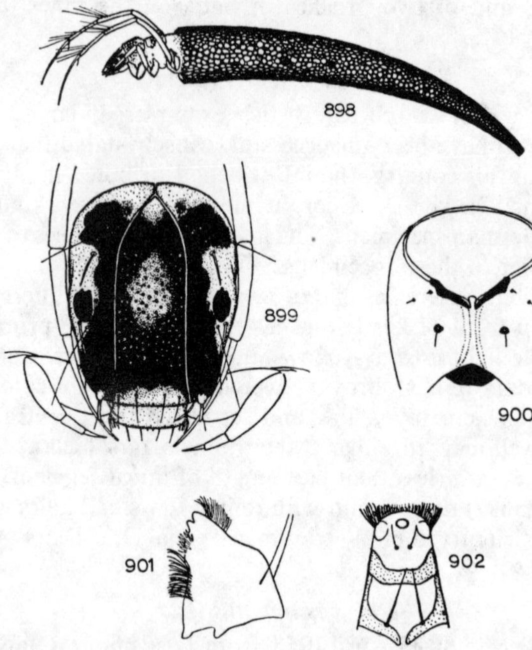

Figs. 898–902. Larva of *Beraeodes minutus*. 898. Larva in case; 899. Head from the front, showing position of prothoracic and mesothoracic legs; 900. Head from below; 901. mandible; 902. Labium.

Mandibles (Fig. 901). Stout with three brushes of hair, the median brush being composed of shorter but stouter hairs.

Labium (Fig. 902). Furnished with hairs at the tip, silk gland orifice large, basal lobe of labial palpi not differentiated.

Pronotum (Fig. 903). Light golden-yellow with black reticulated pattern occupying anterior two-thirds, a fine hair arising from each reticulation; posterior third contains a few but larger black marks; mesonotum with a light brown partially sclerotized area with a few dark brown marks; prosternal horn absent.

Legs. Relative size shown in Fig. 898, and shape of tarsal claws in Figs. 904, 905 and 906.

Abdomen. Bright apple green, gills present in small clusters from first to fifth segments; lateral line absent but sclerotized tubercles present from third to eighth segment.

Anal claws. Basal segments slender with dark brown slender sclerite apparently articulating with distal segment; claw bears two auxiliary claws (not shown in figure).

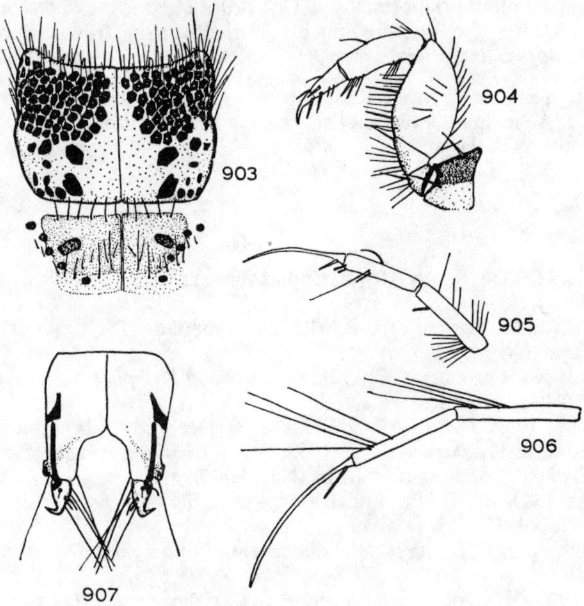

Figs. 903–907. Larva of *Beraeodes minutus*. 903. Pro- and mesonota; 904. Prothoracic leg; 905. Tibia and tarsus of mesothoracic leg; 906. Tibia and tarsus of metathoracic leg; 907. Anal segment.

Key to the Larvae of the British **Beraeidae**

Ulmer's (1909) key is modified to include *Ernodes articularis*.

1. Head oval. Gills present in bunches, brush of hairs present on outer aboral edge of mandible. Head bright yellow, heavily marked with black. Clypeus narrow with parallel sides. Black reticulate pattern present on pronotum—in anterior two-thirds many small marks, in posterior third fewer but some larger marks. Tarsal claws of metathoracic legs as long as or longer than tarsal segment. Dorsal anal lobes absent *Beraeodes minutus* (L.)

— Head round. Gills absent, brush of hairs on outer aboral edge of mandible absent. Head reddish, generally unicolorous. Clypeus transverse, triangular. Black reticulate pattern on pronotum absent, but wing-like lateral extensions prominent. Tarsal claws of metathoracic legs not as long as tarsal segment. Dorsal anal lobes present 2

2. Ridge on dorsal surface of lateral projection of pronotum straight, well marked and dark in colour. Anal claw furnished with a single auxiliary claw. All bristles on head light in colour *Ernodes articularis* (Pictet)

— Ridge on dorsal surface of lateral projection of pronotum concave, not well marked, light in colour. Anal claw with two auxiliary claws. Bristles on anterior margin of clypeus black 3

3. Head and pronotum bright brick-red. One brush of hairs only present on inner face of mandibles. Two or three long black bristles arising from each of the two anal lobes dorsal to the anal claws *Beraea maurus* (Curtis)

— Head and pronotum orange-red. Two brushes of hairs present on inner face of mandibles. A single long black bristle arising from each of the two anal lobes dorsal to the anal claws *Beraea pullata* (Curtis)

BARNARD, K. H. 1934. South African Caddis-flies (Trichoptera). *Trans. R. Soc. S.Afr.* **21**, 291–394.

HICKIN, N. E. 1959. Larvae of the British Trichoptera—The BERAEIDAE. *Proc. R. ent. Soc. Lond.* (A) **34**, 83–9.

LESTAGE, J. A., *in* ROUSSEAU, E. 1921. *Les Larves et Nymphes Aquatiques des Insectes d'Europe*, 1950. Bruxelles.

MOSELY, M. E. 1939. *The British Caddis-flies* (*Trichoptera*), 133. London.

MOSELY, M. E. & KIMMINS, D. E. 1953. *The Trichoptera* (*Caddis-flies*) *of Australia and New Zealand*. London: Brit. Mus. (Nat. Hist.).

NIELSEN, A. 1942. Über die Entwicklung und Biologie der Trichopteren. *Arch. Hydrobiol.* Suppl. **17**, 414.

THIENEMANN, A. 1905. Biologie der Trichopteren-Puppe. *Zool. Jb.* (*Syst.*) **22**, 489–574, figs. 52–56.

ULMER, G. 1903. Über die Metamorphose der Trichopteren. *Abh. naturw., Hamburg* **18**, 96.

ULMER, G. 1909. *Die Süsswasserfauna Deutschlands* **5–6**. *Trichoptera*, 244.

WIGGINS, G. B. 1954. The Caddis-fly Genus *Beraea* in North America (Trichoptera). *Contr. R. Ont. Mus. Zool.* **39**, 1–13.

19

SERICOSTOMATIDAE

The British species in the family SERICOSTOMATIDAE are moderate to moderately small in size. The largest species, *Sericostoma personatum*, has an anterior wing length of 15 mm. in the female whilst in the male of *Crunoecia irrorata* it is only 6 mm. Usually the anterior wing length is about 9 or 10 mm. in this family. The fairly stout antennae are about as long as the wings or only slightly shorter. The basal antennal segment is very hairy and stout, and often longer than the head. Ocelli are absent in all the British species. The maxillary palpi show strong sexual differences and in the male the form and number of segments vary in the different genera. They are, however, always turned upwards. In the female the long and hairy palpi are normal in form. The labial palpi of both sexes are nearly alike, the basal segment being short whilst the second and third segments are long and equal in length. The wings are usually heavily pubescent but *Brachycentrus subnubilus* is exceptional. On the wings of the male there may be modified hairs, scales or folds or grooves. The tibial spur formula is 2, 2, 4 in the SERICOSTOMATINAE, 2, 4, 4 in the GOERINAE and LEPIDOSTOMATINAE, whilst in the BRACHYCENTRINAE it is 2, 3, 3.

The family SERICOSTOMATIDAE is divided into four sub-families, each of which is represented in Britain. Ten species placed in eight genera as in the following check-list are known from Britain.

KIMMINS (1966) ranks the subfamilies given here as families.

The *Lepidostoma fimbriatum* of Mosely's handbook is ranked as a synonym of *L. hirtum*.

SERICOSTOMATINAE

SERICOSTOMA

1. *personatum* (Spence)

NOTIDOBIA

2. *ciliaris* (Linné)

GOERINAE

GOERA

3. *pilosa* (Fabricius)

SILO

4. *nigricornis* (Pictet)
5. *pallipes* (Fabricius)

BRACHYCENTRINAE

BRACHYCENTRUS
6. *subnubilus* Curtis

LEPIDOSTOMATINAE

CRUNOECIA
7. *irrorata* (Curtis)

LEPIDOSTOMA
8. *hirtum* (Fabricius) =*fimbriatum* (Pictet)

LASIOCEPHALA
9. *basalis* (Kolenati)

Distribution

Sericostoma personatum. Streams (M.E.M.); Berks. (M.I.C. *et al.*); Glos. (A.F.P.); Cardigan (J.R.E.J.); Northumberland (G.N.P.); Yorks. (P.F.H.).

Notidobia ciliaris. Large rivers (M.E.M.).

Goera pilosa. Widely distributed and very abundant, rivers and lakes (M.E.M.); Glos. (A.F.P.); Cardigan (J.R.E.J.); Northumberland (G.N.P.); Herts. (W.V.B.).

Silo nigricornis. Equally widely distributed and abundant as *pallipes* (M.E.M.); Berks. (M.I.C. *et al.*); Herts. (W.V.B.).

Silo pallipes. Widely distributed, fairly abundant, rather smaller streams than *nigricornis* (M.E.M.); Hants. (M.I.C. *et al.*); Glos. (A.F.P.); Northumberland (G.N.P.).

Brachycentrus subnubilus. Widely distributed and emerges from some rivers, e.g. Kennet in Berks, in countless numbers (M.E.M.).

Crunoecia irrorata. Very widely distributed but seldom abundant, small springs and waterfalls with species of *Tinodes* or *Wormaldia* (M.E.M.).

Lepidostoma hirtum. Widely distributed, comes to light (M.E.M.); Cardigan (J.R.E.J.); Northumberland (G.N.P.); Yorks. (P.F.H.) as *L. fimbriatum*. Lough Conn., Co. Mayo, Ireland (M.E.M.).

Lasiocephala basalis. Rather local but very abundant where it occurs, fairly fast running water (M.E.M.).

General Description of Larvae

A general picture of the larval taxonomy of the SERICOSTOMATIDAE as a whole is rather difficult to give, on account of the wide variation shown by this heterogeneous group. This presented an even greater problem to Lestage because at that time the HELICOPSYCHINAE (now a separate family, the HELICOPSYCHIDAE, no species of which have yet been found in Britain), the BERAEINAE (now a separate family, the BERAEIDAE) and

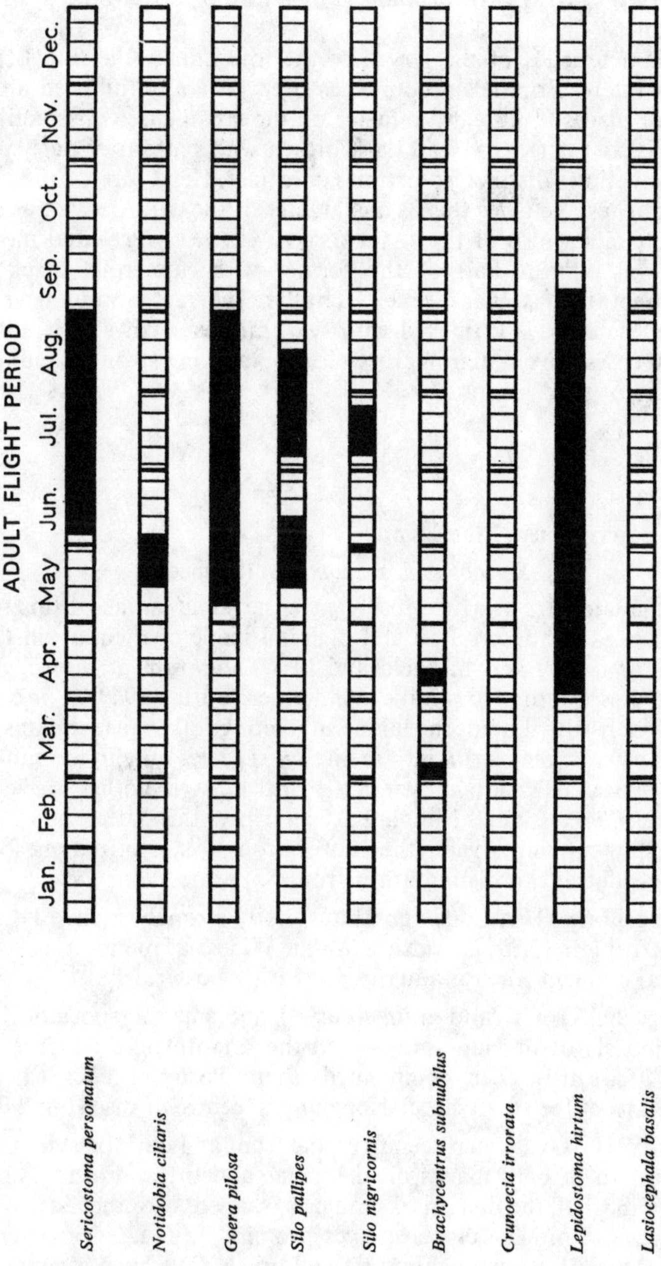

SERICOSTOMATIDAE. *Adult Flight Period taken from published records of individual dates.*

Thremma (not now included in the SERICOSTOMATIDAE and no species of which genus have yet been found in Britain) were all included in the SERICOSTOMATIDAE. His account is therefore of little value to us at the present time.

The characteristics of the larvae of the four subfamilies are therefore given separately. Brindle's account has been generally followed although this author places too great a reliance on the larval case construction.

SERICOSTOMATINAE. Head dark brown with pattern of pale spots or 'flecks', sometimes distinct, sometimes indistinct. Head is retractile into the prothorax. (Lestage gives this as a character of the GOERINAE but it is an obvious character also of the SERICOSTOMATINAE.) Pro- and mesonota dark brown anterior half of the former with numerous black setae. The mesonotum has black setae laterally. The metanotum never with sclerotized patches, is furnished with two transverse rows of black setae. The larval cases are circular in cross-section, tapering strongly, curved and constructed of sand-grains.

SERICOSTOMATINAE

Detailed Descriptions of the Larvae

Sericostoma personatum (Spence)

In 1936 the late Mr. Martin Mosely gave me some larvae of this species. I have checked the description and figures I made of them against larvae obtained for me by Mr. S. Nield and Mr. Thompson of the Freshwater Biological Association from Lake Windermere during 1949. *S. personatum* is fairly widely distributed, the larvae inhabiting rather fast streams, rivers and lakes where streams run into them. Two species only in the sub-family SERICOSTOMATINAE are known in the British list, the other species being *Notidobia ciliaris* Linné. Hanna has described this latter larva, which inhabits slow-moving rivers, the more strongly sclerotized mesonotum should be sufficient to distinguish it from *S. personatum*.

Larva. Eruciform. Head orthognathous, body strongly arched (Figs. 908 and 910). Length 12 mm., width 2.5 mm. Widest at metanotum and first abdominal segment, then gradually decreasing posteriorly.

Case (Fig. 909A). Of sand-grains, curved and tapering posteriorly. The delicate mosaic of the sand-grains and the smooth outer surface of this case has frequently been commented upon. Posterior end sealed with secretion except for small circular opening in centre of disc (Fig. 909B).

Head (Fig. 911). Oval when seen from the front and dark brown in colour, with a pattern of light marks on the genae and the aboral region of the clypeus; almost all the light marks are pear-shaped, with the narrow region directed away from the clypeus. Eyes prominent, situated in pale areas. Antennae small, but each bearing a small bristle. The head is considerably retracted into the prothorax and the anterior apices of the pronotum reach to a position just posterior to the eyes, and the bristles which arise from these apices appear to play a part usually played by bristles arising from the head.

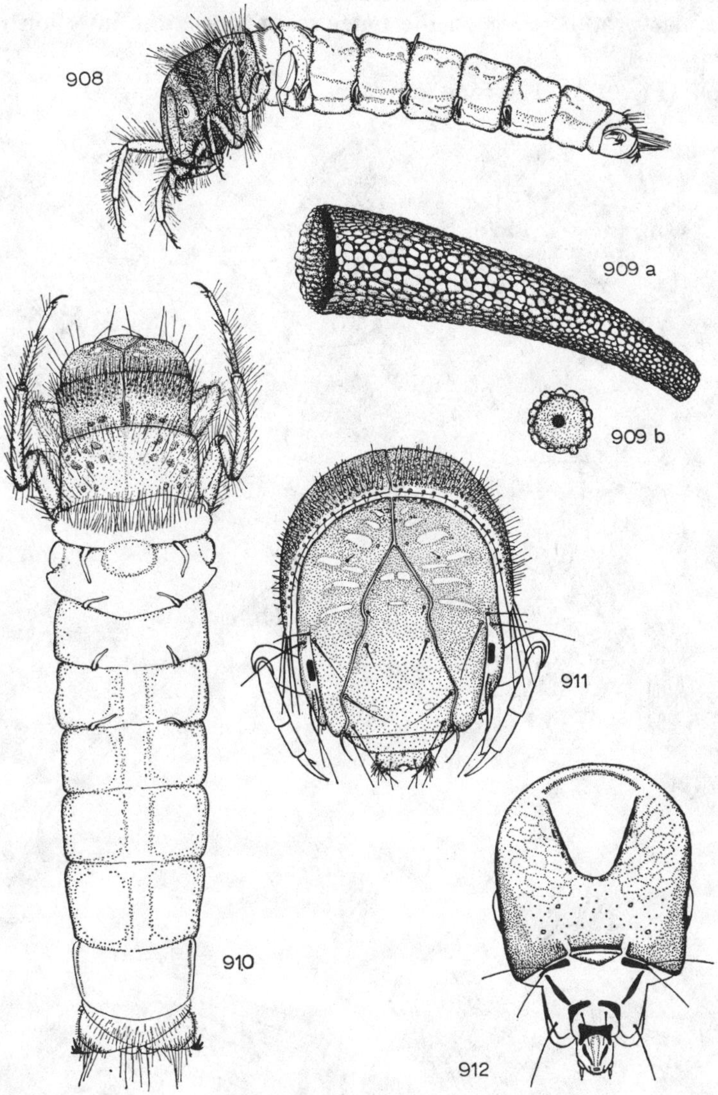

Figs. 908–912. *Sericostoma personatum*. 908. Lateral view of larva; 909. a. larval case; b. posterior end of case from behind; 910. Dorsal view of larva; 911. Head with view of pronotum and prothoracic legs from the front; 912. Head from behind, maxillae and mandibles not drawn.

Gular sclerite not defined by sutures as is usual in the TRICHOPTERA, only the oral region being differentiated; the aboral region is fused with the genae (Fig. 911). Pattern of fine reticulated markings on ventral surface of genae.

Mandibles (Fig. 914). Four teeth. Brush of hairs on the inner edge and a pair of bristles on the base of the outer edge present on each mandible. Triangular protuberance on the outer margin near the insertion of the bristles.

Labrum (Fig. 913). The four transverse bristles long, curved and light in

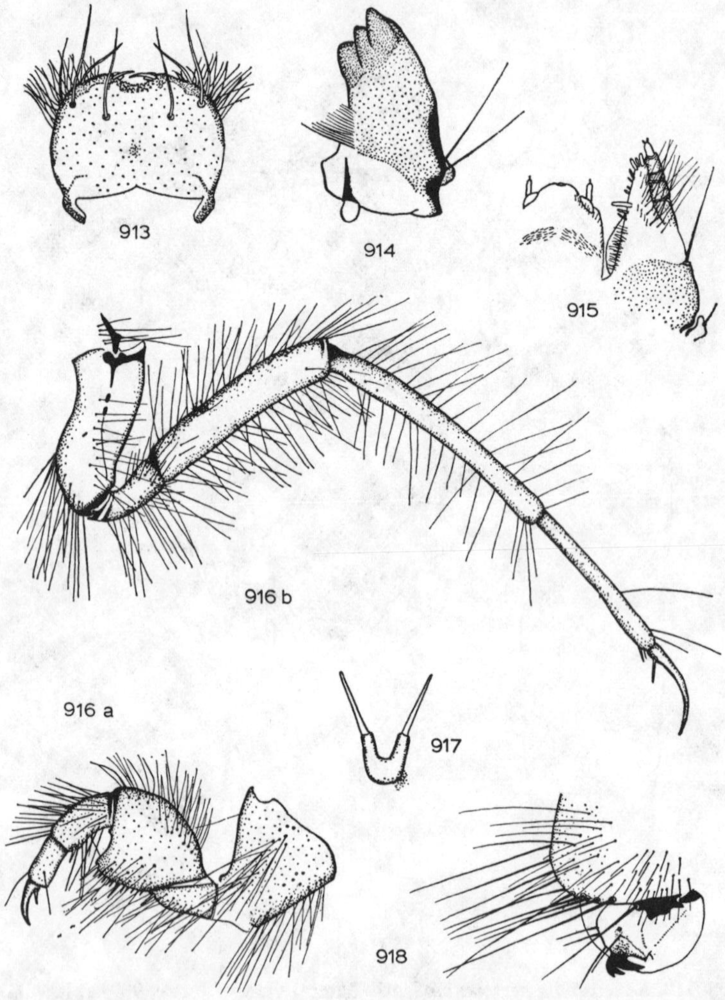

Figs. 913–918. *Sericostoma personatum.* 913. Labrum; 914. Right mandible; 915. Maxilla and labium; 916. a. prothoracic leg; b. metathoracic leg; 917. Sclerotized abdominal tubercle; 918. Anal claw.

colour. The posterior hairs (on the lateral margins) of the bunches longer than the inner, anterior hairs. Anterior excision deep, flanked by pair of curved overlapping bristles lying in front of peg-like projections.

Thorax (Fig. 910). Pronotum heavily sclerotized and dark greyish-brown in colour. Anterior half very hairy with a few dark marks in the posterior half. Anterior margin pale and furnished with a series of pale spine-like hairs. Mesonotum much lighter in colour and not nearly so heavily sclerotized as the pronotum. Indeed some authors have described the mesonotum as membranous, and the two transverse bands of dark marks in the anterior region have been described as sclerotized patches. Metanotum membranous, with two transverse bands of black hairs. Prosternal horn absent.

Prothoracic legs (Fig. 916a). Short, thick and very hairy (except tarsus), *metathoracic legs* long, thin and hairy (except tarsus); *meso- legs* intermediate in length.

Abdomen. Cream in colour, tapering posteriorly. Lateral line only shown by sclerotized tubercles present on third to seventh segments; they are minute, however, except on the seventh segment, where a row of fourteen occur on each side. They are reddish brown in colour, each with the finger-like processes (Fig. 917). Sclerite on eighth abdominal segment absent but many hairs present. Protuberances on first abdominal segments flattened and disc-like. Gills present, either simple filiform or in small bunches. They arise from the anterior border of the segments, although Silfvenius considered that post-segmental gills also occur. An examination of the gills in the larvae in my collection shows them to be much more variable in number than Silfvenius indicates. Anal claws (Fig. 918) consist of three superimposed barbs.

Notidobia ciliaris (L.)

Hanna gives the following description made from larvae collected from near Manchester.

Case (Fig. 919, a and b). Curved and tapered up to 15 mm. in length and 4 mm. wide at anterior end. Posterior end has dark brown silken plate, in the centre a small hole. Built of grains of sand, very smooth outer surface.

Larvae. Eruciform larvae measured up to 13 mm. long and 3 mm. wide.

Head (Figs. 920, 921). Hypognathous, broad, short and dark brown, with lighter pattern on anterior surfaces of genae and aboral end of fronto-clypeus. Posterior surfaces of genae dark brown except area bordering gular sclerite, pale yellow. Gular sclerite chestnut brown and concave ventral margin, markedly convex dorsal margin. No genal suture; two genae are fused together above gular sclerite.

Labrum (Fig. 926). Ventral margin almost straight, with small central

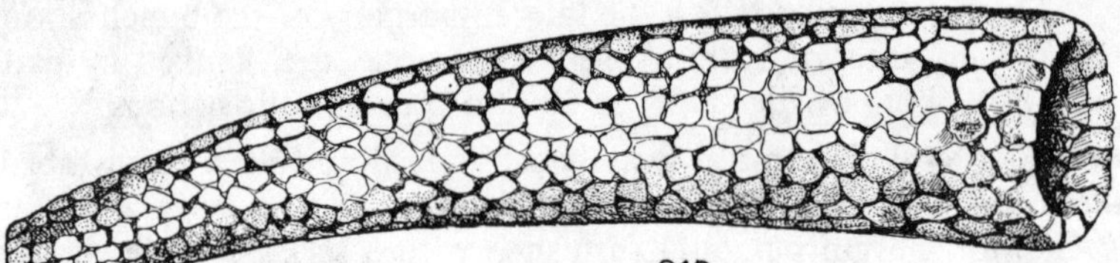

919 a

919.b

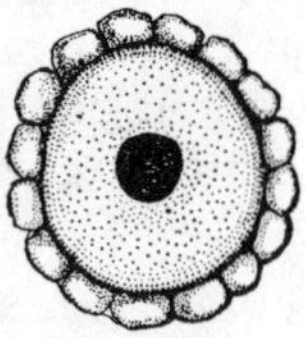

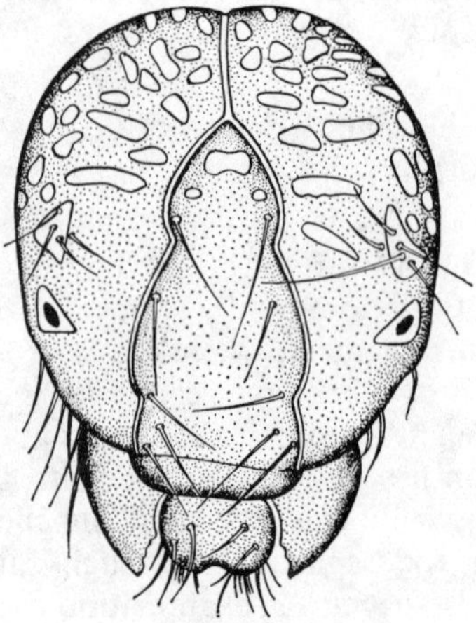

920

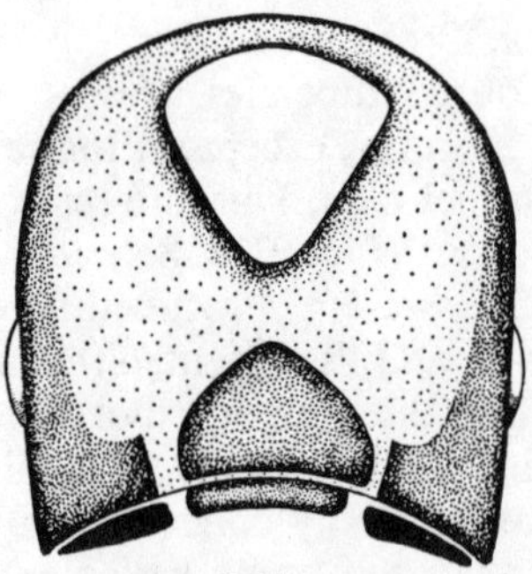

921

Figs. 919–921. *Notidobia ciliaris.* 919. a. larval case; b. posterior end of case from behind; 920. Head from the front; 921. Head from behind. Hanna.

protuberance not always present. On each side three long setae and some smaller setae and hairs. Tormae long, heavily sclerotized, bent inwards.

Mandibles (Figs. 928, 929). Asymmetric, bear two long setae near base. Right three blunt teeth, left four teeth of which outermost is smaller. Each mandible has brush on inner surfaces.

Maxilla (Fig. 930). Cardo and stipes bear two setae. Small additional sclerite between cardo and stipes. Maxillary palp five segments.

Thorax (Fig. 925). Pronotum dark brown, entirely sclerotized, with median longitudinal suture. Antero-lateral margin of the pronotum has a prolongation. Anterior two-thirds of pronotum and its lateral margins carry long hairs. Mesonotum sclerotized except for narrow posterior region, median longitudinal suture. Anterior half of mesonotum darker than posterior. In some specimens mesonotum dark brown except for yellowish brown posterolateral margins. Metanotum soft and bears two transverse rows long dark hairs. No prosternal horn or prosternal sclerite. Propleuron dark brown, with dark longitudinal band. Meso- and metapleura transparent and each bears dark longitudinal band at right-angles to dark transverse band at base of coxa.

Legs (Figs. 922–4). Prothoracic short, robust, with much flattened femur.

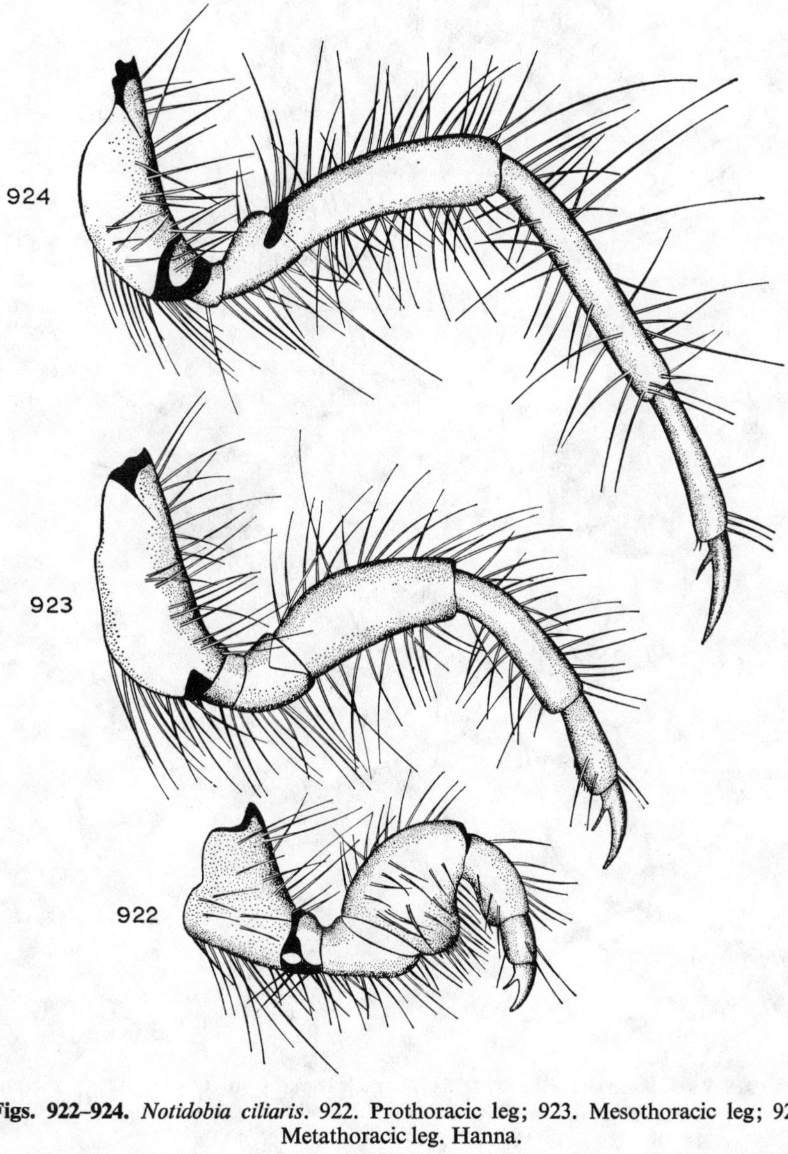

Figs. 922–924. *Notidobia ciliaris*. 922. Prothoracic leg; 923. Mesothoracic leg; 924. Metathoracic leg. Hanna.

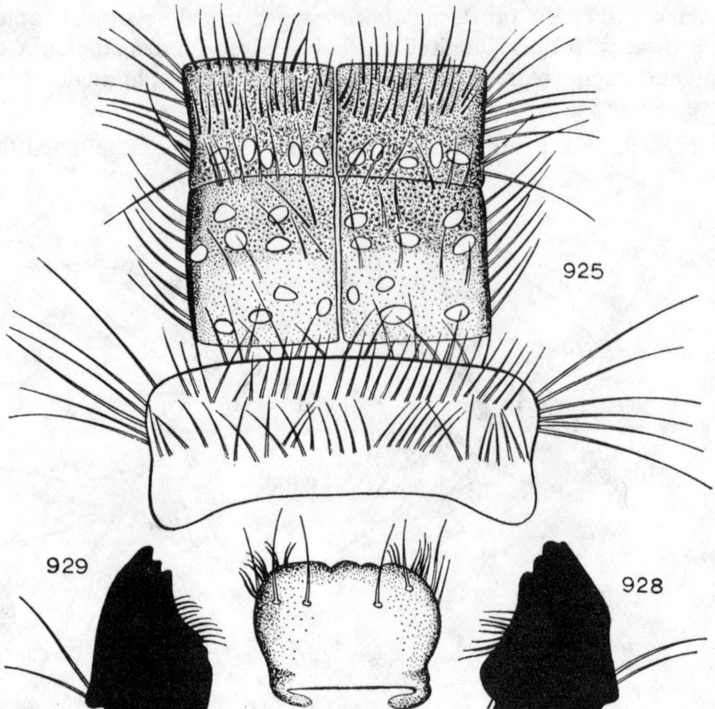

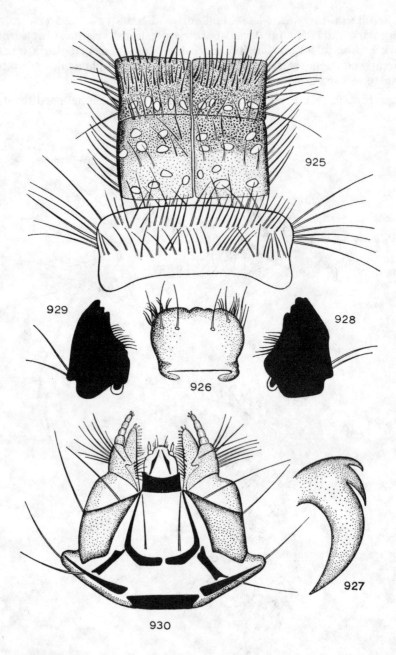

Figs. 925–930. *Notidobia ciliaris*. 925. Thoracic nota from above; 926. Labrum; 927. Anal claw; 928. Left mandible; 929. Right mandible; 930. Labium and maxillae. Hanna.

Inner surface of trochanter, femur and tibia bear small spines. Mesothoracic leg shorter than metathoracic leg. All legs carry many hairs, except on tarsi where they are reduced in number.

Abdomen. On first segment three protuberances of which dorsal is broad and devoid of setae, while each of lateral ones bears single seta. In some specimens the lateral protuberances were devoid of setae and in place of them arose small spines at their margins. Abdominal gills present on segments one to eight. Lateral line represented by small tubercles on the third to the eighth abdominal segments. On eighth abdominal segment the sclerotized tubercles are larger than those on other segments, and are nineteen to twenty in number. Anal sclerite absent. Many long fine hairs on the tergum of the ninth abdominal segment. Anal claw is formed of three superimposed claws.

The following key will easily differentiate between the larvae of *Sericostoma personatum* and *Notidobia ciliaris*.

1. Aboral area of gular sclerite indistinct. Longitudinal suture of mesonotum indistinct. Mesonotum less heavily sclerotized. Length 13–14 mm. Case 15–16 mm. Fast-flowing water but also lake shores where there is wave action. *Sericostoma personatum*

— Aboral area of gular sclerite distinct. Longitudinal suture of mesonotum distinct. Mesonotum more heavily sclerotized. Length 15 mm. Case 17–18 mm. Usually in slow-running water or canals *Notidobia ciliaris*

GOERINAE

In this sub-family the head of the larva is strongly retractile into the prothorax. The anterior angles of the pronotum are produced forwards. The mandibles bear a brush of hairs on the inner margins. A longitudinal suture is present on the mesonotum and the metanotum is sclerotized only in part. There may be three or four pairs of sclerites present on the metanotum. One dorsal and two lateral protuberances are present on the first abdominal segment and a line of hairs is present on the ventral side directed anteriorly. The anal claws bear a single dorsal auxiliary claw. A lateral line is present. The tubular case is constructed of sand-grains with two or three small stones attached laterally. The posterior opening is covered with a membrane pierced by a circular hole.

In addition to the descriptions of Hanna and Hickin, Baker has examined larvae of the three species and has called attention to discrepancies and errors in previous publications.

Goera pilosa (Fabricius)

Abundant larvae of *Goera pilosa* were collected from a stream at Temple Balsall, Warwickshire, where they were crawling over the stones and small pieces of rock in the centre of the stream. The current flowed strongest at

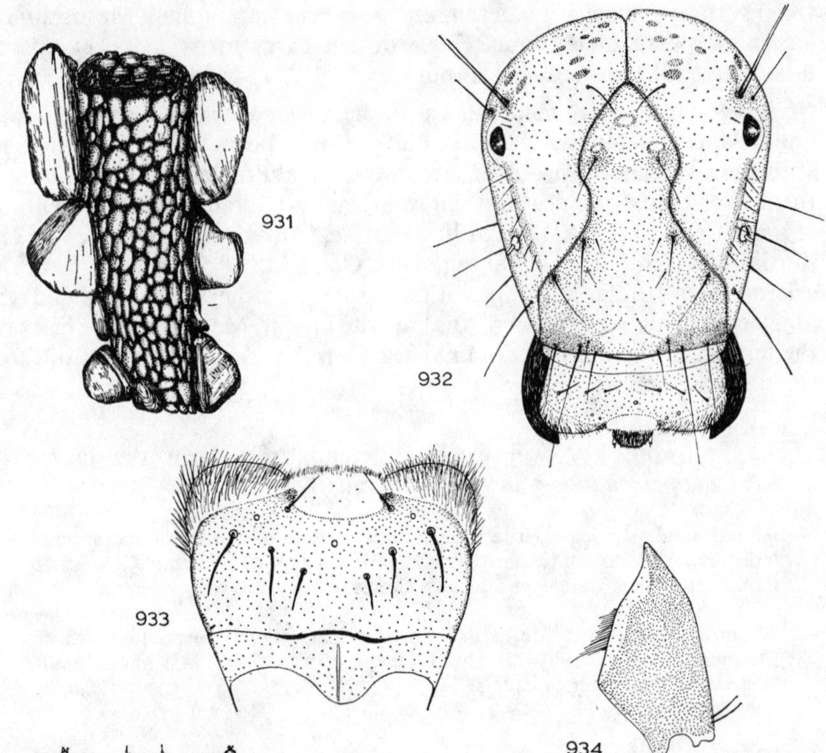

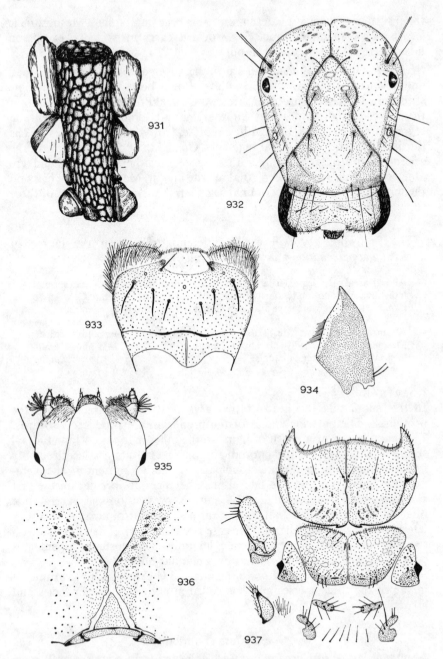

Figs. 931–937. *Goera pilosa*. 931. Case (ventral view); 932. Head; 933. Labrum; 934. Mandible; 935. Maxilla and labium (ventral view); 936. Gular sclerite; 937. Thoracic sclerites.

this point and they were in company with *Potamophylax latipennis* and *Agapetes fuscipes*. Some of the larvae were reared to the adult stage and the determination of the species confirmed.

Larva. Eruciform, head orthocentrous. Sides and frontal region of genae partially covered by extensions of the pronotum. Metanotum only partially sclerotized. Length 13 mm., width 3 mm.

Case (Fig. 931). Consists of central cylindrical part of sand-grains or small rock fragments cemented together with small pebbles or larger rock fragments at the sides, forming wing-like extensions. Cylindrical part slightly curved and flattened on the ventral surface. Anterior opening oblique. Length of case 15 mm., width 4.5 mm., width overall 10 mm.

Head (Fig. 932). Surface of head covered with minute pustules. Genae converging towards labrum. A ridge runs parallel with the genae on each side between the outer face of the genae and the clypeus. Antennae very small, of two parts, the proximal bulbous, the distal filiform and slightly curved. Chaetotaxy as in figure.

Gular sclerite (Fig. 936). Triangular, with anterior transverse bar.

Labrum (Fig. 933). Anterior membranous part quite distinct and separated into three convex lobes, the lateral lobes very hairy.

Mandible (Fig. 934). Triangular, base of outer edge furnished with two bristles, inner edge clothed with a brush of hairs, sclerotization at apex bifid.

Maxillae (Fig. 935). Palp four-segmented, golden brown, short and wide with brush of outwardly directed hairs. (Ulmer gives maxillary palp five-segmented.) Maxillary lobe short, bearing many hairs directed towards the labium, and three fairly large bulbosities.

Labium (Fig. 935). Covered with small spinulose protuberances, labial palp two segmented, proximal segment large and membranous, sensory papilla on distal segment carrying a stout bristle. (Ulmer gives one segment, Lestage *in* Rousseau two segments.) Pair of partially sclerotized quadrangular plates on ventral surface of mentum with a heavily sclerotized black bar stretching from the anterior outer angle of each plate to the base of labium.

Thorax (Fig. 937). Prothorax sclerotized, almost capsular, with anteriorly projecting angles into which the head is recessed. Median longitudinal suture slightly bent where it traverses a small area of the tergum which is not so heavily sclerotized as the remainder. A partial lateral suture occurs contiguous with a horny longitudinal ridge apparently as though a separate sclerite at the posterior outer angle has become fused to form a single sclerite at each side. Mesothoracic notum consists of two pairs of sclerotized plates. The larger inner pair are roughly quadrangular, are divided by the median suture, and have the anterior margin beset with very short spines; there is a small dark mark near the inner anterior angle in each sclerite, and the posterior margin is darker in colour. Outer pair of sclerites roughly

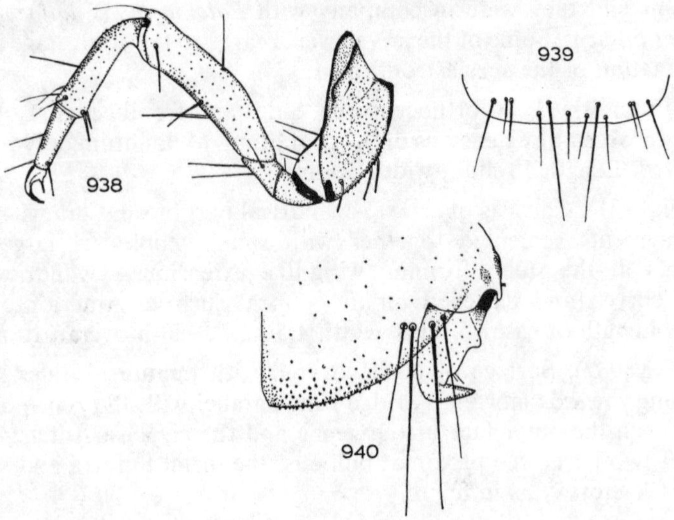

Figs. 938–940. *Goera pilosa.* 938. Prothoracic leg; 939. Position of bristles on anal segment; 940. Anal claw from above.

triangular but with a heavily sclerotized black beak-like projection on the outer margin. A characteristic feature of the mesothorax is the large spinous scoop-like flap projecting from the coxal sclerite and fitting round the proximal part of the coxa. Metathorax partially sclerotized. Two pairs of sclerites on the tergum and relatively large coxal sclerites.

Legs (Fig. 938). Approximately equal. Tibia and tarsus short. Tarsal claw strongly curved and arising from a swollen base which bears a short stout spine.

Abdomen. Cylindrical, diminishing slightly in width towards the posterior end, creamy white in colour. Lateral and dorsal protuberances present on first abdominal segment. Filiform gills present on abdominal segments two to seven, united at their bases into groups of three (sometimes two). These groups occur pre- and post-segmentally on the dorsal and ventral surfaces, and an additional single group occurs laterally on the ventral surface of segment two. Lateral line present and commencing at the middle of the third segment and terminating at the end of the eighth. Series of sclerotized pustules dorsal to the lateral line also present, from two to six in each segment. Anal sclerite not heavily sclerotized but with series of bristles as shown in Figs. 939 and 940. The anal claws of the specimens I examined appeared to be without auxiliary claws but Baker has stated that a single auxiliary claw is present. Four long bristles arise near the base of the sclerite supporting the claw, of which the middle pair are the longest.

Silo nigricornis (Pictet)

Baker examined last instar larvae of this species from Hertfordshire. The average length was 9.06 mm., whilst the cases were 11.5 mm. in length and 2.9 mm. wide. He stated that his specimens agreed closely with Nielsen's description except that the brush of hairs on the inner margins of the mandibles was less extensive (Fig. 952). Baker also found that gills were present on the seventh abdominal segment and that the anal claws were furnished with a single dorsal hook, although difficult to see. Lestage in his key gives *S. nigricornis* as having no gills on the seventh abdominal segment and the anal claws as lacking hooks.

Larvae were collected by Hickin from a small, rapidly flowing stream at the edge of a beechwood at Cranham, Gloucestershire, where they were very plentiful. The stream was bordered with thick vegetation, and the bottom covered with small rocks and stones of limestone. A few larvae were reared to maturity.

Case. A tubular portion of sand-grains with small pebbles cemented along two sides giving dorsal and ventral surfaces to the case, ventral surface slightly flattened (Fig. 941). Length 8 mm., breadth 4.5 mm. Shortly before

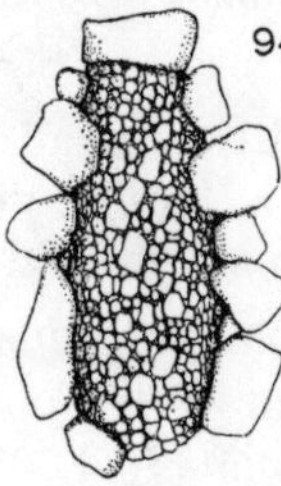
941

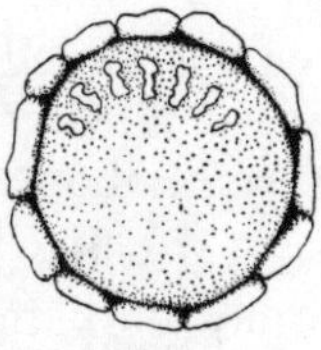
942

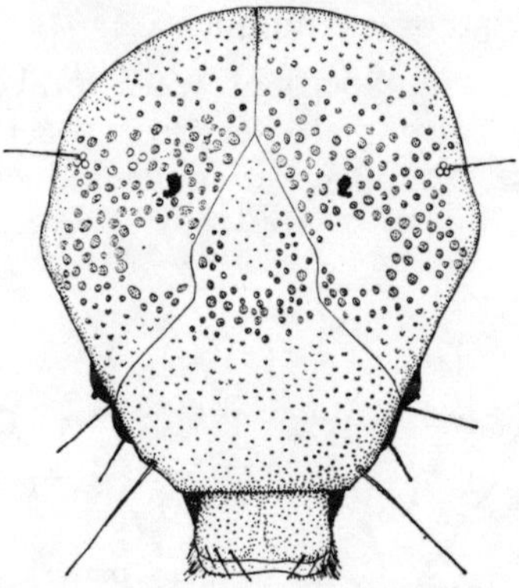
943

Figs. 941–943. *Silo pallipes*. 941. Case; 942. View of posterior end of pupal case; 943. Head.

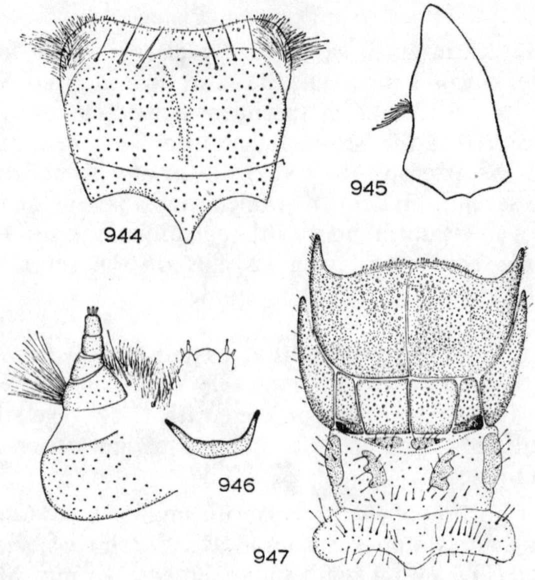

Figs. 944–947. *Silo pallipes*. 944. Labrum; 945. Mandible; 946. Maxilla and labium; 947. Thorax and first abdominal segment.

pupation the anterior end is closed with a small stone, whilst at the posterior end the opening is filled with a dark brown secretion with the exception of a small lattice of about seven small slits arranged in an arc.

Larva. Eruciform, cylindrical with abdomen gradually tapering distally. Sclerotized parts uniformly dark brown. Pro- and mesothoracic nota sclerotized (Fig. 947), forming a globular structure distinct from the partially sclerotized metathoracic notum.

Head. Hypognathous, ovoid, tapering towards oral end (Fig. 943). Angles on genae fairly sharp. Eyes small, set close together at front of head. Clypeus wide at the base. Antennae very much reduced. Head covered with

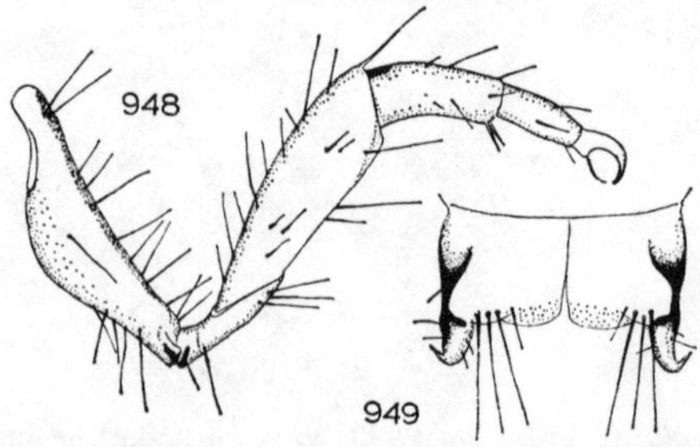

Figs. 948–949. *Silo pallipes*. 948. Prothoracic leg; 949. Anal claws.

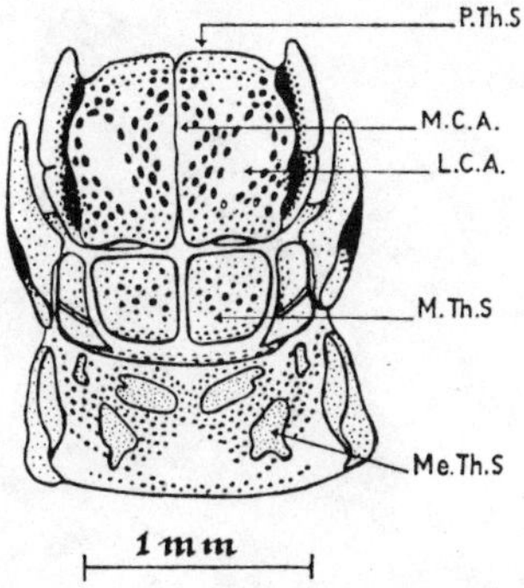

Fig. 950. *Silo pallipes:* dorsal view of thorax. Baker.

[*For key to figs. 950–953 see p. 441.*]

small dark tubercles, except a small area at aboral end of clypeus and an area on each gena adjacent to this. The size of the tubercles decreases towards the top of the head, on the frons, and on the clypeus towards the labrum. Very few bristles present, one large pair at the base of the clypeus and three smaller pairs. One pair of the latter are lateral to the eyes and the other two pairs are lateral to the base of the clypeus.

Mouthparts. Labrum of two parts, a proximal sclerotized part, with a median prolongation at the base and bearing three pairs of bristles transversely near the anterior margin, and a distal unsclerotized part having a brush of fine hairs on each side. Mandibles acute, with a bunch of fine hairs at about the middle of the inner edge a little nearer the base (Fig. 945).

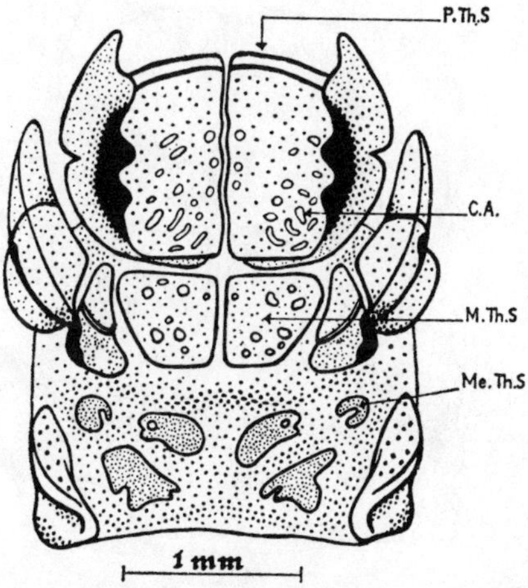

Fig. 951. *Silo nigricornis:* dorsal view of thorax. Baker.

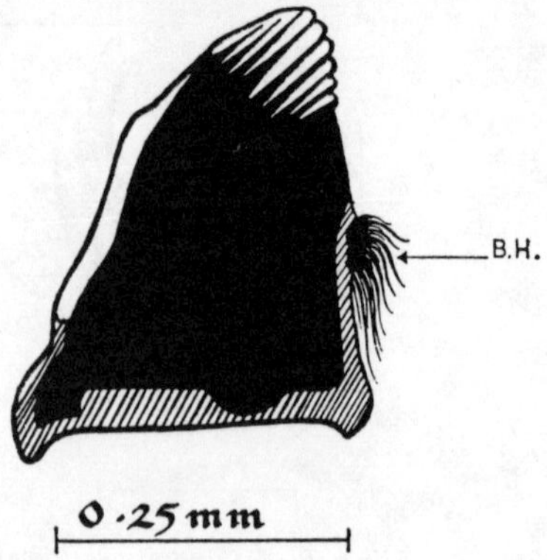

Fig. 952. *Silo nigricornis:* left mandible. Baker.

Maxillary palps four-segmented with basal segments bulbous and brush of fine hairs emerging from the basal segment. Galea covered with fine hairs and bearing two bristles. Labium with fine hairs at the side, labial palps two-segmented with a tubercle on each segment. Bow-shaped sclerite at base of labium (Fig. 946).

Thorax. Pro- and mesothoracic nota sclerotized, covered with small dark coloured tubercles. Pronotum dark reddish-brown with three light patches, one median, two lateral. Pronotum with median suture, convex, anterior margin fringed with small hairs. Pro- and mesonota have lateral wing-like extensions extending anteriorly (Fig. 947). Mesonotum lighter in colour

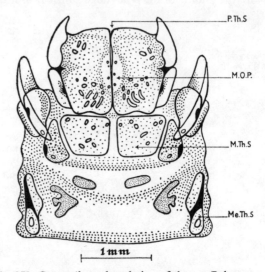

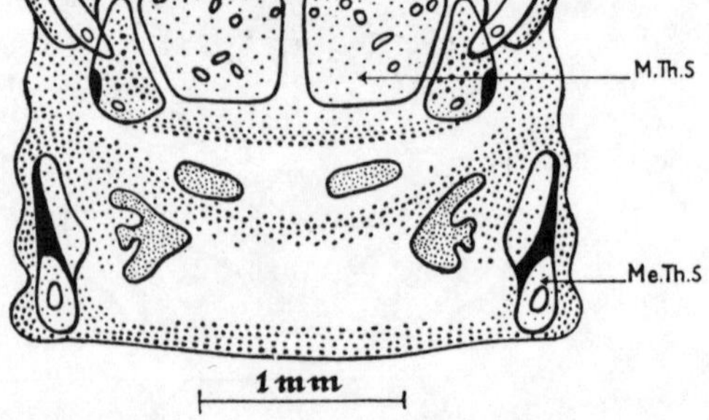

Fig. 953. *Goera pilosa:* dorsal view of thorax. Baker.

than pronotum, divided into ten parts, by a median suture, two lateral longitudinal sutures and two transverse sutures on each side. The hinder of the outer pair of sclerites is not visible from the dorsal surface. Metanotum with six sclerotized patches; two small pear-shaped patches lie transversely near the mesonotum, with a pair of slightly larger patches lying posteriorly with sinuate margins. The largest patches occupy a position in the pleural region. All six sclerotized patches bear bristles.

Legs. Equal, tarsal claws strongly curved and with a basal spine (Fig. 948). Anterior edge of femur with continuous dark mark. Prosternal horn present.

Abdomen. First segment bulbous laterally and in centre of posterior margin. Branched gills present on the second to seventh segments inclusive. Lateral line reaching to the beginning of the eighth segment. Anal claws supported by single large sclerites having small extension pointing towards and almost meeting one small and three large bristles (Fig. 949).

Baker, commenting on the above description by Hickin, considered that the sclerotized parts of the larva varied between light brown and almost black. He stated also that the eyes were large, well separated and situated behind the middle of the head. Instead of six sclerotized patches present on the metanotum as given above. Baker states that there are eight, this being a characteristic feature of the genus *Silo*.

Key to Species of **Goerinae**

The key given below to separate larvae of the three species of GOERINAE is that due to Baker.

1. Pronotum less convex, with raised median ovoid patch (Fig. 953). Sclerotized patches adjacent to the median sclerotized patches on the mesothorax not divided. Metathorax with six sclerotized patches. Thoracic sclerites as shown in Fig. 953. No chitinous support to the ventral side of the lateral protuberances on the first abdominal segment — *Goera pilosa* (F.)

— Pronotum more convex, without raised median ovoid patch. Sclerotized patches adjacent to the median sclerotized patches on the mesothorax divided. Metathorax with eight sclerotized patches. Thoracic sclerites not as in Fig. 953. Chitinous support present on the ventral side of the lateral protuberances of the first abdominal segment — 2

2. Prothorax with three clear areas, one median and two lateral. Thoracic sclerites as in Fig. 950 — *Silo pallipes* (F.)

— Prothorax with more than three clear areas. Thoracic sclerites as in Fig. 950 — *Silo nigricornis* (Pict.)

Explanation of Figure Lettering

B.H.:	brush of hairs.
P.Th.S.:	prothoracic sclerite.
C.A.:	clear areas.
M.Th.S.:	mesothoracic sclerite.
Me.Th.S.:	metathoracic sclerite.
M.C.A.:	median clear area.
L.C.A.:	lateral clear area.
M.O.P.:	median ovoid patch.

The figures do not show the chaetotaxy.

BRACHYCENTRINAE

In this sub-family there is a single British species, *Brachycentrus subnubilus* Curtis. It is widely distributed in slowly flowing rivers where it is often present in enormous numbers.

Brachycentrus subnubilus Curtis

This species occurs in rivers and streams. The description of the fully grown larvae is taken from specimens from the River Kennet, Berks., given to me by the late Mr. M. E. Mosely, whilst the description of first instar larvae is taken from specimens hatched from an egg-mass laid by a captured female. These unfortunately died within a day or two of emergence.

Case (Fig. 954). The young larva constructs a case of vegetable debris, cut leaves (arranged transversely) and secretion, rectangular in cross-section, but older larvae construct the case from secretion only; it is circular in cross-section and the posterior end is attached to the roots of water-weeds or stones and rocks, often very many together. The tail end of the case is

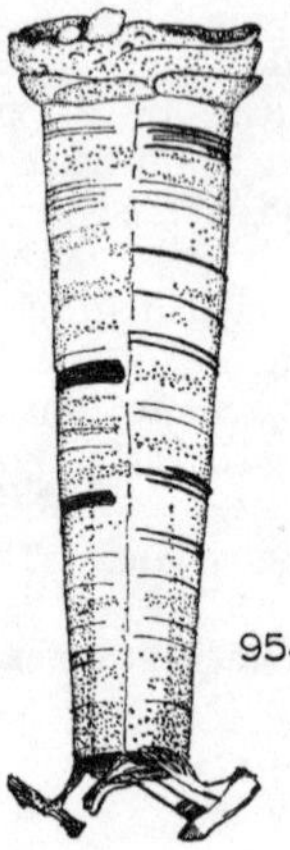

954

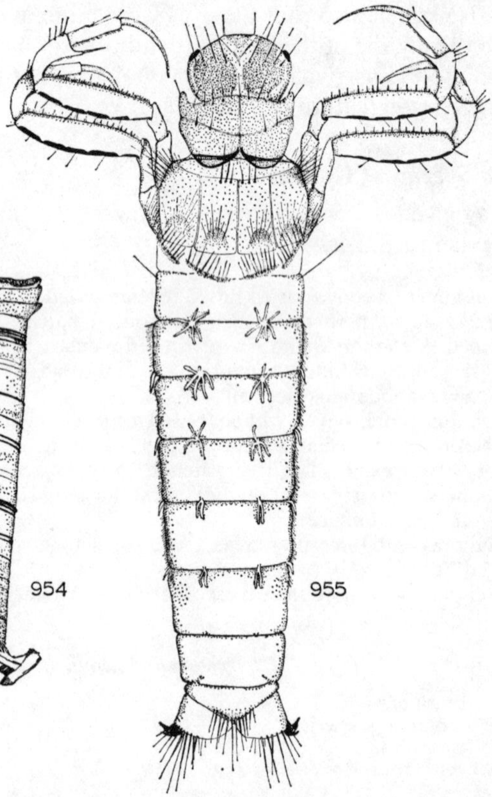

955

Figs. 954–955. *Brachycentrus subnubilus*. 954. Larval case; 955. Larva, full grown.

partly sealed by secretion, leaving a circular orifice. Transverse dark markings are very often a feature of the case. Length 12 mm., width 2–3 mm. *Larva* (Figs. 955, 956). Eruciform, head orthocentrous. The legs are bunched up beneath the thorax and head. Length 10–12 mm., width 2.5 mm.

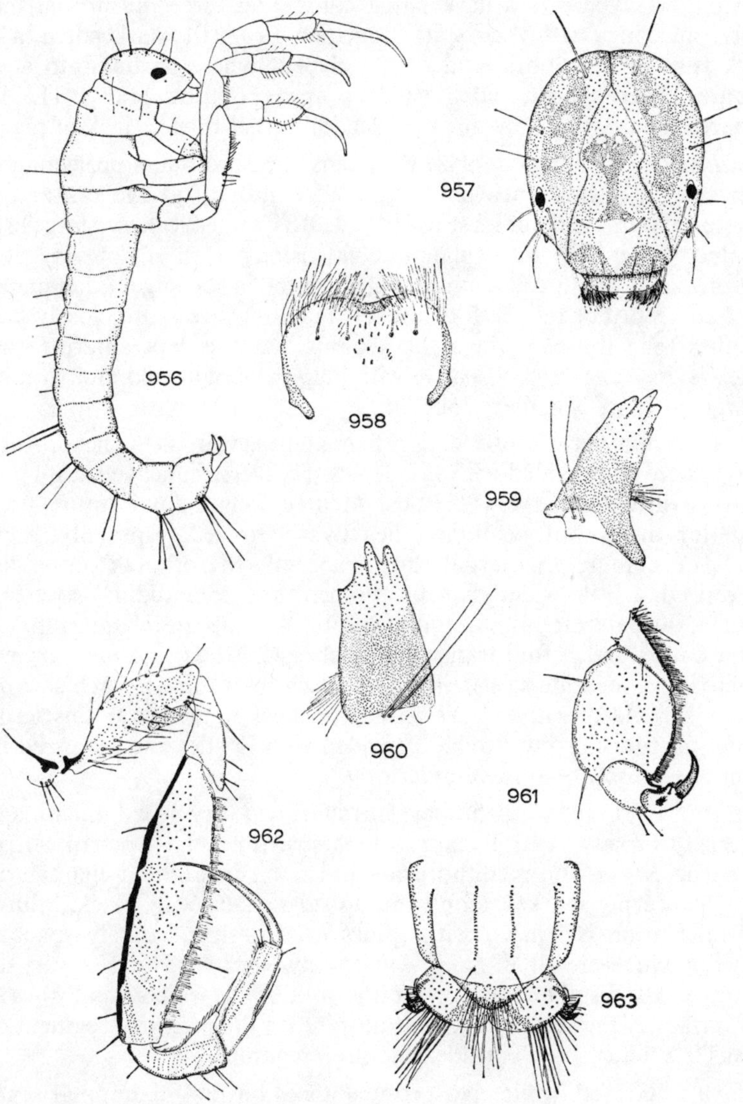

Figs. 956–963. *Brachycentrus subnubilus*. 956. Larva, 1st instar; 957. Head; 958. Labrum; 959. Left mandible; 960. Right mandible; 961. Prothoracic leg; 962. Mesothoracic leg; 963. Anal segment.

Head (Fig. 957). Golden brown with a darker area at the anterior margin of the clypeus (which is provided with a row of four bristles). Back of the head darker also. Dark bands run longitudinally over the genae just skirting the clypeus on each side. Four light-coloured spots are contained in each band anterior to a prominent pair of bristles. Within the aboral area of the clypeus is a dark patch containing three mushroom-shaped lighter markings with their stalks together. From the dark patch, a dark mark runs to the aboral end of the clypeus where it runs into a light-coloured heart-shaped patch with its apex pointing orally. The light-coloured area is probably due to reflection rather than to lack of pigment.

Mouthparts. Labrum (Fig. 958) with dense brush of hairs projecting from each side of a small heavily sclerotized concavity at the centre of the anterior margin. Mandibles (Figs. 959, 960) asymmetrical, two large bristles situated at the base on a plate-like extension which is not very heavily sclerotized. A brush of hairs is situated on the inside edge. Each mandible has four teeth but the teeth of the left mandible are more closely packed together than those of the right. Colour black, except anterior quarter which is deep chestnut. Maxillae with palp five-segmented and with brush of hairs at base. Maxillary lobe hairy.

Thorax. Prothorax sclerotized, approximately as wide as the head. Anterior margin convex provided with bristles. A transverse concave suture present also provided with a few bristles. Median longitudinal suture present. Posterior margin of prothorax heavily sclerotized, especially laterally, black in colour. Prosternal horn absent. Mesothorax only lightly sclerotized, a little wider than prothorax; three longitudinal sutures, one median, two lateral; more heavily sclerotized at the lateral anterior vertices of the notum and at four transverse patches near the posterior margin each situated in a separate sclerite (as divided by the sutures). Each sclerotized area is provided with a few strong bristles. Metanotum unsclerotized except for four narrow patches provided with bristles and arranged in the form of a crescent, concave anteriorly.

Legs (Figs. 961, 962). Prothoracic legs short and very broad, the tibia being especially broad. Ventral margin of tarsus, tibia and trochanter fringed with hair. Meso- and metathoracic legs similar in size and characteristics, the tibia deeply marked along the dorsal surface with black, whilst the ventral margin is fringed with spines of two sizes, equally spaced and between which are small hairs. Tarsal claw very long and on the dorsal surface of the distal tarsal segment is a stout bristle which usually lies along the surface of the segment. A ventral spine on the proximal segment of the tarsus is situated on a considerable prolongation.

Abdomen. Greyish-white. No protuberances on first abdominal segment. Gills present on the second to seventh segments. A pair of tufted gills lie on the dorsal surface of the segments near the posterior margin and become progressively reduced. The gills on the seventh abdominal segment may be very much reduced. A 'lateral line' of fine hairs in the pleural region occurs from segments three to seven.

The young larva on being hatched from an egg-mass has the legs all similar in length and shape, well developed and large. The thoracic nota are not differentiated into sclerites. The tarsal claws are extremely long. Each segment of the abdomen bears two or more large spines.

LEPIDOSTOMATINAE

The two species which I have examined have cases of square cross-section at least when the larva is well grown. The typical case of *Lasiocephala* according to Lestage, however, is tubular with a strong curve near the posterior end.

The ventral margins of the genae where they bound the gular sclerite appear to be thrown into folds to a greater (*Lepidostoma hirtum*) or lesser (*Crunoecia irrorata*) extent, but it is not known how far this feature is characteristic of the sub-family.

Crunoecia irrorata (Curtis)

This is one of the species described in detail by Nielsen (1942) from Danish material. Only those characters have been described, therefore, which are required to enable an identification to be made. I am indebted to Mr. A. Peacey, who kindly sent me larvae of the species, which he had collected on 15 January 1954, from a small fast stream near his home at Brimscombe, near Stroud, Glos. They appeared to be feeding on dead leaves and rotting wood. Mr. Peacey had previously reared this species from larvae collected from a stream near Bisley, Glos., in 1953. A further batch of this series was reared to adult stage, emergence taking place on 7 June 1954.

Larvae. Length 6–7 mm., width 1–6 mm. When extracted from case abdomen assumes variety of shapes quite unlike any other caddis larva I have examined.

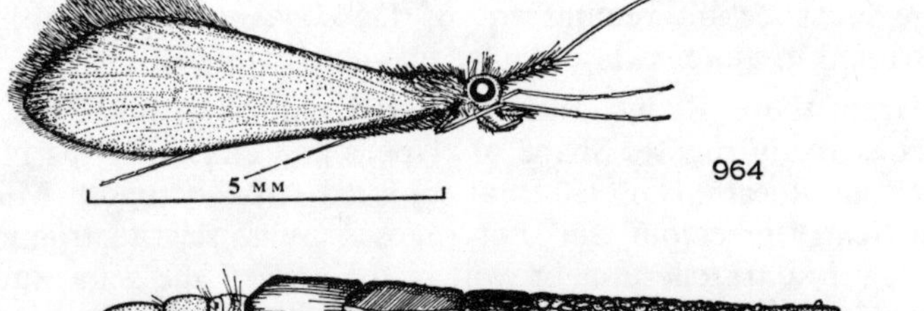

964

965

Figs. 964–965. *Crunoecia irrorata*. 964. Side view of adult; 965. Larva in case.

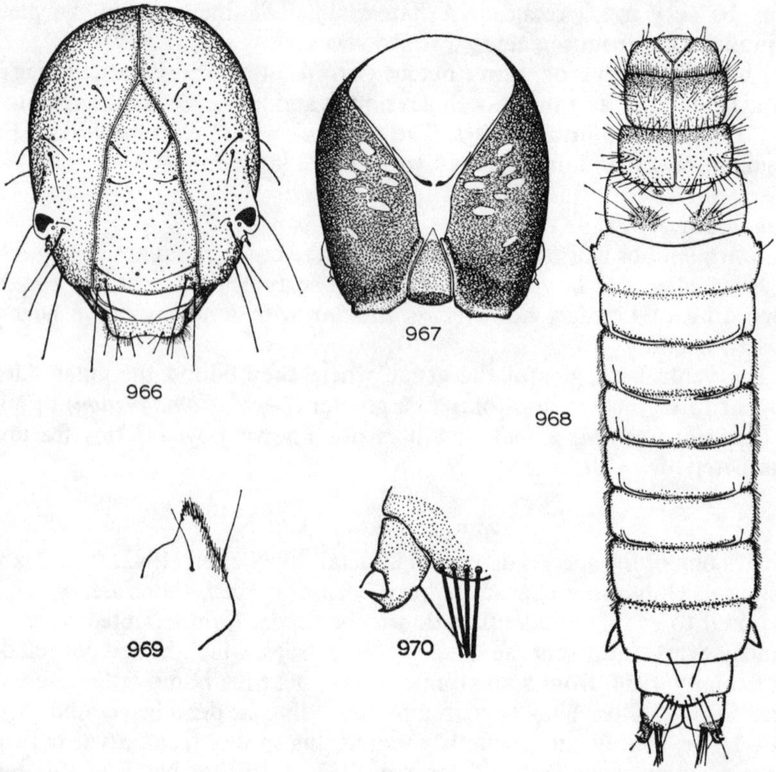

Figs. 966–970. *Crunoecia irrorata.* 966. Head of larva from the front; 967. Head from below; 968. Body of larva from above; 969. Lateral protuberance, abdominal segment I; 970. Anal claw.

Case. Length 7–9 mm., width 2 mm., constructed first of sand-grains, then fragments of rotting wood, dead leaves and other vegetable matter, to give square cross-section reminiscent of *Lepidostoma hirtum*, although not constructed in quite such a regular manner (Fig. 965).

Head from above. Rich chestnut-coloured, reddish to naked eye, devoid of marks, strongly rugose. Shape of clypeus and chaetotaxy as in Fig. 966. *Head* from beneath. Dark chestnut-coloured, strongly rugose. Muscle-spots much lighter in colour and not rugose. Gular sclerite triangular, oral margin very dark chestnut brown; in the area of the genal suture genae membraneous and light cream in colour with two black points at their aboral end. Metanotum of two distinct parts, small anterior part in which a pair of single bristles is situated, and large posterior part with two large groups of bristles. Dorsal protuberance on first abdominal segment absent, lateral protuberances acute, covered with short bristles and hairs. Pairs of single filiform gills present arising from posterior edge of second to sixth abdominal segment on dorsal and ventral surfaces. Lateral line inconspicuous, consisting of short light-coloured hairs, and extending from the third abdominal segment to the beginning of the eighth. On the eighth segment a pair of lappets directed posteriorly and somewhat eversible. Anal claw with auxiliary claw (Fig. 970).

Brindle states of this species that it is found in rivers but that in the north it is more typical of small rivulets in woodland or streams on higher ground or often in marshes where a current of water is passing.

Lepidostoma hirtum (F.)

A large number of larvae were collected from a small stagnant flood pool at the side of the River Rea at Cleobury Mortimer, Shropshire. The larvae were clustered on submerged grass stems. Several were reared through to the adult stage for the species determination to be made.

Larva. Eruciform, cylindrical, head and thoracic sclerites dark golden-brown, abdomen white. Length 11 mm., width 2 mm.

Case (Fig. 971). Of vegetable debris. Square cross-section tapering distally. The vegetable debris of which the case is composed is hardened, apparently by the secretion by means of which it is cemented together. Young larvae construct cases of sand-grains. Length 17 mm., width 2.5 mm.

Head. Almost spherical, dark golden-brown in colour, finely sculptured. Eyes situated in white patches. Oral end of clypeus narrow, not extending to the full width of the head. Pattern of light-coloured patches as shown in Figs. 972 and 973, within the clypeus at the aboral end and on the genae.

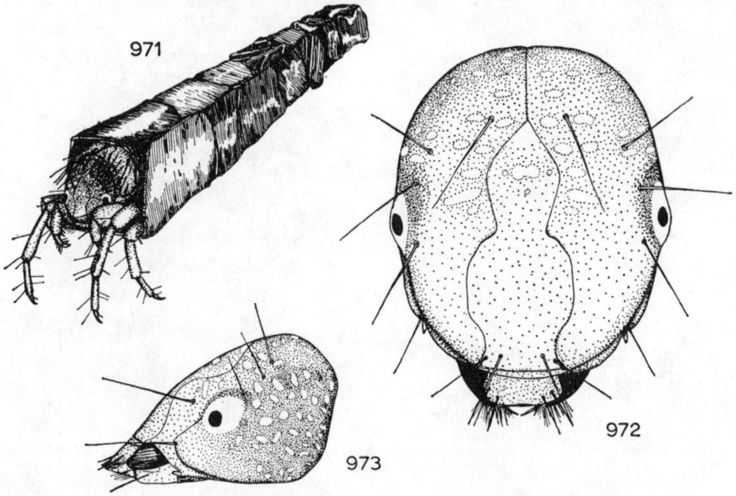

Figs. 971–973. *Lepidostoma hirtum*. 971. Larva in case; 972. Head (from the front); 973. Head (from the side).

Antennae very small. Aboral end of gular sclerite obtuse. Wing-like appendages of genae adjacent to gular sutures (Fig. 974).

Mouthparts. *Labrum* (Fig. 975). Centre of anterior margin sclerotized, transverse, not concave, sclerotized lateral margin curves inwards. Transverse row of robust bristles across centre of labrum.

Mandible (Fig. 976). Three-toothed with tuft of fine hairs at the base on the inside edge.

Maxilla (Fig. 977). Palp four-segmented, hairy at base. Mala obtuse, hairy with a few papillae at the distal extremity.

Labium. Bulbous, covered with small hairs. Labial palps situated on bulbous bases.

Thorax (Fig. 978). Prosternal horn very small. Prothorax sclerotized with longitudinal suture. Dark golden-brown with slightly darker transverse

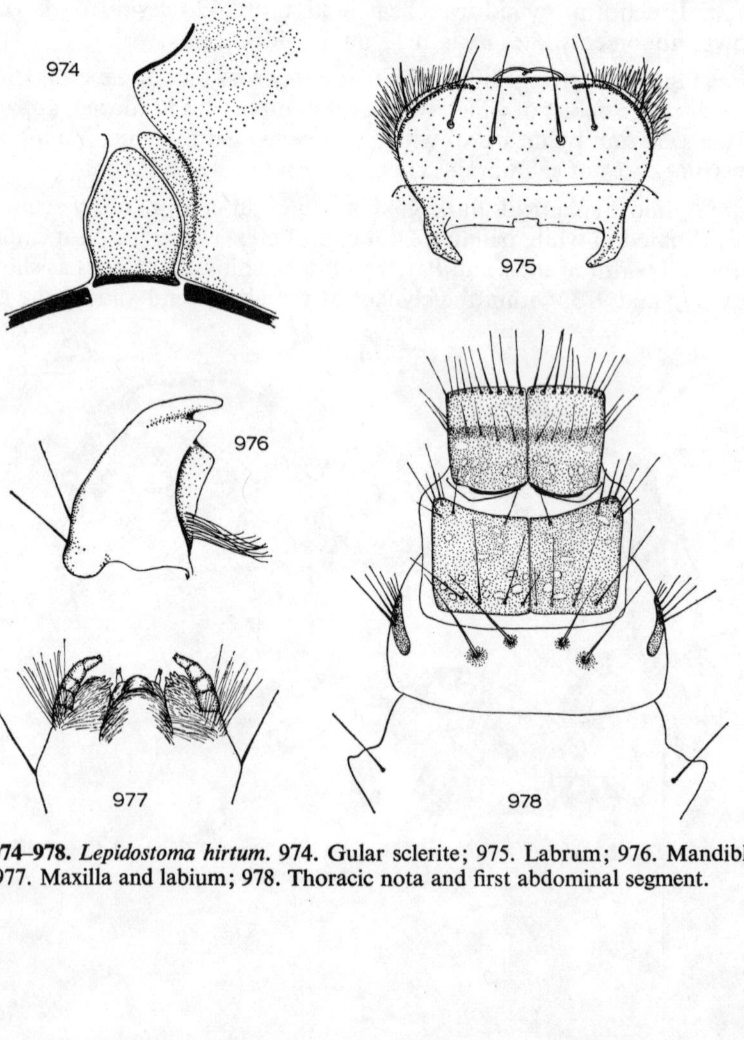

Figs. 974–978. *Lepidostoma hirtum*. 974. Gular sclerite; 975. Labrum; 976. Mandible; 977. Maxilla and labium; 978. Thoracic nota and first abdominal segment.

band. Row of bristles along anterior margin with a second row transversely approximately in middle of notum. Pattern of light marks in posterior half of notum. Posterior margin black. Mesothoracic notum sclerotized, rectangular, shield-like, with longitudinal suture. Dark golden-brown, wider than prothorax. Pattern of light marks as shown in figure. Several bristles at lateral position near anterior margin, and a transverse row of strong bristles near posterior margin. Metathorax wider than mesothorax, unsclerotized except for six small patches. Two pairs of the sclerites are very small and situated on the dorsal surface of the metathorax. These sclerites occur at the base of bristles. The anterior pair are closer together than the posterior pair. The remaining pair are larger, lunate and situated laterally. A few bristles are situated at the anterior end of these latter sclerites.

Legs (Fig. 979, a, b and c). Prothoracic legs short and thick, mesothoracic legs slightly longer than the metathoracic. Chaetotaxy as in the figures.

Abdomen. Lateral protuberances on first abdominal segment pointed and directed anteriorly. Median dorsal protuberance absent. Lateral line indistinct but groups of small chestnut-brown pustules present near lateral line in anterior half of segments three to eight. Gills present, two single

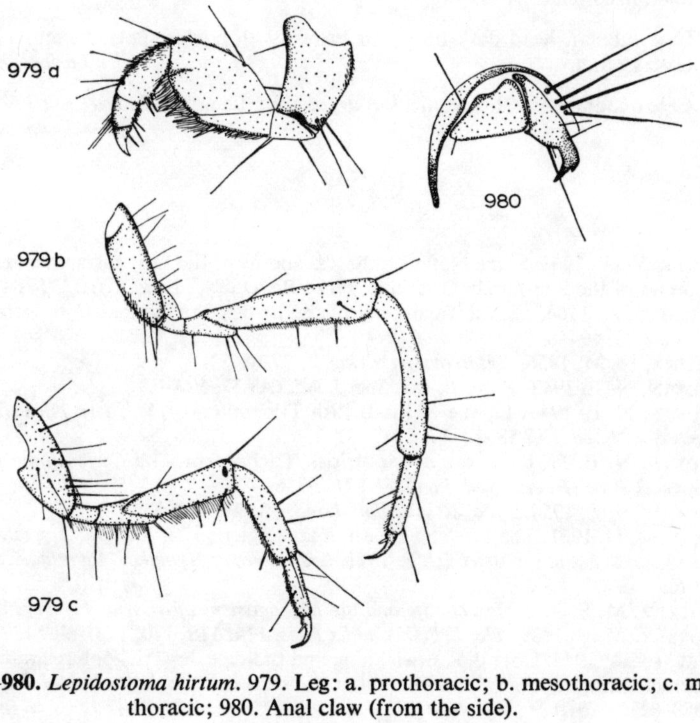

Figs. 979–980. *Lepidostoma hirtum*. 979. Leg: a. prothoracic; b. mesothoracic; c. metathoracic; 980. Anal claw (from the side).

gills on dorsal surface of segment two, and two on ventral surface arising from near posterior margin. In segments three to six, there are four single gills arising from near anterior margin, two dorsal, two ventral, and four single gills arising from near posterior margin, two dorsal and two ventral. In segment seven there are four single gills, all arising from near posterior margin, two dorsal and two ventral. Each anal claw (Fig. 980) with auxiliary claw. Sclerite supporting claw triangular. Four long black bristles.

Lasiocephala basalis (Kolenati)

This species is rather local in Britain but is very abundant where it occurs, which is in fairly fast running water. The case is of sand-grain, circular in cross-section with a fairly abrupt curve near the posterior end.

Key to Larvae of the **Lepidostomatinae**

1. Mesonotum completely sclerotized, metanotum with three pairs of small sclerotized plates. Length 11 mm. Case 16 mm. *Lepidostoma hirtum*

— Mesonotum only partially sclerotized, metanotum with less than three pairs of sclerotized plates 2

2. Case tubular, head dark brown or brown with lighter spots. Length 11 mm. Case 15 mm. *Lasiocephala basalis*

— Case quadrangular, head unicolorous, reddish. Length 6–7 mm. Case 7–9 mm. *Crunoecia irrorata*

BAKER, W. V. 1963. Some Notes on the Taxonomy of the Last Instar Larvae of British Species of the Sub-family Goerinae. *Proc. R. ent. Soc. Lond.* (B) **32**, 171–4.

BRINDLE, A. 1964. Larval Taxonomy of the British Trichoptera. *Entomologist's Rec. J. Var.* **76**, 194–8.

HANNA, H. M. 1956. *Entomologist's Gaz.* **7**, 77–81.

HICKIN, N. E. 1942. *Proc. R. ent. Soc. Lond.* (A) **17**, 123–6.

HICKIN, N. E. 1943. Larvae of the British Trichoptera—14. *Goera pilosa* F. *Proc. R. ent. Soc. Lond.* (A) **18**, 15–17, 75–7, 81–3.

HICKIN, N. E. 1951. Larvae of the British Trichoptera—32. *Sericostoma personatum* Spence. *Proc. R. ent. Soc. Lond.* (A) **26**, 93–6.

HICKIN, N. E. 1954. *Proc. R. ent. Soc. Lond.* (A) **29**, 172–3.

JACQUES, D. 1961. The Life and Death of an Angler's Fly. *Anglers' Annual.*

LESTAGE, J. A., *in* ROUSSEAU, E. 1921. *Les Larves et Nymphes Aquatiques des Insectes d'Europe.*

MOSELY, M. E. 1926. *Insect-Life and the Management of a Trout Fishery.* Pl. XV.

MOSELY, M. E. 1939. *The British Caddis Flies*, 114, 118, 120, 123, 126, 127.

NIELSEN, A. 1942. Über die Entwicklung und Biologie der Trichopteren mit besonderer Berücksichtigung der Quelltrichopteren Himmerlands. *Arch. Hydrobiol.* Suppl. **17**, 465, 588–625.

SILFVENIUS, A. J. 1905–6. Beiträge zur Metamorphose der Trichopteren. *Acta Soc. Fauna Flora fenn.* **27** (6), 10, 13, 19, 21 (fig. 6), 29 (fig. 8).

SILFVENIUS, A. J. 1905–6. *Acta Soc. Fauna Flora fenn.* **28** (6), 15, fig. 4.

THIENEMANN, A. 1905. *Zool. Jb.* (*Syst.*) **22** (5), 29–33.

ULMER, G. 1903. *Allg. Z. Ent.* **8**, 346.

ULMER, G. 1903. *Metamorph. Trichopt.*, 80, 83, 84 (figs. 5, 18, 29, 40), 87 (figs. 6, 17, 28, 57), 90 (fig. 55), 91.

ULMER, G. 1909. *Die Süsswasserfauna Deutschlands* **5–6**, 273, 275 (figs. 411–12), 278, 279, 280.

WESENBERG-LUND, C. 1908. *Int. Revue ges. Hydrobiol. Hydrogr.* **1**, 856.

END-PIECE

'I have held you too long about these caddis'

IZAAK WALTON, 1653

A continuation of publication references dealing with the life cycle of *Trichoptera* from page 54.

BLICKLE, R. L., and MORSE, W. J., 1955, New and little-known *Polycentropus* (TRICHOPTERA). *Bulletin* of the *Brooklyn Entomological Society*. **50** (4): 95–98.

BLICKLE, R. L., and MORSE, W. J., 1957, New HYDROPTILIDAE (TRICHOPTERA) from New Hampshire. *Bulletin* of the *Brooklyn Entomological Society*. **52** (2): 48–50.

BLICKLE, R. L., 1961, New Species of HYDROPTILIDAE (TRICHOPTERA). *Bulletin* of the *Brooklyn Entomological Society*. **56** (5): 131–134.

BLICKLE, R. L., 1962, HYDROPTILIDAE (TRICHOPTERA) of Florida. *Florida Entomologist* **45** (3): 153–155.

BLICKLE, R. L., and MORSE, W. J., 1963, New Species of HYDROPTILIDAE (TRICHOPTERA) *Bulletin* of the *Brooklyn Entomological Society*. **58** (1): 17–22.

CARPENTER, F. M., 1933, Trichoptera from the mountains of North Carolina and Tennessee. *Psyche* **40**: 32–47.

CORBET, P. S., 1966, Parthenogenesis in Trichoptera. *Can. J. Zool.* **44**: 981–982.

CORBET, P. S., SCHMID, F., and AUGUSTIN, C. L., 1966, The Trichoptera of St. Helen's Island, Montreal. 1. The Species present and their relative abundance at Light. *Can. Ent.* **98** (12): 1284–1298.

DA COSTA LIMA, A., 1943, Insetos do Brasil. **4**, chapter 27, pp. 109–133 deal with the caddis flies of Brazil. *Escola Nacional de Agronomia*. Série Didática.

DENNING, D. G., 1941, Descriptions and notes of new and little-known species of Trichoptera. *Ann. Ent. Soc. Amer.* **34**: 195–203.

DENNING, D. G., 1947, New species of Trichoptera from the United States. *Ent. News*, **58**: 249–257.

DENNING, D. G., 1947, New Species and records of Nearctic HYDROPTILIDAE (TRICHOPTERA). *Bulletin* of the *Brooklyn Entomological Society*, **62** (5): 145–158.

DENNING, D. G., 1948, A Review of the RHYACOPHILIDAE (TRICHOPTERA). Vol. of *Can. Entomologist*.

DENNING, D. G., 1949, New and little-known species of Caddis Flies. *Amer. Midl. Nat.* **42** (1): 112–122.

DENNING, D. G., 1949, New Species of Nearctic Caddis Flies. *Bulletin* of the *Brooklyn Entomological Society*. **44** (2): 37–48.

DENNING, D. G., 1950, Records and descriptions of Nearctic Caddis Flies. *Bulletin* of the *Brooklyn Entomological Society*. **65** (4): 97–104.

DENNING, D. G., 1951, Pest Trichoptera at Fort Erie, Ontario. *Can. Entomologist*, **83** (3).

DENNING, D. G., 1952, Descriptions of several new Species of Caddis Flies. *Can. Entomologist*, **84** (1).

DENNING, D. G., 1953, A new Genus of LIMNEPHILIDAE (TRICHOPTERA). *Pan-Pacific Entomologist*, **29** (3): 165–169.

DENNING, D. G., in USINGER, R. L., 1956, Aquatic Insects of California. *Univ.* of *California Press*, *Berkeley*. Chapter 10, pp. 237–270 deals with the Trichoptera. References are given for years 1945–1956, Ross 1944 having given a more or less complete bibliography up to that year. Denning in 1956 considered that seventeen families, one hundred and thirty-five genera and nine hundred and eighty species of Caddis Flies occurred in North America north of Mexico.

DENNING, D. G., 1965, New HYDROPSYCHIDAE. *Journ. Kansas Ent. Soc.* **38**: 75–84.

EDWARDS, S. W., The Immature Stages of *Xiphocentron mexico* (TRICHOPTERA). *Texas Journal of Science*. **13** (1): 51–56.

FISCHER, F. C. J., 1960, Trichopterorum Catalogus. *Nederlandsche Entomologische Vereeniging*. Amsterdam. **1**. NECROTAULIDAE. PROSEPIDIDONTIDAE. RHYACOPHILIDAE. pp. 1–168.

FISCHER, F. C. J., 1961, Trichopterorum Catalogus. *Nederlandsche Entomologische Vereeniging*. Amsterdam. **2**. PHILOPOTAMIDAE. HYDROPTILIDAE. STENOPSYCHIDAE. pp. 1–189 + errata.

FISCHER, F. C. J., 1962, Trichopterorum Catalogus. *Nederlandsche Entomologische Vereeniging*. Amsterdam. **3**. POLYCENTROPODIDAE. PSYCHOMYIIDAE. pp. 1–236 + errata.

FISCHER, F. C. J., 1963, Trichopterorum Catalogus. *Nederlandsche Entomologische Vereeniging*. Amsterdam. **4**. HYDROPSICHIDAE. ARCTOPSYCHIDAE. pp. 1–225 + errata.

FISCHER, F. C. J., 1964, Trichopterorum Catalogus. *Nederlandsche Entomologische Vereeniging*. Amsterdam. **5**. PHRYGANEIDAE. LIMNOCENTROPODIDAE. MOLANNIDAE. pp. 1–213 + errata.

FISCHER, F. C. J., 1965, Trichopterorum Catalogus. *Nederlandsche Entomologische Vereeniging*. Amsterdam. **6**. CALAMOCERATIDAE. PHILORHEITHRIDAE. ODONTOCERIDAE. LEPTOCERIDAE pars 1. pp. 1–242.

FISCHER, F. C. J., 1966, Trichopterorum Catalogus. *Nederlandsche Entomologische Vereeniging*. Amsterdam. **7**. LEPTOCERIDAE pars 2. pp. 1–163.

FISCHER, F. C. J., 1967, Trichopterorum Catalogus. *Nederlandsche Entomologische Vereeniging*. Amsterdam. **8**. GEORIDAE. LIMNEPHILIDAE pars 1. pp. 1–263.

FLINT, O. S., 1947, Description of the Immature Stages of *Drusinus uniformis* Betten (TRICHOPTERA: LIMNEPHILIDAE). *Bulletin* of the *Brooklyn Entomological Soc*. **52** (1): 1–4.

FLINT, O. S., 1956, The Life History and Biology of the Genus *Frenesia* (TRICHOPTERA: LIMNEPHILIDAE). *Bulletin* of the *Brooklyn Entomological Soc*. **51** (4 & 5): 93–108.

FLINT, O. S., 1958, Descriptions of several species of TRICHOPTERA. *Bulletin of the Brooklyn Entomological Soc*. **53** (1): 21–24.

FLINT, O. S., 1958, The Larva and Terrestrial Pupa of *Ironoquia parvula* (TRICHOPTERA: LIMNEPHILIDAE). *Journal* of the *New York Entomological Soc*. **66**: 59–62.

FLINT, O. S., 1960, Taxonomy and Biology of Nearctic Limnephilid Larvae (TRICHOPTERA) with special reference to species in Eastern United States. *Ent. Amer*. **40** (N.S.): 1–117.

FLINT, O. S., 1961, and WIGGINS, G. B., Records and Descriptions of North American species in the Genus *Lepidostoma*, with a Revision of the *Vernalis* group (TRICHOPTERA: LIMNEPHILIDAE). *Can. Entomologist*, **93** (4): 279–297.

FLINT, O. S., 1961, The Immature Stages of the Arctopsychinae occurring in Eastern North America (TRICHOPTERA: HYDROPSYCHIDAE). *Ann. ent. Soc. Amer*. **54** (1): 5–11.

FLINT, O. S., 1962, Larvae of the Caddis Fly Genus *Rhyacophila* in Eastern North America (TRICHOPTERA: RHYACOPHILIDAE). *Proc*. of *U. S. National Museum, Smithsonian Institute*, **113**: 465–493.

FLINT, O. S., 1963, Studies of Neotropical Caddis Flies, I: RHYACOPHILIDAE and GLOSSOSOMATIDAE (TRICHOPTERA). *Proc. U. S. National Museum, Smithsonian Institute*, **114** (3473): 453–477.

FLINT, O. S., 1964, Notes on some Nearctic PSYCHOMYIIDAE with special reference to their larvae. (TRICHOPTERA). *Proc. U. S. National Museum, Smithsonian Institute*, **115**, (3491): 467–481.

FLINT, O. S., 1964, Taxonomy and Life Histories of the Caddis Flies of the Lesser Antilles. American Philosophical Society Year Book. pp. 262–263.

FLINT, O. S., 1964, Two species of LIMNEPHILIDAE new to North America. *Ent. Soc. Washington*. **66** (1).

FLINT, O. S., 1965, New species of TRICHOPTERA from the United States. *Proceedings* of the *Entomological Society* of *Washington*. **69** (1): 55.

FLINT, O. S., 1966, Studies of Neotropical Caddis Flies III. Types of some species described by Ulmer and Brauer. *Proceedings* of the *United States National Museum, Smithsonian Institution*. Washington, D. C. **120** (No. 3559): 1–20.

FLINT, O. S., 1966, On the Identity of *Clymene Aeger fasciella* Chambers (TRICHOPTERA: HYDROPTILIDAE). *Proc. Ent. Soc. Wash.* **68** (2): 135.

FLINT, O. S., 1966, Notes on certain Nearctic Trichoptera in the Museum of Comparative Zoology. *Proc. United States Nat. Mus., Smithsonian Institution*, Washington. **118** (No. 3530): 373–390.

FLINT, O. S., 1967, A new Synonymy in the TRICHOPTERA. *Proc.* of the *Ent. Soc.* of Washington. **69** (1): 55.

FLINT, O. S., 1967, TRICHOPTERA collected by Prof. Dr. J. Illies in the Chilean subregion. Studies of Neotropical Caddis Flies II. Beitrage zur Neotropischen Fauna. Gustav Fischer, Verlag Stuttgart. **5** (1): 45–68.

FLINT, O. S., 1967, Studies of Neotropical Caddis Flies, VI: on a collection from North Western Mexico. *Proc. Ent. Soc.* of Washington, **69** (2): 162–176.

FLINT, O. S., 1967, Studies of Neotropical Caddis Flies. IV. New species from Mexico and Central America. *Proc.* of the *U. S. Nat. Museum., Smithsonian Institution*, D. C. **123** (No. 3608): 1–24.

FLINT, O. S., 1967, Studies of Neotropical Caddis Flies, V. Types of the species described by Banks and Hagen. *Proc.* of the *U. S. Nat. Museum, Smithsonian Institution, Washington*, D. C. **123** (No. 3619): 1–37.

FLINT, O. S., 1967, The First Record of the *Paduniellini* in the New World (TRICHOPTERA: PSYCHOMYIIDAE). *Proc. ent. Soc. Wash.*, **69** (4): 310–311.

LEONARD, J. W., and LEONARD, F. A., 1949, Noteworthy Records of Caddis Flies from Michigan with descriptions of New Species. *Occ. Pap. Mus. Zool. Univ. Mich.*, No. 520, pp. 1–8, 5 pls.

MARSHALL, A. C., 1939, A qualitative and quantitative study of the TRICHOPTERA of Western Lake Erie (as indicated by light-trap material). *Ann. ent. Soc. Amer.* **32**: 665–688.

MILNE, L., 1934, Studies in North American TRICHOPTERA, pt. 1, pp. 1–19, Cambridge, Mass.

MORSE, W. J., and BLICKLE, R. L., 1953, A Check List of the TRICHOPTERA (Caddies Flies) of New Hampshire. *Ent. News*, **64** (3 and 4): 68–102.

MORSE, W. J., and BLICKLE, R. L., 1957, Additions and Corrections to the List of New Hampshire TRICHOPTERA. *Ent. News*. **58** (5): 127–131.

MOSELY, M. E., 1949, New Trichoptera and a redescription of *Leptocellodes flaveola* Ulmer. *Proc. Roy. ent. Soc. Lond.* (*B*) **18** (3–4): 37–41.

MOSELY, M. E., 1954, The *Protoptila* group of the GLOSSOSOMATINAE (TRICHOPTERA: RHYACOPHILIDAE). *Bull. Brit. Mus.* (*Nat. Hist.*). **3** (9): 315–346, 85 figs.

MUNROE, E., 1951, Pest Trichoptera at Fort Erie, Ontario, *Can. Ent.* **83**: 69–72.

NIMMO, A., 1966, A List of TRICHOPTERA taken at Montreal and Chambly, Quebec, with descriptions of three new species. *Can. Ent.* **98**: 688–693.

PROVANCHER, L., 1877, Petite faune entomologique du Canada. TRICHOPTERES. Le Naturaliste Canadien. **9**: 212–7, 241–4, 257–69.

ROBERT, A., 1960, Les Trichopteres de la Region du lac Monroe, Parc du Mont Tremblant, Que. *Ann. Soc. ent. Quebec*. **4**: 47–61.

ROSS, H. H., 1938, Descriptions of Nearctic Caddis Flies (TRICHOPTERA) with special reference to the Illinois species. Bull. of the Div. of the Natural History Survey. **21**, art. 4, pp. 101–183. Urbana, Illinois.

ROSS, H. H., 1944, The Caddis Flies, or TRICHOPTERA, of Illinois. 23 art. 1. of Bull. of the Div. of the Natural History Survey. Urbana, Illinois. pp. 1–326, 961 figs. One of the more important works on the North American Caddis Flies. Contains a check list of the Nearctic species.

ROSS, H. H., 1949, *Xiphocentronidae*, a new family of TRICHOPTERA. *Ent. News*. **60** (1).

ROSS, H. H., 1956, Evolution and classification of the mountain Caddis Flies. *Univ. Illinois Press, Urbana*. pp. 1–213.

ROSS, H. H., 1959, Chapt. 39. TRICHOPTERA in Edmondson, W. T., (Ed.), Ward and Whipple: *Freshwater Biology*, John Wiley, New York, Second Edition.

ROSS, H. H., 1962, Three new species of TRICHOPTERA from Eastern North America. *Ent. News* **73** (5): 129–133.

ROSS, H. H., 1964, The Evolution of Caddisworm cases and nets. *Am. Zoologist*, **4**: 209–2220.

ROSS, H. H., 1965, The Evolutionary History of *Phylocentropus* (TRICHOPTERA: PSYCHOMYIIDAE). *Journal* of the *Kansas Entomological Society*. **38** (4): 398–400.

ROSS, H. H., 1965, New species of the caddis fly genus *Polycentropus* from Eastern North America (TRICHOPTERA: PSYCHOMYIIDAE) *Proc. Biol. Soc. Wash.*, **78**: 241–245.

ROSS, H. H., 1965, The *Micraserna rusticum* group of Caddisflies. (BRACHYCENTRIDAE: TRICHOPTERA). *Proc. Biol. Soc. Wash.*, **78**: 251–258.

SCHMID, F., 1968, La Famille de Arctopsychides (TRICHOPTERA). *Soc. Ent. Quebec.* No. 1.

WIGGINS, G. B., 1954, The Caddisfly Genus BERAEA in North America (TRICHOPTERA). Contributions of the *Royal Ontario Museum* of *Zoology* and *Palaeontology*, No. 39. pp. 1–14.

WIGGINS, G. B., 1956, The KITAGAMIIDAE, A family of Caddis Flies new to North America (TRICHOPTERA). *Roy. Ontario Museum Div. Zool. Palaeontol.* Contribution No. 44.

WIGGINS, G. B., 1956, A Revision of the North American Caddisfly Genus BANKSIOLA (TRICHOPTERA: PHRYGANEIDAE). *Roy. Ont. Museum, Div. Zool. Palaeontol.* No. 43. pp. 1–12.

WIGGINS, G. B., 1957, The Caddisfly Genus OLIGOTRICHA in Japan with the description of a New species. (TRICHOPTERA: PHRYGANEIDAE). Contributions of the *Roy. Ont. Museum. Div. Zool. Palaeontol.* No. 47. pp. 1–8.

WIGGINS, G. B., 1959, A Method of Rearing Caddis Flies (TRICHOPTERA). *Can. Entomologist*, **91** (7): 402–405.

WIGGINS, G. B., 1960, A Preliminary systematic study of the North American Larvae of the Caddisfly family PHRYGANEIDAE (TRICHOPTERA). *Can. J. Zool.* **38**: 1153–1170.

WIGGINS, G. B., 1961, The Rediscovery of an unusual North American Phryganeid, with some additional records of Caddis Flies from Newfoundland (TRICHOPTERA). *Can. Entomologist*, **93** (8): 695–702.

WIGGINS, G. B., 1962, A New Sub-Family of PHRYGANEID Caddis Flies from Western North America (TRICHOPTERA: PHRYGANEIDAE). *Can. J. Zoology*, **40**: 879–891.

WIGGINS, G. B., and ANDERSON, N. H., 1968, Contributions to the systematics of the caddisfly genera *Pseudostenophylax* and *Philocasca* with special reference to the immature stages (TRICHOPTERA: LIMNEPHILIDAE). Canadian Journal of Zoology. **46**: 61–75.

ZIMMERMAN, E. C., 1957, Insects of Hawaii. Vol. **6**, pp. 171–175. *Univ. of Hawaii Press, Honolulu.* Deals with the sole member of the Caddis Fly fauna—an introduced species.

INDEX

Page references in italic indicate an illustration or a diagram